Birkhäuser

Applied and Numerical Harmonic Analysis

Ovidiu Calin
Der-Chen Chang
Kenro Furutani
Chisato Iwasaki

Heat Kernels for Elliptic and Sub-elliptic Operators

Methods and Techniques

Ovidiu Calin
Department of Mathematics
Eastern Michigan University
Ypsilanti, MI 48197, USA
ocalin@emich.edu

Kenro Furutani
Department of Mathematics
Science University of Tokyo
2641 Yamazaki, Noda
Chiba 278-8510, Japan
furutani@ma.noda.sut.ac.jp

Der-Chen Chang
Department of Mathematics
and Statistics
Georgetown University
Washington, DC 20057, USA
chang@georgetown.edu

Chisato Iwasaki
Department of Mathematical Science
University of Hyogo
2167 Shosha Himeji
671-2201, Japan
chisatoiwasaki@hotmail.com

ISBN 978-0-8176-4994-4 e-ISBN 978-0-8176-4995-1
DOI 10.1007/978-0-8176-4995-1
Springer New York Dordrecht Heidelberg London

Library of Congress Control Number: 2010937587

Mathematics Subject Classication (2010): 22E25, 35B40, 35Q80, 35S05, 35R03, 35P10, 35K08, 35K30, 35K65, 35K05, 35F21, 35H10, 35J08, 49L99, 53B20, 53C15, 58J65, 53C17

Printed on acid-free paper

www.birkhauser-science.com

ANHA Series Preface

The *Applied and Numerical Harmonic Analysis (ANHA)* book series aims to provide the engineering, mathematical, and scientific communities with significant developments in harmonic analysis, ranging from abstract harmonic analysis to basic applications. The title of the series reflects the importance of applications and numerical implementation, but richness and relevance of applications and implementation depend fundamentally on the structure and depth of theoretical underpinnings. Thus, from our point of view, the interleaving of theory and applications and their creative symbiotic evolution is axiomatic.

Harmonic analysis is a wellspring of ideas and applicability that has flourished, developed, and deepened over time within many disciplines and by means of creative cross-fertilization with diverse areas. The intricate and fundamental relationship between harmonic analysis and fields such as signal processing, partial differential equations (PDEs), and image processing is reflected in our state-of-the-art *ANHA* series.

Our vision of modern harmonic analysis includes mathematical areas such as wavelet theory, Banach algebras, classical Fourier analysis, time-frequency analysis, and fractal geometry, as well as the diverse topics that impinge on them.

For example, wavelet theory can be considered an appropriate tool to deal with some basic problems in digital signal processing, speech and image processing, geophysics, pattern recognition, biomedical engineering, and turbulence. These areas implement the latest technology from sampling methods on surfaces to fast algorithms and computer vision methods. The underlying mathematics of wavelet theory depends not only on classical Fourier analysis, but also on ideas from abstract harmonic analysis, including von Neumann algebras and the affine group. This leads to a study of the Heisenberg group and its relationship to Gabor systems, and of the metaplectic group for a meaningful interaction of signal decomposition methods. The unifying influence of wavelet theory in the aforementioned topics illustrates the justification for providing a means for centralizing and disseminating information from the broader, but still focused, area of harmonic analysis. This will be a key role of *ANHA*. We intend to publish with the scope and interaction that such a host of issues demands.

Along with our commitment to publish mathematically significant works at the frontiers of harmonic analysis, we have a comparably strong commitment to publish major advances in the following applicable topics in which harmonic analysis plays a substantial role:

Antenna theory	*Prediction theory*
Biomedical signal processing	*Radar applications*
Digital signal processing	*Sampling theory*
Fast algorithms	*Spectral estimation*
Gabor theory and applications	*Speech processing*
Image processing	*Time-frequency and*
Numerical partial differential equations	*time-scale analysis*
	Wavelet theory

The above point of view for the *ANHA* book series is inspired by the history of Fourier analysis itself, whose tentacles reach into so many fields.

In the last two centuries Fourier analysis has had a major impact on the development of mathematics, on the understanding of many engineering and scientific phenomena, and on the solution of some of the most important problems in mathematics and the sciences. Historically, Fourier series were developed in the analysis of some of the classical PDEs of mathematical physics; these series were used to solve such equations. In order to understand Fourier series and the kinds of solutions they could represent, some of the most basic notions of analysis were defined, e.g., the concept of "function." Since the coefficients of Fourier series are integrals, it is no surprise that Riemann integrals were conceived to deal with uniqueness properties of trigonometric series. Cantor's set theory was also developed because of such uniqueness questions.

A basic problem in Fourier analysis is to show how complicated phenomena, such as sound waves, can be described in terms of elementary harmonics. There are two aspects of this problem: first, to find, or even define properly, the harmonics or spectrum of a given phenomenon, e.g., the spectroscopy problem in optics; second, to determine which phenomena can be constructed from given classes of harmonics, as done, for example, by the mechanical synthesizers in tidal analysis.

Fourier analysis is also the natural setting for many other problems in engineering, mathematics, and the sciences. For example, Wiener's Tauberian theorem in Fourier analysis not only characterizes the behavior of the prime numbers, but also provides the proper notion of spectrum for phenomena such as white light; this latter process leads to the Fourier analysis associated with correlation functions in filtering and prediction problems, and these problems, in turn, deal naturally with Hardy spaces in the theory of complex variables.

Nowadays, some of the theory of PDEs has given way to the study of Fourier integral operators. Problems in antenna theory are studied in terms of unimodular trigonometric polynomials. Applications of Fourier analysis abound in signal processing, whether with the fast Fourier transform (FFT), or filter design, or the adaptive modeling inherent in time-frequency-scale methods such as wavelet theory. The coherent states of mathematical physics are translated and modulated Fourier transforms, and these are used, in conjunction with the uncertainty principle, for dealing with signal reconstruction in communications theory. We are back to the raison d'être of the *ANHA* series!

University of Maryland
College Park

John J. Benedetto
Series Editor

Preface

The Fourier transform is known as one of the most powerful and useful methods for finding fundamental solutions of operators with constant coefficients. However, in the general case this method has its own limitations. This book presents several other methods that can be used concurrently with the Fourier transform method to obtain heat kernels for elliptic and sub-elliptic operators. The text contains a large number of examples which facilitate understanding.

An Overview for the Reader. The theory of parabolic operators describes the distribution of heat on a given manifold as well as evolution phenomena and diffusion processes. The solution of an initial value problem for a parabolic partial differential equation depends on its heat kernel, which is the fundamental solution of the associated parabolic operator. Hence the importance of finding explicit formulas for these kernels.

This monograph presents several theories for finding explicit formulas for heat kernels for both elliptic and sub-elliptic operators. These methods are treated in distinct chapters. We shall find heat kernels for classical operators by several different methods. Some methods come from stochastic processes, others come from quantum physics, and others are purely mathematical. Depending on the symmetry, geometry and ellipticity, some methods are more suited for certain operators rather than others.

This book is a perfect reference material for graduate students, researchers in pure and applied mathematics as well as theoretical physicists interested in understanding different ways of approaching evolution operators.

Scientific Outline. Heat kernels arise naturally from probabilistic properties of stochastic processes. The transition density of a stochastic process provides the heat kernel for the associated generator operator, which is a second-order PDE operator. For instance, the generator of a Brownian motion in a plane is the operator $\frac{1}{2}(\partial_x^2 + \partial_y^2)$. On the other side, since the x- and y-components of the Brownian motion are independent, the joint transition density, given that the motion starts at (x_0, y_0) at $t = 0$, is the product of two transition densities:

$$\frac{1}{\sqrt{2\pi t}} e^{-\frac{(x-x_0)^2}{2t}} \times \frac{1}{\sqrt{2\pi t}} e^{-\frac{(x-x_0)^2}{2t}} = \frac{1}{2\pi t} e^{-\frac{1}{2t}[(x-x_0)^2+(y-y_0)^2]},$$

which is the heat kernel for the aforementioned generator operator.

One of the large classes of operators studied in this book is the sum-of-squares operators. These operators might be either elliptic or sub-elliptic. The methods for finding the heat kernel depend on the commutativity condition of the operators. If the operators commute, then the heat kernel is the product of the heat kernels of the operators. If the operators do not commute, then the Trotter formula applies and the heat kernel is computed using the path integral method. This method was borrowed from quantum physics, and it was used to compute propagators for the Schrödinger equation. This technique of path integrals was initiated by Feynman in the early 1940s.

Another class of operators investigated by this book is the sum between a second partial differential operator and a smooth potential. The case of linear and quadratic potentials can be solved explicitly by the path integral method, by van Vleck's formula, or by geometric methods that encounter classical action and volume function. Finally, they can also be solved by means of pseudo-differential operators. For instance, the aforementioned operator $\frac{1}{2}(\partial_x^2 + \partial_y^2)$ has a heat kernel of the type

$$K(x_0, y_0, x, y; t) = V(t)e^{-S_{cl}(x_0,y_0,x,y;t)}, \tag{0.0.1}$$

where $V(t) = \frac{1}{2\pi t}$ describes the density of geodesics emerging from (x_0, y_0), and

$$S_{cl}(x_0, y_0, x, y; t) = \frac{1}{2t}[(x - x_0)^2 + (y - y_0)^2] = \frac{1}{2t}\text{dist}^2\big((x_0, y_0), (x, y)\big)$$

is the classical action on the Euclidean plane from (x_0, y_0) to (x, y) at time t. The key idea here is that the heat propagates mainly along the geodesics of the associated (sub-)Riemannian space. Then the heat density at a certain point in the space, which is the heat kernel of a certain operator, depends on the action along the geodesic and the density of geodesics at that point, given that all geodesics start at a given initial point. The density of geodesics is described by the volume function, which in certain special cases is given by a van Vleck determinant.

This idea can be applied in general for elliptic operators to obtain locally closed-form expressions for the heat kernels for certain operators. However, there are some limitations to the applicability of the method, and it depends on the nature of the potential function. The operators involving a power potential of degree greater than 2 do not in general have a closed-form solution. Geometrically speaking, this reduces to infinitely many geodesics between two points whose energies cannot be obtained in a closed-form solution in terms of the boundary points. None of the methods presented in this book can be applied to obtain exact solutions for these types of operators. For instance, the well-known problem of nonsolvability of the quartic oscillator problem is one of them.

In the case of sub-elliptic operators, formula (0.0.1) no longer holds, due to the degeneracy of the operator, and another formula will take its place. This new formula will require a fiber integration along the characteristic variety (which for elliptic operators degenerate to a point). We note that pseudo-differential techniques and the path integration method do not, in general, provide easy ways of obtaining heat kernels for sub-elliptic operators.

The novelty of this work is the diversity of methods aimed at computing heat kernels for elliptic and sub-elliptic operators. It is interesting that apparently distinct branches of mathematics, such as stochastic processes, differential geometry, special functions, quantum mechanics and PDEs, have all a common concept – the heat kernel. This concept unifies the aforementioned domains of mathematics, and hence deserves us dedicating our study to it.

It is worth noting the relation of the material of this book with other previous books on the subject. One of the long-standing textbooks in the field is the well-known book of David Widder [111], which appeared in the mid-1970s. This treats the heat equation mainly in dimension 1, discussing boundary value problems, Green's functions, integral transforms, theta-functions, the Huygens property, series expansions, heat polynomials, and other miscellaneous topics. Unlike the classical tract of the aforementioned reference, the present monograph covers the heat equation in several other contexts, such as geometric, stochastic, and quantic. However, the present work does not intend to replace Widder's book, but to complement it with newer facts regarding the geometric and analytic contexts.

This book contains most of the heat kernels computable by means of elementary functions. Future developments in this field can consider the possibility of closed-form expressions of heat kernels involving elliptic functions and hyperelliptic functions. These types of special functions have already appeared in the explicit computation of geodesics on certain sub-Riemannian manifolds, and we treated them in the monograph [27]. Similar types of functions also appeared in considering the heat kernel of operators with polynomial potential of degree greater than 2, one of the most famous being the "quartic oscillator" example. In general, the heat kernels of sub-elliptic operators associated with some sub-Riemannian manifolds of step larger than 2 may lead to the need for special functions. However, our present restrictive knowledge of the subject of hyperelliptic function theory indicates today's limit of explicit computability of these types of heat kernels.

The material was prepared as follows: Chaps. 1–8 were prepared by Ovidiu Calin; Chaps. 9–11 were written by Kenro Furutani, Chaps. 12–14 are attributed to Der-Chen Chang, while Chap. 15 was worked out by Chisato Iwasaki. Two of the authors (Calin and Chang) reside in the United States, while the others (Furutani and Iwasaki) live in Japan.

Acknowledgments We wish to express our gratitude to Prof. Peter Greiner, who introduced us to the subject of this book. The monograph was written in 2009–2010 while the first author was on a sabbatical leave from Eastern Michigan University, and he was also partially supported by NSF Grant no. 0631541 and a fund from Tokyo University of Science during his stay there in the summer of 2009. The second author was partially supported by Hong Kong RGC Grant no. 600607, the Norwegian Research Council Research Grant no. 180275/D15 and a competitive research grant at Georgetown University. The third author was partially supported by the JSPS-grant and Grants-in-Aid for Scientific Research (C) no. 20540218, and the fourth author was partially supported by the Grants-in-Aid for Scientific Research (C) no. 21540194. Finally, we would like to express our thanks to the Birkhäuser and ANHA editors, especially to J. J. Benedetto, in making this project possible.

Tokyo, 2009

The chapters' diagram

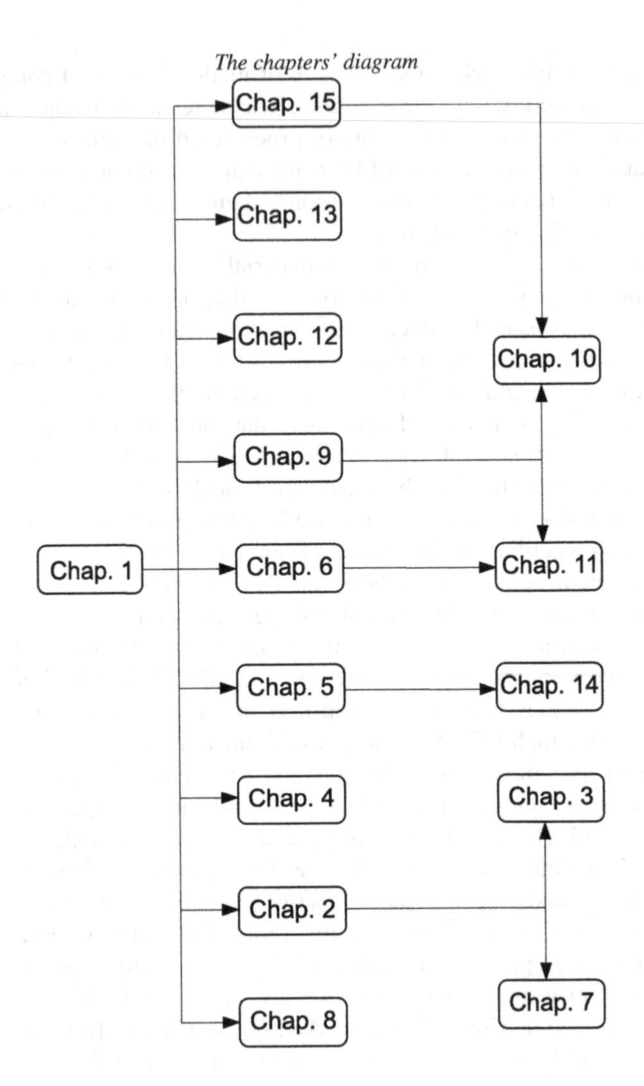

Contents

Part I
Traditional Methods for Computing Heat Kernels

Chapter 1
Introduction

1.1 Physical Significance of the Heat Equation

Consider the problem of finding the *temperature* of a gas contained in a volume V, which is supposed to be a bounded, open and connected set in $\mathbb{R}^3$. Denote by $u(x,t)$ the gas temperature at the point $x \in V$ at time t. Let Ω be a subdomain of V with smooth boundary $\partial\Omega$. Let ν be the unit normal to $\partial\Omega$; see Fig. 1.1a. In order to write the equation of temperature evolution, we shall consider the *law of conservation of thermal energy* on the domain Ω, under the assumption that there is no external absorption or application of heat:

The rate of change of the energy with respect to time in Ω is equal to the net flow of energy across the boundary $\partial\Omega$.

The total thermal energy (heat) in Ω at time t is

$$U(t) = \int_{\Omega} u(x,t)\,dx, \tag{1.1.1}$$

and the net flow of energy across the boundary is given by the *Fourier law*

$$Q(t) = \int_{\partial\Omega} \frac{\partial u(x,t)}{\partial \nu}\,d\sigma,$$

where $d\sigma$ is the area element on $\partial\Omega$. The normal derivative

$$\frac{\partial u(x,t)}{\partial \nu} = \langle \nabla_x u(x,t), \nu(x) \rangle, \qquad x \in \partial\Omega,$$

is the flux density normal to the boundary $\partial\Omega$. If $x \in \partial\Omega$, the only gas particles in a small neighborhood $\mathcal{U}$ of x that hit $\partial\Omega$ are those which have a normal direction to $\partial\Omega$. The other particles in $\mathcal{U}$ that do not have a perpendicular direction to $\partial\Omega$ do not have a contribution toward the term $Q(t)$, since they are colliding with other particles in $\mathcal{U}$ and are bouncing back: see Fig. 1.1b.

O. Calin et al., *Heat Kernels for Elliptic and Sub-elliptic Operators*,
Applied and Numerical Harmonic Analysis, DOI 10.1007/978-0-8176-4995-1_1,

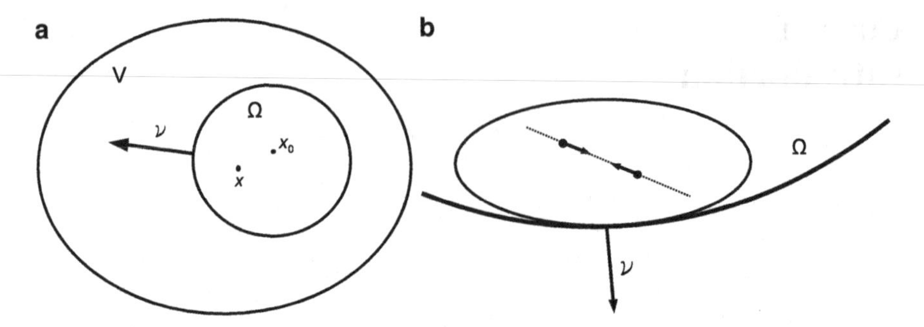

Fig. 1.1 (**a**) The domain Ω and its normal ν. (**b**) The particles which do not have normal direction do not contribute to the heat exchanged through $\partial\Omega$

Applying Gauss's formula, the heat changed through the boundary becomes

$$Q(t) = \int_\Omega \operatorname{div}\nabla_x u(x,t)\,dx = \int_\Omega \Delta_x u(x,t)\,dx. \tag{1.1.2}$$

The aforementioned conservation of energy can be written as

$$\frac{\partial}{\partial t}U(t) = Q(t);$$

using (1.1.1) and (1.1.2), it becomes

$$\int_\Omega \frac{\partial}{\partial t}u(x,t)\,dx = \int_\Omega \Delta_x u(x,t)\,dx.$$

Since Ω is an arbitrary domain of V, it follows that

$$\frac{\partial}{\partial t}u(x,t) = \Delta_x u(x,t), \qquad x \in V,\ t \geq 0,$$

which is called the *heat equation* for the Laplacian Δ_x on $\mathbb{R}^3$.

A similar approach can be applied in the case when (V, g_{ij}) is a Riemannian manifold and $\big(\Omega, (x_1, \ldots, x_n)\big)$ is a local chart on V. In this case the measure dx is replaced by the volume element $d\,v_g = \sqrt{\det(g_{ij})}dx_1 \wedge \cdots \wedge dx_n$ and the divergence and the gradient are given by

$$\operatorname{div} X = \frac{1}{\det(g_{ij})} \sum_{i=1}^n \frac{\partial}{\partial x_i}\Big(\det(g_{ij})X^j\Big),$$

$$\nabla_x u = \sum_{i,j} g^{ij}\frac{\partial u}{\partial x_j}\frac{\partial}{\partial x_i}.$$

The Laplacian in this case becomes the elliptic operator

$$\mathbb{L} = \sum_{i,j} g^{ij} \frac{\partial^2}{\partial x_i \partial x_j} - \sum_k \Gamma^k_{ij} \frac{\partial}{\partial x_k},$$

where $\Gamma^k_{ij} = \frac{1}{2} \sum_\ell g^{k\ell} \left(\frac{\partial g_{i\ell}}{\partial x_j} + \frac{\partial g_{j\ell}}{\partial x_i} - \frac{\partial g_{ij}}{\partial x_\ell} \right)$ are the Christoffel symbols.

1.2 The Fundamental Solution

Consider the following *heat problem:* Given two distinct points x_0, x on a Riemannian manifold (V, g), assume that a unit volume of heat is applied at the point x_0 at time $t = 0$. What is the heat distribution at any time $t > 0$? Equivalently, what is volume of heat $K(x_0, x; t)$ that transmits from x_0 to x after time t?

$K(x_0, x; t)$ is defined on $V \times V \times (0, \infty)$ and is called the *fundamental solution* of the heat operator, or the *heat kernel*, if it satisfies the equations

$$\begin{aligned} \frac{\partial}{\partial t} K(x_0, x; t) &= \mathbb{L} K(x_0, x; t), \qquad x \in V, \ t \geq 0, \\ \lim_{t \searrow 0} K(x_0, x; t) &= \delta_{x_0}. \end{aligned}$$

Since in the compact case (with no boundary) no heat is lost, the total amount of heat is preserved:

$$\int K(x_0, x; t)\, dx = \int \delta_{x_0}(x) dx = 1.$$

This is the reason why $K(x_0, x; t)$ can be considered as a probability density function for any $t > 0$ for a certain random process discussed in Chap. 2. For instance, in the one-dimensional case, at $t = 0$ the density is the Dirac distribution δ_{x_0} centered at x_0, while at $t > 0$, the distribution is Gaussian; see Fig. 1.2a, b.

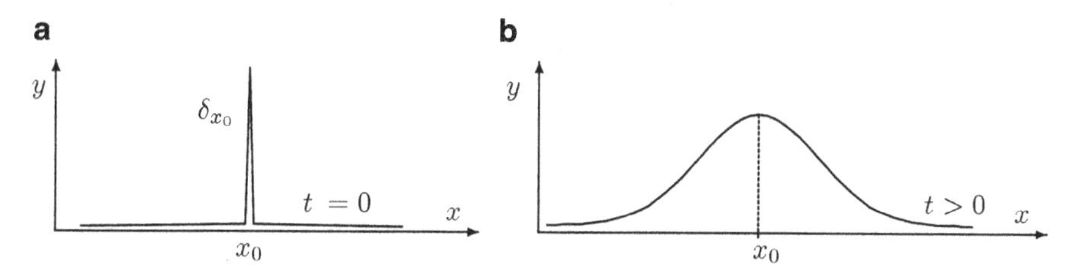

Fig. 1.2 (**a**) The Dirac distribution δ_{x_0}. (**b**) The one-dimensional Gaussian $\frac{1}{\sqrt{2\pi t}} e^{\frac{-(x-x_0)^2}{2t}}$

The main use of the heat kernel is to find the temperature evolution in V, provided the initial distribution of temperature is given. For any continuous, bounded function ϕ, Cauchy's problem

$$\frac{\partial}{\partial t}u(x,t) = \mathbb{L}u(x,t), \qquad x \in V, \; t \geq 0,$$
$$u(x,0) = \phi(x),$$

has the solution

$$u(x,t) = \int_V K(x,y;t)\phi(y)\,dv_g.$$

Since $K(x,y;t)\phi(y)$ represents the volume of heat transmitted from y to x after time t, given the initial temperature $\phi(x)$, the kernel $K(x,y;t)$ is sometimes also called a *propagator*.

1.3 The Heat Equation in Its Setting

The heat equation is a particular case of a more general linear second-order partial differential equation of the type

$$A\partial_x^2 + B\partial_x\partial_t + C\partial_t^2 + D\partial_x + E\partial_t + F = 0,$$

where the coefficients A, B, C, D, E, F are functions on x and t only. The following classification is familiar from undergraduate courses on differential equations:

Condition	Type	Example	Name
$B^2 < 4AC$	Elliptic	$\partial_x^2 u + \partial_t^2 u = 0$	Laplace
$B^2 > 4AC$	Hyperbolic	$\partial_x^2 u - \partial_t^2 u = 0$	Wave
$B^2 = 4AC$	Parabolic	$\partial_x^2 u - \partial_t u = 0$	Heat

The type is suggested by the corresponding conic curve defined by the principal symbol. The Laplace equation describes static equations such as minimal surfaces, gravitational potentials, electric potentials, etc., having the Riemannian geometry as its underlined geometry. The wave operators describe the propagation of mechanical, radio and electro-magnetic waves, and radiation. It is associated with a geometry of Lorentz type. The heat equation describes diffusion phenomena in classical physics, Schrödinger-type equations in quantum mechanics, and the Black–Scholes equation in finance. The reader can find an elementary treatment of these equations, for instance, in Folland [44].

1.4 Methodology

The first part of this book provides several methods for computing heat kernels for both elliptic and sub-elliptic operators using the following methods:

- Stochastic analysis of diffusion processes
- The Path integral method
- The Geometric method involving the action along the geodesics
- The Fourier transform method
- The eigenfunction expansion method

The stochastic analysis method associates with each Ito diffusion a generator operator. The heat kernel of this operator is given by a transition density for the associated Ito diffusion. Several methods for computing transition densities are provided.

The path integral method stems from the computation of propagators in quantum mechanics and has the following approaches:

- By direct computation
- By using Trotter's formula
- By using van Vleck's formula
- By using Feynman–Kac's formula

The eigenfunction expansion is a method which can be applied as long as we are able to compute the eigenfunctions and eigenvalues of the operator. However, closed-form solutions can be provided only if one of the following bilinear generating formulas is used:

- Mehler's formula
- Hille–Hardy's formula
- Poisson's summation formula
- Any other bilinear generating function

The geometric method expresses the heat kernel in terms of the action and volume element. It is a method which can be applied successfully as long as we are able to find the explicit formula for the action and solve the transport equation. It can be applied for both elliptic and sub-elliptic operators.

1.5 Theta-Functions

A *theta-function* is a special function that describes the evolution of temperature on a segment domain subject to certain boundary conditions. We shall use them in the sequel when several heat kernels will be expressed in terms of theta-functions.

Consider the heat equation

$$\frac{\partial \theta}{\partial t} - \kappa \frac{\partial^2 \theta}{\partial x^2} = 0 \tag{1.5.3}$$

on $[0, \pi]$, with constant diffusivity κ. There are two important boundary-type conditions which lead to several types of theta-functions.

Dirichlet boundary condition. The heat is maintained at zero temperature at the endpoints at all times t:

$$\theta(0, t) = \theta(\pi, t) = 0.$$

The separation-of-variables method provides the solution of (1.5.3) in the form

$$\theta_1(x, t) = 2 \sum_{n \geq 0} (-1)^n e^{-(2n+1)^2 \kappa t} \sin(2n + 1)x,$$

also called the *first theta-function.* We note the temperature will decrease to zero in the long run, so $\lim_{t \to \infty} \theta_1(x, t) = 0$.

Neumann boundary condition. There is no heat exchange at the endpoints $x = 0, \pi$. In this case we say the endpoints are insulated and the following condition holds at the boundary:

$$\frac{\partial \theta}{\partial x}(0, t) = \frac{\partial \theta}{\partial x}(\pi, t) = 0.$$

In this case the solution of (1.5.3) is

$$\theta_4(x, t) = 1 + 2 \sum_{n \geq 1} (-1)^n e^{-4n^2 \kappa t} \cos 2nx,$$

called the *fourth theta-function.* Since the heat cannot escape the domain, in the long run the solution tends to a constant equilibrium solution.

A translation of θ_1 and θ_4 with $\pi/2$ in the direction of x yields two other theta-functions:

$$\theta_2(x, t) = \theta_1(x + \pi/2, t) = 2 \sum_{n \geq 0} e^{-4(n+1/2)^2 \kappa t} \cos(2n + 1)x,$$

$$\theta_3(x, t) = \theta_4(x + \pi/2, t) = 1 + 2 \sum_{n \geq 1} e^{-4n^2 \kappa t} \cos 2nx.$$

The theta-functions θ_1 and θ_2 are periodic with period 2π, and the functions θ_3 and θ_4 are periodic with period π.

A few variations of the definition of theta-functions can be found in the literature. If we let $q = e^{-4\kappa t}$, then the aforementioned formulas become

$$\theta_1(x,q) = 2\sum_{n\geq 0}(-1)^n q^{(n+1/2)^2}\sin(2n+1)x,$$

$$\theta_2(x,q) = 2\sum_{n\geq 0} q^{(n+1/2)^2}\cos(2n+1)x,$$

$$\theta_3(x,q) = 1+2\sum_{n\geq 1} q^{n^2}\cos 2nx,$$

$$\theta_4(x,q) = 1+2\sum_{n\geq 1}(-1)^n q^{n^2}\cos 2nx.$$

The graphs of the theta-functions with $q = 0.7$ and $x \in [0, 12]$ can be seen in Fig. 1.3. In the aforementioned formulas the variable q can also be considered complex with $|q| < 1$. In this case, if we let $q = e^{i\pi\tau}$, with $\tau \in \mathbb{C}$, and $Im(\tau) > 0$, we consider the notation

$$\theta_3(x|\tau) = 1+2\sum_{n\geq 1} e^{i\pi n^2\tau}\cos 2nx,$$

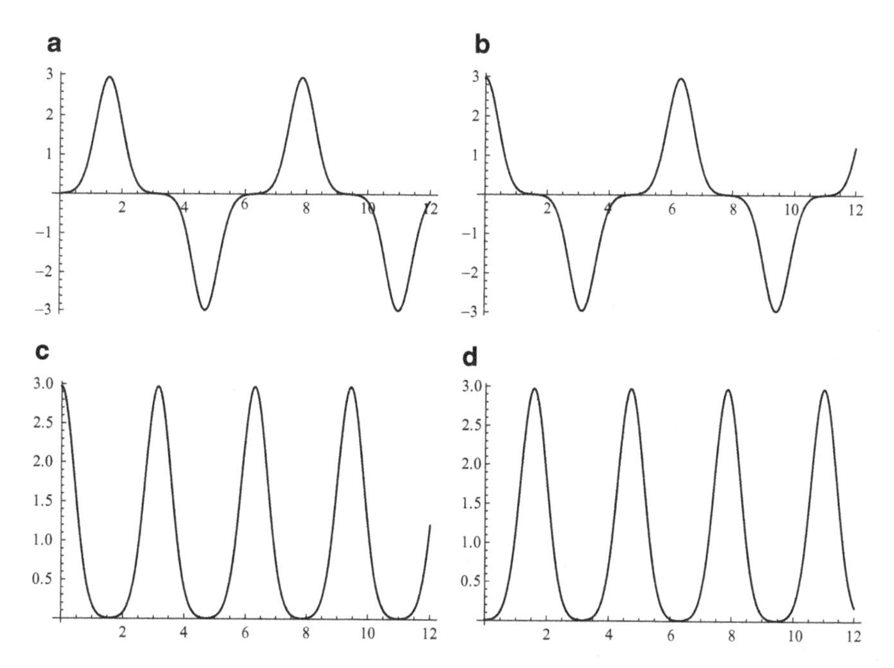

Fig. 1.3 The theta-functions plotted for $q = 0.7$: **(a)** $\theta_1(x, 0.7)$, **(b)** $\theta_2(x, 0.7)$, **(c)** $\theta_3(x, 0.7)$ and **(d)** $\theta_4(x, 0.7)$

which is sometimes useful to be written as

$$\theta_3(x|i\tau) = 1 + 2\sum_{n\geq 1} e^{-\pi n^2 \tau} \cos 2nx. \tag{1.5.4}$$

Using Euler's formula and the fact that sine is an odd function, we have the equivalent definition

$$\theta_3(x|\tau) = \sum_{n=-\infty}^{\infty} q^{n^2} e^{2nxi} = \sum_{n=-\infty}^{\infty} e^{i\pi n^2 \tau} e^{2nxi}.$$

Similar considerations can be carried out for the other theta-functions. More properties of the theta-functions can be found in Chap. 1 of [86].

1.6 Some Useful Integrals

When computing heat kernels, one often gets into integrals which can be reduced to one given by the following result.

Proposition 1.6.1. *We have*

$$\int e^{-ay^2+by}\, dy = \sqrt{\frac{\pi}{a}}\, e^{\frac{b^2}{4a}}, \qquad \forall a \in \mathbb{R}^*,\ b \in \mathbb{R}, \tag{1.6.5}$$

$$\int e^{-ay^2+i\xi y}\, dy = \sqrt{\frac{\pi}{a}}\, e^{\frac{-\xi^2}{4a}}, \qquad \forall a \in \mathbb{R}^*,\ \xi \in \mathbb{R}. \tag{1.6.6}$$

Proof. Completing the square and using the Gaussian integral $\int e^{-v^2}\, dv = \sqrt{\pi}$ yields

$$\int e^{-ay^2+by}\, dy = e^{\frac{b^2}{4a}} \int e^{-\left(\sqrt{a}y - b/(2\sqrt{a})\right)^2} dy = e^{\frac{b^2}{4a}} \frac{1}{\sqrt{a}} \int e^{-v^2}\, dv = \sqrt{\frac{\pi}{a}}\, e^{\frac{b^2}{4a}}.$$

The second integral is obtained by formally replacing b by $i\xi$. Next we shall present a complete proof of this result. Completing the square yields

$$\begin{aligned}\int e^{-ay^2+i\xi y} dy &= e^{-\frac{\xi^2}{4a}} \int e^{-\left(\sqrt{a}y - i\xi/(2\sqrt{a})\right)^2} dy = e^{-\frac{\xi^2}{4a}} \frac{1}{\sqrt{a}} \int e^{-\left(u - i\xi/(2\sqrt{a})\right)^2} du \\ &= \frac{1}{\sqrt{a}} e^{-\frac{\xi^2}{4a}} \int_{Im\, z = -\xi/(2\sqrt{a})} e^{-z^2}\, dz,\end{aligned} \tag{1.6.7}$$

where $Im\, z$ denotes the imaginary part of the complex number z. The integral can be brought down to an integral on the real line as follows. Since the function e^{-z^2} is holomorphic, its integral on the contour $\{(-R,0), (R,0), (R, -\xi/(2\sqrt{a})),$

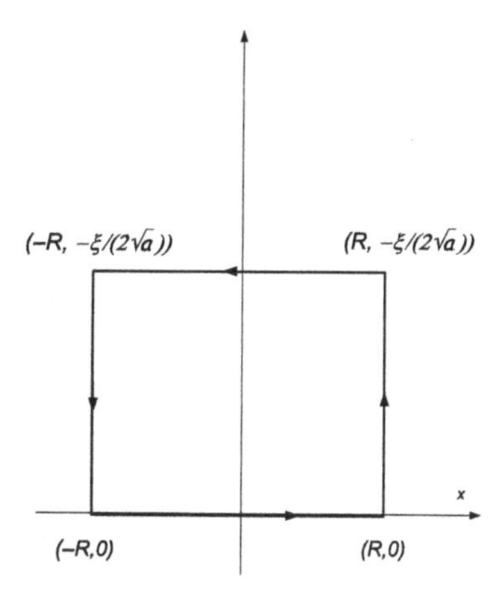

Fig. 1.4 The rectangular contour $\{(-R,0), (R,0), (R,-\xi/(2\sqrt{a})), (-R,-\xi/(2\sqrt{a}))\}$

$(-R,-\xi/(2\sqrt{a}))\}$ vanishes; see Fig. 1.4. Since the limit of the integral of e^{-z^2} on the vertical edges of the aforementioned rectangle vanishes as $R \to \infty$, it follows that

$$\int_{Im\, z=-\xi/(2\sqrt{a})} e^{-z^2}\, dz = \int_{\mathbb{R}} e^{-x^2}\, dx = \sqrt{\pi}.$$

Substituting in (1.6.7) yields the desired result. ∎

Chapter 2
A Brief Introduction to the Calculus of Variations

The Lagrangian and Hamiltonian formalisms will be useful in the following chapters when the heat kernel will be computed using the path integral and geometric variational methods. In the following we shall present a brief overview of the variational theory needed in the sequel.

2.1 Lagrangian Mechanics

In classical mechanics a moving particle is completely described at any instance of time t by its *position* $x(t)$ and its *velocity* $\dot{x}(t)$. The position x belongs to the *coordinate space*, which is, in general, a Riemannian manifold with the metric defined by the kinetic energy. The space of the positions and velocities $(x, \dot{x})$ is called the *phase space*, and it is identified with the tangent bundle TM of the coordinate space M. The pair $(x, \dot{x})$ is called the *state* of the particle.

A real-valued function defined on the tangent bundle $L : TM \to \mathbb{R}$ is called *Lagrangian*. In classical mechanics the usual Lagrangian is given by the difference between the *kinetic energy* and the *potential energy* of the particle:

$$L(x, \dot{x}) = K(\dot{x}) - U(x),$$

where $K(\dot{x}) = \frac{1}{2}\sum_{i=1}^{n} \dot{x}_i^2$ and $U(x)$ is usually a polynomial function of x. The trajectory of a particle $x(t)$ in the coordinate space is a curve parameterized by the time parameter t. The Lagrangian $L(x, \dot{x}, t)$ describes the dynamics of the particle, in the sense that the particle moves on a trajectory $x(t)$ such that the following action integral

$$S(x_0, x, \tau) = \int_0^\tau L(x(t), \dot{x}(t), t)\, dt \tag{2.1.1}$$

is locally minimized under small variations of the path $x(t)$. We shall investigate this problem in the next few sections.

O. Calin et al., *Heat Kernels for Elliptic and Sub-elliptic Operators*,
Applied and Numerical Harmonic Analysis, DOI 10.1007/978-0-8176-4995-1_2,

2.1.1 The First Variation

For our study it would be sufficient to investigate the case of Lagrangians that do not depend explicitly on time t. Let $x(t)$ be a smooth curve joining $x_0 = x(0)$ and $x = x(\tau)$. The action along $x(t)$ is given by

$$S = S\big(x(t)\big) = \int_0^\tau L\big(x(s), \dot{x}(s)\big)\, ds.$$

The main problem of Lagrangian formalism is formulated below:

Given the fixed endpoints x_0 and x, find the path $x(t)$ for which the functional

$$x(t) \to S\big(x(t)\big)$$

(a) Has a "critical point"
(b) Has a minimum

Customarily, this problem is approached by considering a fixed path $x(t)$ and a variation

$$x_\epsilon(t) = x(t) + \epsilon\eta(t)$$

with fixed endpoints

$$x_\epsilon(0) = x(0) = x_0, \qquad x_\epsilon(\tau) = x(\tau) = x_0.$$

The vector field $\eta(t)$ defined on $[0, \tau]$ is an arbitrary *variation vector field* with homogeneous boundary conditions

$$\eta(0) = \eta(\tau) = 0. \tag{2.1.2}$$

Expanding S about the path $x(t)$ in powers of ϵ yields

$$\begin{aligned} S\big(x_\epsilon(t)\big) &= S\big(x(t)\big) + \epsilon \int_0^\tau \left(\left\langle \frac{\partial L}{\partial \dot{x}}, \dot{\eta} \right\rangle + \left\langle \frac{\partial L}{\partial x}, \eta \right\rangle \right) dt \\ &\quad + \frac{\epsilon^2}{2!} \int_0^\tau \left(\left\langle \frac{\partial^2 L}{\partial \dot{x}^2} \dot{\eta}, \dot{\eta} \right\rangle + 2 \left\langle \frac{\partial^2 L}{\partial \dot{x} \partial x} \dot{\eta}, \eta \right\rangle + \left\langle \frac{\partial^2 L}{\partial x^2} \eta, \eta \right\rangle \right) dt + \mathcal{O}(\epsilon^3) \\ &= S + \epsilon\, \delta S + \frac{\epsilon^2}{2!} \delta^2 S + \mathcal{O}(\epsilon^3), \end{aligned}$$

The path $x(t)$ is a "critical point" for the action functional $S\big(x(t)\big)$ if and only if $\delta S = 0$. Integrating by parts and using the boundary conditions (2.1.2) yields

$$0 = \delta S = \int_0^\tau \left(\left\langle \frac{\partial L}{\partial \dot{x}}, \dot{\eta} \right\rangle + \left\langle \frac{\partial L}{\partial x}, \eta \right\rangle \right) dt = \int_0^\tau \left(\left\langle -\frac{d}{dt} \frac{\partial L}{\partial \dot{x}} + \frac{\partial L}{\partial x}, \eta \right\rangle \right) dt.$$

Since the vector field η is arbitrary, we obtain the variational equations, which are the famous *Euler–Lagrange equations*

$$\frac{d}{dt}\frac{\partial L}{\partial \dot{x}} = \frac{\partial L}{\partial x}, \tag{2.1.3}$$

where $\dot{x} = (\dot{x}_1, \dots, \dot{x}_n)$ and $x = (x_1, \dots, x_n)$ denote the velocity and position of the particle, respectively. A solution of (2.1.3) is called a *classical path* and will play a central role in the rest of the book.

2.1.2 The Second Variation

Assume $x(t)$ is a solution of the Euler–Lagrange equations; i.e., it is a classical path. If the second variation is positive definite along $x(t)$, i.e.,

$$\langle \delta^2 S\, \dot{\eta}(t), \eta(t)\rangle > 0, \qquad \forall t \in (0, \tau), \tag{2.1.4}$$

for any variation vector field $\eta(t)$ with $\eta(0) = \eta(\tau) = 0$, then the path $x(t)$ minimizes the action functional between x_0 and $x(\tau)$. In order to have relation (2.1.4) satisfied, it suffices to prove that

$$\min_{\eta}\{\langle \delta^2 S\, \dot{\eta}(t), \eta(t)\rangle, \ \forall t \in (0, \tau)\} > 0.$$

Let η be a variation vector field for which the previous minimum is reached. Then η will satisfy the Euler–Lagrange equations for the associated Lagrangian

$$\overline{L}(\eta, \dot{\eta}) = \langle \delta^2 S\, \dot{\eta}(t), \eta(t)\rangle,$$

which are given by

$$\frac{d}{dt}\left(\frac{\partial^2 L}{\partial \dot{x}^2}\dot{\eta}\right) + \left[\frac{d}{dt}\left(\frac{\partial^2 L}{\partial x \partial \dot{x}}\right) - \frac{\partial^2 L}{\partial x^2}\right]\eta + \left[\frac{\partial^2 L}{\partial \dot{x}\partial x} - \frac{\partial^2 L}{\partial x \partial \dot{x}}\right]\dot{\eta} = 0, \tag{2.1.5}$$

which is called the *Jacobi equation*. A solution $\eta(t)$ of (2.1.5) is called a *Jacobi vector field*. If there is a value $t_1 \in (0, \tau)$ such that a Jacobi vector field $\eta(t_1) = 0$, then inequality (2.1.4) fails at $t = t_1$. A well-known result of variational calculus states that as long as the Jacobi vector field doesn't vanish, the classical path is still minimizing the action. The classical path ceases to be a minimizer as soon as the Jacobi vector field vanishes the first time. The vanishing points of a Jacobi vector field are called *conjugate points*. This is equivalent to saying that the classical path is minimizing between two consecutive conjugate points and is no longer minimizing after that.

2.1.3 Geometrical Interpretation

Next we shall deal with the geometrical significance of the Jacobi vector fields. Consider the classical path starting at x_0 and all the neighboring, classical paths $x(p,t)$ starting at the same point and parameterized by their initial momenta p; this is $x(p,0) = x_0$; see Fig. 2.1. For the time being we neglect the fixed endpoint condition at $t = \tau$. The separation between two classical paths is given by

$$x(p+\epsilon,t) = x(p,t) + \epsilon J(p,t) + \mathcal{O}(\epsilon^2),$$

with $J(p,t) = \big(J_{ij}\big)(p,t)$, where

$$J_{ij}(p,t) = \frac{\partial x_i(p,t)}{\partial p_j}, \qquad i,j = 1,\dots,n. \tag{2.1.6}$$

Since $x(p,t)$ are classical paths, the Euler–Lagrange equations are satisfied:

$$\frac{d}{dt}\left(\frac{\partial L}{\partial \dot{x}_r}\right) = \frac{\partial L}{\partial x_r}, \qquad r = 1,\dots,n.$$

Using the chain rule

$$\begin{aligned}\frac{\partial L}{\partial p_k} &= \frac{\partial L}{\partial x_i}\frac{\partial x_i}{\partial p_k} + \frac{\partial L}{\partial \dot{x}_j}\frac{\partial \dot{x}_j}{\partial p_k} \\ &= \frac{\partial L}{\partial x_i} J_{ik} + \frac{\partial L}{\partial \dot{x}_j}\dot{J}_{jk}\end{aligned}$$

and differentiating in the Euler–Lagrange equations yields

$$\frac{d}{dt}\left[\frac{\partial^2 L}{\partial \dot{x}_r \partial \dot{x}_i}\dot{J}_{ik}\right] + \left[\frac{d}{dt}\frac{\partial^2 L}{\partial \dot{x}_r \partial x_i} - \frac{\partial L}{\partial x_r \partial x_i}\right] J_{ik} + \left[\frac{\partial^2 L}{\partial \dot{x}_r \partial x_i} - \frac{\partial^2 L}{\partial x_r \partial \dot{x}_i}\right]\dot{J}_{ik} = 0. \tag{2.1.7}$$

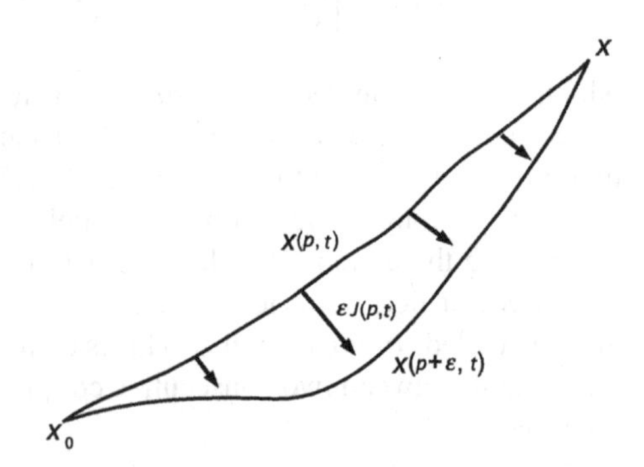

Fig. 2.1 The Jacobi vector field along a classical path between two conjugate points

A standard result of ODEs shows that the second-order-system (2.1.7) together with the following $2n^2$ initial conditions

$$J_{ik}(p,0) = \frac{\partial x_i(p,0)}{\partial p_k} = 0,$$
$$\dot{J}_{ik}(p,0) = \frac{\partial \dot{x}_i(p,0)}{\partial p_k} = \delta_{ik},$$

has a unique solution $J_{ij}(p,t) \neq 0$, for $t \in (0,\tau)$, with $\tau > 0$ small enough.

Let $p = \dot{x}(0)$ be fixed. For any vector $v = (v^1, \dots, v^n) \neq 0$, we can define the vector field $\eta(t) = J(t)v$ along the classical path $x(t)$, by $\eta^i(t) = J_{ik}(t)v^k$. If there is a $T > 0$ such that $\eta(T) = 0$, then $\det J_{ik}(T) = 0$. The point $x(T)$ is conjugate with $x_0 = x(0)$, in the sense of the definition given in the previous section. This fact will be shown next.

Multiplying (2.1.7) by v^k and summing over k yields the equation

$$\frac{d}{dt}\left[\frac{\partial^2 L}{\partial \dot{x}_r \partial \dot{x}_i}\dot{\eta}^i\right] + \left[\frac{d}{dt}\frac{\partial^2 L}{\partial \dot{x}_r \partial x_i} - \frac{\partial^2 L}{\partial x_r \partial x_i}\right]\eta^i + \left[\frac{\partial^2 L}{\partial \dot{x}_r \partial x_i} - \frac{\partial^2 L}{\partial x_r \partial \dot{x}_i}\right]\dot{\eta} = 0,$$

which is exactly the Jacobi equation (2.1.5). The vector field $\eta(t)$ becomes a Jacobi vector field. Since J_{ik} is nondegenerate for $t \in (0,T)$, the set of Jacobi vectors forms an n-dimensional space at every point along the path $x(t)$. The first point where one Jacobi vector (and hence all of them) vanishes is a conjugate point with x_0. The classical path $x(t)$ minimizes the action as long as it does not pass through a conjugate point.

In the following we shall write the matrix J, which satisfies (2.1.7), in terms of the action S. Let x_0 and x be two fixed endpoints and let $x(t)$ be a classical path such that $x(0) = x_0$ and $x(\tau) = x$. The momentum along $x(t)$ at time t is denoted by $p(x_0, x; t)$. Let $\widetilde{x}(t)$ be the reverted curve defined by $\widetilde{x}(t) = x(\tau - t) = x(\tilde{t})$. The curves $\widetilde{x}(t)$ and $x(t)$ have the same endpoints, $x_0 = \widetilde{x}(\tau) = x(0)$ and $x = \widetilde{x}(0) = x(\tau)$. The following relation between the momenta along the curves $x(t)$ and $\widetilde{x}(t)$ holds:

$$p(x_0, x; t) = -p(x, x_0; \tau - t) = -p(x, x_0; \tilde{t}).$$

Using the well-known relation $\frac{\partial S}{\partial x} = p$ along $x(t)$ at instances t and τ yields

$$\frac{\partial S(x_0, x(t); t)}{\partial x(t)} = p(x_0, x(t); t), \qquad \frac{\partial S(x_0, x; \tau)}{\partial x} = p(x_0, x; \tau).$$

A similar argument for the curve $\widetilde{x}(t)$ yields

$$\frac{\partial S(x, x_0; \tilde{t})}{\partial x_0} = p(x, x_0; \tilde{t}) = -p(x_0, x; t).$$

Differentiating with respect to x yields

$$\frac{\partial S}{\partial x \partial x_0} = -\frac{\partial p(x_0, x; t)}{\partial x} = -\frac{1}{J(p, x)}. \tag{2.1.8}$$

This means that $\frac{\partial S}{\partial x^i \partial x_0^k}$ and $-J_{ik}$ are inverse matrices. Since at conjugate points $\det J_{ik} = 0$, it follows that at conjugate points

$$D = \det\Big(-\frac{\partial S}{\partial x \partial x_0}\Big) = \pm\infty.$$

One of the main properties of D is that it satisfies a *continuity equation* of the following type (see [54]):

$$\frac{\partial D}{\partial t} + \sum_k \frac{\partial}{\partial x_k}(\dot{x}_k\, D) = 0, \tag{2.1.9}$$

which means that D can be interpreted as a density of paths. This fact will be useful in the future chapters that deal with path integration, van Vleck's formula and the geometric method of computing heat kernels.

2.1.4 The Case of Riemannian Geometry

Consider the Lagrangian $L(x, \dot{x}) = \frac{1}{2}\sum_{i,j=1}^n g_{ij}(x)\dot{x}_i\dot{x}_j$, with g_{ij} a positive definite and non-degenerate matrix at each point $x \in \mathbb{R}^n$. This Lagrangian can also be considered on the tangent bundle of an n-dimensional Riemannian manifold (M, g). However, since we are studying local properties, we can make the simplifying assumption $M = \mathbb{R}^n$.

In this case the Euler–Lagrange equations become the familiar equations of geodesics

$$\ddot{x}_k(t) + \sum_{i,j} \Gamma_{ij}^k(x(t))\dot{x}_i(t)\dot{x}_j(t) = 0, \qquad k = 1, \ldots, n.$$

The classical paths are called geodesics and satisfy the following local result.

Theorem 2.1.1. *Given a point x_0 on a Riemannian manifold (M, g), there is a neighborhood $\mathcal{V}$ of x_0 such that for any $x \in \mathcal{V}$, there is a unique geodesic joining the points x_0 and x.*

The aforementioned result does not necessarily hold globally for any Riemannian manifold. However, it holds on compact manifolds, and in general on metrically complete manifolds, as the Hopf–Rinov theorem states; see [79].

Moreover, any geodesic is locally minimizing the action functional. The distance on the Riemannian manifold (M, g) is defined by

$$d(A, B) = \inf\{\ell(x) \mid x \text{ geodesic}, \ x(0) = A, x(\tau) = B\},$$

where the length of $x(t)$ is

$$\ell(x) = \int_0^\tau \sqrt{\sum_{i,j} g_{ij}(x(t))\dot{x}_i(t)\dot{x}_j(t)\, dt.}$$

The classical action in this case is given by the formula

$$S(x_0, x; t) = \frac{d^2(x_0, x)}{2t}.$$

Along a geodesic we have $\nabla S = \dot{x}(t)$, where ∇ is the gradient in the metric g:

$$g(U, \nabla S) = dS(U),$$

for any tangent vector field U; see [24].

The Jacobi equation (2.1.5) on Riemannian manifolds takes the form

$$\ddot{\eta}(t) = R\big(\eta(t)\dot{x}(t)\big)\dot{x}(t), \tag{2.1.10}$$

where R is the Riemannian curvature tensor induced by the Levi–Civita connection D:

$$R(U, V)W = D_{[U,V]}W - [D_U, D_V]W.$$

If the manifold has constant curvature K, then

$$R(U, V)W = K\big(g(W, U)V - g(W, V)U\big).$$

In this case, under the additional hypotheses that $x(t)$ is unit speed and the Jacobi vector field η is normal to the geodesic $x(t)$, the Jacobi equation (2.1.10) can be written in the suggestive form

$$\ddot{\eta}(t) = -K\eta(t), \tag{2.1.11}$$

with the initial condition $\eta(0) = 0$. Solving, we distinguish the following cases:

(1) Euclidean case: $K = 0$, $\eta(t) = ct$, c constant. See Fig. 2.2a.
(2) Elliptic case: $K = k^2 > 0$, $\eta(t) = c\sin(kt)$. See Fig. 2.2b.
(3) Hyperbolic case: $K = -k^2 < 0$, $\eta(t) = c\sinh(kt)$. See Fig. 2.2c.

In cases (1) and (3) there are no conjugate points, while in case (2) there are infinitely many conjugate points to $x(0)$ that occur at $t_n = n\pi/k$, $n = 1, 2, \ldots$.

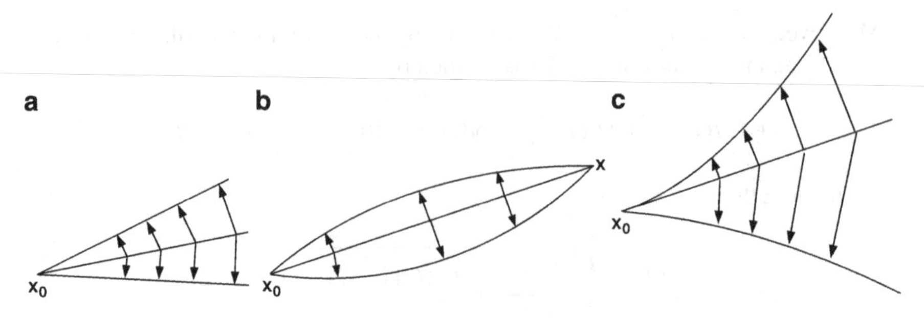

Fig. 2.2 (**a**) Euclidean case: $K = 0$; (**b**) elliptic case: $K > 0$; (**c**) hyperbolic case: $K < 0$

This behavior, for instance, occurs on a sphere. In general, all manifolds in situation (2) are compact. Just for the record, we include here a generalization of this case. Recall the notation for the diameter of a manifold as $\text{diam}(M) = \sup\{d(p,q); p,q \in M\}$.

Theorem 2.1.2 (Myers). *Let (M,g) be a complete, connected n-dimensional Riemannian manifold. If there is a positive number $k > 0$ such that*

$$Ric \geq (n-1)k^2 g, \tag{2.1.12}$$

where Ric denotes the Ricci tensor of M, then the following relations hold:

(i) $diam(M) \leq \pi/k$
(ii) M is compact

As a special case of the previous theorem, we have

Corollary 2.1.3. *If (M,g) is a complete, connected n-dimensional Riemannian manifold with constant curvature K satisfying $K \geq k^2 > 0$, then*

(i) $diam(M) \leq \pi/k$
(ii) M is compact

References for Riemannian geometry and its variational methods are the books [24, 79, 94].

2.1.5 Examples of Lagrangian Dynamics

In the following examples we shall assume that the initial and final positions of the particle $x_0 = x(0)$ and $x = x(\tau)$ are given and we shall determine its trajectory between these endpoints.

Example 2.1.4 (The natural Lagrangian). Let $U(x)$ be a smooth potential, and consider the Lagrangian $L = \frac{1}{2}\dot{x}^2 - U(x)$. The Euler–Lagrange equation is

$$\ddot{x}(t) = -\frac{\partial U}{\partial x}. \tag{2.1.13}$$

Given the boundary conditions $x(0) = x_0$, $x(\tau) = x$, (2.1.13) might not always have a unique solution, regardless of how close the boundary points x_0 and x are: see [24]. If the potential is linear or quadratic, the aforementioned boundary value equation has a unique solution, and in this case the action is well defined. The Jacobi equation for the previous Lagrangian takes the form

$$\ddot{\eta}(t) = -\frac{\partial^2 U(x)}{\partial x^2}\bigg|_{x(t)} \eta(t). \tag{2.1.14}$$

We note that if $U(x)$ is linear or quadratic in x, then the previous equation becomes similar to (2.1.11).

Example 2.1.5 (The free particle). If the Lagrangian is $L(\dot{x}, x) = \frac{1}{2}|\dot{x}|^2 = \frac{1}{2}\sum_j \dot{x}_j^2$, then the variational equation (2.1.13) is $\ddot{x}(t) = 0$, and the trajectories are the straight lines

$$x(t) = x_0 + \frac{t}{\tau}\big(x - x_0\big). \tag{2.1.15}$$

The Jacobi equation (2.1.14) is $\ddot{\eta}(t) = 0$. Using the initial conditions $\eta(0) = 0$, $\dot{\eta}(0) = 1$ yields the Jacobi vector field $\eta(t) = t$. The classical action is obtained by integrating the Lagrangian along the classical path:

$$S(x_0, x; \tau) = \int_0^\tau \frac{1}{2}\Big(\frac{x - x_0}{\tau}\Big)^2\, dt = \frac{(x - x_0)^2}{2\tau}.$$

We recuperate formula (2.1.8) by taking the mixed derivative of the action

$$\frac{\partial^2 S}{\partial x \partial x_0} = -\frac{1}{\tau} = -\frac{1}{\eta(\tau)} \neq 0.$$

Hence there are no conjugate points to x_0 in this case. Next we shall check the continuity equation (2.1.9). We have $D = \frac{1}{t}$, $\dot{x}^k(t) = \frac{x^k - x_0^k}{t}$, and then

$$\begin{aligned}\frac{\partial D}{\partial t} + \sum_{k=1}^n \partial_{x_k}(v_k D) &= \frac{\partial}{\partial t}\frac{1}{t} + \sum_{k=1}^n \frac{\partial}{\partial x_k}\left(\frac{x^k - x_0^k}{t^2}\right)\\ &= -\frac{1}{t^2} + \frac{1}{t^2} = 0.\end{aligned}$$

Example 2.1.6 (Particle in constant gravitational field). The Lagrangian describing the dynamics of a free-falling particle under the action of a constant gravitational force is given by $L(\dot{x}, x) = \frac{1}{2}\dot{x}^2 - kx$, $k > 0$. The trajectory $x(t)$ satisfies Galileo's equation $\ddot{x} = -k$ with the solution

$$x(t) = -\frac{k}{2}t^2 + (x - x_0)\frac{t}{\tau} + \frac{k}{2}t\tau + x_0.$$

Since the Jacobi equation is $\ddot{\eta}(t) = 0$ with the Jacobi vector field $\eta(t) = t \neq 0$, for $t > 0$, there are no conjugate points along the trajectory. We leave the computation of the classical action as an instructive exercise to the reader.

Example 2.1.7 (The linear oscillator). If $L(\dot{x}, x) = \frac{1}{2}\dot{x}^2 - \frac{k}{2}x^2$, $k > 0$, then the Euler–Lagrange equation is $\ddot{x}(t) = kx$, with the solution

$$x(t) = \left[x - x_0 \cosh(\sqrt{k}\tau)\right]\frac{\sinh(\sqrt{k}t)}{\sinh(\sqrt{k}\tau)} + x_0 \cosh(\sqrt{k}t).$$

This represents the trajectory of a particle under an elastic force $F(x) = kx$. The Jacobi equation (2.1.14) can be written as $\ddot{\eta}(t) = k\eta(t)$, and the Jacobi vector field is given by $\eta(t) = \frac{1}{\sqrt{k}} \sinh(\sqrt{k}t) \neq 0$, for $t > 0$. Then there are no conjugate points to x_0 along the classical path $x(t)$. A tedious computation shows that the classical action in this case is

$$S(x_0, x; \tau) = \frac{\sqrt{k}}{2\sinh(\sqrt{k}\,\tau)}\left[(x^2 + x_0^2)\cosh(\sqrt{k}\,\tau) - 2xx_0\right].$$

Differentiating yields

$$\frac{\partial^2 S}{\partial x \partial x_0} = -\frac{\sqrt{k}}{\sinh(\sqrt{k}\,\tau)} = -\frac{1}{\eta(t)},$$

which is (2.1.8).

Example 2.1.8 (The simple pendulum). The dynamics of a unit mass pendulum bob with unit length pendulum string under the action of the gravitational force is modeled by the Lagrangian $L(\dot{x}, x) = \frac{1}{2}\dot{x}^2 - k(1 - \cos x)$, $k > 0$. Its variational equation is given by the second-order differential equation $\ddot{x}(t) = -k \sin x$ that cannot be solved using elementary functions. A solution using elliptic functions can be found in [24], p. 38. In the case of small oscillations the dynamics is approximated by the linearized pendulum equation $\ddot{x}(t) = -kx$, with the solution

$$x(t) = \left[x - x_0 \cos(\sqrt{k}\,\tau)\right]\frac{\sin(\sqrt{k}\,t)}{\sin(\sqrt{k}\tau)} + x_0 \cos(\sqrt{k}\,t).$$

The Jacobi equation (2.1.14) becomes

$$\ddot{\eta}(t) = -k\eta(t).$$

Under the standard initial conditions, the Jacobi vector field is $\eta(t) = \frac{1}{\sqrt{k}}\sin(\sqrt{k}\,t)$. Hence the conjugate points to x_0 will occur along $x(t)$ at instances $t_n = n\pi/\sqrt{k}$, $n = 1, 2, \ldots$. The classical action is

$$S(x_0, x; \tau) = \frac{\sqrt{k}}{2\sin(\sqrt{k}\,\tau)}\big[(x^2 + x_0^2)\cos(\sqrt{k}\,\tau) - 2xx_0\big], \qquad \tau \notin \{n\pi/\sqrt{k}\}.$$

Example 2.1.9 (Particle in a constant electric field). The Lagrangian of a particle in a one-dimensional electric potential $U(x) = k/x$, $x > 0$, is $L(\dot{x}, x) = \frac{1}{2}\dot{x}^2 - k/x$, $k > 0$. The trajectories satisfy the equation $\ddot{x} = k/x^2$, which cannot be integrated using elementary functions. The same occurs for the Jacobi equation

$$\ddot{\eta} = -\frac{2k}{x^3}\eta.$$

2.2 Hamiltonian Mechanics

An alternate way of describing the dynamics of a particle is using the *Hamiltonian function.* A real-valued function defined on the cotangent bundle of the coordinate space T^*M is called a Hamiltonian function. There is an intimate relationship between the Lagrangian and the Hamiltonian associated with a moving particle. The Hamiltonian associated with a Lagrangian is obtained using Legendre's transform

$$H(x, p) = p\dot{x} - L(x, \dot{x}, t),$$

where $\dot{x}$ is a function of the momentum p obtained by solving the equation

$$p = \frac{\partial L}{\partial \dot{x}}. \tag{2.2.16}$$

For instance, if the Lagrangian $L(x, \dot{x}) = \frac{1}{2}\dot{x}^2 - x^m$, then the Hamiltonian is

$$H(x, p) = \frac{1}{2}p^2 + x^m.$$

The procedure works vice versa, i.e., for a given Hamiltonian, the associated Lagrangian can be obtained by

$$L(x, \dot{x}, t) = p\dot{x} - H(x, p),$$

with the momentum p given by (2.2.16).

The dynamics of the particle is described in the cotangent space by the Hamiltonian system of equations

$$\begin{aligned} \dot{x} &= \partial_p H, \\ \dot{p} &= -\partial_x H. \end{aligned}$$

A solution $\big(x(t), p(t)\big)$ of the aforementioned system is called a *bicharacteristic.*

If the Hamiltonian H does not depend explicitly on the time variable t, the system is said to be conservative, because in this case the Hamiltonian evaluated along the above solutions is constant. This can be verified by using the chain rule. If $\big(x(t), p(t)\big)$ is a bicharacteristic curve, then

$$\frac{d}{dt}H\big(x(t), p(t)\big) = \dot{x}(t)\,\partial_x H + \dot{p}(t)\,\partial_p H = 0.$$

The value of the Hamiltonian evaluated along the bicharacteristic is called the *total energy* of the particle. This is a first integral of motion.

If $(x(t), p(t))$ is a bicharacteristic curve, then the component $x(t)$ is a solution of the Euler–Lagrange equation. Conversely, if $x(t)$ is a solution of the Euler–Lagrange equations, then $\big(x(t), \frac{\partial L}{\partial \dot{x}}(t)\big)$ is a bicharacteristic curve, called the *lift* of $x(t)$.

2.3 The Hamilton–Jacobi Equation

The *classical action* associated with the Lagrangian $L(x, \dot{x}, t)$ is obtained by integrating the Lagrangian along the solution of the Euler–Lagrange equation

$$S_{cl}(x_0, x; \tau) = \int_0^\tau L\big(x(t), \dot{x}(t), t\big)\,dt. \tag{2.3.17}$$

This assumes that there is only one solution $x(t)$ satisfying the boundary conditions $x(0) = x_0$ and $x(\tau) = x$. The classical action is a solution of the following *Hamilton–Jacobi equation*:

$$\frac{\partial}{\partial \tau}S_{cl} + H(x, \partial_x S_{cl}) = 0. \tag{2.3.18}$$

Since the above equation is nonlinear, there might be more than one solution. In the case of a conservative system, if $E = E(x_0, x, \tau) = H$ denotes the energy along the solution, the equation becomes

$$\frac{\partial}{\partial \tau}S_{cl} = -E,$$

with the solution

$$S_{cl}(x_0, x, \tau) = -E(x_0, x, \tau)\tau + c(x_0, x), \tag{2.3.19}$$

where $c(x_0, x) = S_{cl}(x_0, x, 0)$ is the initial condition.

The action plays a very important role in the geometric methods of finding the heat kernel. To conclude our brief discussion, there are three ways of computing the action.

1. *Starting from the Lagrangian.* After solving the Euler–Lagrange equations, we integrate the Lagrangian along solutions using formula (2.3.17). This is a robust method and works as long as we are able to solve the Euler–Lagrange system of equations explicitly.

2. *Solving the Hamilton–Jacobi equation*. Substitute $p = \partial_x S$ in the expression of the Hamiltonian H and solve (2.3.18). Because it is nonlinear, there are no standard methods to solve the aforementioned equation. Depending on the problem, one may try different strategies. For instance, we can try to look for particular solutions of the type $S(x,t) = a(t)+b(x)$. Substituted in (2.3.18), the functions $a(t)$ and $b(x)$ satisfy the separable equation

$$a'(t) + H(x, \partial_x b) = 0,$$

so there is a constant C such that

$$\begin{aligned} a'(t) &= -C \Longrightarrow a(t) = -Ct + a(0), \\ H(x, \partial_x b) &= C. \end{aligned} \tag{2.3.20}$$

If the Hamiltonian H does not depend explicitly on the variable x, (2.3.20) can sometimes be reduced to an *eikonal equation*. For instance, if $H = \frac{1}{2}|p|^2$, then the function $b(x)$ satisfies

$$|\partial_x b|^2 = C/2, \qquad C > 0.$$

This equation has infinitely many solutions of the type

$$b(x) = \sqrt{\frac{2}{C}}|x - x_0| = \sqrt{\frac{2}{C}\sum_{i=1}^{n}(x_i - x_i^0)^2}, \qquad x_0 \in \mathbb{R}^n.$$

3. *The conservative Hamiltonian case*. If the Hamiltonian does not depend explicitly on the time parameter t, the action function is given by (2.3.19), where the energy should be expressed in terms of the endpoints $x(0)$ and $x(\tau)$.

Next we shall work out an example. The aforementioned methods will be used to find the action function for the *free particle case*; see Example 2.1.5. Differentiating in (2.1.15) yields

$$\dot{x} = \frac{1}{\tau}\big(x(\tau) - x(0)\big).$$

The Lagrangian along the solution is

$$L(x, \dot{x}) = \frac{1}{2}\dot{x}^2 = \frac{|x(\tau) - x(0)|^2}{2\tau^2},$$

and using (2.3.17) we obtain the action

$$S(x_0, x; \tau) = \int_0^\tau \frac{|x(\tau) - x(0)|^2}{2\tau^2}\, dt = \frac{|x(\tau) - x(0)|^2}{2\tau} = \frac{|x - x_0|^2}{2\tau}.$$

If we try to solve the Hamilton–Jacobi equation, using $\partial_x S = p = \frac{\partial L}{\partial \dot{x}} = \dot{x}$, (2.3.18) becomes

$$\frac{\partial}{\partial \tau} S + \frac{1}{2} |\dot{x}(t)|^2 = 0 \Longleftrightarrow \frac{\partial}{\partial \tau} S = -\frac{1}{2} \frac{|x - x_0|^2}{\tau^2} \Longleftrightarrow$$
$$\frac{\partial}{\partial \tau} S = \frac{\partial}{\partial \tau} \left(\frac{|x - x_0|^2}{2\tau} \right).$$

Integrating yields the aforementioned relation for the action.

The reader can find more advanced topics on the calculus of variations in [6, 24, 52].

Chapter 3
The Geometric Method

This chapter deals with a construction of heat kernels from the geometric point of view. Each operator will be associated with a geometry. Investigating the geodesic flow in this geometry, one can describe the heat kernels for a large family of operators. The idea behind this method is that the heat flow propagates along the geodesics of the associated geometry. The "density" of the heat flow is described by a volume function that satisfies a transport equation which is an analog of the continuity equation from fluid dynamics. This corresponds to the density of paths given by the van Vleck determinant in the path integral approach. This method works for elliptic operators with or without potentials or linear terms. The method can be modified to work even in the case of sub-elliptic operators, as the reader will become familiar with in Chaps. 9 and 10. This method was initially applied for the Heisenberg Laplacian; see, for instance [28].

3.1 Heat Kernel for $\mathbb{L} = \frac{1}{2}\sum_{i,j=1}^{n} a_{ij}\,\partial_{x_i}\partial_{x_j}$

Consider the elliptic differential operator $\mathbb{L} = \frac{1}{2}\sum_{i,j=1}^{n} a_{ij}\,\partial_{x_i}\partial_{x_j}$. Since the matrix a_{ij} is symmetric, nondegenerate and positive definite at each point, then $(\mathbb{R}^n, a^{ij})$ becomes a Riemannian space, where $(a_{ij})^{-1} = (a^{ij})$. The study of the geometry of this space provides a method for finding the heat kernel of the operator $\mathbb{L}$.

The heat kernel models the physical phenomenon of heat propagation from a point x_0 to another point x within time t. The main idea is that the heat flows along the geodesics of the space $(\mathbb{R}^n, a^{ij})$ from the initial point x_0 to the final point x. For the sake of simplicity, we shall assume in Theorems 3.1.1 and 3.4.3 that the Riemannian space $(\mathbb{R}^n, a^{ij})$ has the property that any two points can be joined by a *unique* geodesic, i.e., the space is *geodesically complete*, and there are no *conjugate* or *cut points* along the geodesics. This condition is always satisfied locally on a Riemannian manifold, so the aforementioned theorems provide a local expression for the heat kernels. However, in some cases, for instance in the case of spaces with negative or zero curvature, this property is globally satisfied, and hence the expression for the heat kernels is globally defined.

O. Calin et al., *Heat Kernels for Elliptic and Sub-elliptic Operators*,
Applied and Numerical Harmonic Analysis, DOI 10.1007/978-0-8176-4995-1_3,

The case when the cut locus of x_0 is not empty, i.e., there is more than one geodesic joining the points x_0 and x, the analysis is more complicated and shall be treated separately. The geometric method described in this chapter is based on finding geodesics.

Obtaining the geodesics. The *geodesics* of the Riemannian space $(\mathbb{R}^n, a^{ij})$ can be obtained in two ways. Consider the following Hamiltonian function obtained by taking the principal symbol of the operator $\mathbb{L}$:

$$H(x,p) = \frac{1}{2}\sum_{i,j=1}^{n} a_{ij}\, p_i\, p_j. \tag{3.1.1}$$

One way to define the geodesics is by considering the x-component of the bicharacteristic curve that solves the Hamiltonian system

$$\dot{x} = H_p,$$
$$\dot{p} = -H_x$$

with the boundary conditions

$$x(0) = x_0, \qquad x(t) = x.$$

The solution will be a geodesic joining the points x_0 and x. For a large class of ODE systems, the solution of the previous boundary value problem is unique.

The alternate approach for finding the geodesics is by solving the Euler–Lagrange system for the Lagrangian associated with the Hamiltonian (3.1.1).

Assuming that we were able to find the geodesic $x(s)$ in one way or another, we can measure its length by the formula

$$\ell(x) = \int_0^t \sqrt{\sum_{i,j=1}^{n} a^{ij}(x(t))\dot{x}^i(t)\dot{x}^j(t)}\, dt;$$

Denote by $d(x_0, x)$ the Riemannian distance between the points x_0 and x measured in the metric a_{ij}. This is the length of the shortest geodesic $x(s)$ which joins $x_0 = x(0)$ with $x = x(t)$. This length is independent of the parameter t, in the sense that it is invariant when the geodesic changes its parameterization. One may show that the associated action is

$$S(x_0, x; t) = \frac{d^2(x_0, x)}{2t};$$

see, for instance, [24]. The aforementioned action satisfies the Hamilton–Jacobi equation

$$\partial_t S + H(\nabla S) = 0; \tag{3.1.2}$$

where H is the Hamiltonian (3.1.1). There is an alternate way of finding the action by integrating the Lagrangian along the geodesic

$$S(x_0, x; t) = \int_0^t \frac{1}{2} L\big(x(s), \dot{x}(s)\big)\, ds,$$

where $x : [0, t] \to \mathbb{R}^n$ is the geodesic which satisfies $x(0) = x_0$ and $x(t) = x$.

There is an intimate relationship between the geometric method and the path integration technique. In Chap. 7 the previous action is called the *classical action* and is denoted by S_{cl}. It will be shown that the heat kernel can be represented as a path integral by

$$K(x_0, x; t) = \int_{\mathfrak{P}_{x_0,x;t}} e^{-S(\phi,t)}\, d\mathfrak{m}(\phi) = e^{-S_{cl}(x_0,x;t)} \int_{\mathfrak{P}_{0,0;t}} e^{-S(\psi,t)}\, d\mathfrak{m}(\psi).$$

We shall see that in several cases the path integral $\int_{\mathfrak{P}_{0,0;t}} e^{-S(\psi,t)}\, d\mathfrak{m}(\psi)$ is a function of t only. In this case the heat kernel takes the friendly form

$$K(x_0, x; t) = V(t) e^{-S_{cl}(x_0,x;t)}, \qquad t > 0, \tag{3.1.3}$$

so we may say that the geometric method is a particular case of the path integral method. More precisely, we have the following result.

Theorem 3.1.1. *Consider the operator $\mathbb{L} = \frac{1}{2}\sum_{i,j=1}^{n} a_{ij}\partial_{x_i}\partial_{x_j}$ and assume that $\mathbb{L}(S_{cl})$ is a function of t only. If there is only one geodesic joining the points x_0 and x within time t, then the heat kernel of $\mathbb{L}$ is given by (3.1.3), where $V(t)$ satisfies the following transport equation:*

$$V'(t) + V(t)\mathbb{L}(S_{cl}) = 0, \tag{3.1.4}$$

with the initial condition

$$V(0) = \frac{1}{\lim\limits_{t \searrow 0} \int e^{-S_{cl}(x,x_0;0)}\, dx}. \tag{3.1.5}$$

Proof. Let $S = S_{cl}(x_0, x; t)$ denote the classical action. Since

$$\partial_{x_i}\partial_{x_j} e^{-S} = \partial_{x_i}\big(-e^{-S}\partial_{x_j} S\big) = e^{-S}(\partial_{x_i} S\, \partial_{x_j} S - \partial_{x_i}\partial_{x_j} S),$$

then

$$\mathbb{L}(e^{-S}) = \frac{1}{2}\sum_{i,j=1}^{n} a_{ij}\partial_{x_i}\partial_{x_j} e^{-S} = e^{-S}\left(H(\nabla S) - \mathbb{L}(S)\right). \tag{3.1.6}$$

On the other hand,

$$\begin{aligned}\partial_t\left(e^{-S}V(t)\right) &= \partial_t e^{-S}\,V(t) + e^{-S}\partial_t V(t)\\ &= e^{-S}\left(\partial_t V(t) - V(t)\partial_t S\right).\end{aligned} \tag{3.1.7}$$

Subtracting (3.1.6) and (3.1.7) yields

$$\begin{aligned}(\partial_t - \mathbb{L})\left(e^{-S}V(t)\right) &= \partial_t\left(e^{-S}V(t)\right) - \mathbb{L}\left(e^{-S}V(t)\right)\\ &= e^{-S}\left\{\partial_t V + V(t)\,\mathbb{L}(S) - V(t)\underbrace{(\partial_t S + H(\nabla S))}_{=0}\right\}\\ &= e^{-S}\left(\partial_t V + V(t)\,\mathbb{L}(S)\right)\\ &= 0,\end{aligned}$$

where we used the Hamilton–Jacobi equation (3.1.2) and the transport equation (3.1.4).

The solution $V(t)$ of the transport equation is unique up to a multiplicative constant. Relation (3.1.5) is fixing this constant. The reason behind relation (3.1.5) is the *normalization condition* for the heat kernel

$$\int K(x, x_0; t)\,dx = 1.$$

■

Remark 3.1.2. Since the differential operator $\mathbb{L}$ is quadratic, the term $\mathbb{L}(S)$ does not depend on x in the cases when the action S depends on x linearly or quadratically. Formula (3.1.3) resembles van Vleck's formula with van Vleck's determinant $V(t)$.

The function $V(t)$ shows how the geodesics spread away from the point x_0 as t increases. For this reason we shall refer to V as the *volume function*.

3.2 Interpretation of the Volume Function

One of the well-known conservation equations in fluid dynamics is the *continuity equation*

$$\frac{\partial \rho}{\partial t} + \operatorname{div}(\rho\, v_k) = 0,$$

where $v = (v_1, \ldots, v_n)$ denotes the velocity of the fluid and ρ is the fluid density. If the fluid follows the direction of the geodesic flow of a Riemannian manifold, then the velocity of the flow is related to the action by

$$v = \nabla S_{cl}.$$

Assuming the fluid is homogeneous, i.e., it has the same density $\rho(t)$ at each point, the aforementioned equation becomes

$$\rho'(t) + \rho(t)\mathrm{div}\nabla S_{cl} = 0.$$

Substituting $V(t) = \sqrt{\rho(t)}$, we get

$$V'(t) + V(t)\mathbb{L}S_{cl} = 0,$$

where $\mathbb{L} = \frac{1}{2}\Delta$. This is exactly the transport equation (3.1.4) in the case of the Laplace–Beltrami operator. Hence the square of the volume function describes the "density" of the geodesic flow, i.e., the way the geodesics spread away from a given geodesic. We also note the relation with the Jacobi vector field emphasized in Chap. 2.

In the following we shall deal with explicit computations in a few particular cases.

3.3 The Operator $\mathbb{L} = \frac{1}{2}\sum_{i=1}^{n} b_i \partial_{x_i}^2$

Consider the operator $\mathbb{L} = \frac{1}{2}\sum_{i=1}^{n} b_i \partial_{x_i}^2$, with $b_i > 0$ constants. Associate the Hamiltonian function

$$H(p) = \frac{1}{2}\sum_{i=1}^{n} b_i p_i^2.$$

Solving the associated Hamiltonian system

$$\begin{aligned} \dot{x}_i(s) &= b_i p_i, \\ \dot{p}_i(s) &= 0, \\ x(0) &= x_0, \quad x(t) = x, \end{aligned}$$

yields

$$x_i(s) = b_i p_i s + x_0 = \frac{x_i - x_i^0}{t} s + x_i^0, \quad i = 1, \ldots, n,$$

where $x(t) = x = (x_1, \ldots, x_n)$, $x(0) = x^0 = (x_1^0, \ldots, x_n^0)$, and the p_i are constants.

The associated Lagrangian is

$$L(x, \dot{x}) = \frac{1}{2}\sum_{i=1}^{n} \frac{1}{b_i}\dot{x}_i^2,$$

and the action becomes

$$S(x_0, x; t) = \int_0^t L(x(s), \dot{x}(s))\, ds = \frac{1}{2}\sum_{i=1}^{n} \frac{1}{b_i}\int_0^t \left(\frac{x_i - x_i^0}{t}\right)^2 ds$$
$$= \sum_{i=1}^{n} \frac{(x_i - x_i^0)^2}{2b_i t}.$$

Since

$$\mathbb{L}S = \frac{1}{2}\sum_{j=1}^{n} b_j \partial_{x_j}^2 \left(\sum_{i=1}^{n} \frac{(x_i - x_i^0)^2}{2b_i t}\right) = \frac{n}{2t},$$

the transport equation becomes

$$V'(t) + \frac{n}{2t}V(t) = 0.$$

Integrating yields

$$V(t) = ct^{-n/2}.$$

By Theorem 3.1.1, the heat kernel of $\frac{1}{2}\sum_{i=1}^{n} b_i \partial_{x_i}^2$ is

$$K(x_0, x; t) = ct^{-n/2} e^{-\sum_{i=1}^{n} \frac{(x_i - x_i^0)^2}{2b_i t}}.$$

The constant c is determined from the relation

$$\int_{\mathbb{R}^n} K(x_0, x; t)\, dx = 1, \qquad (3.3.8)$$

by using the well-known formula

$$\int_{\mathbb{R}^n} e^{-\frac{1}{2}\sum_{i,j=1}^{n} M_{ij} u_i u_j}\, du_1 \cdots du_n = \frac{(2\pi)^{n/2}}{(\det M)^{1/2}}.$$

In our case $u_i = x_i - x_i^0$ and $M_{ij} = \frac{1}{b_i t}\delta_{ij}$, with $\det M = t^{-n}(\prod_{i=1}^{n} b_i)^{-1}$. Integrating in (3.3.8) yields

$$1 = \int_{\mathbb{R}^n} K(x_0, x; t)\, dx = ct^{-n/2} \int_{\mathbb{R}^n} e^{-\frac{1}{2}\sum_{i=1}^{n} \frac{(x_i - x_i^0)^2}{b_i t}}\, dx_1 \cdots dx_n$$
$$= ct^{-n/2} \frac{(2\pi)^{n/2}}{\left(\frac{1}{t^n \prod_{i=1}^{n} b_i}\right)^{1/2}} = c(2\pi)^{n/2} \left(\prod_{i=1}^{n} b_i\right)^{1/2}.$$

Hence

$$c = (2\pi)^{-n/2}\left(\prod_{i=1}^{n} b_i\right)^{-1/2}.$$

We have arrived at the following result.

Theorem 3.3.1. *Let $b_i > 0$. Then the heat kernel of $\mathbb{L} = \frac{1}{2}\sum_{i=1}^{n} b_i \partial_{x_i}^2$ is given by*

$$K(x, x_0; t) = (2\pi t)^{-n/2}\left(\prod_{i=1}^{n} b_i\right)^{-1/2} e^{-\sum_{i=1}^{n}\frac{(x_i - x_i^0)^2}{2b_i t}}, \quad t > 0.$$

When $b_i = 1$, we obtain the familiar formula for the heat kernel of the Laplacian.

Corollary 3.3.2. *The heat kernel for $\mathbb{L} = \frac{1}{2}\sum_{i=1}^{n} \partial_{x_i}^2$ is given by*

$$K(x, x_0; t) = (2\pi t)^{-n/2} e^{-\sum_{i=1}^{n}\frac{(x_i - x_i^0)^2}{2t}}, \quad t > 0.$$

3.4 The Generalized Transport Equation

In Theorem 3.1.1 we had assumed that $L(S_{cl})$ depends on t only. In this case the transport equation (3.1.4) can easily be solved by integration. However, there are some cases when $L(S_{cl})$ depends on both variables t and x. In these cases the transport equation is replaced by the *generalized transport equation*. This equation describes how the heat density evolves along the geodesics.

We shall start with an example. One of the operators for which $\mathbb{L}(S_{cl})$ depends on both x and t is $\mathbb{L} = \frac{1}{2}x\partial_x^2$. The following discussion applies to the domain $\{x; x > 0\}$, where the operator is elliptic. The associated Hamiltonian function is

$$H(x, p) = \frac{1}{2}xp^2. \tag{3.4.9}$$

The Hamiltonian system is

$$\begin{aligned}\dot{x} &= H_p = xp,\\ \dot{p} &= -H_x = -\frac{1}{2}p^2,\end{aligned}$$

with boundary conditions $x(0) = x_0$, $x(t) = x$.

First we solve for p. Integrating in the equation

$$-\frac{\dot{p}}{p^2} = \frac{1}{2}$$

yields

$$p = \frac{1}{s_0 + \frac{1}{2}s},$$

with $s_0 \in \mathbb{R}$. Substituting in the first Hamiltonian equation, we get

$$\frac{\dot{x}}{x} = \frac{1}{s_0 + \frac{1}{2}s} \Longrightarrow \ln x(s) = \int \frac{ds}{s_0 + \frac{1}{2}s} = \ln\left(s_0 + \frac{1}{2}s\right)^2 + C,$$

and hence

$$x(s) = e^C \left(s_0 + \frac{1}{2}s\right)^2.$$

It is worth noting that $x(s) > 0$ for $s > 0$ and hence only points on the positive semi-axis can be joined by a geodesic, i.e., no heat propagates in the negative semi-axis, a fact which agrees with the domain $\{x; x > 0\}$ considered before.

Given the endpoints $x_0, x > 0$, we shall determine the unique constants C and s_0 such that $x_0 = x(0)$ and $x = x(t)$. This will show that the geodesic between x_0 and x within time t is unique.

Setting $s = 0$ and $s = t$, we arrive at the system

$$\begin{aligned} x_0 &= e^C s_0^2, \\ x &= e^C \left(s_0 + \frac{1}{2}t\right)^2. \end{aligned}$$

The elimination method provides

$$s_0 = \frac{t}{2(\sqrt{x/x_0} - 1)}, \qquad e^C = \frac{4}{t^2}(x + x_0 - \sqrt{x x_0}). \tag{3.4.10}$$

The Lagrangian associated with the Hamiltonian (3.4.9) is obtained by applying the Legendre transform

$$\begin{aligned} L(x, \dot{x}) = p\dot{x} - H &= \frac{\dot{x}}{x}\dot{x} - \frac{1}{2}x\frac{\dot{x}^2}{x^2} \\ &= \frac{1}{2}\frac{\dot{x}^2}{x}. \end{aligned}$$

Integrating along the geodesic and using (3.4.10), we obtain the classical action

$$\begin{aligned} S_{cl}(x_0, x; t) &= \int_0^t \frac{1}{2}\frac{\dot{x}^2(s)}{x(s)}\,ds = \int_0^t \frac{1}{2}\frac{\left(e^C\left(s_0 + \frac{1}{2}s\right)\right)^2}{e^C\left(s_0 + \frac{1}{2}s\right)^2}\,ds = \frac{1}{2}e^C t \\ &= \frac{2}{t}(x + x_0 - \sqrt{x x_0}). \end{aligned}$$

A straightforward computation shows

$$\mathbb{L}(S_{cl}) = \frac{1}{2}x\partial_x^2 S_{cl} = \frac{1}{4t}\left(\frac{x_0}{x}\right)^{1/2}.$$

Since the function V also depends on x, the hypothesis of Theorem 3.1.1 is not satisfied, and therefore the transport equation (3.1.4) does not hold.

In the following we shall find the equation satisfied by the volume function $V(t,x)$. This will be called the *generalized transport equation.*

Assume the heat kernel has the form $K = V(t,x)e^{-S_{cl}}$. Since

$$\begin{aligned}
\partial_x K &= e^{-S_{cl}}(\partial_x V - V\,\partial_x S_{cl}),\\
\partial_x^2 K &= e^{-S_{cl}}\left(\partial_x^2 V - 2\partial_x S_{cl}\,\partial_x V + V(\partial_x S_{cl})^2 - V\,\partial_x^2 S_{cl}\right),\\
\partial_t K &= e^{-S_{cl}}\left(\partial_t V - \partial_t S_{cl}\,V\right),
\end{aligned}$$

we obtain

$$\begin{aligned}
\mathbb{L}K &= \partial_t K - \frac{1}{2}x\partial_x^2 K\\
&= e^{-S_{cl}}\left\{\partial_t V - V\underbrace{\left(\partial_t S_{cl} + \frac{1}{2}x(S_x)^2\right)}_{=0} - \frac{1}{2}x\left(\partial_x^2 V - 2\partial_x S_{cl}\,\partial_x V - V\partial_x^2 S_{cl}\right)\right\}\\
&= e^{-S_{cl}}\left\{\partial_t V - \frac{1}{2}x\left(\partial_x^2 V - 2\partial_x S_{cl}\,\partial_x V - V\partial_x^2 S_{cl}\right)\right\},
\end{aligned}$$

where we used that S_{cl} satisfies the Hamilton–Jacobi equation

$$\partial_t S_{cl} + \frac{1}{2}x(S_x)^2 = 0.$$

Since $\mathbb{L}K = 0$ for any $t > 0$, it follows that $V(t,x)$ satisfies the following *generalized transport equation*:

$$\partial_t V - \frac{1}{2}x\left(\partial_x^2 V - 2\partial_x S_{cl}\,\partial_x V - V\partial_x^2 S_{cl}\right) = 0. \tag{3.4.11}$$

An explicit computation of the coefficients of $\partial_x V$ and V in the above equation yields

$$\partial_t V - \frac{1}{2}x\partial_x^2 V + \frac{1}{t}(2x - \sqrt{xx_0})\partial_x V + \frac{1}{4t}\sqrt{\frac{x_0}{x}}V = 0. \tag{3.4.12}$$

This equation does not have a straightforward solution. The function V should be looked for as a product between an exponential term and a modified Bessel function, as in (8.20.79), where the operator is associated with a squared Bessel process.

However, an easy explicit solution can be obtained if $x_0 = 0$. In this particular case it is not hard to check that a solution is

$$V(t,x) = \frac{cx}{t^2},$$

with c constant. In this case the classical action is $S_{cl}(0,x;t) = \frac{2}{t}x$. By Theorem 3.1.1, the heat kernel for $\frac{1}{2}x\partial_x^2$ from the origin is

$$K(0,x;t) = V(t,x)e^{-S_{cl}} = \frac{cx}{t^2}e^{-\frac{2}{t}x}.$$

The constant c is determined from the condition

$$\int_{x>0} K(0,x;t)\,dx = 1.$$

Since integration by parts shows that

$$\int_0^\infty \frac{cx}{t^2}e^{-\frac{2}{t}x}\,dx = \frac{c}{4},$$

it follows that $c = 4$. If $x_0 = 0$, then $s_0 = 0$, and $e^c = 4x/t^2$, and the geodesics joining $x_0 = 0$ and x at time t is $x(s) = xs^2/t^2$.

Proposition 3.4.1. *The heat kernel for $\frac{1}{2}x\partial_x^2$ from the origin is given by*

$$K(0,x;t) = \frac{4x}{t^2}e^{-\frac{2}{t}x}, \qquad t>0, x>0.$$

Proof. We have shown that $\frac{1}{2}x\partial_x^2 K = 0$ for $t > 0$. We still need to check that $\lim_{t\searrow 0} K(0,x;t) = \delta_0$ in the distribution sense. For any test function ϕ, we have

$$\begin{aligned}
\lim_{t\searrow 0} K(0,x;t)(\phi) &= \lim_{t\searrow 0}\int_0^\infty K(0,x;t)\phi(x)\,dx \\
&= \lim_{t\searrow 0}\int_0^\infty \frac{4x}{t^2}e^{-\frac{2}{t}x}\phi(x)\,dx \quad \left(\text{make } u = \frac{2x}{t}\right) \\
&= \lim_{t\searrow 0}\int_0^\infty \frac{4}{t^2}\frac{tu}{2}e^{-u}\phi\left(\frac{tu}{2}\right)\frac{t}{2}\,du = \lim_{t\searrow 0}\int_0^\infty ue^{-u}\phi\left(\frac{tu}{2}\right)du \\
&= \phi(0)\underbrace{\int_0^\infty ue^{-u}\,du}_{=1} = \phi(0) = \delta_0(\phi).
\end{aligned}$$

■

If $x_0 > 0$, $x > 0$, then the heat travels in infinitely many ways between x_0 and x since it is reflected at the wall $x = 0$; the kernel will be given by a series in this case.

Remark 3.4.2. The change of variable $x = r^2$ transforms the operator $\frac{1}{2}x\partial_x^2$ into an operator that looks just like a two-dimensional Bessel operator with a changed sign

$$\frac{1}{2}x\partial_x^2 = \frac{1}{8}\left(\partial_r^2 - \frac{1}{r}\partial_r\right).$$

The above calculations also hold in the general case. This is given by the following result.

Theorem 3.4.3. *Consider the operator $\mathbb{L} = \frac{1}{2}\sum_{i,j=1}^n a_{ij}\partial_{x_i}\partial_{x_j}$ with (a_{ij}) a symmetric, non-degenerate matrix. Assume $\mathbb{L}(S_{cl})$ depends on both variables x and t. If there is a unique geodesic joining the points x_0 and x within time t, then the heat kernel of $\mathbb{L}$ is given by*

$$K(x_0, x; t) = V(t, x)e^{-S_{cl}(x_0,x;t)}, \tag{3.4.13}$$

where $V(t)$ satisfies the following generalized transport equation:

$$\partial_t V + \sum_{i,j} a_{ij}\partial_{x_i} S\,\partial_{x_j} V - \mathbb{L}V + V\,\mathbb{L}S = 0, \tag{3.4.14}$$

with

$$\lim_{t\searrow 0}\int V(t, x)e^{-S_{cl}(x,x_0;t)}\,dx = 1. \tag{3.4.15}$$

Proof. To simplify notations, we shall denote the classical action by $S = S_{cl}(x_0, x; t)$. Differentiating in (3.4.13) yields

$$\begin{aligned}
\partial_{x_j}\partial_{x_i} K &= e^{-S}\left(V\,\partial_{x_i} S\,\partial_{x_j} S + \partial_{x_i}\partial_{x_j} V - V\partial_{x_i}\partial_{x_j} S\right.\\
&\quad \left.-(\partial_{x_i} S\partial_{x_j} V + \partial_{x_i} V\partial_{x_j} S)\right),\\
\mathbb{L}K &= e^{-S}\left\{VH(\nabla S) + \mathbb{L}V - V\,\mathbb{L}S - a(\nabla S, \nabla V)\right\},\\
\partial_t K &= e^{-S}\{\partial_t V - V\,\partial_t S\},
\end{aligned}$$

where

$$\begin{aligned}
H(\nabla S) &= \frac{1}{2}\sum a_{ij}\partial_{x_i} S\,\partial_{x_j} S,\\
a(\nabla S, \nabla V) &= \sum a_{ij}\partial_{x_i} S\,\partial_{x_j} V.
\end{aligned}$$

Since S satisfies the Hamilton–Jacobi equation

$$\partial_t S + H(\nabla S) = 0,$$

and V satisfies the generalized transport equation (3.4.14), we get

$$(\partial_t - \mathbb{L})K = e^{-S}\left\{\partial_t V + a(\nabla S, \nabla V) - \mathbb{L}V + V\,\mathbb{L}S\right\} = 0. \qquad \blacksquare$$

Remark 3.4.4. If V does not depend on x, the generalized transport equation (3.4.14) becomes the usual transport equation (3.1.4).

The first two terms of the generalized transport equation (3.4.11) can be written as a derivative along the classical path $x(t)$:

$$\frac{d}{dt}V\big(t,x(t)\big) = \frac{\partial V}{\partial t} + \sum \frac{\partial V}{\partial x_k}\dot{x}_k = \frac{\partial V}{\partial t} + a(\nabla V, \dot{x})$$
$$= \frac{\partial V}{\partial t} + a(\nabla V, \nabla S).$$

Hence we obtain another form of the generalized transport equation,

$$\frac{d}{dt}V\big(t,x(t)\big) + V\,\mathbb{L}S = \mathbb{L}V.$$

3.5 Solving the Generalized Transport Equation

The generalized transport equation (3.4.14) is in general hard to solve explicitly. Its solution $V(t,x)$ might be represented as an integral formula rather than being an elementary function. The following computation holds in the case of an elliptic operator and uses a series expansion for t small. A parametrix expansion can also be found in [84]. We shall consider

$$V^{(k)}(t,x_0,x) = (2\pi t)^{-n/2}\big(\psi_0(x_0,x) + \psi_1(x_0,x)t + \cdots + \psi_k(x_0,x)t^k\big),$$

which approximates the volume function locally for t small, as a correction to the Euclidean volume element. Then

$$K^{(k)}(x_0,x;t) = V_k(t,x_0,x)e^{-S_{cl}(x_0,x;t)}$$

is an approximation of the heat kernel

$$K(x_0,x;t) = K^{(k)}(x_0,x;t) + O(t^k), \quad t \sim 0.$$

We need to solve for ψ_m recursively for small t. A computation shows

$$\partial_t V^{(k)} = (2\pi t)^{-n/2}\left(-\frac{n}{2t}\psi_0 + \left(1-\frac{n}{2}\right)\psi_1 + t\left(2-\frac{n}{2}\right)\psi_2 \cdots + t^{k-1}\left(k-\frac{n}{2}\right)\psi_k\right). \tag{3.5.16}$$

Since the classical action on a Riemannian manifold is related to the Riemannian distance by

$$S_{cl}(x_0,x;t) = \frac{d^2(x_0,x)}{2t},$$

then $\nabla S_{cl} = \frac{1}{2t}\nabla d^2(x_0,x)$. Hence

$$a(\nabla S_{cl}, \nabla V^{(k)}) = (2\pi t)^{-n/2}\frac{1}{2t}\sum_{j=0}^{k} a\left(\nabla d^2(x_0,x), \nabla\psi_j\right)t^j. \tag{3.5.17}$$

We also have

$$\mathbb{L}V^{(k)} = (2\pi t)^{-n/2} \sum_{j=0}^{k} \mathbb{L}\psi_j(x_0, x)\, t^j \tag{3.5.18}$$

and

$$V^{(k)}\,\mathbb{L}S_{cl} = (2\pi t)^{-n/2} \frac{1}{2t} \mathbb{L}\big(d^2(x_0, x)\big) \sum_{j=0}^{k} \psi_j(x_0, x) t^j. \tag{3.5.19}$$

Using (3.5.16)–(3.5.19) and equating the coefficients of t^j with 0 in the generalized transport equation for $V^{(k)}$,

$$\partial_t V^{(k)} + a\big(\nabla S, \nabla V^{(k)}\big) - \mathbb{L}V^{(k)} + V^{(k)}\,\mathbb{L}S = \mathcal{O}(t^k), \tag{3.5.20}$$

yields the following system in the unknown functions $\psi_0, \psi_1, \ldots$:

$$\begin{aligned}
\big(\mathbb{L}d^2 - n\big)\psi_0 + a(\nabla d^2, \nabla\psi_0) &= 0, \\
\big(\mathbb{L}d^2 - n + 2\big)\psi_1 + a(\nabla d^2, \nabla\psi_1) &= 2\mathbb{L}\psi_0, \\
\big(\mathbb{L}d^2 - n + 4\big)\psi_2 + a(\nabla d^2, \nabla\psi_2) &= 2\mathbb{L}\psi_1, \\
\vdots \qquad\qquad &= \ \vdots \\
\big(\mathbb{L}d^2 - n + 2k\big)\psi_k + a(\nabla d^2, \nabla\psi_k) &= 2\mathbb{L}\psi_{k-1},
\end{aligned}$$

together with the initial conditions $\psi_j(x_0, x_0) = 0$, $j = 1, \ldots, k$.

In general, solving the above system is as hard as solving the generalized transport equation. The system can be solved only in simple cases, as we shall see in the next example.

Example 3.5.1. Let $\mathbb{L} = \frac{1}{2}\partial_x^2$. Then $d^2(x_0, x) = (x - x_0)^2$, $\mathbb{L}\big(d^2(x_0, x)\big) = 1$, and $\nabla d^2(x_0, x) = 2(x - x_0)$. The first equation of the above system becomes

$$2(x - x_0)\psi_0' = 0, \tag{3.5.21}$$

with the solution $\psi_0 = c_0$, a constant. Then the second equation becomes

$$2\psi_1 + 2(x - x_0)\psi_1' = 0,$$

with the solution

$$\psi_1 = \frac{c_1}{x - x_0} = 0,$$

because $\psi_1(x_0) = 0$. Inductively we obtain all $\psi_j = 0$, $j \geq 1$.

In the following we shall discuss a few more examples.

3.6 The Operator $\mathbb{L} = \frac{1}{2}y^m(\partial_x^2 + \partial_y^2)$, $m \in \mathbb{N}$

We shall discuss the cases $m = 0, 1$ and 2.

Case $m = 0$. In this case the operator is the usual two-dimensional Laplacian $\mathbb{L} = \frac{1}{2}(\partial_x^2 + \partial_y^2)$ with the heat kernel

$$K(x_0, x; t) = \frac{1}{(2\pi t)} e^{-\frac{(x-x_0)^2+(y-y_0)^2}{2t}}.$$

Case $m = 1$. The operator becomes $\mathbb{L} = \frac{1}{2}y(\partial_x^2 + \partial_y^2)$. We shall consider it in the domain $\{(x, y); y > 0\}$ where the operator is elliptic. The Hamiltonian function in this case is

$$H = \frac{1}{2}y(p_1^2 + p_2^2), \tag{3.6.22}$$

and the associated Lagrangian is given by

$$L = \frac{1}{2y}(\dot{x}^2 + \dot{y}^2).$$

We are interested in the geodesics and the action from (x_0, y_0) to (x, y) within time t. From the Hamiltonian system we have

$$\begin{aligned}
\dot{x} &= H_{p_1} = yp_1, \\
\dot{y} &= H_{p_2} = yp_2 \Longrightarrow p_2 = \frac{\dot{y}}{y}, \\
\dot{p}_1 &= -H_x = 0 \Longrightarrow p_1 = k \text{ constant}, \\
\dot{p}_2 &= -H_y = \frac{1}{2}p_1^2 + \frac{1}{2}p_2^2.
\end{aligned}$$

Since the Hamiltonian function H does not depend explicitly on the parameter s, it is preserved along the solutions; i.e., there is a constant $C > 0$, which depends on the boundary points (x_0, y_0), (x, y) and t, such that

$$H(x, y, p_1, p_2) = \frac{1}{2}C^2.$$

Using the expression of the Hamiltonian (3.6.22) yields the following equation in y:

$$\left(k^2 + \frac{\dot{y}^2}{y^2}\right) = C^2. \tag{3.6.23}$$

Solving for $\dot{y}$ yields

$$\dot{y}^2 = C^2 y - k^2 y^2.$$

Completing the above to a square, we obtain

$$\dot{y}^2 = k^2 \left\{ \frac{C^4}{4k^4} - \left(y - \frac{1}{2}\frac{C^2}{k^2} \right)^2 \right\}.$$

With the substitution $u = y - \frac{1}{2}\frac{C^2}{k^2}$, the previous equation becomes

$$\dot{u}^2 = k^2 \left\{ \left(\frac{C^2}{2k^2} - u^2 \right)^2 \right\} \Longrightarrow \dot{u} = \pm k \sqrt{\frac{C^2}{2k^2} - u^2}.$$

Separating and integrating yields

$$\begin{aligned} \int_{u_0}^{u(s)} \frac{du}{\sqrt{\frac{C^2}{2k^2} - u^2}} &= \pm \int_0^s k \Longleftrightarrow \\ \arcsin\left(\frac{2k^2}{C^2} u(s) \right) &= \arcsin\left(\frac{2k^2}{C^2} u(0) \right) \pm ks \Longleftrightarrow \\ u(s) &= \frac{C^2}{2k^2} \sin(\phi \pm ks), \end{aligned}$$

where

$$\phi = \arcsin\left(\frac{2k^2}{C^2} u(0) \right).$$

Going back to the variable y, we obtain

$$y(s) = \frac{1}{2}\frac{C^2}{k^2}\left(1 + \sin(\phi \pm ks)\right). \tag{3.6.24}$$

Integrating in the first Hamiltonian equation $\dot{x} = ky$ yields

$$\begin{aligned} x(s) &= x_0 + k \int_0^s y \\ &= x_0 + \frac{k}{2}\frac{C^2}{k^2}\left(s - \frac{\cos(\phi \pm ks)}{\pm k} \right). \end{aligned} \tag{3.6.25}$$

Sign convention. If we choose the "plus" sign in front of k, it means that $y(s)$ is increasing, and hence $y > y_0$. If we choose the "negative" sign in front of k, it means that $y(s)$ is decreasing, and hence $y < y_0$.

Relations (3.6.25) and (3.6.24) provide the equations of the geodesics. In the following we shall eliminate their trigonometric part. Rewrite the equations as

$$\begin{aligned} x(s) - x_0 - \frac{k}{2}\frac{C^2}{k^2}s &= -\frac{1}{\pm 2}\frac{C^2}{k^2}\cos(\phi \pm ks), \\ y(s) - \frac{1}{2}\frac{C^2}{k^2} &= \frac{1}{2}\frac{C^2}{k^2}\sin(\phi \pm ks). \end{aligned}$$

Summing the squares yields the following implicit equation:

$$\left(x(s) - x_0 - \frac{k}{2}\frac{C^2}{k^2}s\right)^2 + \left(y(s) - \frac{1}{2}\frac{C^2}{k^2}\right)^2 = \frac{1}{4}\frac{C^4}{k^4}. \tag{3.6.26}$$

Making $s = 0$ yields

$$\left(y - \frac{1}{2}\frac{C^2}{k^2}\right)^2 = \frac{1}{4}\frac{C^4}{k^4} \Longleftrightarrow$$
$$y_0\left(y_0 - \frac{C^2}{k^2}\right) = 0.$$

Since $y_0 \neq 0$, it follows that

$$\frac{C^2}{k^2} = y_0. \tag{3.6.27}$$

Then (3.6.26) can be rewritten as

$$\left(x(s) - x_0 - \frac{k}{2}y_0 s\right)^2 + \left(y(s) - \frac{1}{2}y_0\right)^2 = \frac{1}{4}y_0^2 \Longleftrightarrow$$
$$\left(x(s) - x_0 - \frac{k}{2}y_0 s\right)^2 + y(s)\big(y(s) - y_0\big) = 0. \tag{3.6.28}$$

From (3.6.28) it follows that on the half-plane $\{(x, y); y > 0\}$ the solution $y(s)$ is a decreasing function of s. If by contradiction we assume that $y(s)$ is increasing, then $0 < y_0 < y(s)$ and hence $y(y - y_0) > 0$ and (3.6.28) cannot hold. Hence the sign in front of k in formulas (3.6.24)–(3.6.25) must be negative.

A particular family of geodesics is obtained when both terms of (3.6.28) vanish:

$$\begin{aligned} x(s) &= x_0 + \frac{k}{2}y_0 s \\ &= x_0 + \frac{x - x_0}{t}s, \\ y(s) &= y_0. \end{aligned}$$

In this case the geodesics are lines and the action is

$$\begin{aligned} S\big((x_0, y_0), (x, y_0); t\big) &= \int_0^t \frac{1}{2y(s)}(\dot{x}(s)^2 + \dot{y}(s)^2)\, ds \\ &= \int_0^t \frac{1}{2y_0}\left(\frac{x - x_0}{t}\right)^2 ds = \frac{1}{2y_0}\frac{(x - x_0)^2}{t}. \end{aligned}$$

Since

$$\mathbb{L}S = \frac{1}{2}y(\partial_x^2 + \partial_y^2)\left(\frac{1}{2y_0}\frac{(x-x_0)^2}{t}\right) = \frac{1}{2t},$$

the transport equation is

$$V'(t) + \frac{1}{2t}V(t) = 0,$$

with the solution $V(t) = \frac{c}{t^{1/2}}$, c a constant. Then the heat kernel between (x_0, y_0) and (x, y_0) is given by Theorem 3.1.1:

$$K\big((x_0, y_0), (x, y_0); t\big) = \frac{c}{t^{1/2}}e^{-\frac{1}{2y_0}\frac{(x-x_0)^2}{t}}, \qquad t > 0.$$

Finding the constant c.

$$\begin{aligned} 1 &= \int K\,dx = \frac{c}{t^{1/2}}\int e^{-\frac{1}{2y_0}\frac{(x-x_0)^2}{t}}\,dx \\ &= \frac{c}{t^{1/2}}\cdot(2\pi y_0 t)^{1/2} = c(2\pi y_0)^{1/2} \Longrightarrow c = \frac{1}{(2\pi y_0)^{1/2}}. \end{aligned}$$

Hence

$$K\big((x_0, y_0), (x, y_0); t\big) = \frac{1}{(2\pi y_0 t)^{1/2}}e^{-\frac{1}{2y_0}\frac{(x-x_0)^2}{t}}, \qquad t > 0.$$

The general case. From the Hamiltonian equation $\dot{x} = ky$ and $y > 0$, it follows that if $k > 0$, then $x(s)$ is increasing and hence $x_0 < x$. And if $k < 0$, then $x(s)$ is decreasing and then $x_0 < x$. Making $s = t$ in (3.6.28), we obtain

$$\left(x - x_0 - \frac{k}{2}y_0 t\right)^2 + y(y - y_0) = 0,$$

and then solving for k, we get

$$k = \begin{cases} \frac{2}{y_0 t}\big(x - x_0 + \sqrt{y(y_0 - y)}\big), & \text{if } x > x_0; \\ \frac{2}{y_0 t}\big(x - x_0 - \sqrt{y(y_0 - y)}\big), & \text{if } x < x_0. \end{cases}$$

We note that $y_0 \geq y$ always, since $y(s)$ is decreasing.

Finding the action between (x_0, y_0) and (x, y) within time t. Since the Lagrangian along the geodesics is

$$\begin{aligned} L &= \frac{1}{2y}(\dot{x}^2 + \dot{y}^2) = \frac{1}{2y}(y^2 p_1^2 + y^2 p_2^2) \\ &= \frac{1}{2}y(p_1^2 + p_2^2) = H = \frac{1}{2}C^2, \end{aligned}$$

integrating yields the action

$$S\big((x_0,y_0),(x,y);t\big) = \int_0^t L = \frac{1}{2}C^2 t = \frac{1}{2}\frac{C^2}{k^2}k^2 t = \frac{1}{2}y_0 t k^2$$
$$= \begin{cases} \frac{2}{y_0 t}\big(x - x_0 + \sqrt{y(y_0-y)}\big)^2, & \text{if } x > x_0; \\ \frac{2}{y_0 t}\big(x - x_0 - \sqrt{y(y_0-y)}\big)^2, & \text{if } x < x_0. \end{cases}$$

Since $\mathbb{L}S = \frac{1}{2}y(\partial_x^2 + \partial_y^2)S$ depends on t, x and y, the heat kernel is given in this case by Theorem 3.4.3:

$$K\big((x_0,y_0),(x,y);y\big) = \begin{cases} V(t,x,y)e^{-\frac{2}{y_0 t}\left(x-x_0+\sqrt{y(y_0-y)}\right)^2}, & \text{if } x > x_0; \\ V(t,x,y)e^{-\frac{2}{y_0 t}\left(x-x_0-\sqrt{y(y_0-y)}\right)^2}, & \text{if } x < x_0, \end{cases}$$

where $V(t,x,y)$ satisfies the generalized transport equation (3.4.14).

Case $m = 2$. The operator becomes $\mathbb{L} = \frac{1}{2}y^2(\partial_x^2 + \partial_y^2)$, which is the Laplace–Beltrami operator on the upperhalf-plane $\mathrm{U} = \{(x,y); y > 0\}$, with the metric $ds^2 = \frac{1}{y^2}(dx^2 + dy^2)$. The Hamiltonian function is

$$H = \frac{1}{2}y^2(p_1^2 + p_2^2), \tag{3.6.29}$$

with the associated Lagrangian

Any two points of U can be joined by a unique geodesic, which is either a vertical radius or a half-circle perpendicular on the line $\{y = 0\}$. Hence, by Theorem 3.4.3, the heat kernel can be represented as a product

$$K\big((x_0,y_0),(x,y);t\big) = V(t,x,y)e^{-S_{cl}\left((x_0,y_0),(x,y);t\right)}. \tag{3.6.30}$$

Let d_h denote the hyperbolic distance on the upperhalf-plane. If the points $A(x_0,y_0)$ and $B(x,y)$ belong to the same vertical line, then the distance between them is

$$d_h(A,B) = \left|\ln\frac{AM}{BM}\right| = \left|\ln\frac{y_0}{y}\right| = \left|\ln y - \ln y_0\right|;$$

see Fig. 3.1a. Otherwise,

$$d_h(A,B) = \left|\ln\frac{BN/BM}{AN/AM}\right| = \left|\ln\frac{A'M\cdot NB'}{AA'\cdot BB'}\right| = \left|\ln\frac{(K-x_0+r)(x-K+r)}{y_0 y}\right|,$$

where

$$r^2 = y_0^2 + (K-x_0)^2,$$
$$K = \frac{1}{2}\frac{(x-x_0)^2 + y^2 - y_0^2}{x - x_0};$$

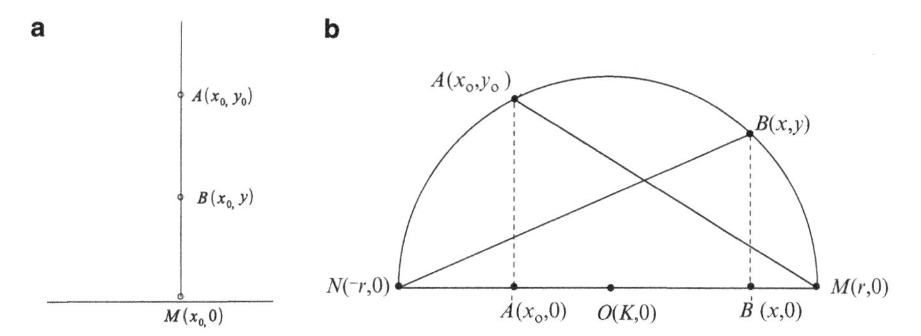

Fig. 3.1 The hyperbolic distance $d_h(A, B)$. (**a**) The case when A and B are on a vertical direction; (**b**) A and B are on a non-vertical direction

see Fig. 3.1b. Since d_h is the associated Riemannian distance on the upper space U, we have

$$S_{cl}\big((x_0, y_0), (x, y); t\big) = \frac{d_h^2\big((x_0, y_0), (x, y)\big)}{2t}.$$

Since $\mathbb{L}S_{cl}$ does not depend on t only, the transport equation is messy and the heat kernel in this case is hard to find in the product form (3.6.30). However, McKean [89] found the integral representation

$$K\big((x_0, y_0), (x, y); t\big) = \frac{\sqrt{2}}{(2\pi t)^{3/2}} e^{-t/2} \int_\rho^\infty \frac{s e^{-\frac{s^2}{2t}}}{\sqrt{\cosh s - \cosh \rho}}\, ds, \tag{3.6.31}$$

where $\rho = d_h\big((x_0, y_0), (x, y)\big)$. It is interesting for further investigations to clarify the relationship between the integral representation (3.6.31) and the product formula (3.6.30). As we shall see in the next section, this relation is obvious in the case of a three-dimensional hyperbolic space.

3.7 Heat Kernels on Curved Spaces

Let Δ_g be the Laplace–Beltrami operator on a Riemannian space (M, g_{ij}) of dimension n. Let $R = g^{ij} R_{ij}$ be the Ricci scalar curvature of the space, which will be assumed constant. Let $d(x_0, x)$ denote the Riemannian distance between the points x_0 and x. The heat kernel for the elliptic operator Δ_g is given by

$$K(x_0, x, t) = \frac{1}{(4\pi t)^{n/2}} g(x)^{-1/4} D^{1/2} g(x_0)^{-1/4} e^{Rt} e^{-\frac{d^2(x_0, x)}{4t}}, \tag{3.7.32}$$

where

$$D = \det\left[-\frac{\partial^2 S_{cl}}{\partial x_0 \partial x}\right]$$

is the van Vleck determinant; see Schulman [102], Chap. 24.

Applying the aforementioned formula for the classical spaces with constant curvatures $0, -1, 1$, we arrive at the following classical results.

The three-dimensional hyperbolic space. The Laplace–Beltrami operator on the three-dimensional hyperbolic space, i.e., the domain $\{(x_1, x_2, x_3)|x_3 > 0\}$ endowed with the metric $ds^2 = (dx_1^2 + dx_2^2 + dx_3^2)/(x_3)^2$, is given by

$$\Delta = x_3^2(\partial_{x_1}^2 + \partial_{x_2}^2 + \partial_{x_3}^2) - x_3\partial_{x_3}.$$

The heat kernel of Δ is obtained from formula (3.7.32) by letting scalar curvature $R = -1$ and hyperbolic distance $d(x_0, x) = \rho$,

$$K = \frac{1}{(4\pi t)^{3/2}} \frac{\rho}{\sinh \rho} e^{-t} e^{-\frac{\rho^2}{4t}}.$$

This relation was obtained by a direct method in [39], p. 396.

The three-dimensional unit sphere S^3. The Laplace–Beltrami operator on S^3 in spherical coordinates is given by

$$\Delta = \frac{\partial^2}{\partial\theta^2} + \cot\theta \frac{\partial}{\partial\theta} + \frac{1}{\sin^2\theta}\left(\frac{\partial^2}{\partial\theta^2} + \frac{\partial^2}{\partial\psi^2} - 2\cos\theta \frac{\partial^2}{\partial\phi\partial\psi}\right).$$

In this case $R = 1$ and the heat kernel becomes

$$K(x_0, x; t) = \frac{1}{(4\pi t)^{3/2}} \frac{\rho}{\sin\rho} e^{t} e^{-\frac{\rho^2}{4t}}, \tag{3.7.33}$$

where $\rho = d_{S^3}(x_0, x)$. This was obtained for $SU(2)$ by Schulman [101], who also conjectured that this formula works in general for Lie groups.

The three-dimensional Euclidean space. In the case of the Laplace operator

$$\Delta = \frac{\partial^2}{\partial x_1^2} + \frac{\partial^2}{\partial x_2^2} + \frac{\partial^2}{\partial x_3^2},$$

making the curvature $R = 0$ in (3.7.32) yields the familiar formula for the Euclidean heat kernel

$$K(x_0, x; t) = \frac{1}{(4\pi t)^{3/2}} e^{-\frac{|x-x_0|^2}{4t}}.$$

3.8 Heat Kernel at the Cut-Locus

The point x belongs to the *cut-locus* of x_0 if there is more than one geodesic between the points x_0 and x in time t, and this number is finite. In this case the heat propagates from x_0 to x in more than one way, each geodesic having its own contribution toward the heat kernel. If there are k geodesics between x_0 and x, parameterized by $[0,t]$, with the corresponding volume element V_k and classical action S_k, then the formula for the heat kernel is the sum of all contributions:

$$K(x_0, x; t) = \sum_{j=1}^{k} V_k(t, x_0, x) e^{-S_k}. \tag{3.8.34}$$

The above sum has only one term in the case of elliptic operators. In the case of sub-elliptic operators the sum may become an infinite series, as in the case of the Grushin operator.

3.9 Heat Kernel at the Conjugate Locus

If the points x_0 and x are conjugate, there is a smooth variation with geodesics of the same length with fixed endpoints x_0 and x. Consider the space

$$\Gamma_{x_0,x;t} = \{c : [0,t] \to M; c(0) = x_0, c(t) = x, c \text{ geodesic}\}.$$

Each element of the above space contributes to the propagation of heat between x_0 and x. In this case the sum (3.8.34) becomes an integral over $\Gamma_{x_0,x;t}$:

$$K(x_0, x; t) = \int_{\Gamma_{x_0,x;t}} e^{-S(\phi)}\, d\nu(\phi),$$

where $d\nu(\phi)$ is the volume measure obtained by a limit process from V_k. This recovers a refined concept of path integral.

In the particular case of homogeneous spaces, the volume element depends only on the Riemannian distance between x_0 and x and does not depend in an explicit way on the end points. This can be seen in the cases of the Euclidean space and the three-dimensional hyperbolic space.

3.10 Heat Kernel on the Half-Line

We interested in finding the heat kernel of the Laplacian $\frac{1}{2}\frac{d^2}{dx^2}$ on the domain $[0,\infty)$, subject to the boundary condition $\frac{\partial}{\partial x}u(t,0) = 0$. This means that no heat can leak across the point $x = 0$. Let $x_0, x > 0$. The heat can travel from the point x_0 to the point x in time t in two different ways:

- On the shortest path between the points, of length $d_1 = |x - x_0|$, given by

$$x(s) = x_0 + \frac{(x - x_0)s}{t}, \qquad s \in [0,t].$$

- On a piecewise path that hits the origin at time τ and then is reflected toward x:

$$x(s) = \begin{cases} (1 - s\tau)x_0, & 0 \le s \le \tau, \\ x(s-\tau)/(t-\tau), & \tau \le s \le t. \end{cases}$$

of length $d_2 = x_0 + x$.

The heat kernel is the amount of heat received by x at time t, which is the sum of the aforementioned amounts of heat:

$$\begin{aligned} K(x_0, x; t) &= \frac{1}{\sqrt{2\pi t}} e^{-\frac{d_1^2}{2t}} + \frac{1}{\sqrt{2\pi t}} e^{-\frac{d_2^2}{2t}} \\ &= \frac{1}{\sqrt{2\pi t}} e^{-\frac{(x-x_0)^2}{2t}} + \frac{1}{\sqrt{2\pi t}} e^{-\frac{(x+x_0)^2}{2t}}. \end{aligned}$$

The following result will be needed in the next two applications.

Lemma 3.10.1. *For any $\xi \in \mathbb{R}$, $T > 0$, $i = \sqrt{-1}$, we have*

$$\sum_{n\ge 1} e^{-\frac{1}{2t}(2nT+\xi)^2} + \sum_{n\ge 1} e^{-\frac{1}{2t}(2nT-\xi)^2} = 2e^{-\frac{\xi^2}{2t}} \sum_{n\ge 1} e^{-\frac{2n^2T^2}{t}} \cos\left(\frac{2nT\xi}{t} i\right).$$

Proof. Expanding the binomial in the exponent and using Euler's formula, we have

$$\begin{aligned} &\sum_{n\ge 1} e^{-\frac{1}{2t}(2nT+\xi)^2} + \sum_{n\ge 1} e^{-\frac{1}{2t}(2nT-\xi)^2} \\ &= \sum_{n\ge 1} e^{-\frac{1}{2t}(4n^2T^2+\xi^2+4nT\xi)} + \sum_{n\ge 1} e^{-\frac{1}{2t}(4n^2T^2+\xi^2-4nT\xi)} \\ &= e^{-\frac{\xi^2}{2t}} \sum_{n\ge 1} e^{-\frac{4n^2T^2}{2t}} \left(e^{-\frac{2nT\xi}{t}} + e^{\frac{2nT\xi}{t}}\right) = 2e^{-\frac{\xi^2}{2t}} \sum_{n\ge 1} e^{-\frac{2n^2T^2}{t}} \cosh \frac{2nT\xi}{t} \\ &= 2e^{-\frac{\xi^2}{2t}} \sum_{n\ge 1} e^{-\frac{2n^2T^2}{t}} \cos\left(\frac{2nT\xi}{t} i\right). \end{aligned}$$

■

3.11 Heat Kernel on S^1

Let X_0 and X be two points on the unit circle S^1 with angular arguments s_0 and s. The heat kernel $K(X_0, X; t)$ depends on all geodesics that join the points X_0 and X. The square of the lengths of both clockwise and counterclockwise geodesics joining X_0 and X are

$$\ell_0^2 = (s - s_0)^2, \qquad \ell_n^2 = \big(2n\pi + (s - s_0)\big)^2, \qquad \ell_{-n}^2 = \big(2n\pi - (s - s_0)\big)^2, \quad n \in \mathbb{N}.$$

Since the unit circle is locally Euclidean and one-dimensional, the volume element is $V(t) = 1/\sqrt{2\pi t}$, and the heat kernel has the following expression:

$$K(X_0, X; t) = \sum_{n\in\mathbb{Z}} V(t)e^{-\frac{\ell_k^2}{2t}} = \frac{1}{\sqrt{2\pi t}} \sum_{n=-\infty}^{\infty} e^{-\frac{1}{2t}\left(2n\pi+(s-s_0)\right)^2}$$

$$= \frac{1}{\sqrt{2\pi t}}\left(e^{-\frac{(s-s_0)^2}{2t}} + \sum_{n\geq 1} e^{-\frac{1}{2t}\left(2n\pi+(s-s_0)\right)^2} + \sum_{n\geq 1} e^{-\frac{1}{2t}\left(2n\pi-(s-s_0)\right)^2}\right).$$

Applying Lemma 3.10.1 with $T = \pi$ and $\xi = s - s_0$, we obtain

$$K(X_0, X; t) = \frac{1}{\sqrt{2\pi t}} e^{-\frac{(s-s_0)^2}{2t}} \left(1 + 2\sum_{n\geq 1} e^{-\frac{2\pi^2 n^2}{t}} \cos\left(\frac{2n\pi}{t}(s-s_0)i\right)\right).$$

Choosing $z = \pi(s - s_0)i/t$ and $\tau = 2\pi/t$ in the definition of the theta-function

$$\theta_3(z|i\tau) = 1 + 2\sum_{n\geq 1} e^{-\pi n^2 \tau}\cos(2nz)$$

[see (1.5.4)], we arrive at the following formula for the heat kernel on S^1:

$$K(X_0, X; t) = \frac{1}{\sqrt{2\pi t}} e^{-\frac{(s-s_0)^2}{2t}} \theta_3\left(\frac{\pi(s-s_0)i}{t}\,\Big|\,\frac{2\pi i}{t}\right). \tag{3.11.35}$$

Since there is no generating formula for θ_3, this shows that even in the simplest compact case, the heat kernel does not have a neat formula. For an approach using eigenfunctions expansion and the Poisson summation formula, see Sect. 6.4.

3.12 Heat Kernel on the Segment [0, T]

The heat kernel of concern here is the one corresponding to the Neumann boundary conditions $\frac{\partial}{\partial x}u(t,0) = \frac{\partial}{\partial x}u(t,T) = 0$. This corresponds to the case when $x = 0$ and $x = T$ are perfectly insulated reflecting walls. The distances traveled by heat from x_0 to x in a direct way or using reflections in the walls are

$$\begin{aligned} &x - x_0, && x + x_0, \\ &2nT + x - x_0, && 2nT - x + x_0, \\ &2nT + x + x_0, && 2nT - x - x_0, \ldots, \end{aligned}$$

with $n = 1, 2, \ldots$. These are the distances from x_0 to the images of x seen in two parallel plane mirrors situated at coordinates 0 and T. Each of these distances has a contribution to the heat kernel as follows:

$$
\begin{aligned}
K(x_0, x; t) &= \sum_{n \geq 0} V(t) e^{-\frac{\ell_n^2}{2t}} \\
&= \frac{1}{\sqrt{2\pi t}} \Bigg[e^{-\frac{(x-x_0)^2}{2t}} + e^{-\frac{(x+x_0)^2}{2t}} \\
&\qquad + \sum_{n\geq 1} e^{-\frac{1}{2t}(2nT+x-x_0)^2} + \sum_{n\geq 1} e^{-\frac{1}{2t}(2nT-x+x_0)^2} \\
&\qquad + \sum_{n\geq 1} e^{-\frac{1}{2t}(2nT+x+x_0)^2} + \sum_{n\geq 1} e^{-\frac{1}{2t}(2nT-x-x_0)^2} \Bigg].
\end{aligned}
$$

Lemma 3.10.1 enables us to write the previous expression as

$$
\begin{aligned}
K(x_0, x; t) &= \frac{1}{\sqrt{2\pi t}} e^{-\frac{(x-x_0)^2}{2t}} \left[1 + 2 \sum_{n\geq 1} e^{-\frac{2n^2T^2}{t}} \cos\left(\frac{2nT}{t}(x - x_0)i \right) \right] \\
&\quad + \frac{1}{\sqrt{2\pi t}} e^{-\frac{(x+x_0)^2}{2t}} \left[1 + 2 \sum_{n\geq 1} e^{-\frac{2n^2T^2}{t}} \cos\left(\frac{2nT}{t}(x + x_0)i \right) \right],
\end{aligned}
$$

which, after using the definition of the θ_3-function, can be expressed as

$$
K(x_0, x; t) = \frac{1}{\sqrt{2\pi t}} e^{-\frac{(x-x_0)^2}{2t}} \theta_3 \left(\frac{T}{t}(x - x_0)i \,\Big|\, \frac{2T^2 i}{\pi t} \right) \tag{3.12.36}
$$

$$
+ \frac{1}{\sqrt{2\pi t}} e^{-\frac{(x+x_0)^2}{2t}} \theta_3 \left(\frac{T}{t}(x + x_0)i \,\Big|\, \frac{2T^2 i}{\pi t} \right). \tag{3.12.37}
$$

3.13 Heat Kernel on the Cylinder

The cylinder is two-dimensional and locally Euclidean, so the volume element is $V(t) = \frac{1}{2\pi t}$. Let (x_0, y_0) and (x, y) be two points on the cylinder. The geodesic joining the points is a spiral which makes a constant angle θ with the vertical. If the geodesic winds n times, the vertical distance is the sum of n equal increments plus a remainder amount (see Fig. 3.2):

$$
y - y_0 = 2n\pi \tan\theta + (x - x_0)\tan\theta.
$$

This shows that the angle $\theta = \theta_n$ must be quantized by the relation

$$
\tan\theta_n = \frac{y - y_0}{2n\pi + (x - x_0)}, \qquad n \in \mathbb{Z}.
$$

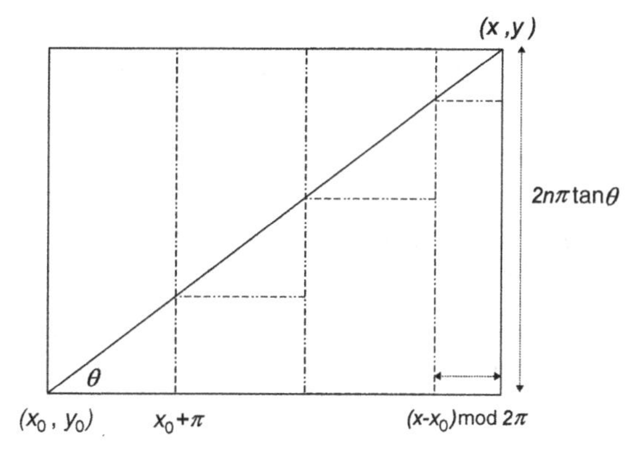

Fig. 3.2 The geodesic between the points (x_0, y_0) and (x, y) on the cylinder $\mathbb{R} \times S^1$

By convention, we consider the integer n positive if the winding is counterclockwise and negative otherwise. The length of the geodesic is the sum of $n+1$ oblique segments:

$$\ell_n = \frac{2\pi n}{\cos\theta_n} + \frac{x - x_0}{\cos\theta_n}.$$

Expressing cosine in terms of tangent function and using the aforementioned formula yields

$$\ell_n = = \sqrt{1 + \tan^2\theta_n}(2n\pi + x - x_0) = \sqrt{\big(2n\pi + (x - x_0)\big)^2 + (y - y_0)^2}.$$

Then the heat kernel is given by the contribution of all geodesics from (x_0, y_0) and (x, y):

$$\begin{aligned}
&K\big((x_0, y_0), (x, y); t\big) \\
&\quad = \frac{1}{2\pi t}\sum_{n\in\mathbb{Z}} e^{-\frac{\ell_n^2}{2t}} = \frac{1}{2\pi t}\sum_{n\in\mathbb{Z}} e^{-\frac{1}{2t}\big((2n\pi + x - x_0)^2 + (y - y_0)^2\big)} \\
&\quad = \frac{1}{2\pi t} e^{-\frac{(x-x_0)^2+(y-y_0)^2}{2t}} \sum_{n\in\mathbb{Z}} e^{-\frac{1}{2t}\big(4n^2\pi^2 + 4n\pi(x - x_0)\big)} \\
&\quad = \frac{1}{2\pi t} e^{-\frac{(x-x_0)^2+(y-y_0)^2}{2t}} \sum_{n\in\mathbb{Z}} e^{-\frac{2n^2\pi^2}{t}} \cos\left(\frac{2n\pi(x - x_0)i}{t}\right) \\
&\qquad + \frac{i}{2\pi t} e^{-\frac{(x-x_0)^2+(y-y_0)^2}{2t}} \sum_{n\in\mathbb{Z}} e^{-\frac{2n^2\pi^2}{t}} \sin\left(\frac{2n\pi(x - x_0)i}{t}\right) \\
&\quad = \frac{1}{2\pi t} e^{-\frac{(x-x_0)^2+(y-y_0)^2}{2t}} \left(1 + 2\sum_{n\geq 1} e^{-\frac{2n^2\pi^2}{t}} \cos\left(\frac{2n\pi(x - x_0)i}{t}\right)\right) \\
&\quad = \frac{1}{2\pi t} e^{-\frac{(x-x_0)^2+(y-y_0)^2}{2t}} \theta_3\left(\frac{(x - x_0)i}{t} \,\Big|\, \frac{2\pi i}{t}\right).
\end{aligned}$$

It is worth noting that the previous heat kernel is the direct product of the heat kernels on $\mathbb{R}$ and S^1. This formula could have been written directly since the cylinder is the product of these two manifolds.

3.14 Operators with Potentials

This is an extension of the method described in the previous sections to operators that have a potential. We start the presentation with the heat kernel of the operator $\frac{1}{2}\frac{d}{dx}^2 + U(x)$, where $U(x)$ is a function of the variable x, called the *potential*. Since the heat kernel will depend on the classical action between any two given points, in order to obtain closed-form formulas, we need to consider only those potentials $U(x)$ for which the classical action can be computed explicitly. This will occur in several cases, for instance, in the case of linear and quadratic potentials. In these cases there is a unique solution joining any two given points x_0 and x within time t. The classical action cannot be obtained explicitly for potentials of the type $U(x) = a^2x^m$, with $m \geq 3$.

Consider the operator $L = \frac{1}{2}\frac{d}{dx}^2 + U(x)$ with smooth potential $U(x)$. We associate the Hamiltonian function

$$H(p,x) = \frac{1}{2}p^2 + U(x). \tag{3.14.38}$$

Hamilton's equations are

$$\begin{aligned} &\dot{x} = H_p = p, \\ &\dot{p} = -H_x = -U'(x), \quad \text{and hence } \ddot{x} = \dot{p} = -U'(x). \end{aligned}$$

For any two given points x_0 and x, the classical path joining them is obtained by solving the equation

$$\begin{cases} \ddot{x} = -U'(x), \\ x(0) = x_0, \\ x(t) = x. \end{cases} \tag{3.14.39}$$

We shall assume the potential $U(x)$ is such that the previous system has a unique solution. Since the Hamiltonian (3.14.38) does not depend explicitly on the variable t, it will be preserved along the solutions of (3.14.39):

$$H = \frac{\dot{x}^2}{2} + U(x) = E, \tag{3.14.40}$$

where $E = E(x_0, x, t)$ is the constant of energy. Hence $x(s)$ verifies the integral equation

$$\int_{x_0}^{x(s)} \frac{dw}{\sqrt{2E - 2U(w)}} = \pm s,$$

where the positive (negative) sign is taken on the right-hand side if $x > x_0$ $(x < x_0)$. The energy $E = E(x_0, x, t)$ satisfies the equation

$$\int_{x_0}^{x} \frac{dw}{\sqrt{2E - 2U(w)}} = \pm t,$$

with the same sign convention. The action S is verifying the Hamilton–Jacobi equation $\partial_t S = -E(x_0, x, t)$. Since along the solutions we have $p = \partial_x S$, using $\dot{x} = p$, we get $\dot{x} = \partial_x S$, and hence (3.14.40) becomes

$$(\partial_x S)^2 = 2E - 2U(x). \tag{3.14.41}$$

We shall look for a fundamental solution of the type $K = V(t, x)e^{-S}$, where $S = S_{cl}$ is the classical action. A computation provides

$$\begin{aligned} \partial_t K &= K\left(\frac{V'}{V} + E\right), \\ \partial_x K &= K\left(\frac{\partial_x V}{V} + \partial_x S\right), \\ \partial_x^2 K &= K\left(\frac{\partial_x^2 V}{V} - 2\frac{\partial_x V}{V}\partial_x S + (\partial_x S)^2 - \partial_x^2 S\right). \end{aligned}$$

Let $P = \partial_t - \frac{1}{2}\partial_x^2 - U(x)$. Using the above formulas, we obtain

$$\begin{aligned} PK &= K\left(\frac{\partial_t V}{V} + E - \frac{1}{2}\frac{\partial_x^2 V}{V} + \frac{\partial_x V \partial_x S}{V} - \frac{1}{2}(\partial_x S)^2 + \frac{1}{2}\partial_x^2 S - U(x)\right) \\ &= K\left(\frac{\partial_t V - \frac{1}{2}\partial_x^2 V + \partial_x V \partial_x S}{V} + \frac{1}{2}\partial_x^2 S\right), \end{aligned}$$

where we used (3.14.41). Hence $PK = 0$ if and only if V satisfies the generalized transport equation

$$\partial_t V - \frac{1}{2}\partial_x^2 V + \partial_x S\, \partial_x V + \frac{1}{2}\partial_x^2 S\, V = 0. \tag{3.14.42}$$

Using that $\dot{x} = \partial_x S$, we have

$$\frac{d}{dt}V(t, x) = \partial_t V + \dot{x}\partial_x V = \partial_t V + \partial_x S\, \partial_x V,$$

and hence (3.14.42) becomes

$$\frac{d}{dt}V(t, x) = \frac{1}{2}\partial_x^2 V(t, x) - \frac{1}{2}\partial_x^2 S\, V(t, x). \tag{3.14.43}$$

If $\partial_x^2 S$ depends on t only, then it makes sense to look for a function $V = V(t)$ satisfying the simplified transport equation

$$\partial_t V + \frac{1}{2}\partial_x^2 S \; V = 0. \tag{3.14.44}$$

The volume function $V(t,x)$ should satisfy the initial condition

$$\lim_{t \searrow 0} \int V(t,x) e^{-S_{cl}} \, dx = 1. \tag{3.14.45}$$

We conclude the above calculations with the following method for finding explicit formulas for heat kernels:

1. Assume the boundary value problem (3.14.39) satisfies the uniqueness condition; let $x(s)$ be its solution.
2. Let $S = S_{cl}(x_0, x; t)$ be the classical action along solution $x(s)$ between x_0 and x in time t.
3. Consider the solution $V(t,x)$ of the transport equation (3.14.42) satisfying (3.14.45).
4. Then $K(x_0, x; t) = V(t,x)e^{-S}$ is the heat kernel for the operator $\frac{1}{2}\partial_x^2 + U(x)$.

Unfortunately, there are potential functions $U(x)$ for which the boundary value problem (3.14.39) has a unique solution. This happens, for instance, in the case when $U(x) = x^4$; see Calin and Chang [24]. Even if the solution is unique and the action function is found, the transport equation might be difficult to solve.

It is worth noting that the potential function $U(x)$ does not appear explicitly in the expression of the generalized transport equation. It is even more interesting to observe that (3.14.42) is a particular form of the multidimensional generalized transport equation (3.4.11) with $(a_{ij}) = \mathbf{1}$ (one-dimensional matrix).

In the following se shall consider the case of linear and quadratic potentials, for which all the computations can be done explicitly.

3.15 The Linear Potential

Let $U(x) = -ax$, with a a real constant. The operator is $\mathbb{L} = \frac{1}{2}\frac{d}{dx}^2 - ax$ and its associated Hamiltonian function is

$$H(p,x) = \frac{1}{2}p^2 - ax.$$

The Hamiltonian system becomes

$$\dot{x} = H_p = p,$$
$$\dot{p} = -H_x = a.$$

The associated Lagrangian function is $L(x,\dot{x}) = \frac{1}{2}\dot{x}^2 + ax$. The Euler–Lagrange equation $\ddot{x} = a$ with the boundary conditions $x(0) = x_0$, $x(t) = x$ has the unique solution

$$x(s) = \frac{1}{2}as^2 + bs + x_0, \tag{3.15.46}$$

with

$$b = \frac{x - x_0}{t} - \frac{at}{2}. \tag{3.15.47}$$

The Lagrangian along the solution (7.9.43) is

$$\begin{aligned} L\big(x(s),\dot{x}(s)\big) &= \frac{1}{2}\dot{x}(s)^2 + ax(s) \\ &= \frac{1}{2}(as+b)^2 + a\left(\frac{1}{2}as^2 + bs + x_0\right) \\ &= a^2s^2 + 2abs + ax_0 + \frac{1}{2}b^2. \end{aligned}$$

The classical action is obtained by integrating the Lagrangian along the solution:

$$\begin{aligned} S(x_0,x;t) &= \int_0^t L\big(x(s),\dot{x}(s)\big)\,ds \\ &= \frac{1}{3}a^2t^3 + abt^2 + ax_0t + \frac{1}{2}b^2t. \end{aligned}$$

Substituting b from (7.10.67) yields

$$\begin{aligned} S(x_0,x;t) &= \frac{1}{3}a^2t^3 + at^2\left(\frac{x-x_0}{t} - \frac{at}{2}\right) + ax_0t + \frac{1}{2}t\left(\frac{x-x_0}{t} - \frac{at}{2}\right)^2 \\ &= \frac{(x-x_0)^2}{2t} + \frac{1}{2}a(x+x_0)t - \frac{1}{24}a^2t^3. \end{aligned} \tag{3.15.48}$$

There is an alternate method for finding the action using the Hamilton–Jacobi equation. For this we need first to find the energy along the solution as a function of the boundary points. In our case (3.14.40) becomes

$$\frac{1}{2}\dot{x}(s)^2 - ax(s) = E.$$

Using (7.9.43) yields

$$E = \frac{1}{2}(as+b)^2 - a\left(\frac{1}{2}as^2 + bs + x_0\right), \qquad \forall s \in [0,t].$$

Taking the particular value $s = 0$ yields

$$
\begin{aligned}
E &= \frac{1}{2}b^2 - ax_0 \\
&= \frac{1}{2}\left(\frac{x - x_0}{t} - \frac{at}{2}\right)^2 - ax_0 \\
&= \frac{(x - x_0)^2}{2t^2} - \frac{1}{2}a(x + x_0) + \frac{a^2t^2}{8}.
\end{aligned}
$$

It is not hard to check that the Hamilton–Jacobi equation

$$
\begin{aligned}
\partial_t S &= -E \\
&= -\frac{(x - x_0)^2}{2t^2} + \frac{1}{2}a(x + x_0) - \frac{a^2t^2}{8},
\end{aligned}
$$

with the initial condition $S_{|t=0} = 0$, has the solution

$$
S(x_0, x; t) = \frac{(x - x_0)^2}{2t} + \frac{1}{2}a(x + x_0)t - \frac{1}{24}a^2t^3. \tag{3.15.49}
$$

In the following we shall find the volume function V. Since $\frac{1}{2}\partial_x^2 S = \frac{1}{2t}$, then $V = V(t)$ and the transport equation is

$$
V'(t) + \frac{1}{2t}V(t) = 0,
$$

with the solution

$$
V(t) = \frac{c}{t^{1/2}}.
$$

The constant c can be found using (3.14.45):

$$
\begin{aligned}
1 &= \lim_{t \searrow 0} \int K(x_0, x; t)\, dx = \lim_{t \searrow 0} V(t) \int e^{-S(x_0,x;t)}\, dx \\
&= \lim_{t \searrow 0} \frac{c}{t^{1/2}} \int e^{-\frac{(x-x_0)^2}{2t} - \frac{1}{2}a(x+x_0)t + \frac{1}{24}a^2t^3}\, dx \\
&= \lim_{t \searrow 0} \frac{c}{t^{1/2}} e^{-\frac{1}{2}ax_0t + \frac{1}{24}a^2t^3} \int e^{-\frac{(x-x_0)^2}{2t} - \frac{1}{2}a(x-x_0)t}\, dx \\
&= \lim_{t \searrow 0} \frac{c}{t^{1/2}} e^{-\frac{1}{2}ax_0t + \frac{1}{24}a^2t^3} (2\pi t)^{1/2} e^{\frac{1}{8}a^2t^3} \\
&= c(2\pi)^{1/2},
\end{aligned}
$$

and hence $c = (2\pi)^{-1/2}$ and the volume function is $V(t) = 1/(2\pi t)^{1/2}$.

We have arrived at the following result:

Theorem 3.15.1. *Let $a \in \mathbb{R}$. The heat kernel of the operator $\mathbb{L} = \frac{1}{2}\frac{d}{dx}^2 - ax$ is given by*

$$K(x, x_0, t) = \frac{1}{\sqrt{2\pi t}} e^{-\frac{(x-x_0)^2}{2t} - \frac{1}{2}a(x+x_0)t + \frac{1}{24}a^2t^3}, \quad t > 0.$$

Remark 3.15.2. In the case when $a = 0$ we recover the well-known heat kernel of $\frac{1}{2}\partial_x^2$, which is $\frac{1}{\sqrt{2\pi t}} e^{-\frac{(x-x_0)^2}{2t}}$.

Without much effort, the previous result can be generalized to the n-dimensional operator $\mathbb{L} = \frac{1}{2}\sum_{i=1}^n \partial_{x_i}^2 - \sum_{i=1}^n a_i x_i$. In this case the Hamiltonian function is

$$H(p, x) = \frac{1}{2}\sum_{i=1}^n p_i^2 - \sum_{i=1}^n a_i x_i,$$

and the Hamiltonian equations are

$$\begin{aligned} \dot{x}_j &= p_j, \\ \dot{p}_j &= -a_j, \qquad j = 1, \dots, n. \end{aligned}$$

The solution joining the points x_0 and x is

$$x^j(s) = \frac{1}{2}a_j s^2 + b_j s + x_0^j, \tag{3.15.50}$$

with

$$b_j = \frac{x^j - x_0^j}{t} - \frac{a_j t}{2}, \qquad j = 1, \dots, n. \tag{3.15.51}$$

The Lagrangian

$$L(x, \dot{x}) = \frac{1}{2}|\dot{x}_i|^2 + \langle a, x \rangle = \frac{1}{2}\sum_{i=1}^n \dot{x}_i^2 + \sum_{i=1}^n a_i x_i$$

evaluated along the previous solution is

$$L\big(x(s), \dot{x}(s)\big) = |a|^2 s^2 + 2\langle a, b \rangle s + \langle a, x_0 \rangle + \frac{1}{2}|b|^2.$$

Integrating the Lagrangian along the solution yields the following classical action:

$$S(x_0, x; t) = \frac{|x - x_0|^2}{2t} + \frac{1}{2}\langle a, x + x_0 \rangle t - \frac{1}{24}|a|^2 t^3. \tag{3.15.52}$$

Similar calculations show that the generalized transport equation in this case is

$$\partial_t V - \frac{1}{2}\sum_{i=1}^{n} \partial_{x_i}^2 V + \langle \nabla S, \nabla V \rangle + \frac{1}{2}\left(\sum_{i=1}^{n} \partial_{x_i}^2 S\right) V = 0.$$

Since $\frac{1}{2}\sum_{i=1}^{n} \partial_{x_i}^2 S = \frac{n}{2t}$ does not depend on x, it makes sense to look for a volume function of the form $V = V(t)$, in which case the previous equation takes the simplified form

$$V'(t) + \frac{n}{2t} V(t) = 0.$$

The solution is $V(t) = ct^{-n/2}$. Using a similar method as before, we find $c = (2\pi)^{-n/2}$. We conclude with the following result:

Theorem 3.15.3. *Let $a \in \mathbb{R}^n$. The heat kernel of the operator $\mathbb{L} = \frac{1}{2}\sum_{i=1}^{n} \partial_{x_i}^2 - \sum_{i=1}^{n} a_i x_i$ is given by*

$$K(x, x_0, t) = \frac{1}{(2\pi t)^{n/2}} e^{-\frac{|x-x_0|^2}{2t} - \frac{1}{2}\langle a, x+x_0\rangle t + \frac{1}{24}|a|^2 t^3}, \quad t > 0.$$

3.16 The Quadratic Potential

We consider the operator $\mathbb{L} = \frac{1}{2}(\frac{d}{dx}^2 - a^2x^2)$, with a constant, called the *Hermite operator*. The associated Hamiltonian is given by $H(p, x) = \frac{1}{2}p^2 - \frac{1}{2}a^2x^2$ and the Hamiltonian system of equations is

$$\dot{x} = H_p = p,$$
$$\dot{p} = -H_x = a^2 x.$$

The geodesic between x_0 and x within time t satisfies

$$\ddot{x} = a^2 x,$$
$$x(0) = x_0, \; x(t) = x. \tag{3.16.53}$$

The solution is

$$x(s) = \frac{x - x_0 \cosh(at)}{\sinh(at)} \sinh(as) + x_0 \cosh(as). \tag{3.16.54}$$

The Lagrangian associated with the previous Hamiltonian is

$$L(x, \dot{x}) = p\dot{x} - H = \frac{1}{2}\dot{x}^2 + \frac{1}{2}a^2x^2. \tag{3.16.55}$$

Integrating the solution (7.9.51) along the Lagrangian (7.9.52) and performing the same computations as in Sect. 7.8 yields

$$S(x_0, x, t) = \frac{a}{2\sinh(at)}\left((x^2 + x_0^2)\cosh(at) - 2xx_0\right). \tag{3.16.56}$$

There is an alternate way of finding the action using the Hamilton–Jacobi equation. The conservation of energy law for (7.9.50) is

$$\frac{1}{2}\dot{x}^2(s) - \frac{1}{2}a^2x^2(s) = E,$$

where E is the energy constant. This can also be written as

$$\frac{dx}{ds} = \sqrt{2E + a^2x^2} \Longrightarrow \frac{dx}{\sqrt{2E + a^2x^2}} = ds.$$

Integrating between $s = 0$ and $s = t$, with $x(0) = x_0$ and $x(t) = x$, and solving for the energy yields

$$2E = \frac{a^2(x - x_0\cosh(at))^2}{\sinh(at)^2} - a^2x_0^2 = \frac{a^2\left(x^2 + x_0^2 - 2xx_0\cosh(at)\right)}{\sinh(at)^2}.$$

The Hamilton–Jacobi equation becomes

$$\begin{aligned}\partial_t S &= -H(\nabla S) = -E \\ &= -\frac{a^2\left(x^2 + x_0^2 - 2xx_0\cosh(at)\right)}{2\sinh(at)^2} \\ &= \frac{\partial}{\partial t}\left[\frac{a}{2}(x^2 + x_0^2)\coth(at) - \frac{axx_0}{\sinh(at)}\right].\end{aligned}$$

Hence

$$\begin{aligned}S(x_0, x, t) &= \frac{a}{2}\left[(x^2 + x_0^2)\coth(at) - \frac{2xx_0}{\sinh(at)}\right] \\ &= \frac{a}{2}\frac{1}{\sinh(at)}\left[(x^2 + x_0^2)\cosh(at) - 2xx_0\right].\end{aligned} \tag{3.16.57}$$

Since $\partial_x^2 S = a\coth(at)$ is a function of t only, the transport equation becomes

$$V'(t) + \frac{1}{2}a\coth(at)V(t) = 0.$$

Integrating, we obtain the solution

$$V(t) = \frac{c}{\sqrt{\sinh(at)}},$$

where c is a constant which will be determined later. Using $K = V(t)e^{-S}$, we get the fundamental solution

$$K(x_0, x, t) = \frac{c}{\sqrt{\sinh(at)}} e^{-\frac{a}{2}\frac{1}{\sinh(at)}[(x^2+x_0^2)\cosh(at)-2xx_0]}.$$

In order to determine the constant c we write

$$K(x_0, x, t) = \frac{c}{\sqrt{at}} \sqrt{\frac{at}{\sinh(at)}}\, e^{-\frac{1}{2t}\cdot\frac{at}{\sinh(at)}[(x^2+x_0^2)\cosh(at)-2xx_0]}.$$

Since $at/\sinh(at) \to 1$, for $a \to 0$, we can write

$$K(x_0, x, t) \sim \frac{c}{\sqrt{at}}\, e^{\frac{1}{2t}(x-x_0)^2}.$$

By comparison with the fundamental solution for the usual heat operator $1/\sqrt{2\pi t}\, e^{\frac{1}{2t}(x-x_0)^2}$, we obtain $c = \sqrt{a/2\pi}$. To conclude, we have the following result.

Theorem 3.16.1. *Let a be a constant. The heat kernel for the operator $\frac{1}{2}(\frac{d}{dx}^2 - a^2x^2)$ is*

$$K(x_0, x, t) = \frac{1}{\sqrt{2\pi t}} \sqrt{\frac{at}{\sinh(at)}}\, e^{-\frac{1}{2t}\frac{at}{\sinh(at)}[(x^2+x_0^2)\cosh(at)-2xx_0]}, \quad t > 0.$$

Using the substitutions $a = -i\alpha$ and $\cosh(i\alpha t) = \cos(\alpha t)$ and $\sinh(2i\alpha t) = i\sin(2\alpha t)$, we get

Corollary 3.16.2. *Let α be a constant. The heat kernel for the operator $\frac{1}{2}(\frac{d}{dx}^2 + \alpha^2x^2)$ is*

$$K(x_0, x, t) = \frac{1}{\sqrt{2\pi t}} \sqrt{\frac{2\alpha t}{\sin(2\alpha t)}}\, e^{-\frac{1}{2t}\frac{2\alpha t}{\sin(2\alpha t)}[(x^2+x_0^2)\cos(2\alpha t)-2xx_0]}, \quad t > 0.$$

3.17 The Operator $\frac{1}{2}\sum \partial_{x_i}^2 \pm \frac{1}{2}a^2|x|^2$

Consider the operator

$$\frac{1}{2}\Delta_n - \frac{1}{2}a^2|x|^2 = \frac{1}{2}(\partial_{x_1}^2 + \cdots + \partial_{x_n}^2) - \frac{1}{2}a^2(x_1^2 + \cdots + x_n^2), \quad a \geq 0.$$

The associated Hamiltonian is

$$H = \frac{1}{2}(\xi_1^2 + \cdots + \xi_n^2) - \frac{1}{2}a^2(x_1^2 + \cdots + x_n^2),$$

with the Hamiltonian system

$$\begin{cases} \dot{x}_j = H_{\xi_j} = \xi_j, \\ \dot{\xi}_j = -H_{x_j} = a^2 x_j, \quad j = 1, \ldots, n. \end{cases}$$

The geodesic $x(s)$ starting at $x_0 = (x_1^0, \ldots, x_n^0)$ and having the final point $x = (x_1, \ldots, x_n)$ satisfies the equations

$$\begin{cases} \ddot{x}_j = a^2 x_j, \\ x_j(0) = x_j^0, \\ x_j(t) = x_j, \quad j = 1, \ldots, n. \end{cases}$$

As in the one-dimensional case, we have the law of conservation of energy

$$\dot{x}_j^2(s) - a^2 x_j^2(s) = 2E_j, \quad j = 1, \ldots, n,$$

where E_j is the energy constant for the jth component. The total energy, which is the Hamiltonian, is given by

$$H = \sum_{j=1}^{n} \left(\frac{1}{2}\dot{x}_j^2 - \frac{1}{2}a^2 x_j^2 \right) = E_1 + \cdots + E_n = E(\text{constant}).$$

Since the energy for the one-dimensional case is

$$E_j = \frac{a^2[x_j^2 + (x_j^0)^2 - 2x_j x_j^0 \cosh(at)]}{2\sinh^2(at)},$$

we get the total energy

$$H = E = \sum_{j=1}^{n} E_j = \frac{a^2[|x|^2 + |x_0|^2 - 2\langle x, x_0 \rangle \cosh(at)]}{2\sinh^2(at)},$$

where $|x|^2 = \sum_{j=1}^n x_j^2$ and $\langle x, x_0 \rangle = \sum_{j=1}^n x_j x_j^0$.

Computing the classical action. The action between x_0 and x within time t satisfies the equation $\frac{\partial}{\partial t} S = -E$ or

$$\begin{aligned} \frac{\partial}{\partial t} S &= -\frac{a^2[|x|^2 + |x_0|^2 - 2\langle x, x_0 \rangle \cosh(at)]}{2\sinh^2(at)} \\ &= \frac{\partial}{\partial t}\left[\frac{a}{2}(|x|^2 + |x_0|^2)\coth(at) - \frac{a\langle x, x_0 \rangle}{\sinh(at)} \right] \end{aligned}$$

and hence

$$S_{cl}(x_0, x; t) = \frac{a}{2} \frac{1}{\sinh(at)} \left[(|x|^2 + |x_0|^2) \cosh(at) - 2\langle x, x_0 \rangle \right]. \tag{3.17.58}$$

We are looking for a kernel of the form

$$K(x_0, x, t) = V(t) e^{k S_{cl}(x_0, x, t)}, \qquad k \in \mathbb{R}. \tag{3.17.59}$$

Since the Lagrangian is at most quadratically in x and $\dot{x}$, the function $V(t)$ is given by the van Vleck formula (see also Chap. 7)

$$V(t) = \sqrt{\det \left(-\frac{1}{2\pi} \frac{\partial^2 S_{cl}}{\partial x \, \partial x_0} \right)}.$$

Since in this case

$$\frac{\partial^2 S_{cl}}{\partial x \, \partial x_0} = \frac{-a}{\sinh(at)} \mathbb{I}_n,$$

we obtain

$$V(t) = \left(\frac{a}{2\pi \sinh(at)} \right)^{n/2}.$$

Hence the heat kernel is

$$K(x_0, x, t) = \left(\frac{a}{2\pi \sinh(at)} \right)^{n/2} e^{-\frac{a}{2} \frac{1}{\sinh(at)} \left[(|x|^2 + |x_0|^2) \cosh(at) - 2\langle x, x_0 \rangle \right]}$$

for $t > 0$.

In a similar way, one can find the heat kernel for the operator $\frac{1}{2}\Delta_n + \frac{1}{2}a^2|x|^2$. Formally, this reduces to changing sinh into sin and cosh into cos in the previous expression.

An alternate method. Another method for finding the heat kernel of

$$L = \frac{1}{2} \sum_{i=1}^{n} \partial_{x_i}^2 - 2a|x|^2$$

is described next by following reference [11]. Using the rotational symmetry of the operator L, it makes sense to state the following.

Anzatz: The heat kernel of L starting at the origin is of the form

$$K(x, a; t) = \phi_a(t) e^{-\frac{1}{2}\alpha_a(t)|x|^2}, \qquad t > 0.$$

Substituting in equation $(\partial_t - L)K = 0$ yields

$$\left(-\frac{\phi'(t)}{\phi(t)} - \frac{n}{2}\alpha(t) + \frac{1}{2}\left(\alpha^2(t) + \alpha'(t) - a^2\right)|x|^2\right) K = 0.$$

We shall look for functions ϕ and α such that

$$\alpha'(t) = a^2 - \alpha^2(t),$$
$$\frac{\phi'(t)}{\phi(t)} = -\frac{n}{2}\alpha(t).$$

Using the substitution $\alpha = \beta'/\beta$ yields the equation $\beta'' = a^2\beta$, with the solution

$$\beta(t) = A\cosh(at) + B\sinh(at).$$

Using $\lim_{t\searrow 0+} K(x, a; t) = \delta_x$ yields

$$\infty = \lim_{t\searrow 0+} \alpha(t) = \frac{\beta'(0)}{\beta(0)} = \frac{2aB}{A},$$

so $A = 0$. Hence $\alpha(t) = a\coth(at)$.

Integrating in $\frac{\phi'(t)}{\phi(t)} = -\frac{n}{2}a\coth(at)$ yields

$$\phi(t) = \frac{c_n}{\sinh(at)^{n/2}}.$$

The constant c_n may be obtained from the condition

$$\lim_{t\searrow 0+} \int \frac{c_n}{\sinh(at)^{n/2}} e^{-\frac{1}{2}\alpha_a(t)|x|^2}\,dx = 1 \iff$$
$$\lim_{t\searrow 0+} \frac{c_n}{\sinh(at)^{n/2}} \left(\frac{2\pi}{\alpha(t)}\right)^{n/2} = 1 \iff$$
$$\lim_{t\searrow 0+} \left(\frac{2\pi}{a\cosh(at)}\right)^{n/2} = c_n^{-1} \iff$$
$$c_n = \left(\frac{a}{2\pi}\right)^{n/2}.$$

Hence we obtain the heat kernel of L:

$$\begin{aligned} K(x, a; t) &= \phi_a(t) e^{-\frac{1}{2}\alpha_a(t)|x|^2} \\ &= \left(\frac{a}{2\pi\sinh(at)}\right)^{n/2} e^{-\frac{1}{2}a\coth(at)|x|^2}, \qquad t > 0. \end{aligned}$$

3.18 The Operator $L = \frac{1}{2}\partial_x^2 + \frac{\lambda}{x^2}$

The operator is defined on the domain $\{x > 0\}$. This is an example where the transport equation reduces to a modified Bessel equation. This operator has been considered from the geometric point of view in reference [26]. First we find the classical action. The associated Hamiltonian is $H = \frac{1}{2}p^2 + \frac{\lambda}{x^2}$, λ a real constant. The geodesic joining the points x_0 and x within time t satisfies the Euler–Lagrange equation associated with the Lagrangian $L = \frac{1}{2}\dot{x}^2 - \frac{\lambda}{x^2}$:

$$\ddot{x} = \frac{2\lambda}{x^3},$$
$$x(0) = x_0, \quad x(t) = x.$$

Since the regions $\{x < 0\}$ and $\{x > 0\}$ are separated, in order to have connectivity, we have to assume that either $x_0, x > 0$ or $x_0, x < 0$. We can show that the energy is a first integral of motion, so that

$$\frac{1}{2}\dot{x}^2(s) + \frac{\lambda}{x^2(s)} = E, \qquad s \in [0, t].$$

Under the assumption $x_0, x > 0$, the previous relation becomes

$$\dot{x}(s)x(s) = \sqrt{2Ex^2(s) - 2\lambda}.$$

Let $u(s) = x^2(s)$, $u_0 = x_0^2$, $u_t = x^2$. Then $u(s)$ verifies the ODE

$$\dot{u} = 2\sqrt{2Eu - 2\lambda},$$
$$u(0) = u_0, \quad u(t) = u_t.$$

Integrating yields

$$\int_{u_0}^{u_t} \frac{du}{\sqrt{2Eu - 2\lambda}} = 2t \iff$$
$$\sqrt{2Eu_t - 2\lambda} - \sqrt{2Eu_0 - 2\lambda} = 2Et.$$

Eliminating the square roots, we obtain

$$\left((u_0 + u_t) - 2Et^2\right)^2 = 4(u_0 u_t - 2\lambda t^2), \tag{3.18.60}$$

assuming the following condition:

$$\lambda < \frac{x_0^2 x^2}{2t^2}.$$

Solving for E in (3.18.60) yields

$$E = \frac{x_0^2 + x^2}{2t^2} - \frac{\sqrt{x_0^2 x^2 - 2\lambda t^2}}{t^2}. \tag{3.18.61}$$

The classical action S_{cl} satisfies the following Hamilton–Jacobi equation:

$$\partial_t S_{cl} = -E,$$

with E given by (3.18.61). We can write $S_{cl} = S_0 + S_1$, where

$$\partial_t S_0 = -\frac{x_0^2 + x^2}{2t^2} \Longrightarrow S_0 = \frac{x_0^2 + x^2}{2t}, \tag{3.18.62}$$

$$\partial_t S_1 = \frac{\sqrt{x_0^2 x^2 - 2\lambda t^2}}{t^2}. \tag{3.18.63}$$

We shall solve (3.18.63) as a homogeneous equation. Let $\tau = \frac{t}{x_0 x}$, and $S_2(\tau) = S_1(\frac{t}{x_0 x})$. Then

$$\frac{d}{d\tau} S_2(\tau) = \frac{\sqrt{1 - 2\lambda\tau^2}}{\tau^2}.$$

With the substitution $\tau = \frac{1}{\sqrt{2\lambda}} \sin\phi$, integrating yields

$$\begin{aligned} S_2(\tau) &= \int \frac{\sqrt{1 - 2\lambda\tau^2}}{\tau^2}\, d\tau = \sqrt{2\lambda} \int \cot^2\phi\, d\phi \\ &= \sqrt{2\lambda} \int \big(- 1 - \cot'\phi \big)\, d\phi = -\sqrt{2\lambda}(\phi + \cot\phi) \\ &= -\sqrt{2\lambda} \left\{ \sin^{-1}(\sqrt{2\lambda}\tau) + \frac{\sqrt{1 - 2\lambda\tau^2}}{\sqrt{2\lambda}\tau} \right\} \end{aligned}$$

and hence

$$S_1(x_0, x, \tau) = \frac{-2\sqrt{x_0^2 x^2 - \lambda t^2}}{2t} - \sqrt{2\lambda} \sin^{-1}\left(\sqrt{2\lambda} \frac{t}{x_0 x} \right). \tag{3.18.64}$$

From (3.18.64) and (3.18.62), we obtain the classical action

$$S_{cl}(x_0, x, t) = \frac{x_0^2 + x^2}{2t} - \frac{2\sqrt{x_0^2 x^2 - \lambda t^2}}{2t} - \sqrt{2\lambda}\sin^{-1}\left(\sqrt{2\lambda}\frac{t}{x_0 x}\right). \quad (3.18.65)$$

By the general theory, or by a direct computation, the classical action S_{cl} satisfies the Hamilton–Jacobi equation

$$\partial_t S_{cl} + \frac{1}{2}(\partial_x S_{cl})^2 + \frac{\lambda}{x^2} = 0. \quad (3.18.66)$$

The transport equation. We shall assume the heat kernel for $\mathbb{L}$ of the type

$$K(x_0, x, t) = V(x, t)e^{-S(x_0, x, t)}.$$

Then a computation shows

$$\partial_t K = e^{-S}\left(\partial_t V - V\,\partial_t S\right),$$
$$\partial_x^2 K = e^{-S}\left(\partial_x^2 V - 2\partial_x V\,\partial_x S + V(\partial_x S)^2 - V(\partial_x^2 S)\right),$$

and hence

$$(\partial_t - \mathbb{L})K = \left(\partial_t - \frac{1}{2}\partial_x^2 - \frac{\lambda}{x^2}\right)K$$
$$= e^{-S}\Bigg\{ - V\Bigg[\underbrace{\partial_t S + \frac{1}{2}(\partial_x S)^2 + \frac{\lambda}{x^2}}_{=0}\Bigg] \quad \text{[by (3.18.66)]}$$
$$+\partial_t V - \frac{1}{2}\partial_x^2 V + \partial_x V \partial_x S + \frac{1}{2}V(\partial_x^2 S)\Bigg\}.$$

We shall ask V to satisfy the following transport equation:

$$\partial_t V - \frac{1}{2}\partial_x^2 V + \partial_x V \partial_x S + \frac{1}{2}V(\partial_x^2 S) = 0. \quad (3.18.67)$$

Equation (3.18.67) might be hard to solve since the action S and its derivatives are complicated. In the following we shall consider a shortcut for these computations. We note that the action is the sum $S = S_0 + S_1$, where the term S_1 is a function of $\frac{x_0 x}{t}$. Then

$$e^{-S} = W\left(\frac{x_0 x}{t}\right)e^{-\frac{x_0^2 + x^2}{2t}}.$$

Then it makes sense now to look for a heat kernel of the type

$$K(x_0, x, t) = V(x,t)e^{-S_0} = V(x,t)e^{-\frac{x_0^2+x^2}{2t}},$$

where $V(x,t) = \frac{1}{\sqrt{t}}Z(\frac{x_0 x}{t})$ satisfies the *extended transport equation*

$$\partial_t V - \frac{1}{2}\partial_x^2 V + \partial_x V \partial_x S_0 + \frac{1}{2}V(\partial_x^2 S_0) - V\left[\partial_t S_0 + \frac{1}{2}(\partial_x S_0)^2 + \frac{\lambda}{x^2}\right] = 0. \tag{3.18.68}$$

In the following we shall solve (3.18.68). Let $\tau = \frac{x_0 x}{t}$. Then we have

$$\begin{aligned} V &= t^{-\frac{1}{2}} Z(\tau), \\ \partial_t V &= -t^{-\frac{3}{2}}\left(\frac{1}{2}Z(\tau) + \tau Z'(\tau)\right), \\ \partial_x V &= t^{-\frac{3}{2}} Z'(\tau)\, x_0, \\ \partial_x^2 V &= t^{-\frac{3}{2}} Z''(\tau)\, \frac{x_0^2}{t}. \end{aligned}$$

Since

$$\partial_t S = -\frac{x_0^2 + x^2}{2t^2}, \qquad \partial_x S = \frac{x}{t}, \qquad \partial_x^2 S = \frac{1}{t},$$

(3.18.68) becomes, after cancelations,

$$-\frac{1}{2}t^{-\frac{1}{2}}Z''(\tau)\left(\frac{x_0}{t}\right)^2 - t^{-\frac{1}{2}}Z(\tau)\left[\frac{\lambda}{x^2} - \frac{1}{2}\left(\frac{x_0}{t}\right)^2\right] = 0.$$

Multiplying by $-2x^2 t^{-\frac{1}{2}}$ yields

$$\tau^2 Z''(\tau) + Z(\tau)[2\lambda - \tau^2] = 0. \tag{3.18.69}$$

Let $U(\tau) = \tau^{-\frac{1}{2}}Z(\tau)$. A computation shows

$$\begin{aligned} \tau^2 U''(\tau) &= \frac{3}{4}\tau^{-\frac{1}{2}}Z(\tau) - \tau^{\frac{1}{2}}Z'(\tau) + \tau^{\frac{3}{2}}Z''(\tau), \\ \tau U'(\tau) &= -\frac{1}{2}\tau^{-\frac{1}{2}}Z(\tau) + \tau^{\frac{1}{2}}Z'(\tau), \end{aligned}$$

and using (3.18.69), we have

$$\begin{aligned}\tau^2U''(\tau) + \tau U'(\tau) &= \tau^{-\frac{1}{2}}\left(\tau^2 Z''(\tau) + \frac{1}{4}Z(\tau)\right)\\ &= \tau^{-\frac{1}{2}}\left(Z(\tau)(\tau^2 - 2\lambda) + \frac{1}{4}Z(\tau)\right)\\ &= U(\tau)\left(\tau^2 - 2\lambda + \frac{1}{4}\right).\end{aligned}$$

Hence $U(\tau)$ satisfies the modified Bessel equation

$$\tau^2 U'' + \tau U' + (-\tau^2 - \gamma^2)U = 0,$$

with $\gamma = \frac{1}{2}\sqrt{1-8\lambda}$. The general solution can be written as a linear combination

$$U(\tau) = \alpha I_\gamma(\tau) + \beta J_\gamma(\tau), \qquad \alpha, \beta \in \mathbb{R},$$

where $I_\gamma(\tau)$ and J_γ are the modified Bessel functions of the first and second types. Hence the general solution of (3.18.69) is given by

$$Z(\tau) = \sqrt{\tau}U(\tau) = \alpha\sqrt{\tau}I_\gamma(\tau) + \beta\sqrt{\tau}J_\gamma(\tau),$$

where

$$I_\gamma(\tau) \sim \sqrt{\frac{1}{2\pi\tau}}e^\tau, \qquad J_\gamma(\tau) \sim \sqrt{\frac{\pi}{2\tau}}\frac{1}{e^\tau} \quad \text{as } \tau \to \infty; \tag{3.18.70}$$

see, for instance, [41].

Consequently, the solution of the extended transport equation (3.18.68) will be given by

$$V(x_0, x, t) = t^{-1/2}Z(\tau) = \frac{\sqrt{xx_0}}{t}\left(\alpha I_\gamma\left(\frac{xx_0}{t}\right) + \beta J_\gamma\left(\frac{xx_0}{t}\right)\right), \tag{3.18.71}$$

with $\alpha, \beta \in \mathbb{R}$.

Theorem 3.18.1. *The heat kernel for the operator* $\mathbb{L} = \frac{1}{2}\partial_x^2 + \frac{\lambda}{x^2}$, *with* $0 < \lambda < 1/8$ *and* $x > 0$, *is*

$$K(x_0, x; t) = \frac{\sqrt{x_0 x}}{t} I_\gamma\left(\frac{x_0 x}{t}\right) e^{-\frac{x_0^2 + x^2}{2t}}, \qquad t > 0,$$

where I_γ *is the nonsingular modified Bessel function of order* $\gamma = \frac{1}{2}\sqrt{1-8\lambda}$.

Proof. We have shown already in the previous section that

$$(\partial_t - \mathbb{L})K(x_0, x, t) = 0, \qquad t > 0,$$

with

$$K(x_0, x, t) = V(x_0, x, t)e^{-\frac{x_0^2 + x^2}{2t}},$$

and V given by (3.18.71). We need to choose the constants α and β such that

$$\lim_{t \searrow 0} K(x_0, x, t) = \delta_{x_0}$$

in the distributions sense. Let $K = K_1 + K_2$ with

$$K_1(x_0, x, t) = \alpha \frac{\sqrt{x x_0}}{t} I_\gamma \left(\frac{x x_0}{t}\right) e^{-\frac{x^2 + x_0^2}{2t}},$$

$$K_2(x_0, x, t) = \beta \frac{\sqrt{x x_0}}{t} J_\gamma \left(\frac{x x_0}{t}\right) e^{-\frac{x^2 + x_0^2}{2t}}.$$

Then

$$\begin{aligned}
\lim_{t \searrow 0} K_1(x_0, x, t) &= \lim_{t \searrow 0} \alpha \frac{\sqrt{x x_0}}{t} I_\gamma \left(\frac{x x_0}{t}\right) e^{-\frac{x x_0}{t}} e^{-\frac{(x - x_0)^2}{2t}} \\
&= \alpha \lim_{\tau \to \infty} \sqrt{\tau} I_\gamma(\tau) e^{-\tau} \lim_{t \searrow 0} \frac{1}{\sqrt{t}} e^{-\frac{(x - x_0)^2}{2t}} \\
&= \alpha \frac{1}{\sqrt{2\pi}} \lim_{t \searrow 0} \frac{1}{\sqrt{t}} e^{-\frac{(x - x_0)^2}{2t}} \\
&= \alpha \delta_{x_0},
\end{aligned}$$

where we have used the first relation of (3.18.70). Hence we shall choose $\alpha = 1$.

A similar computation, using the second relation of (3.18.70) yields

$$\begin{aligned}
\lim_{t \searrow 0} K_2(x_0, x, t) &= \lim_{t \searrow 0} \beta \frac{\sqrt{x x_0}}{t} J_\gamma \left(\frac{x x_0}{t}\right) e^{-\frac{x x_0}{t}} e^{-\frac{(x - x_0)^2}{2t}} \\
&= \beta \lim_{\tau \to \infty} \sqrt{\tau} J_\gamma(\tau) e^{-\tau} \lim_{t \searrow 0} \frac{1}{\sqrt{t}} e^{-\frac{(x - x_0)^2}{2t}} \\
&= \beta \sqrt{\pi} \lim_{\tau \to \infty} e^{-2\tau} \lim_{t \searrow 0} \frac{1}{\sqrt{2\pi t}} e^{-\frac{(x - x_0)^2}{2t}} \\
&= 0.
\end{aligned}$$

Hence

$$\lim_{t \searrow 0} K(x_0, x, t) = \lim_{t \searrow 0} K_1(x_0, x, t) + \lim_{t \searrow 0} K_2(x_0, x, t) = \alpha \delta_{x_0},$$

so we need to choose $\alpha = 1$. In order to find β, we shall consider the limit $\lambda \to 0$, in which case we recover the Gaussian kernel

$$\begin{aligned}\frac{1}{\sqrt{2\pi t}} e^{-\frac{(x-x_0)^2}{2t}} &= \lim_{\lambda \searrow 0} K(x_0, x, t)\\ &= \frac{1}{\sqrt{t}} \sqrt{\tau} I_{1/2}(\tau) e^{-\tau} e^{-\frac{(x-x_0)^2}{2t}} + \beta \frac{1}{\sqrt{t}} \sqrt{\tau} J_{1/2}(\tau) e^{-\tau} e^{-\frac{(x-x_0)^2}{2t}}\\ &= \frac{1}{\sqrt{2\pi t}} e^{-\frac{(x-x_0)^2}{2t}} + \beta \frac{1}{\sqrt{t}} \sqrt{\tau} J_{1/2}(\tau) e^{-\tau} e^{-\frac{(x-x_0)^2}{2t}},\end{aligned}$$

since we take $\gamma = 1/2$ in

$$I_\gamma(\tau) = \frac{1}{\sqrt{2\pi\tau}} e^{\tau - \frac{\frac{1}{2}(\gamma^2 - \frac{1}{4})}{\tau}} + O(1/\tau^2),$$

and hence we need to choose $\beta = 0$ in relation (3.18.71). ■

Chapter 4
Commuting Operators

4.1 Commuting Operators

If X and Y are two vector fields, or in general, two operators which commute, i.e., $XY = YX$, then it is obvious that their squares also commute. If $\mathbb{L} = \frac{1}{2}(X^2 + Y^2)$ with $[X^2, Y^2] = 0$, then the problem of finding the heat kernel for $\mathbb{L}$ is reduced to the same problem for the operators X^2 and Y^2, with

$$e^{t(X^2+Y^2)} = e^{tX^2} e^{tY^2}.$$

The aforementioned formula is reminiscent from stochastic calculus, where if the operators $\frac{1}{2}X^2$ and $\frac{1}{2}Y^2$ are generators for certain independent Ito diffusion processes, then their joint density function is a product of their individual density functions. A similar formula holds in the general case of n commuting vector fields.

The nontrivial problem occurs when X^2 and Y^2 do not commute, in which case the heat kernel can be obtained either by using Trotter's formula and path integrals (see Chap. 7), or by using the geometric method (see Chap. 3). We shall provide a few examples below that can be reduced to a commuting sum of squares of two vector fields.

4.1.1 The Operator $\mathbb{L} = \frac{1}{2}(\partial_x^2 + \partial_y^2)$

This is an elementary example of reducing the two-dimensional Laplacian problem to a one-dimensional Laplacian problem. Since the operators ∂_x^2 and ∂_y^2 commute, we can write

$$\begin{aligned} e^{t\mathbb{L}} &= e^{\frac{1}{2}t\partial_x^2} e^{\frac{1}{2}t\partial_y^2} = \frac{1}{\sqrt{2\pi t}} e^{-\frac{(x-x_0)^2}{2t}} \frac{1}{\sqrt{2\pi t}} e^{-\frac{(y-y_0)^2}{2t}} \\ &= \frac{1}{2\pi t} e^{-\frac{(x-x_0)^2+(y-y_0)^2}{2t}}. \end{aligned}$$

O. Calin et al., *Heat Kernels for Elliptic and Sub-elliptic Operators*,
Applied and Numerical Harmonic Analysis, DOI 10.1007/978-0-8176-4995-1_4,

4.1.2 The Operator $\mathbb{L} = \frac{1}{2}(\partial_x^2 + y\partial_y^2)$

Let $(x, y) \in \mathbb{R}\times(0, \infty)$. Since the operators ∂_x^2 and $y\partial_y^2$ commute, using Proposition 3.4.1, the heat kernel for $\mathbb{L}$ between $(x_0, 0)$ and (x, y) within time t is

$$K\big((x_0,0),(x,y);t\big) = e^{t\mathbb{L}} = e^{\frac{t}{2}(\partial_x^2+y\partial_y^2)} = e^{\frac{t}{2}\partial_x^2}e^{\frac{t}{2}y\partial_y^2}$$
$$= \frac{1}{(2\pi t)^{1/2}}e^{-\frac{(x-x_0)^2}{2t}}\frac{4y}{t^2}e^{-\frac{2}{t}y}$$
$$= 2\sqrt{\frac{2}{\pi}}\frac{y}{t^{3/2}}e^{-\frac{1}{2t}\left(4y-(x-x_0)^2\right)}, \qquad t > 0.$$

4.1.3 The Operator $\mathbb{L} = \frac{1}{2}(x\partial_x^2 + y\partial_y^2)$

The operator is defined on the domain $(x, y) \in (0, \infty)\times(0, \infty)$. Since the operators $x\partial_x^2$ and $y\partial_y^2$ commute, using Proposition 3.4.1 we have

$$K\big((0,0),(x,y);t\big) = e^{t\mathbb{L}} = e^{\frac{t}{2}x\partial_x^2}e^{\frac{t}{2}y\partial_y^2}$$
$$= \frac{4x}{t^2}e^{-\frac{2}{t}x}\,\frac{4y}{t^2}e^{-\frac{2}{t}y} = \frac{4xy}{t^4}e^{-\frac{2}{t}(x+y)}, \qquad t > 0.$$

The graph of this heat kernel for $t = 0.25$ and $x > 0$, $y > 0$ is given by Fig. 4.1a.

4.1.4 Sum of Squares of Linear Potentials

The following example will be useful in the sequel. It is obtained by applying the Fourier transform to the Heisenberg operator. Consider the noncommutative

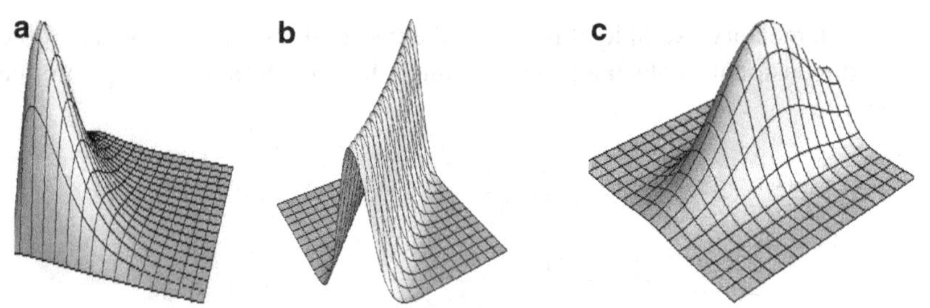

Fig. 4.1 Heat kernel profiles for different operators. **(a)** $\frac{1}{2}(x\partial_x^2 + y\partial_y^2)$; **(b)** $\frac{1}{2}\Delta_2 + \frac{1}{2}\omega + \frac{1}{2}|x|^2$; **(c)** $\frac{1}{2}(x^2 + y^2)(\partial_x^2 + \partial_y^2)$

operators $L_1 = \partial_{x_1} + ax_2$ and $L_2 = \partial_{x_2} - ax_1$, with $a \in \mathbb{R}$ constant. Construct the operator

$$\begin{aligned} L &= \frac{1}{2}(L_1^2 + L_2^2) \\ &= \frac{1}{2}(\partial_{x_1}^2 + \partial_{x_2}^2) + \frac{a}{2}(x_2\partial_{x_1} - x_1\partial_{x_2}) + \frac{a^2}{2}(x_1^2 + x_2^2) \\ &= \frac{1}{2}\Delta_2 + \frac{a}{2}\omega + \frac{a^2}{2}|x|^2, \end{aligned}$$

where ω stands for the angular momentum. We note that $e^{t\frac{a}{2}\omega}\delta_{(0,0)} = \delta_{(0,0)}$. Since the operators ω and $\frac{1}{2}\Delta + \frac{a}{2}|x|^2$ commute, the heat kernel for L is

$$\begin{aligned} K(x, y; t) &= e^{\frac{t}{2}(L_1^2+L_2^2)}\delta_{(0,0)} = e^{t\left(\frac{1}{2}\Delta_2+\frac{a}{2}\omega+\frac{a^2}{2}|x|^2\right)}\delta_{(0,0)} \\ &= e^{t\left(\frac{1}{2}\Delta_2+\frac{a^2}{2}|x|^2\right)}e^{t\frac{a}{2}\omega}\delta_{(0,0)} = e^{t(\frac{1}{2}\Delta_2+\frac{a^2}{2}|x|^2)}\delta_{(0,0)} \\ &= e^{\frac{t}{2}(\Delta_2+a^2|x|^2)}\delta_{(0,0)} \\ &= \frac{1}{2\pi t}\frac{2at}{\sin(2at)}e^{-\frac{1}{2t}\frac{2at}{\sin(2at)}[(|x|^2+|y|^2)\cos(2at)-2\langle x, y\rangle]}, \qquad t > 0. \end{aligned}$$

The shape of the heat kernel for the values $a = 1$ and $t = 0.05$ is given in Fig. 4.1b.

4.1.5 *The Operator* $\mathbb{L} = \frac{1}{2}(x^2 + y^2)(\partial_x^2 + \partial_y^2)$

Assume $(x, y), (x_0, y_0) \neq (0, 0)$. Using the representation in polar coordinates $x = r\cos\theta$, $y = r\sin\theta$, the operator can be written as a sum of two operators in variables r and θ:

$$(x^2 + y^2)(\partial_x^2 + \partial_y^2) = (r^2\partial_r^2 + r\partial_r) + \partial_\theta^2.$$

Using formulas which will be proved in Sect. 6.3, we have

$$\begin{aligned} K(x_0, y_0, x, y; t) &= e^{\frac{t}{2}(x^2+y^2)(\partial_x^2+\partial_y^2)} = e^{\frac{t}{2}(r^2\partial_r^2+r\partial_r)}e^{\frac{t}{2}\partial_\theta^2} \\ &= \frac{1}{2t}e^{-\frac{r_0^2+r^2}{2t}}I_0\Big(\frac{r_0 r}{t}\Big)\frac{1}{\sqrt{2\pi t}}e^{-\frac{(\theta-\theta_0)^2}{2t}} \\ &= \pi^{-1/2}(2t)^{-3/2}I_0\Big(\frac{r_0 r}{t}\Big)e^{-\frac{1}{2t}\left(r_0^2+r^2+(\theta-\theta_0)^2\right)}, \qquad t, r, r_0 > 0, \end{aligned}$$

where I_0 is the Bessel function of the first kind of order zero. The profile of the heat kernel with $r_0 = 1$, $\theta_0 = 0$ and $t = 0.2$ is depicted in Fig. 4.1c. In the view of the geometric method (see Chap. 3), the term $\pi^{-1/2}(2t)^{-3/2} I_0\left(\frac{r_0 r}{t}\right)$ represents the volume element, while $d^2\big((r_0, \theta_0), (r, \theta)\big) = r_0^2 + r^2 + (\theta - \theta_0)^2$ is the square of the associated Riemannian distance.

Chapter 5
The Fourier Transform Method

The Fourier transform has been known as one of the most powerful and useful methods of finding fundamental solutions for operators with constant coefficients. Sometimes the application of a partial Fourier transform might be more useful than the full Fourier transform. In this chapter, by the application of the partial Fourier transform, we shall reduce the problem of finding the heat kernel of a complicated operator to a simpler problem involving an operator with fewer variables. After solving the problem for this simple operator, the inverse Fourier transform provides the heat kernel for the initial operator represented under an integral form. In general, this integral cannot be computed explicitly, but in certain particular cases it actually can be worked out. We shall also apply this method to some degenerate operators.

5.1 The Algorithm

If $\mathbb{L}$ is a partial differential operator in variables $(x, y) \in \mathbb{R}^n \times \mathbb{R}$, then the heat kernel $K(x_0, y_0, x, y; t) = v(x, y; t)$ satisfies

$$\begin{aligned} \partial_t v &= \mathbb{L}_{xy} v, \\ \lim_{t \searrow 0} v(x, y; t) &= \delta_{x_0}(x) \otimes \delta_{y_0}(y), \end{aligned}$$

where δ_a is the Dirac distribution centered at a. Let $u(x, \xi; t) = \mathcal{F}_y(v)(x, \xi; t)$, where $\mathcal{F}_y$ denotes the partial Fourier transform with respect to y:

$$(\mathcal{F}_y v)(x, \xi; t) = \int_{\mathbb{R}} e^{-iy\xi} v(x, y; t)\, dy.$$

Then u satisfies the equation

$$\begin{aligned} \partial_t u &= \mathbb{P}_x u, \\ u(x, \xi; 0) &= \delta_{x_0}(x) \otimes \mathbb{I}_\xi, \end{aligned}$$

O. Calin et al., *Heat Kernels for Elliptic and Sub-elliptic Operators*,
Applied and Numerical Harmonic Analysis, DOI 10.1007/978-0-8176-4995-1_5,

where $\mathbb{P}_x$ is obtained from $\mathbb{L}_{xy}$ by replacing ∂_y^2 by $i\xi$. In order to recover v, we need to apply the inverse Fourier transform on u:

$$v(x,y;t) = (\mathcal{F}_\xi^{-1}u)(x,\xi;t).$$

The following formulas will be useful in the sequel:

$$\mathcal{F}^{-1}(u) = \frac{1}{2\pi}\mathcal{F}(\check{u}), \text{ where } \check{u}(x) = u(-x),$$

$$\mathcal{F}(e^{-a\xi^2})(x) = \left(\frac{\pi}{a}\right)^{1/2} e^{-x^2/(4a)}, \qquad a > 0,$$

$$\mathcal{F}(\delta) = \mathbb{I}, \qquad \mathcal{F}(\mathbb{I}) = 2\pi\delta.$$

5.2 Heat Kernel for the Grushin Operator

The two-dimensional, two-step Grushin operator is defined as the sum of squares

$$\Delta_G = \frac{1}{2}(X^2 + Y^2) = \frac{1}{2}(\partial_x^2 + x^2\partial_y^2),$$

where $X = \partial_x$ and $Y = x\partial_y$. The heat kernel satisfies

$$\partial_t v = \Delta_G v, \qquad t > 0,$$

$$\lim_{t\searrow 0} v(x,y;t) = \delta_{x_0}(x) \otimes \delta_{y_0}(y).$$

Let $u = \mathcal{F}_y(v)$ be the partial Fourier transform of v with respect to y. Then u satisfies

$$\partial_t u = \frac{1}{2}(\partial_x^2 u - x^2\xi^2 u), \qquad t > 0,$$

$$u(x,\xi;0) = \delta_{x_0}(x) \otimes \mathbb{I}_\xi.$$

The solution is

$$u(x,\xi;t) = e^{\frac{t}{2}(\partial_x^2 - x^2\xi^2)}(\delta_{x_0}(x) \otimes \mathbb{I}_\xi). \tag{5.2.1}$$

By Theorem 3.16.1, the kernel of the Hermite operator $\frac{1}{2}\partial_x^2 - \frac{1}{2}\xi^2x^2$ is

$$K(x_0,x;t) = \frac{1}{\sqrt{2\pi t}}\sqrt{\frac{\xi t}{\sinh(\xi t)}}\, e^{-\frac{1}{2t}\frac{\xi t}{\sinh(\xi t)}\left((x^2+x_0^2)\cosh(\xi t) - 2xx_0\right)}, \qquad t > 0.$$

Applying an inverse Fourier transform with respect to ξ in (5.2.1) yields

$$\begin{aligned}
v(x,y;t) &= \frac{1}{2\pi}\int e^{iy\xi}u(x,\xi;t)\,d\xi \\
&= \frac{1}{2\pi}\int e^{iy\xi}\left\{\frac{1}{\sqrt{2\pi t}}\sqrt{\frac{\xi t}{\sinh(\xi t)}}\,e^{-\frac{1}{2t}\frac{\xi t}{\sinh(\xi t)}\left((x^2+x_0^2)\cosh(\xi t)-2xx_0\right)}\right\}d\xi \\
&\qquad\qquad\qquad\qquad\qquad\qquad\qquad\qquad\qquad\qquad\qquad(\text{let } \tau = \xi t) \\
&= \frac{1}{(2\pi t)^2}\int\sqrt{\frac{\tau}{\sinh\tau}}\,e^{iy\tau/t-\frac{\tau/t}{2\sinh\tau}\left((x^2+x_0^2)\cosh(\tau)-2xx_0\right)}\,d\tau \\
&= \frac{1}{(2\pi)^{3/2}\sqrt{t}}\int\sqrt{\frac{\tau}{\sinh\tau}}\,e^{\{iy\tau-\frac{1}{2}(x^2+x_0^2)\tau\coth\tau-\frac{\tau}{\sinh\tau}xx_0\}/t}\,d\tau.
\end{aligned}$$

If we let

$$f(x_0,x,y;\tau) = -iy\tau + \frac{1}{2}(x^2+x_0^2)\tau\coth\tau + \frac{\tau}{\sinh\tau}xx_0, \tag{5.2.2}$$

$$V(\tau) = \sqrt{\frac{\tau}{\sinh\tau}}, \tag{5.2.3}$$

we obtain the following form:

$$v(x,y;t) = \frac{1}{(2\pi)^{3/2}\sqrt{t}}\int e^{-f(x_0,x,y;\tau)/t}\,V(\tau)\,d\tau, \qquad t>0,$$

where f is called the *modified complex action* (see also [16]), and $V(\tau)$ is the *volume element*. Formula (5.2.2) cannot be reduced to just an elementary function, since the Grushin operator does not have a heat kernel of the function type. This result is true for most sub-elliptic operators given as a sum of squares.

5.3 Heat Kernel for $\partial_x^2 + x\partial_x\partial_y$

The heat kernel of the operator $\mathbb{L} = \partial_x^2 + x\partial_x\partial_y$ satisfies

$$\begin{aligned}
\partial_t v &= \partial_x^2 v + x\partial_x\partial_y v, \qquad t>0, \\
\lim_{t\searrow 0} v(x,y;t) &= \delta_{x_0}(x)\otimes\delta_{y_0}(y).
\end{aligned}$$

Let $u = \mathcal{F}_y(v)$ be the partial Fourier transform of v with respect to y. Then u satisfies

$$\begin{aligned}
\partial_t u &= \partial_x^2 - i\xi x\partial_x u, \qquad t>0, \\
u(x,\xi;0) &= \delta_{x_0}(x)\otimes\mathbb{I}_\xi.
\end{aligned}$$

The solution is

$$u(x,\xi;t) = e^{t(\partial_x^2 - i\xi x\partial_x)}(\delta_{x-x_0} \otimes \mathbb{I}_\xi). \tag{5.3.4}$$

By Theorem 10.29 of [24], the fundamental solution of $\partial_t - \partial_x^2 + i\xi x\partial_x$ is

$$K(x_0,x;t) = \frac{1}{\sqrt{4\pi t}}\sqrt{\frac{\xi t}{\sin(\xi t)}}\, e^{-\frac{1}{4t}\frac{\xi t}{\sin(\xi t)}\left((x^2+x_0^2)\cos(\xi t)-2xx_0\right)}, \qquad t>0.$$

Applying the inverse Fourier transform, we obtain

$$\begin{aligned}
v(x,y;t) &= \frac{1}{2\pi}\int e^{iy\xi}u(x,\xi;t)\,d\xi \\
&= \frac{1}{2\pi}\int e^{iy\xi}\left\{\frac{1}{\sqrt{4\pi t}}\sqrt{\frac{\xi t}{\sin(\xi t)}}\, e^{-\frac{1}{4t}\frac{\xi t}{\sin(\xi t)}\left((x^2+x_0^2)\cos(\xi t)-2xx_0\right)}\right\} d\xi \\
&\qquad\qquad\qquad\qquad\qquad\qquad\qquad\qquad\qquad\qquad (\text{let } \tau = \xi t) \\
&= \frac{1}{4(\pi t)^{3/2}}\int \sqrt{\frac{\tau}{\sin\tau}}\, e^{i\frac{y\tau}{t}-\frac{1}{4t}\left(\tau(x^2+x_0^2)\cot\tau - \frac{2xx_0}{\sin\tau}\right)}\,d\tau \\
&= \frac{1}{4(\pi t)^{3/2}}\int V(\tau)e^{-f/t}\,d\tau,
\end{aligned} \tag{5.3.5}$$

with

$$\begin{aligned}
V(\tau) &= \sqrt{\frac{\tau}{\sin\tau}}, \\
f(x_0,x,y;\tau) &= -iy\tau + \frac{1}{4}\left(\tau(x^2+x_0^2)\cot\tau - \frac{2xx_0}{\sin\tau}\right).
\end{aligned}$$

It is remarkable that heat kernel formulas (5.2.2) and (5.3.5) are similar; just the modified complex actions and the volume elements are different. This leads to the question of whether this is a universal formula for sub-elliptic-type operators.

5.4 The Formula of Beals, Gaveau and Greiner

In this section we shall consider a sub-elliptic operator with one degenerate direction of the type

$$\mathbb{L} = \frac{1}{2}\sum_{i,j=1}^{n} a_{ij}\partial_{x_i}\partial_{x_j} + \phi(x)\partial_y.$$

In general, heat kernels in this case can be represented as an integral, which usually does not integrate to yield a function-type kernel. The degenerate direction can be eliminated by applying a partial Fourier transform in the y-variable. Let v be the fundamental solution for the sub-elliptic operator $\mathbb{L}$:

$$\mathbb{L}v = 0, \qquad t > 0,$$
$$\lim_{t \searrow 0} v(x, y; t) = \delta_x \otimes \delta_y.$$

Let $u(x, \xi; t) = \mathcal{F}_y v(x, y; t)$ be the partial Fourier transform of v with respect to y. Then u satisfies

$$\mathbb{P}u = 0, \qquad t > 0,$$
$$\lim_{t \searrow 0} u(x, \xi; t) = \delta_x \otimes \mathbb{I}_\xi,$$

where

$$\mathbb{P} = \frac{1}{2} \sum_{i,j=1}^{n} a_{ij} \partial_{x_i} \partial_{x_j} - i\xi\phi(x)$$

is an elliptic operator with potential $U(x) = i\xi\phi(x)$. If we treat ξ as a real parameter, then u becomes the heat kernel for the operator $\mathbb{P}$. Now we shall assume that the potential is at most quadratic, i.e., $\phi(x) = \sum b_{ij} x_i x_j + \sum c_j x_j + k$, so the geodesic joining any two points is unique. The geometric method or the path integral method yields in this case a fundamental solution of the type

$$u(x, \xi, t) = V(t, \xi) e^{-S_{cl}},$$

where S_{cl} is the classical action associated with the operator $\mathbb{P}$. This is obtained by integrating the Lagrangian

$$L = \frac{1}{2} \sum_{i,j=1}^{n} a_{ij} p_i p_j + i\xi\phi(x)$$

along the geodesics. The parameter ξ can be considered as a Lagrange multiplier that depends on the endpoints of the geodesic.

Taking the inverse Fourier transform yields the heat kernel for the sub-elliptic operator $\mathbb{L}$:

$$v(x, y, t) = \mathcal{F}_\xi^{-1} u(x, \xi; t) = \frac{1}{2\pi} \int e^{iy\xi} V(t, \xi) e^{-S_{cl}(x_0, x, t; \xi)} \, d\xi. \quad (5.4.6)$$

We shall assume that in the formula of S_{cl} the variables t and ξ come as a product, so we can write $\tau = t\xi$. Let

$$g(x_0, x, \tau) = t S_{cl}(x_0, x, t; \xi).$$

Under this hypothesis, the variables t and ξ will also come as a product in the formula of V. This follows from the van Vleck formula

$$V = \sqrt{\det\Big(-\frac{1}{2\pi}\frac{\partial^2 S_{cl}}{\partial x\, \partial x_0}\Big)}.$$

Consider the function $W(\tau) = V(t, \xi)$. Then formula (5.4.6) becomes

$$\begin{aligned} v(x, y; t) &= \frac{1}{2\pi}\int W(\tau)e^{iy(\tau/t)}e^{-S_{cl}(x_0,x,\tau)}\, d(\tau/t) \\ &= \frac{1}{2\pi t}\int W(\tau)e^{[iy\tau - g(x_0,x,\tau)]/t}\, d(\tau) \\ &= \frac{1}{2\pi t}\int W(\tau)e^{-f(x_0,x,\tau)/t}\, d(\tau), \end{aligned} \tag{5.4.7}$$

where

$$f(x_0, x, \tau) = -iy\tau + g(x_0, x, \tau)$$

is called the *modified complex action function*. The interested reader can find more details about finding the complex action function by methods of complex Hamiltonian mechanics in Chap. 5 of [28]. Formula (5.4.7) was first shown in the case of the Heisenberg operator in the work of Beals, Gaveau and Greiner [16, 48].

5.5 Kolmogorov Operators

This section deals with the case of a degenerate operator for which the integral formula (5.4.7) actually does integrate exactly, providing a closed-form formula for the heat kernel. The operator

$$\mathbb{L} = \frac{1}{2}\partial_x^2 - x\partial_y \tag{5.5.8}$$

was first studied by Kolmogorov [80], who constructed its heat kernel as a transition probability of a diffusion process. We note that the operator is degenerate in the y-direction and is not a sum of squares of vector fields. This section is based on the results obtained in Calin and Chang [27].

Let K be the heat kernel for $\mathbb{L}$, and denote by $u = \mathcal{F}_y K$ the partial Fourier transform with respect to the y-variable. Then

$$\begin{aligned} \partial_t u &= \frac{1}{2}\partial_x^2 u - i\xi x u, \\ u(x, y; t) &= \delta_{x_0}(x) \otimes \mathbb{I}_\xi. \end{aligned}$$

Denote $a = i\xi$, and consider it a parameter. Then u is the heat kernel for the operator

$$\mathbb{P} = \frac{1}{2}\partial_x^2 - ax.$$

Applying Theorem 3.15.1 yields

$$\begin{aligned} u(x_0, x; t) &= \frac{1}{\sqrt{2\pi t}} e^{-\frac{(x-x_0)^2}{2t} - \frac{1}{2}a(x+x_0)t + \frac{1}{24}a^2t^3} \\ &= \frac{1}{\sqrt{2\pi t}} e^{-\frac{(x-x_0)^2}{2t} - \frac{1}{2}i\xi(x+x_0)t - \frac{1}{24}\xi^2 t^3}. \end{aligned}$$

The heat kernel of $\mathbb{L}$ can be obtained by taking the inverse Fourier transform:

$$\begin{aligned} K(x_0, y_0, x, y; t) &= \frac{1}{2\pi}\int e^{i(y-y_0)\xi} u(x, \xi; t)\, d\xi \\ &= \frac{1}{2\pi}\int \frac{1}{\sqrt{2\pi t}} e^{-\frac{(x-x_0)^2}{2t} - \frac{1}{24}\xi^2t^3 + i\xi[(y-y_0) - \frac{1}{2}(x+x_0)t]}\, d\xi \\ &= \frac{1}{2\pi\sqrt{2\pi t}} e^{-\frac{(x-x_0)^2}{2t}} \int e^{-\frac{1}{24}\xi^2t^3 + i\xi[(y-y_0) - \frac{1}{2}(x+x_0)t]}\, d\xi. \end{aligned} \tag{5.5.9}$$

Let

$$A = A(x_0, x, y_0, y, t) = (y - y_0) - \frac{1}{2}(x + x_0)t.$$

If we complete the square, the exponent becomes

$$\begin{aligned} -\frac{1}{24}\xi^2t^3 + i\xi\Big[(y-y_0) - \frac{1}{2}(x+x_0)t\Big] &= -\frac{1}{24}\xi^2t^3 + i\xi A = \frac{t^3}{24}\Big\{(i\xi)^2 + i\xi\frac{24}{t^3}A\Big\} \\ &= \frac{t^3}{24}\Big\{\Big(i\xi + \frac{12}{t^3}A\Big)^2 - \Big(\frac{12A}{t^3}\Big)^2\Big\} \\ &= \frac{t^3}{24}\Big(i\xi + \frac{12}{t^3}A\Big)^2 - \frac{6A^2}{t^3}. \end{aligned}$$

The integral can be written as an integral along a line:

$$\int e^{-\frac{1}{24}\xi^2t^3 + i\xi A}\, d\xi = e^{-\frac{6A^2}{t^3}} \int e^{-\frac{t^3}{24}(\xi - i\frac{12}{t^3}A)^2}\, d\xi = e^{-\frac{6A^2}{t^3}} \int_{Im\, z=b} e^{-\frac{t^3}{24}z^2}\, dz, \tag{5.5.10}$$

where $b = \frac{-12}{t^3}A$. Let $R > 0$. Using that $f(z) = e^{-t^3 z^2/24}$ is holomorphic, the integral along the rectangular contour $(-R,0), (-R,b), (R,b), (R,0)$ vanishes, so

$$\int_{Im\, z=b} e^{-\frac{t^3}{24}z^2}\, dz = \lim_{R\to\infty}\left\{\int_{-R}^{R} e^{-\frac{t^3}{24}y^2}\, dy + \int_0^b e^{-\frac{t^3}{24}(R+i\eta)^2} i\, d\eta \right.$$
$$\left. + \int_b^0 e^{-\frac{t^3}{24}(R+i\eta)^2} i\, d\eta\right\} = \int e^{-\frac{t^3}{24}y^2}\, dy = \frac{2\sqrt{6\pi}}{t^{3/2}}, \tag{5.5.11}$$

since

$$\left|e^{-\frac{t^3}{24}(R+i\eta)^2}\right| = e^{-\frac{t^3}{24}(R^2-\eta^2)} \to 0, \quad \text{for } R \to \infty.$$

We also note that the computation could be short-cut by the use of Proposition 1.6.1. Using (5.5.10) and (5.5.11), (5.5.9) yields

$$K(x_0, y_0, x, y; t) = \frac{1}{2\pi\sqrt{2\pi t}} e^{\frac{-(x-x_0)^2}{2t} - \frac{6A^2}{t^3}} \frac{2\sqrt{6\pi}}{t^{3/2}}$$
$$= \frac{\sqrt{3}}{\pi t^2} e^{\frac{-(x-x_0)^2}{2t} - \frac{6}{t^3}\left(y - y_0 - \frac{1}{2}(x+x_0)t\right)^2}.$$

Theorem 5.5.1. *The heat kernel of the sub-elliptic operator* $\frac{1}{2}\partial_x^2 - x\partial_y$ *is given by*

$$K(x_0, y_0, x, y; t) = \frac{\sqrt{3}}{\pi t^2} e^{\frac{-(x-x_0)^2}{2t} - \frac{6}{t^3}\left(y - y_0 - \frac{1}{2}(x+x_0)t\right)^2}, \qquad t > 0.$$

One may consider Kolmogorov operators of the type $\frac{1}{2}\partial_x^2 - x^m\partial_y$. The previous method provides heat kernels only for the values $m = 0, 1, 2$.

In the following we shall treat the case $m = 0$. We need the following result.

Lemma 5.5.2. *The heat kernel for the operator* $\frac{1}{2}\partial_x^2 - a$, *with* $a \in \mathbb{C}$, *is*

$$u(x_0, x, y_0, y; t) = \frac{1}{\sqrt{2\pi t}} e^{-\frac{(x-x_0)^2}{2t} - at}.$$

Proof. The associated Hamiltonian and Lagrangian are $H = \frac{1}{2}p - a$ and $L = \frac{1}{2}\dot{x}^2 + a$. The Euler–Lagrange equation is $\ddot{x} = 0$, with the boundary conditions $x(0) = x_0$, $x(t) = x$. The classical action is

$$S_{cl} = \frac{(x - x_0)^2}{2t} + at.$$

The fundamental solution has the form

$$u(x_0, x, y_0, y; t) = \sqrt{\det\Big(-\frac{1}{2\pi}\frac{\partial^2 S_{cl}}{\partial x\,\partial x_0}\Big)}e^{-S_{cl}}$$
$$= \frac{1}{\sqrt{2\pi t}}e^{-\frac{(x-x_0)^2}{2t}-at}.$$

∎

Let K be the heat kernel for the sub-elliptic operator $\frac{1}{2}\partial_x^2 - \partial_y$. Then $u = \mathcal{F}_y K$ is the heat kernel for the operator $\frac{1}{2}\partial_x^2 - i\xi$. Applying Lemma 5.5.2 with $a = i\xi$ yields

$$K = \frac{1}{2\pi}\int e^{i(y-y_0)\xi}u(x_0, x, \xi; t)\,d\xi$$
$$= \frac{1}{2\pi}\int e^{i(y-y_0)\xi}\frac{1}{\sqrt{2\pi t}}e^{-\frac{(x-x_0)^2}{2t}-i\xi t}\,d\xi$$
$$= \frac{1}{(2\pi)^{3/2}}\frac{1}{\sqrt{t}}e^{-\frac{(x-x_0)^2}{2t}}\int e^{i\xi(y-y_0-t)}\,d\xi$$
$$= \frac{1}{\sqrt{2\pi t}}e^{-\frac{(x-x_0)^2}{2t}}\otimes\delta_{y_0+t}(y),$$

where we have used the formula for the Dirac distribution

$$\int e^{i\xi x}\,d\xi = \widehat{1}_{(x)} = 2\pi\delta_0(x).$$

To conclude, we have the following result:

Proposition 5.5.3. *The heat kernel for the operator* $\frac{1}{2}\partial_x^2 - \partial_y$ *is*

$$K(x_0, x, y_0, y; t) = \frac{1}{\sqrt{2\pi t}}e^{-\frac{(x-x_0)^2}{2t}}\otimes\delta_{y_0+t}(y).$$

We shall next find the heat kernel for the operator $\frac{1}{2}\partial_x^2 + ix^2\partial_y$. After applying a partial Fourier transform with respect to y, we obtain the Hermite operator $\frac{1}{2}\partial_x^2 - \xi x^2$ with the heat kernel

$$u(x_0, x, t) = \frac{1}{\sqrt{2\pi}}\sqrt{\frac{a}{\sinh(at)}}\,e^{-\frac{1}{2}\frac{a}{\sinh(at)}[(x^2+x_0^2)\cosh(at)-2xx_0]},\quad t>0,$$

with $a^2 = 2\xi$; see Theorem 3.16.1. The inverse Fourier transform yields the heat kernel for the operator $\frac{1}{2}\partial_x^2 + ix^2\partial_y$:

$$
\begin{aligned}
K &= \frac{1}{2\pi}\int e^{i(y-y_0)\xi}u(x_0,x,\xi;t)\,d\xi \\
&= \frac{1}{(2\pi)^{3/2}}\sqrt{\frac{a}{\sinh(at)}}\int e^{i(y-y_0)\frac{a^2}{2}-\frac{1}{2}\frac{a}{\sinh(at)}[(x^2+x_0^2)\cosh(at)-2xx_0]}\,\frac{1}{2}a\,da \\
&= \frac{1}{2^{5/2}}\Big(\frac{a}{\pi}\Big)^{3/2}\int ae^{i(y-y_0)\frac{a^2}{2}-\frac{1}{2}\frac{a}{\sinh(at)}[(x^2+x_0^2)\cosh(at)-2xx_0]}\,da.
\end{aligned}
$$

We note that in this case the heat kernel has an integral representation that cannot be reduced to a function.

In the rest of the section we shall study the sub-elliptic operator

$$
\mathbb{L} = \frac{1}{2}(\partial_{x_1}^2 + \partial_{x_2}^2) + (x_1 - x_2)\partial_y, \tag{5.5.12}
$$

which can be regarded as a two-dimensional Kolmogorov operator. If K denotes the heat kernel for $\mathbb{L}$, then $u = \mathcal{F}_y K$ is the heat kernel for the operator

$$
\mathbb{P} = \frac{1}{2}(\partial_{x_1}^2 + \partial_{x_2}^2) + i\xi(x_1 - x_2). \tag{5.5.13}
$$

u can be obtained by letting $a_1 = -a_2 = -i\xi$ in Theorem 3.15.3:

$$
\begin{aligned}
u(x_0,x,\xi;t) &= \frac{1}{2\pi t}e^{-\frac{|x-x_0|^2}{2t}-\frac{1}{2}\langle a,x+x_0\rangle t+\frac{1}{24}|a|^2t^3} \\
&= \frac{1}{2\pi t}e^{-\frac{|x-x_0|^2}{2t}-\frac{i\xi}{2}\langle \rho,x+x_0\rangle t-\frac{\xi^2}{12}t^3},
\end{aligned}
$$

with $\rho = (-1,1) \in \mathbb{R}^2$. Let $A = (y - y_0) - \frac{1}{2}\langle \rho, x + x_0\rangle t$. The heat kernel of (5.5.12) is obtained by taking the inverse Fourier transform:

$$
\begin{aligned}
K = \mathcal{F}_y^{-1}u &= \frac{1}{2\pi}\int e^{i(y-y_0)\xi}u(x_0,x,\xi;t)\,d\xi \\
&= \frac{1}{(2\pi)^2 t}e^{-\frac{|x-x_0|^2}{2t}}\int e^{-\frac{\xi^2t^3}{12}+i\xi A}\,d\xi.
\end{aligned} \tag{5.5.14}
$$

The second formula of Proposition 1.6.1 yields

$$
\int e^{-\frac{\xi^2t^3}{12}+i\xi A} = e^{-\frac{3A^2}{t^3}}\frac{2\sqrt{3\pi}}{t^{3/2}},
$$

and substituting in (5.5.14), we arrive at the following result:

Theorem 5.5.4. *The heat kernel of the operator (5.5.12) is*

$$K(x_0, y_0, x, y; t) = \frac{\sqrt{3}}{2\pi^{3/2}\, t^{5/2}}\, e^{-\frac{|x-x_0|^2}{2t} - \frac{3}{t^3}[(y-y_0)-\frac{1}{2}\langle\rho, x+x_0\rangle t]^2}$$

with $t > 0$ *and vector* $\rho = (-1, 1)$.

5.6 The Operator $\frac{1}{2}\partial_x^2 + x\partial_y^2$

The operator $\mathbb{L} = \frac{1}{2}\partial_x^2 + x\partial_y^2$ resembles the Kolmogorov operator, but it has a higher derivative in y. Its heat kernel can be obtained in a similar way, but it is not of a function type.

Applying the Fourier transform with respect to y, we arrive at the operator

$$P = \frac{1}{2}\partial_x^2 - x\xi^2,$$

which according to Theorem 3.15.1 has the heat kernel given by

$$u(x_0, x, t; \xi) = \frac{1}{\sqrt{2\pi t}} e^{\frac{-(x-x_0)^2}{2t} - \frac{1}{2}\xi^2(x+x_0)t + \frac{1}{24}\xi^4 t^3}.$$

The heat kernel for the sub-elliptic operator $\mathbb{L}$ is obtained by applying the inverse Fourier transform:

$$\begin{aligned} K(x_0, x, y_0, y, t) &= \frac{1}{2\pi}\int e^{i(y-y_0)\xi}\, u(x_0, x, t; \xi)\, d\xi \\ &= \frac{1}{(2\pi)^{3/2} t^{1/2}}\, e^{\frac{-(x-x_0)^2}{2t}} \int e^{i(y-y_0)\xi - \frac{1}{2}(x+x_0)t\xi^2 + \frac{1}{24}t^3\xi^4}\, d\xi, \end{aligned}$$

which is an integral that cannot be worked out by elementary functions since the exponent is of the fourth degree.

5.7 The Generalized Grushin Operator $\frac{1}{2}\sum_{i=1}^{n}\partial_{x_i}^2 + 2|x|^2\partial_t^2$

Consider the sub-elliptic operator

$$\mathbb{L}_n = \frac{1}{2}\left(\sum_{k=1}^{n} X_i^2 + Y^2\right) = \frac{1}{2}\sum_{i=1}^{n}\partial_{x_i}^2 + 2|x|^2\partial_t^2, \tag{5.7.15}$$

where $X_k = \partial_{x_k}$, $Y = 2|x|\partial_t$ are vector fields on $\mathbb{R}^{n+1}$. It is worth noting that $\mathbb{L}_1 = \frac{1}{2}\partial_x^2 + 2x^2\partial_t^2$ is the two-dimensional Grushin operator treated in Sect. 5.2.

In the following we shall find the heat kernel for $\mathbb{L}_n$. Applying the Fourier transform in t, we obtain the operator

$$P = \frac{1}{2}\sum_{i=1}^{n} \partial_{x_i}^2 - 2\xi^2|x|^2.$$

Making $a = 2\xi$ in Sect. 3.17, we obtain the following heat kernel for P:

$$u(x_0, x, t; \xi) = \left(\frac{\xi}{\pi \sinh(2\xi t)}\right)^{n/2} e^{-\frac{\xi}{\sinh(2\xi t)}[(|x|^2+|x_0|^2)\cosh(2\xi t) - 2\langle x, x_0\rangle]}, \qquad t > 0.$$

The inverse Fourier transform in t provides the heat kernel for the operator $\mathbb{L}_n$:

$$\begin{aligned} K &= \frac{1}{2\pi}\int e^{i(y-y_0)\xi}\, u(x_0, x, t; \xi)\, d\xi \\ &= \frac{1}{2\pi}\int \left(\frac{\xi}{\pi \sinh(2\xi t)}\right)^{n/2} e^{i(y-y_0)\xi - \frac{\xi}{\sinh(2\xi t)}[(|x|^2+|x_0|^2)\cosh(2\xi t) - 2\langle x, x_0\rangle]}\, d\xi. \end{aligned}$$

Changing the variable $\tau = \xi t$ yields

$$\begin{aligned} K &= \frac{1}{2\pi}\int \left(\frac{2\tau}{2\pi t \sinh(2\tau)}\right)^{n/2} e^{\{i(y-y_0)\tau - \tau(|x|^2+|x_0|^2)\coth(2\tau) + \frac{2\tau}{\sinh(2\tau)}\langle x, x_0\rangle\}/t}\, \frac{1}{t}\, d\tau \\ &= \frac{1}{(2\pi t)^{1+\frac{n}{2}}}\int \left(\frac{2\tau}{\sinh(2\tau)}\right)^{n/2} e^{\{i(y-y_0)\tau - \tau(|x|^2+|x_0|^2)\coth(2\tau) + \frac{2\tau}{\sinh(2\tau)}\langle x, x_0\rangle\}/t}\, d\tau. \end{aligned}$$

We have arrived at the following result:

Theorem 5.7.1. *The heat kernel for the sub-elliptic operator*

$$\mathbb{L}_n = \frac{1}{2}\sum_{i=1}^{n} \partial_{x_i}^2 + 2|x|^2\partial_t^2$$

is given by

$$K(x_0, x, t) = \frac{1}{(2\pi t)^{1+\frac{n}{2}}}\int V(\tau) e^{-f(x_0, x, t, \tau)/t\, d\tau}, \qquad (5.7.16)$$

with

$$V(\tau) = \left(\frac{2\tau}{\sinh(2\tau)}\right)^{n/2},$$

$$f(x_0, x, t, \tau) = -i(y - y_0)\tau + \tau(|x|^2 + |x_0|^2)\coth(2\tau) - \frac{2\tau}{\sinh(2\tau)}\langle x, x_0\rangle.$$

Formula (5.7.16) agrees with the formula obtained by Beals, Gaveau and Greiner for sub-elliptic operators [16], where f and V stand for the *complex modified action* and *volume element*, respectively.

5.8 The Heisenberg Laplacian

Consider on $\mathbb{R}^{2n+1}$ the following $2n$ vector fields:

$$X_j = \partial_{x_j} + 2x_{n+j}\partial_t, \qquad X_{n+j} = \partial_{x_{n+j}} - 2x_j\partial_t, \qquad j = 1, \dots, n,$$

called the *Heisenberg vector fields*. The commutation relations are

$$[X_{n+j}, X_j] = 4\partial_t, \qquad [X_j, X_k] = 0 \text{ for } |k - j| \neq n.$$

The vector fields X_k together with ∂_t are left invariant with respect to the following group law on $\mathbb{R}^{2n+1}$:

$$(x,t) \circ (x', t') = \left(x + x', t + t' + 2\sum_{j=1}^{n}(x_{n+j}x'_j - x_j x'_{n+j}) \right).$$

This group is denoted by $\mathbb{H}_n$ and is called the *Heisenberg group*.

The sub-elliptic operator

$$\Delta_H = \frac{1}{2}\sum_{k=1}^{2n} X_k^2$$

is called the *Heisenberg Laplacian*. In this section we shall determine the heat kernel for Δ_H. A computation yields

$$\begin{aligned}\Delta_H &= \frac{1}{2}\sum_{k=1}^{2n} X_k^2 = \frac{1}{2}\sum_{k=1}^{2n} \partial_{x_k}^2 + 2|x|^2\partial_t^2 + 2\sum_{j=1}^{n}\left(x_{n+j}\partial_{x_j} - x_j\partial_{x_{n+j}}\right)\partial_t \\ &= \mathbb{L}_n + 2\Omega\partial_t,\end{aligned}$$

where $\mathbb{L}_n$ is the Grushin operator (5.7.15) and

$$\Omega = \sum_{j=1}^{n}\left(x_{n+j}\partial_{x_j} - x_j\partial_{x_{n+j}}\right)$$

is the angular momentum operator. A computation shows that $[\mathbb{L}_n, \Omega] = 0$. Using the property of the semigroup generated by the sum of two operators which commute, we have

$$e^{t\Delta_H} = e^{t\mathbb{L}_n + 2t\Omega\partial_t} = e^{t\mathbb{L}_n} e^{2t\Omega\partial_t}.$$

Since $e^{2t\Omega\partial_t} = \mathbb{I}$ at $x = 0$, it follows that $e^{2t\Omega\partial_t}\delta_x = \mathbb{I}\delta_x = \delta_x$. Then the heat kernel of Δ_H at $x_0 = 0$ is

$$K(0, x, t) = e^{t\Delta_H}\delta(x) = e^{t\mathbb{L}_n} e^{2t\Omega\partial_t}\delta(x) = e^{t\mathbb{L}_n}\delta(x) = u(0, x, t),$$

where u is the heat kernel of the operator $\mathbb{L}_n$ at $x_0 = 0$, which is obtained from Theorem 5.7.1. Making $x_0 = y_0 = 0$ and replacing n by $2n$ in formula (5.7.16), we obtain

Theorem 5.8.1. *The heat kernel for the Heisenberg Laplacian Δ_H starting from the origin is given by*

$$K(0, x, t) = \frac{1}{(2\pi t)^{n+1}} \int V(\tau) e^{-f(x,t,\tau)/t\, d\tau}, \tag{5.8.17}$$

with the volume element and complex modified action given by

$$V(\tau) = \Big(\frac{2\tau}{\sinh(2\tau)}\Big)^n,$$

$$f(x, t, \tau) = -i(y)\tau + \tau|x|^2 \coth(2\tau).$$

Chapter 6
The Eigenfunction Expansion Method

Finding the heat kernel of an elliptic operator on a compact manifold using the eigenvalues method is a well-known method in mathematical physics and quantum mechanics. Roughly speaking, the eigenvalues and eigenfunctions of an operator determine its heat kernel. The formula is an infinite series that involves products of eigenfunctions; see Theorem 6.1.1. It is interesting that in several cases this series can be written as an elementary function by using the associated bilinear generating function.

We shall present applications of the bilinear generating formulas of Mehler, Hille–Hardy, and Poisson. The main disadvantage of this method is that in general the bilinear generating formula is missing, in which case the series that represents the heat kernel cannot be represented as a function. This fact is not surprising, since many heat kernels are not of the function type. Since this method involves a great deal of work with special functions, we shall refer the reader to the useful book of Erdélyi [41].

6.1 General Results

Consider the self-adjoint differential operator $\mathbb{L}$ defined on the interval I and let f_i be its eigenfunctions

$$\mathbb{L} f_i = \lambda_i f_i, \qquad i = 0, 1, \ldots,$$

with $\lambda_i \in \mathbb{R}$ eigenvalues. We shall assume that $\{f_i\}_{i \geq 0}$ is a complete orthogonal system of $\mathfrak{L}^2(I) = \{f : I \to \mathbb{C};\ \int_I |f|^2 < \infty\}$, i.e.,

$$\int_I f_n(x) \overline{f_m}(x)\, dx = \delta_{nm},$$

and for any $\phi \in L^2(I)$ with

$$\int_I \phi(x) f_n(x)\, dx = 0, \quad \forall n = 0, 1, \ldots \quad \Longrightarrow \phi = 0.$$

O. Calin et al., *Heat Kernels for Elliptic and Sub-elliptic Operators*,
Applied and Numerical Harmonic Analysis, DOI 10.1007/978-0-8176-4995-1_6,

Then for any $\phi \in \mathfrak{L}^2(I)$, we can write

$$\phi(x) = \sum_{i \geq 0} \langle f_i, \phi \rangle f_i(x),$$

where $\langle f_i, \phi \rangle = \int_I f_i(x)\phi(x)\, dx$.

The next result provides a formal expression for the heat kernel.

Proposition 6.1.1. *Let $\{f_i\}_i$ be an orthogonal complete system of $\mathfrak{L}^2(I)$ of real eigenfunctions of operator $\mathbb{L}$. Then the heat kernel of $\mathbb{L}$ is given by*

$$K(x_0, x, t) = \sum_{i \geq 0} e^{\lambda_i t} f_i(x_0) f_i(x). \tag{6.1.1}$$

Proof. We shall check the definition properties of the heat kernel

$$\begin{aligned}
(\partial_t - \mathbb{L})K(x_0, x, t) &= \sum_{i \geq 0} \Big(\lambda_i e^{\lambda_i t} f_i(x_0) f_i(x) - e^{\lambda_i t} f_i(x_0) \mathbb{L} f_i(x) \Big) \\
&= \sum_{i \geq 0} \Big(\lambda_i e^{\lambda_i t} f_i(x_0) f_i(x) - e^{\lambda_i t} f_i(x_0) \lambda_i f_i(x) \Big) \\
&= 0.
\end{aligned}$$

We also need to check the following limit in the distribution sense:

$$\lim_{t \searrow 0} K(x_0, x, t) = \delta_{x_0}.$$

For any ϕ smooth function with compact support, we have

$$\begin{aligned}
\lim_{t \searrow 0} \int_I K(x_0, x, t)\phi(x)\, dx &= \lim_{t \searrow 0} \sum_{i \geq 0} e^{\lambda_i t} \left(\int_I f_i(x)\phi(x)\, dx \right) f_i(x_0) \\
&= \sum_{i \geq 0} \left(\int_I f_i(x)\phi(x)\, dx \right) f_i(x_0) \\
&= \sum_{i \geq 0} \langle f_i, \phi \rangle f_i(x_0) = \phi(x_0) = \delta_{x_0}(\phi).
\end{aligned}$$

Hence the heat kernel is given by formula (6.1.1). ■

Remark 6.1.2. If the eigenfunctions f_n are complex, the heat kernel of $\mathbb{L}$ is given by

$$K(x_0, x, t) = \sum_{i \geq 0} e^{\lambda_i t} f_i(x_0) \overline{f_i}(x).$$

6.2 Mehler's Formula and Applications

In this section we show how a bilinear generating function involving Hermite polynomials, called *Mehler's formula,* is reducing the series (6.1.1) to a function in the case of the Hermite operator. We shall start first with some prerequisites regarding Hermite polynomials.

The *Hermite polynomial* $H_n(x)$ of order n is defined by

$$H_n(x) = (-1)^n e^{x^2}(d/dx)^n \left(e^{-x^2}\right) = 2^n x^n + \dots.$$

Using repetitive integration by parts, we obtain

$$\int_{\mathbb{R}} H_n(x)H_m(x)e^{-x^2}\,dx = \begin{cases} \displaystyle\int_{\mathbb{R}} H_n(x)(-1)^m(d/dx)^m e^{-x^2}\,dx = 0, & \text{if } n < m, \\ \displaystyle\int_{\mathbb{R}} (H_n)^{(n)}(x)e^{-x^2} = 2^n n!\sqrt{\pi}, & \text{if } n = m. \end{cases}$$

This can be written as

$$\int_{\mathbb{R}} H_n(x)H_m(x)e^{-x^2}\,dx = 2^n n! \pi^{1/2}\delta_{nm},$$

where δ_{nm} denotes the Kronecker delta function. The graphs of a few Hermite polynomials are sketched in Fig. 6.1. Hence

$$f_n(x) = \frac{e^{-x^2/2}H_n(x)}{2^{n/2}\sqrt{n!}\,\pi^{1/4}}, \qquad n = 0, 1, \dots, \tag{6.2.2}$$

forms an orthogonal system in $\mathfrak{L}^2(\mathbb{R})$. One may show that it is also a complete system.

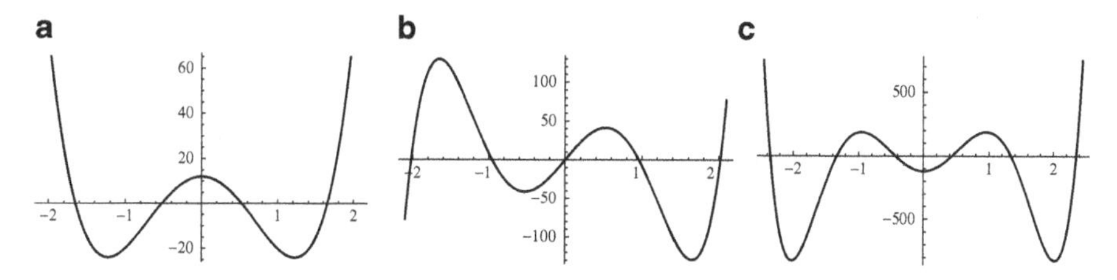

Fig. 6.1 The graphs of a few Hermite polynomials $H_n(x)$: **(a)** $n = 4$; **(b)** $n = 5$; **(c)** $n = 6$

Lemma 6.2.1 (Generating function formula). *We have*

$$\sum_{n\geq 0} H_n(x)\frac{y^n}{n!} = e^{2xy-y^2}.$$

Proof. Taylor's formula for $F(y) = e^{-(x-y)^2}$ yields

$$F(y) = \sum_{n\geq 0} \frac{F^{(n)}(0)}{n!}(-1)^n y^n = e^{-x^2} \sum_{n\geq 0} H_n(x)\frac{y^n}{n!},$$

and hence

$$\sum_{n\geq 0} H_n(x)\frac{y^n}{n!} = e^{x^2} F(y) = e^{x^2-(x-y)^2} = e^{2xy-y^2}.$$

■

The following result deals with the eigenfunctions and eigenvalues of the Hermite operator $\partial_x^2 - x^2$.

Proposition 6.2.2. *The functions given by (6.2.2) satisfy*

$$(\partial_x^2 - x^2) f_n(x) = -(2n+1) f_n(x), \qquad n = 0, 1, \ldots. \tag{6.2.3}$$

Proof. Let $F(y) = e^{2xy-y^2} = \sum_{n\geq 0} \frac{H_n(x)}{n!} y^n$ and $\psi = -\frac{1}{2}x^2 + 2xy - y^2$. Then

$$2y\frac{\partial}{\partial y}e^{\psi} = 2ye^{-\frac{1}{2}x^2} F'(y) = \sum_{n\geq 0} H_n(x)e^{-\frac{1}{2}x^2} 2n\frac{y^n}{n!}. \tag{6.2.4}$$

And applying (6.2.4), we have

$$\begin{aligned}(\partial_x^2 - x^2 + 1)e^{\psi} &= e^{\psi}\left(\psi_x^2 + \psi_{xx} - x^2 + 1\right)\\ &= e^{\psi}\left((-x+2y)^2 - x^2\right) = -2ye^{\psi}(2x-2y)\\ &= -2y\frac{\partial}{\partial y}e^{\psi} = \sum_{n\geq 0} H_n(x)e^{-\frac{1}{2}x^2}(-2n)\frac{y^n}{n!}. \end{aligned} \tag{6.2.5}$$

On the other hand, we have

$$\begin{aligned}(\partial_x^2 - x^2 + 1)e^{\psi} &= \left(\partial_x^2 - x^2 + 1\right)\left[e^{-\frac{1}{2}x^2} F(y)\right]\\ &= \sum_{n\geq 0}\left[\left(\partial_x^2 - x^2 + 1\right)\left(H_n(x)e^{-\frac{1}{2}x^2}\right)\right]\frac{y^n}{n!}. \end{aligned} \tag{6.2.6}$$

Comparing the coefficients of the power series (6.2.5) and (6.2.6) yields

$$(\partial_x^2 - x^2)\left(H_n(x)e^{-\frac{1}{2}x^2}\right) = -(2n+1)H_n(x)e^{-\frac{1}{2}x^2},$$

which is proportional to (6.2.3). ■

Lemma 6.2.3 (Mehler's formula). *If H_n is the nth Hermite polynomial, then*

$$e^{-(x^2+y^2)}\sum_{n=0}^{\infty}\frac{z^n}{2^n n!}H_n(x)H_n(y) = \frac{e^{-(x^2+y^2-2xyz)/(1-z^2)}}{\sqrt{1-z^2}}.$$

For the proof, the interested reader can look in Erdélyi [41], Vol. 2, p. 194.

As an application of Mehler's formula, we shall find a closed-form expression for the heat kernel $K(x_0,x,t)$ of the operator $\partial_x^2 - x^2$. The heat kernel as an eigenfunction expansion is given by Proposition 6.1.1. Using Proposition 6.2.2, the heat kernel is

$$\begin{aligned} K(x_0,x,t) &= \sum_{n\geq 0} e^{\lambda_n t} f_n(x_0) f_n(x) \\ &= \sum_{n\geq 0} e^{-(2n+1)t} e^{-\frac{1}{2}(x_0^2+x^2)} \frac{1}{\sqrt{\pi}}\frac{1}{2^n n!} H_n(x_0)H_n(x) \\ &= \frac{e^{-t}}{\sqrt{\pi}} e^{-\frac{1}{2}(x_0^2+x^2)} \sum_{n\geq 0} \frac{(e^{-2t})^n}{2^n n!} H_n(x_0)H_n(x). \end{aligned} \tag{6.2.7}$$

Making $z = e^{-2t}$ and applying Mehler's formula, (6.2.7) becomes

$$\begin{aligned} K(x_0,x,t) &= \frac{1}{\sqrt{\pi}}\frac{e^{-t}}{\sqrt{1-z^2}} e^{-\frac{1}{2}(x_0^2+x^2)} e^{-\frac{(x_0^2+x^2)z^2-2x_0xz}{1-z^2}} \\ &= \frac{1}{\sqrt{\pi}}\frac{e^{-t}}{\sqrt{1-z^2}} e^{-[\frac{1}{2}(x_0^2+x^2)\frac{1+z^2}{1-z^2} - 2x_0x\frac{z}{1-z^2}]}. \end{aligned} \tag{6.2.8}$$

Since

$$\frac{e^{-t}}{\sqrt{1-z^2}} = \sqrt{\frac{e^{-2t}}{1-e^{-4t}}} = \sqrt{\frac{1}{e^{2t}-e^{-2t}}} = \frac{1}{\sqrt{2\sinh(2t)}},$$

$$\frac{1+z^2}{1-z^2} = \frac{1+e^{-4t}}{1-e^{-4t}} = \frac{e^{2t}+e^{-2t}}{e^{2t}-e^{-2t}} = \coth(2t),$$

$$\frac{z}{1-z^2} = \frac{e^{-2t}}{1-e^{-4t}} = \frac{1}{2\sinh(2t)},$$

formula (6.2.7) becomes the heat kernel of $\partial_x^2 - x^2$:

$$K(x_0, x, t) = \frac{1}{\sqrt{2\pi}} \frac{1}{\sqrt{\sinh(2t)}} e^{-\left[\frac{1}{2}(x_0^2+x^2)\coth(2t) - x_0 x/\sinh(2t)\right]}, \quad t > 0. \quad (6.2.9)$$

6.3 Hille–Hardy's Formula and Applications

In this section we show how a bilinear generating function involving Laguerre polynomials, called *Hille–Hardy's formula*, is reducing the series (6.1.1) to a function in the case of the operator $\mathbb{L} = x\partial_x^2 + \partial_x - \frac{1}{4}(b^2 x + \frac{a^2}{x})$. Then we shall provide a couple of applications regarding the Bessel operator and an operator with gravitational potential in two dimensions. We shall first present some prerequisites regarding Laguerre polynomials.

The *Laguerre polynomial* L_n^a of degree n and parameter a can be defined either by the *Rodrigues formula*

$$L_n^a = \frac{e^x x^{-a}}{n!} \frac{d^n}{dx^n} \left(e^{-x} x^{n+a}\right) = \frac{1}{n!}(-x)^n + \dots,$$

or equivalently, by the generating function formula

$$\sum_{n \geq 0} L_n^a(x) y^n = \frac{1}{(1-y)^{a+1}} e^{xy/(y-1)}. \quad (6.3.10)$$

It is worth noting that for the same degree n the graph does not depend essentially on the value of the parameter a; see Fig. 6.2. Applying integration by parts several times yields

$$\int_0^\infty L_n^a(x) L_m^a(x) e^{-x} x^a \, dx = \begin{cases} 0, & \text{if } n \neq m, \\ \dfrac{\Gamma(a+n+1)}{n!}, & \text{if } n = m. \end{cases}$$

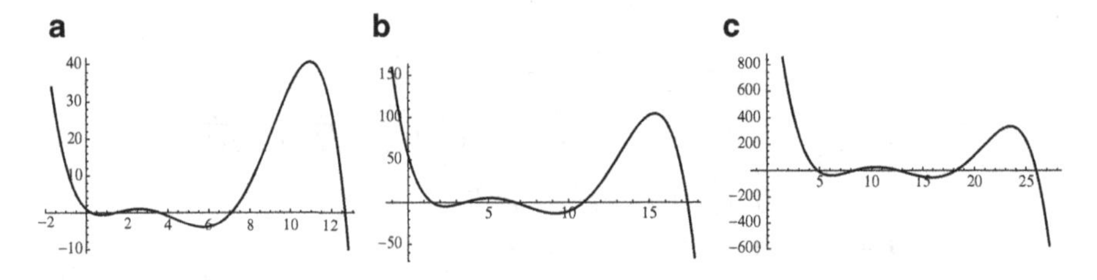

Fig. 6.2 The graphs of the Laguerre polynomial $L_5^a(x)$ for a few distinct values of the parameter a: **(a)** $a = 0$; **(b)** $a = 3$; **(c)** $a = 9$

Hence

$$f_n(x) = \sqrt{\frac{n!}{\Gamma(a+n+1)}} e^{-x/2} x^{a/2} L_n^a(x), \qquad n = 0, 1, \ldots, \tag{6.3.11}$$

forms an orthogonal system for $\mathfrak{L}^2(0, \infty)$. One may show that the system is also complete.

Proposition 6.3.1. *The functions (6.3.11) are eigenfunctions for the operator*

$$\mathbb{L} = x\partial_x^2 + \partial_x - \frac{1}{4}\left(x + \frac{a^2}{x}\right), \tag{6.3.12}$$

with the corresponding eigenvalue $\lambda_n = -\big(n + \frac{a+1}{2}\big)$.

Proof. It comes from the generating formula (6.3.10) and the fact that $y = L_n^a(x)$ verifies the *Laguerre equation*

$$xy'' + (\alpha + a - x)y' + ny = 0.$$

■

The following bilinear generating formula can be found in Erdélyi [41], Vol. 2, p. 189.

Proposition 6.3.2 (Hille–Hardy's formula). *If* L_n^a *denotes the Laguerre polynomial, and* I_a *is the modified Bessel function of first type of order* a*, then for* $|z| < 1$ *we have*

$$\sum_{n=0}^{\infty} \frac{n!}{\Gamma(n+a+1)} z^n (xyz)^{a/2} L_n^a(x) L_n^a(y) = \frac{1}{(1-z)} e^{-\frac{(x+y)z}{(1-z)}} I_a\left(\frac{2\sqrt{xyz}}{1-z}\right).$$

In the following let $z = e^{-t}$ and use that

$$\frac{\sqrt{z}}{1-z} = \frac{1}{2\sinh(t/2)}, \qquad 1 + \frac{2z}{1-z} = \frac{1+e^{-t}}{1-e^{-t}} = \coth(t/2).$$

By Proposition 6.1.1, the heat kernel for the operator (6.3.12) on $(0, \infty)$ is given by

$$\begin{aligned}
K(x_0, x, t) &= \sum_{n \geq 0} e^{\lambda_n t} f_n(x_0) f_n(x) \\
&= \sum_{n \geq 0} e^{-nt} e^{-\frac{a}{2}t} e^{-\frac{t}{2}} \frac{n!}{\Gamma(a+n+1)} L_n^a(x_0) L_n^a(x) x_0^{a/2} e^{-x_0/2} x^{a/2} e^{-x/2} \\
&= e^{-\frac{1}{2}(x_0+x)} \sqrt{z} \sum_{n \geq 0} \frac{n!}{\Gamma(a+n+1)} z^n (x_0 x z)^{a/2} L_n^a(x_0) L_n^a(x)
\end{aligned}$$

$$= e^{-\frac{1}{2}(x_0+x)} \frac{\sqrt{z}}{1-z} e^{-(x_0+x)\frac{z}{1-z}} I_a\left(2\sqrt{x_0 x}\frac{\sqrt{z}}{1-z}\right)$$

(by Hille–Hardy's formula)

$$= \frac{1}{2\sinh(t/2)} e^{-\frac{1}{2}(x_0+x)(1+\frac{2z}{1-z})} I_a\left(\sqrt{x_0 x}\frac{1}{\sinh(t/2)}\right)$$

$$= \frac{1}{2\sinh(t/2)} e^{-\frac{1}{2}(x_0+x)\coth(t/2)} I_a\left(\frac{\sqrt{x_0 x}}{\sinh(t/2)}\right).$$

Hence the heat kernel of the operator

$$\mathbb{L} = x\partial_x^2 + \partial_x - \frac{1}{4}\left(x + \frac{a^2}{x}\right)$$

is given by the formula

$$K(x_0, x, t) = \frac{1}{2\sinh(t/2)} e^{-\frac{1}{2}(x_0+x)\coth(t/2)} I_a\left(\frac{\sqrt{x_0 x}}{\sinh(t/2)}\right), \qquad t > 0.$$

Next we shall present a slight generalization of the previous result. If we define $\mathbb{L}_1 = \mathbb{L}$, then we shall find the kernel of the operator

$$\mathbb{L}_b = x\partial_x^2 + \partial_x - \frac{1}{4}\left(b^2 x + \frac{a^2}{x}\right).$$

If we take $\tilde{x} = bx$ as a new variable, a computation shows that $\mathbb{L}_b = b\tilde{\mathbb{L}}_1$. Then

$$(e^{t\mathbb{L}_b} f)(x_0) = \left(e^{tb\tilde{\mathbb{L}}_1} \tilde{f}\right)(\tilde{x}_0) = \int K(bx_0, bx, bt) f(x_0) b dx_0,$$

and hence the heat kernel of $e^{t\mathbb{L}_b}$ is $bK(bx_0, bx, tb)$; that is,

$$K_b(x_0, x, t) = \frac{b/2}{\sinh(bt/2)} e^{-\frac{b}{2}(x_0+x)\coth(bt/2)} I_a\left(\frac{b\sqrt{x_0 x}}{\sinh(bt/2)}\right). \qquad (6.3.13)$$

Therefore, the heat kernel of the operator $\mathbb{L}_b$ is given by formula (6.3.13). We shall present next two applications.

Heat kernel for the Bessel operator. If $\mathbb{L}_0 = x\partial_x^2 + \partial_x - \frac{a^2}{4x}$, then its heat kernel is obtained by making $b \to 0$ in the above relation:

$$K_0(x_0, x, t) = \frac{1}{t} e^{-\frac{1}{t}(x_0+x)} I_a\left(\frac{2\sqrt{x_0 x}}{t}\right), \qquad (6.3.14)$$

where we used

$$\lim_{b\to 0} \frac{bt/2}{\sinh(bt/2)} = 1.$$

The change of variable $x = r^2$ in $\mathbb{L}_0$ yields a Bessel operator with potential $-a^2/r^2$:

$$x\partial_x^2 + \partial_x - \frac{a^2}{4x} = \frac{1}{4}\left(\partial_r^2 + \frac{1}{r}\partial_r - \frac{a^2}{r^2}\right).$$

Let $\tau = t/2$. Since the following relation holds among the heat kernels

$$e^{t\left(x\partial_x^2+\partial_x-\frac{a^2}{4x}\right)} = e^{\frac{t}{4}\left(\partial_r^2+\frac{1}{r}\partial_r^2-\frac{a^2}{r^2}\right)} = e^{\frac{\tau}{2}\left(\partial_r^2+\frac{1}{r}\partial_r^2-\frac{a^2}{r^2}\right)},$$

then the heat kernel of $\frac{1}{2}(\partial_r^2 + \frac{1}{r}\partial_r^2 - \frac{a^2}{r^2})$ is obtained from the heat kernel of $x\partial_x^2 + \partial_x - \frac{a^2}{4x}$ by making the substitutions $x_0 = r_0^2$, $x = r^2$ and $\tau = t/2$ in formula (6.3.14).

Hence the heat kernel of $\frac{1}{2}\left(\partial_r^2 + \frac{1}{r}\partial_r - \frac{a^2}{r^2}\right)$ is

$$G(r_0, r, \tau) = \frac{1}{2\tau}e^{-\frac{1}{2\tau}(r_0^2+r^2)} I_a\left(\frac{r_0 r}{\tau}\right). \tag{6.3.15}$$

Operator with gravitational potential. In the following we shall present an application of formula (6.3.15) to a two-dimensional operator with gravitational potential; see [26]. We shall look for the heat kernel of the operator

$$\mathbb{L} = \frac{1}{2}\left(\partial_{x_1}^2 + \partial_{x_2}^2\right) - \frac{\lambda^2}{x_1^2 + x_2^2}, \quad \text{with } \lambda \in \mathbb{R}. \tag{6.3.16}$$

In polar coordinates this becomes

$$\mathbb{L} = \frac{1}{2}\left(\partial_r^2 + \frac{1}{r}\partial_r + \frac{1}{r^2}\partial_\theta^2\right) - \frac{\lambda^2}{r^2}.$$

If K is the heat kernel of the above operator, then applying a partial Fourier transform with respect to θ, we obtain that $\hat{K} = \mathfrak{F}_\theta K$ is the heat kernel of the operator

$$\mathbb{P} = \frac{1}{2}\left(\partial_r^2 + \frac{1}{r}\partial_r - \frac{1}{r^2}\xi^2\right) - \frac{2\lambda^2}{2r^2} = \frac{1}{2}\left(\partial_r^2 + \frac{1}{r}\partial_r - \frac{a^2}{r^2}\right),$$

with $a^2 = 2\lambda^2 + \xi^2$. The heat kernel $e^{t\mathbb{P}}$ can be obtained from the formula (6.3.14):

$$\hat{K}(r_0, r, \tau; \xi) = \frac{1}{2\tau}e^{-\frac{1}{2\tau}(r_0^2+r^2)} I_a\left(\frac{r_0 r}{\tau}\right).$$

Applying the inverse Fourier transform yields the heat kernel of the operator (6.3.16):

$$
\begin{aligned}
K(r_0,\theta_0,r,\theta,\tau) &= \frac{1}{2\pi}\int e^{i(\theta-\theta_0)\xi}\frac{1}{2\tau}e^{-\frac{1}{2\tau}(r_0^2+r^2)}I_{(2\lambda^2+\xi^2)^{1/2}}\left(\frac{r_0 r}{\tau}\right)d\xi \\
&= \frac{1}{4\pi\tau}e^{-\frac{1}{2\tau}(r_0^2+r^2)}\int e^{i(\theta-\theta_0)\xi}I_{(2\lambda^2+\xi^2)^{1/2}}\left(\frac{r_0 r}{\tau}\right)d\xi.
\end{aligned}
$$

Remark 6.3.3. In the case $\lambda = 0$ the above formula becomes

$$
K(r_0,\theta_0,r,\theta,\tau) = \frac{1}{4\pi\tau}e^{-\frac{1}{2\tau}(r_0^2+r^2)}\int e^{i(\theta-\theta_0)\xi}I_{\xi}\left(\frac{r_0 r}{\tau}\right)d\xi. \tag{6.3.17}
$$

On the other hand, the heat kernel for (6.3.16) with $\lambda = 0$ is given by

$$
\frac{1}{2\pi t}e^{-\frac{|x-x_0|^2}{2\tau}} = \frac{1}{2\pi t}e^{-\frac{r_0^2+r^2-2rr_0\cos(\theta-\theta_0)}{2\tau}}.
$$

Comparing with (6.3.17) and making $\rho = \frac{r_0 r}{\tau}$ and $u = \theta - \theta_0$ yields

$$
\frac{1}{2}\int e^{iu\xi}I_\xi(\rho)\,d\xi = e^{\rho\cos u}.
$$

6.4 Poisson's Summation Formula and Applications

Under some convergence and regularity conditions on the function f, we have the following summation formula, called *Poisson's summation formula* (see [86]):

$$
\sum_{n=-\infty}^{\infty} f(x+2n\pi) = \frac{1}{2\pi}\sum_{k=-\infty}^{\infty} e^{ikx}\int_{-\infty}^{\infty} e^{-iky}f(y)\,dy.
$$

As a consequence, we have the next result regarding the theta-function, which was introduced in Sect. 1.5. The result deals with the case of the θ_3-function, but similar transformation formulas work for the other theta-functions.

Proposition 6.4.1 (Jacobi's transformation for θ_3). *If $t' = -1/t$, we have*

$$
\theta_3(z|t') = (-it)^{1/2}e^{itz^2/\pi}\theta_3\big(tz|t\big), \tag{6.4.18}
$$

where the third Jacobi theta-function is defined by

$$
\theta_3(z|t) = \sum_{n=-\infty}^{\infty} e^{i\pi t n^2+2nzi}. \tag{6.4.19}
$$

Proof. We note that we can write

$$\theta_3(z|t) = e^{z^2/(i\pi t)} \sum_{n=-\infty}^{\infty} e^{\frac{it}{4\pi}(u+2n\pi)^2},$$

with $u = 2z/t$. Applying Poisson summation formula with $f(u) = e^{itu^2/4\pi}$ yields

$$\begin{aligned}\theta_3(z|t) &= \frac{1}{2\pi} e^{z^2/i\pi t} \sum_{k=-\infty}^{\infty} e^{iku} \int_{-\infty}^{\infty} e^{\frac{ity^2}{4\pi} - iky}\, dy \\ &= \frac{1}{2\pi} e^{z^2/i\pi t} \sum_{k=-\infty}^{\infty} e^{iku} 2\pi(-it)^{-1/2} e^{-i\pi k^2/t} \\ &= (-it)^{-1/2} e^{z^2/i\pi t} \theta_3(z/t \mid -1/t).\end{aligned}$$

Replacing z by tz yields (6.4.18). ■

The interested reader can find details about the theta-functions in reference [86]. Next we shall present a few applications.

6.4.1 Heat Kernel on S^1

In this section we shall compute the heat kernel for the Laplacian on the unit circle S^1 using Jacobi's transformation. If ϕ is the arc length on the unit circle, the eigenfunctions of $\frac{1}{2}\frac{d^2}{d\phi^2}$ on the circle S^1 satisfy

$$\frac{1}{2}\frac{d^2}{d\phi^2} u_k = \lambda_k u_k,$$

where $\lambda_k = -k^2/2$ and $u_k(\phi) = c_k e^{ik\phi}$. The constant c_k can be obtained from the orthogonality condition

$$1 = \int_0^{2\pi} u_k(\phi)\overline{u_k}(\phi)\, d\phi = 2\pi c_k^2 \Longrightarrow c_k = \frac{1}{\sqrt{2\pi}}.$$

The functions $u_k(\phi) = \frac{1}{\sqrt{2\pi}} e^{ik\phi}$ form a complete orthogonal system on $\mathfrak{L}^2(S^1, d\phi)$. Therefore, the heat kernel for $\frac{1}{2}\frac{d^2}{d\phi^2}$ on the circle S^1 is given by

$$\begin{aligned}K(\phi_0, \phi; \tau) &= \sum e^{-k^2\tau/2} u_k(\phi)\overline{u_k}(\phi_0) = \frac{1}{2\pi}\sum e^{-k^2\tau/2} e^{ik(\phi-\phi_0)} \\ &= \frac{1}{2\pi}\sum e^{-k^2\tau/2 + ik(\phi-\phi_0)} = \frac{1}{2\pi}\theta_3\Big(\frac{\phi-\phi_0}{2}\Big|\frac{i\tau}{2\pi}\Big),\end{aligned}$$

by the definition of the θ_3 given in (6.4.19). Denoting

$$z = \frac{\phi - \phi_0}{2}, \quad t' = \frac{i\tau}{2\pi} = -1/t, \quad t = 2\pi i/\tau,$$

Jacobi's transformation formula provides

$$\begin{aligned} K(\phi_0, \phi; \tau) &= \frac{1}{2\pi}\theta_3\Big(\frac{\phi - \phi_0}{2}\,\Big|\,\frac{i\tau}{2\pi}\Big) = \frac{1}{2\pi}\theta_3(z|t') \\ &= \frac{1}{2\pi}(-it)^{1/2} e^{itz^2/\pi}\theta_3(tz|t) \\ &= \frac{1}{\sqrt{2\pi\tau}} e^{-\frac{(\phi-\phi_0)^2}{2\tau}}\,\theta_3\Big(\frac{i\pi}{\tau}(\phi - \phi_0)\,\Big|\,\frac{2\pi i}{\tau}\Big). \end{aligned} \tag{6.4.20}$$

This formula agrees with the geometric formula of the heat kernel for S^1 obtained in Sect. 3.11.

6.4.2 Heat Kernel on the Segment $[0, T]$

We shall determine the heat kernels on the segment $[0, T]$ with the Dirichlet and Neumann boundary conditions, respectively.

Dirichlet boundary condition. In this case the temperature $u(x,t)$ is kept equal to zero at the endpoints $x = 0, T$ all the time. A complete orthonormal system for the Laplacian $\frac{1}{2}\frac{d^2}{dx^2}$ on $[0, T]$ is given by

$$u_k(x) = \sqrt{\frac{2}{T}} \sin \frac{k\pi x}{T},$$

with the corresponding eigenvalues $\lambda_k = -\frac{k^2\pi^2}{2T^2}$, $k = 0, 1, \ldots$. For any two points $x_0, x \in [0, T]$, the heat kernel can be expressed as

$$\begin{aligned} K(x_0, x; \tau) &= \sum_{k\geq 0} e^{\lambda_k \tau} u_k(x_0)\overline{u}_k(x) \\ &= \frac{2}{T}\sum_{k\geq 1} e^{-\frac{k^2\pi^2\tau}{2T^2}} \sin\frac{k\pi x_0}{T} \sin\frac{k\pi x}{T} \\ &= \frac{1}{T}\sum_{k\geq 1} e^{-\frac{k^2\pi^2\tau}{2T^2}} \Big[\cos\frac{k\pi(x - x_0)}{T} - \cos\frac{k\pi(x + x_0)}{T}\Big]. \end{aligned} \tag{6.4.21}$$

Substituting $z = (x \pm x_0)\pi/(2T)$ and $t = \pi\tau/(2T^2)$, the previous series can be written in terms of theta-functions:

$$\sum_{k\geq 1} e^{-\frac{k^2\pi^2\tau}{2T^2}} \cos\frac{k\pi(x\pm x_0)}{T} = \frac{1}{2}\Big[\theta_3\big(z\,|\,it\big) - 1\Big]$$
$$= \frac{1}{2}\Big[\theta_3\Big(\frac{(x\pm x_0)\pi}{2T}\,\Big|\,\frac{\pi i\tau}{2T^2}\Big) - 1\Big],$$

and hence (6.4.21) becomes

$$K(x_0, x; \tau) = \frac{1}{2T}\theta_3\Big(\frac{(x-x_0)\pi}{2T}\,\Big|\,\frac{\pi i\tau}{2T^2}\Big) - \frac{1}{2T}\theta_3\Big(\frac{(x+x_0)\pi}{2T}\,\Big|\,\frac{\pi i\tau}{2T^2}\Big). \quad (6.4.22)$$

Even if (6.4.22) is a perfectly valid formula, we shall still transform it according to Poisson's summation formula, obtaining the familiar term $e^{-\frac{(x\pm x_0)^2}{2\tau}}$ in front of it. Taking $z = \frac{(x\pm x_0)\pi}{2T}$ and $t' = \frac{\pi i\tau}{2T^2}$ in formula (6.4.18) yields

$$\theta_3\left(\frac{(x\pm x_0)\pi}{2T}\,\Big|\,\frac{\pi i\tau}{2T^2}\right) = \theta_3(z|t') = \frac{2T}{\sqrt{2\pi\tau}}e^{-\frac{(x\pm x_0)^2}{2\tau}}\theta_3\left(\frac{(x\pm x_0)Ti}{\tau}\,\Big|\,\frac{2T^2 i}{\pi\tau}\right),$$

and hence the heat kernel becomes

$$K(x_0, x; \tau) = \frac{1}{\sqrt{2\pi\tau}}\Bigg[e^{-\frac{(x-x_0)^2}{2\tau}}\theta_3\left(\frac{(x-x_0)Ti}{\tau}\,\Big|\,\frac{2T^2 i}{\pi\tau}\right)$$
$$-e^{-\frac{(x+x_0)^2}{2\tau}}\theta_3\left(\frac{(x+x_0)Ti}{\tau}\,\Big|\,\frac{2T^2 i}{\pi\tau}\right)\Bigg].$$

Neumann boundary condition. In this case there is no leak of heat at the endpoints of $[0, T]$. This condition can be written as $\frac{\partial}{\partial x}u(x,t) = 0$ for $x = 0, T$, and all $t \geq 0$. Then a complete orthonormal system for $\frac{1}{2}\frac{d^2}{dx^2}$ on $[0, T]$ with respect to the aforementioned boundary conditions is

$$u_k(x) = \sqrt{\frac{2}{T}}\cos\frac{k\pi x}{T},$$

with the corresponding eigenvalues $\lambda_k = -\frac{k^2\pi^2}{2T^2}$, $k = 0, 1, \ldots$. A computation similar to the one before provides the heat kernel:

$$K(x_0, x; \tau) = \sum_{k\geq 0} e^{\lambda_k\tau}u_k(x_0)\overline{u}_k(x)$$
$$= \frac{2}{T} + \frac{2}{T}\sum_{k\geq 1} e^{-\frac{k^2\pi^2\tau}{2T^2}}\cos\frac{k\pi x_0}{T}\cos\frac{k\pi x}{T}$$

$$= \frac{1}{T}\left[\left(1 + \sum_{k\geq 1} e^{-\frac{k^2\pi^2\tau}{2T^2}} \cos \frac{k\pi(x + x_0)}{T}\right)\right.$$
$$\left. + \left(1 + \sum_{k\geq 1} e^{-\frac{k^2\pi^2\tau}{2T^2}} \cos \frac{k\pi(x - x_0)}{T}\right)\right]$$
$$= \frac{1}{T}\theta_3\left(\frac{(x + x_0)\pi}{2T} \,\Big|\, \frac{\pi i\tau}{2T^2}\right) + \frac{1}{T}\theta_3\left(\frac{(x - x_0)\pi}{2T} \,\Big|\, \frac{\pi i\tau}{2T^2}\right). \quad (6.4.23)$$

Applying Poisson's summation formula yields

$$K(x_0, x; \tau) = \frac{1}{\sqrt{2\pi\tau}}\left[e^{-\frac{(x+x_0)^2}{2\tau}}\theta_3\left(\frac{(x + x_0)Ti}{\tau} \,\Big|\, \frac{2T^2 i}{\pi\tau}\right)\right.$$
$$\left. + e^{-\frac{(x-x_0)^2}{2\tau}}\theta_3\left(\frac{(x - x_0)Ti}{\tau} \,\Big|\, \frac{2T^2 i}{\pi\tau}\right)\right],$$

which agrees with the geometric formula (3.12.37).

6.5 Legendre's Polynomials and Applications

Consider the operator

$$\partial_\theta^2 + \cot\theta\, \partial_\theta. \quad (6.5.24)$$

If we make the change of variable $\cos\theta = x$, the operator becomes

$$(1 - x^2)\partial_x^2 - 2x\partial_x, \qquad x \in (-1, 1). \quad (6.5.25)$$

The nth-degree Legendre polynomial P_n is defined by the Rodrigues formula

$$P_n(x) = \frac{1}{2^n n!}\frac{d^n}{dx^n}(x^2 - 1)^n.$$

P_n can also be defined by the following generating function:

$$\sum_{n\geq 0} P_n(x)y^n = \frac{1}{(1 - 2xy + y^2)^{1/2}}.$$

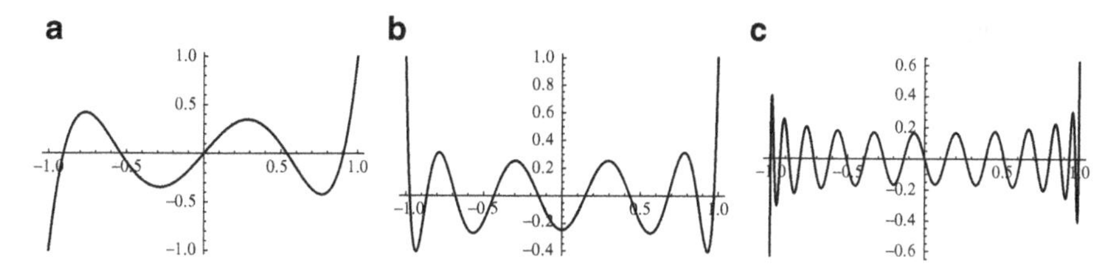

Fig. 6.3 The graphs of a few Legendre polynomials $P_n(x)$: **(a)** $n = 5$; **(b)** $n = 10$; **(c)** $n = 23$

The graphs of three polynomials are contained in Fig. 6.3. One may show that P_n satisfies the equation

$$(1 - x^2)\partial_x^2 P_n(x) - 2x\partial_x P_n(x) = -n(n+1)P_n(x),$$

and hence $P_n(x)$ is an eigenfunction of the operator (6.5.25) with the corresponding eigenvalue $\lambda_n = -n(n+1)$. Since

$$\int_{-1}^{1} P_n(x)P_m(x)\,dx = 0, \qquad n \neq m,$$

$$\int_{-1}^{1} [P_n(x)]^2\,dx = \frac{1}{n+1/2},$$

it follows that

$$f_n(x) = \sqrt{n + \frac{1}{2}}\,P_n(x), \qquad n = 0, 1, 2\ldots,$$

forms an orthogonal system on $\mathfrak{L}^2(-1,1)$. The system is also complete. By Proposition 6.1.1, the heat kernel for (6.5.25) on $(-1,1)$ is

$$K(x,y,t) = \sum_{n\geq 0}(n+1/2)e^{-n(n+1)t}P_n(x)P_n(y),$$

and hence the heat kernel for (6.5.24) is

$$K(\theta,\psi,t) = \sum_{n\geq 0}(n+1/2)e^{-n(n+1)t}P_n(\cos\theta)P_n(\cos\psi).$$

In this case a bilinear generating function is missing, so the operator (6.5.24) does not have a function-type heat kernel.

6.6 Legendre Functions and Applications

In the following we shall deal with the heat kernel for the Laplacian on the sphere $S^2 = \{x \in \mathbb{R}^3; |x| = 1\}$. In spherical coordinates $\theta \in (0, \pi)$, $\phi \in [0, 2\pi)$, the Laplacian is given by

$$\Delta_{S^2} = \frac{1}{\sin\theta}\frac{\partial}{\partial\theta}\Big(\sin\theta\,\frac{\partial}{\partial\theta}\Big) + \frac{1}{(\sin\theta)^2}\frac{\partial^2}{\partial^2\phi^2}.$$

The eigenfunctions are given by the spherical harmonics

$$Y_m^k(\theta,\phi) = c_{mk}\,P_m^{|k|}(\cos\theta)e^{ik\phi},$$

where $m = 0, 1, 2, \ldots$ and $k \in \{0, \pm 1, \ldots, \pm m\}$. P_m^k denotes the Legendre function. Since

$$\Delta_{S^2} Y_m^k = -m(m+1)Y_m^k,$$

the eigenvalue corresponding to Y_m^k is $\lambda_m^k = -m^2 - m$; see, for instance, [63], p. 74. The heat kernel is given by

$$\begin{aligned}
K(\theta_0,\phi_0;\theta,\phi;t) &= \sum e^{\lambda_m^k t}\,Y_m^k(\theta,\phi)\overline{Y_m^k}(\theta_0,\phi_0)\\
&= \sum c_{mk}^2 e^{-m(m+1)t}\,P_m^k(\cos\theta)e^{ik\phi}\,P_m^k(\cos\theta_0)e^{-ik\phi_0}\\
&= \sum_{m\ge 0} e^{-m(m+1)t}\sum_{k=-m}^{m} c_{mk}^2\,P_m^k(\cos\theta)\,P_m^k(\cos\theta_0)e^{ik(\phi-\phi_0)}.
\end{aligned}$$

A neater formula can be obtained if the kernel is represented in terms of the angle ξ between x_0 and x given by $\xi = \cos^{-1}(\langle x_0, x\rangle)$. Even if the inverse cosine function is multi-valued, since the heat kernel is an even, 2π-periodic function of ξ, we may use any value of ξ with $\cos\xi = \langle x_0, x\rangle$; see references [30, 106]. The heat kernel on S^2 can be written as

$$\begin{aligned}
K(x_0, x; t) &= \frac{1}{2\sqrt{\pi t}}e^{t/4}\frac{1}{\sqrt{2\pi t}}\int_\xi^\pi \frac{1}{\sqrt{\cos\xi - \cos\rho}}\\
&\quad\times \sum_{n=-\infty}^{\infty}(-1)^n(\rho - 2\pi n)e^{-\frac{1}{2t}(\xi-2\pi n)^2}\,d\rho.
\end{aligned}$$

It is worth noting that in the case of the three-dimensional sphere there is a simple formula given by (3.7.33).

To conclude this chapter, we note that the eigenfunction expansion technique is a robust method, which works as long as we are able to find an orthonormal set of functions and their eigenvalues. The heat kernel can be expressed in this case as an infinite series. There are only a few known cases when this series has a bilinear generating function, in which case the heat kernel has an elegant expression.

Chapter 7
The Path Integral Approach

7.1 Introducing Path Integrals

In 1948 Feynman [42] provided an informal expression for the propagators of the famous Schrödinger equation, involving an integration over all the continuous paths with respect to a nonexistent infinite-dimensional Lebesgue measure. In fact, this is not really an integral, since there is no measure to give the integral. Since then, a large number of papers have tried to explain the precise mathematical meaning of the *Feynman integral.*

In this chapter we do not attempt this direction. We are concerned only with the immediate applications of the concept of a Feynman integral to obtaining heat kernels for differential operators.

In classical mechanics the particles follow trajectories provided by the Euler–Lagrange equations or Hamiltonian system. Given the initial conditions, standard theorems of ODE theory state the existence and uniqueness of the solution of the above systems of equations. This can be stated by saying that classical particles travel along deterministic trajectories.

Unlike the aforementioned case, in quantum mechanics particles travel on nondeterministic trajectories. This was inferred from the double-slit experiment and was theoretically stated by the *Heisenberg principle* of uncertainty.

Nondeterministic trajectories are better described by the notion of transition probability. Let $P(c \mid a)$ denote the probability that the particle is in state c given that it starts at state a. An application of the conditional probability rule yields that the previous probability depends on all the intermediate states b:

$$P(c \mid a) = \sum_b P(c \mid b) P(b \mid a);$$

see Fig. 7.1b. In quantum mechanics the conditional probability $P(c \mid a)$ is replaced by the probability amplitude function φ_{ca}, in which case the aforementioned rule becomes

$$\varphi_{ca} = \sum_b \varphi_{cb} \varphi_{ba}.$$

O. Calin et al., *Heat Kernels for Elliptic and Sub-elliptic Operators*, Applied and Numerical Harmonic Analysis, DOI 10.1007/978-0-8176-4995-1_7,

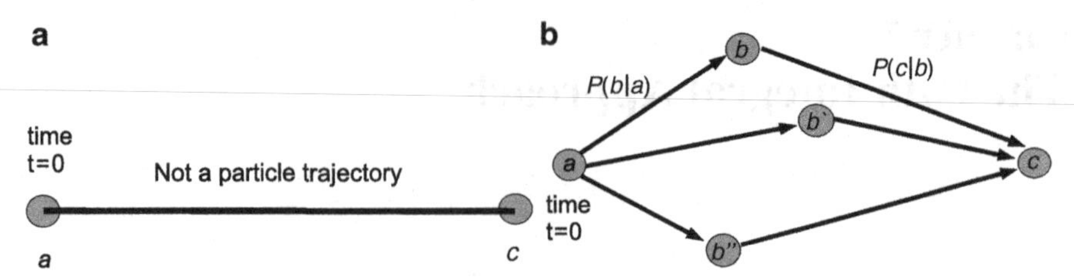

Fig. 7.1 (**a**) Quantum particles do not travel on deterministic paths. (**b**) Nondeterministic trajectories

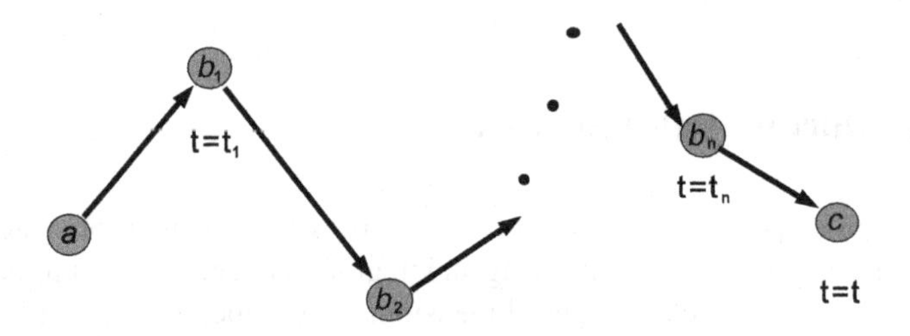

Fig. 7.2 Intermediate states and probability amplitude functions

Assume the particle starts at a at time zero and ends at b at time t. Consider the following n intermediate states between a and c:

$$a = b_0,\ b_1,\ \ldots,\ b_n,\ b_{n+1} = c,$$

such that the particle is in state b_k at time t_k, $k = 0, \ldots, n$; see Fig. 7.2. The probability amplitude function between a and c can be written as

$$\varphi_{ca} = \sum_{b_1,\ldots,b_n} \varphi_{cb_n}\varphi_{b_n b_{n-1}} \cdots \varphi_{b_2 b_1}\varphi_{b_1 a}.$$

When the number of the intermediate states $n \to \infty$, the linear function $ab_1b_2\cdots b_nc$ tends to a continuous path $x(s)$ with $x(0) = a$ and $x(t) = c$. The "transition probability" from a to c at time t is obtained by summing over all the continuous paths between a and c parameterized by $[0, t]$:

$$\varphi_{a,c;t} = \sum_{x(s)} \varphi_{x(s)}.$$

A breakthrough idea came from Dirac, who indicated that $\varphi_{x(s)} = e^{iS/\hbar}$ for t small, where S satisfies the Hamilton–Jacobi equation and $\hbar$ denotes the Planck constant. The idea was continued by Feynman, who provided a heuristical expression for the propagator of Schrödinger's operator in the form

$$K(a,c;t) = C \cdot \sum_{x(s)} e^{iS/\hbar},$$

with $C = (2\pi i\hbar t)^{-n/2}$, for n large. This was expressed later as the path integral written in the form

$$K(a,c;t) = \int_{\mathfrak{P}(a,c;t)} e^{iS(\phi,t)/\hbar}\, dm(\phi), \tag{7.1.1}$$

where $\mathfrak{P}(a,c;t) = \{\phi : [0,t] \to \mathbb{R}^n; \phi(0) = x_0, \phi(t) = x\}$ denotes the space of the continuous paths from x_0 to x within time t. The "propagator" $K(x_0,x;t)$ given by (7.1.1) satisfies Schrödinger's equation

$$i\hbar\frac{\partial}{\partial t}K + \frac{\hbar^2}{2}\Delta_x K = U(x)K, \tag{7.1.2}$$

where $S(\phi,t)$ is the solution of the Hamilton–Jacobi equation

$$\frac{\partial S}{\partial t} + H(\phi, \partial_\phi S) = 0,$$

with the Hamiltonian

$$H(x,p) = \frac{1}{2}|p|^2 + U(x).$$

Feynman's formula for the propagator of the Schrödinger equation can be adapted for the heat kernels by replacing t by $-it$ and $U(x)$ by $-V(x)$ in (7.1.2). If for the sake of simplicity we choose $m = \hbar = 1$, then Schrödinger's equation becomes the following heat equation:

$$\frac{\partial K}{\partial t} = \frac{1}{2}\Delta_x K + V(x)K.$$

In this chapter we shall deal with finding heat kernels for second-order differential operators using path integrals. Since any heat kernel can be expressed formally as a path integral, the problem of finding the heat kernel is to compute the path integral explicitly. As we shall see throughout the chapter, even if the computation is sometimes tedious, this can be done for several examples of operators.

7.2 Paths Integrals via Trotter's Formula

This section deals with the application of Trotter's formula to a formal construction of path integrals as an infinite limit of improper integrals. Consider the heat equation

$$\frac{\partial}{\partial t}\phi = \mathbb{L}\phi, \tag{7.2.3}$$

where $\mathbb{L} = \Delta + V(x)$, with V a smooth function called the *potential*, and with the Laplacian $\Delta = \frac{1}{2}(\partial_{x_1}^2 + \cdots + \partial_{x_n}^2)$. If the initial condition is $\phi_{|t=0} = \phi_0$, then the solution of (7.2.3) is given by

$$\phi(x,t) = K(t)\phi_0(x), \qquad t > 0,$$

where

$$K(t) = e^{t\mathbb{L}} = e^{t(\Delta + V(x))} \tag{7.2.4}$$

is called the *evolution operator* or the *propagator* of the operator $\mathbb{L}$. If $\phi_0 = \delta_0$ is the Dirac distribution centered at $x = 0$, then the *heat kernel* is given by $K(t)\delta_0$.

We need to find an integral representation for the evolution operator $K(t)$ defined by (7.2.4). Before going on, we shall recall a few concepts regarding *integral kernels.*

Let K be an integral operator on $L^2(\mathbb{R}^m)$. Then its integral kernel, denoted by $\widehat{K}$, is defined by

$$K(f)(x) = \int_{\mathbb{R}^m} \widehat{K}(x,y) f(y)\, dy,$$

with $\widehat{K} : \mathbb{R}^m \times \mathbb{R}^m \to \mathbb{R}$.

Example 7.2.1. The Gaussian kernel on $\mathbb{R}^m$ is $\widehat{K_t}(x,y) = \left(\frac{1}{2\pi t}\right)^{m/2} e^{-\frac{|x-y|^2}{2t}}$ and hence

$$(e^{t\Delta} f)(x) = \left(\frac{1}{2\pi t}\right)^{m/2} \int_{\mathbb{R}^m} e^{-\frac{|x-y|^2}{2t}} f(y)\, dy. \tag{7.2.5}$$

The following result deals with the integral kernels of the product of two or more operators.

Proposition 7.2.2. *(1) Let K_1 and K_2 be two integral operators with the integral kernels $\widehat{K_1}$ and $\widehat{K_2}$. Then the integral kernel of the product $K = K_1K_2$ is*

$$\widehat{K}(x,y) = \int_{\mathbb{R}^m} \widehat{K_1}(x,u)\widehat{K_2}(u,y)\, du.$$

(2) In the case of r operators $K_1, K_2, \ldots, K_r$, the integral kernel of the product operator $K = K_1K_2 \cdots K_r$ is

$$\widehat{K}(x,y) = \int_{\mathbb{R}^m} \cdots \int_{\mathbb{R}^m} \widehat{K_1}(x,u_1)\widehat{K_2}(u_1,u_2) \cdots \widehat{K_{r-1}}(u_{r-2},u_{r-1}) \widehat{K_r}(u_{r-1},y)\, du_{r-1} \cdots du_1.$$

Let A and B be two self-adjoint operators on a Hilbert space $\mathcal{H}$. There are two cases:

1. The operators commute. In this case $[A,B] = AB - BA = 0$ and $e^{A+B} = e^A e^B$. Using Proposition 7.2.2, part (1), the integral kernel of $e^{t(A+B)}$ is given by

$$\int_{\mathbb{R}^m} \widehat{e^{tA}}(x,u)\widehat{e^{tB}}(u,y)\,du.$$

Example 7.2.3. Let $A = \frac{1}{2}\partial_x^2$ and $B = \frac{1}{2}\partial_y^2$. Let

$$\phi(x,y) = \left(e^{tB} f\right)(x,y) = \int \frac{1}{\sqrt{2\pi t}} e^{-\frac{(y-v)^2}{2t}} f(x,v)\,dv.$$

Using that $AB = BA$, we have

$$\begin{aligned}
\left(e^{\frac{t}{2}(\partial_x^2+\partial_y^2)}\right) f(x,y) &= \left(e^{t(A+B)}\right) f(x,y) = e^{tA}\left(e^{tB} f\right)(x,y) \\
&= (e^{tA}\phi)(x,y) = \frac{1}{\sqrt{2\pi t}} \int e^{-\frac{(x-u)^2}{2t}} \phi(u,y)\,du \\
&= \frac{1}{\sqrt{2\pi t}} \int e^{-\frac{(x-u)^2}{2t}} \left(\int \frac{1}{\sqrt{2\pi t}} e^{-\frac{(y-v)^2}{2t}} f(u,v)\,dv\right) du \\
&= \frac{1}{2\pi t} \iint e^{-\frac{(x-u)^2+(y-v)^2}{2t}} f(u,v)\,dudv.
\end{aligned}$$

If we let $f(u,v) = \delta_{x_0}(u) \otimes \delta_{y_0}(v)$, we get the heat kernel

$$e^{\frac{t}{2}(\partial_x^2+\partial_y^2)}(\delta_{x_0} \otimes \delta_{y_0}) = \frac{1}{2\pi} e^{-\frac{(x-x_0)^2+(y-y_0)^2}{2t}}.$$

This is just the product of the heat kernels of A and B.

2. The operators do not commute. If $[A,B] \neq 0$, then $e^{A+B} \neq e^A e^B$. In this case the following *Trotter product formula* is needed.

Theorem 7.2.4. *Let A and B be self-adjoint operators on a separable Hilbert space $\mathcal{H}$ such that they are bounded from below and $A+B$ is essentially self-adjoint. Then*

$$e^{A+B} = \lim_{n\to\infty} \left(e^{\frac{t}{n}A} e^{\frac{t}{n}B}\right)^n,$$

with the convergence in the strong operator topology.[1]

[1] $\lim_{n\to\infty} A_n = A$ in the strong operator topology if $\|A_n f - Af\|_{\mathcal{H}} \to 0$, for any $f \in \mathcal{H}$.

In our case the Hilbert space $\mathcal{H}$ is the space of test functions. We also denote

$$\begin{aligned} A &= t\Delta, \\ B &= tV(x) \qquad \text{[we mean multiplication by } tV(x)], \\ K_n &= e^{\frac{t}{n}\Delta} e^{\frac{t}{n}V(x)}. \end{aligned}$$

Using the Trotter product formula, (7.2.4) becomes

$$K(t) = e^{A+B} = \lim_{n\to\infty} K_n \circ \cdots \circ K_n = \lim_{n\to\infty} K_n^n. \tag{7.2.6}$$

In order to compute the integral kernel of the product operator K_n^n, we first need to find the integral kernel for K_n. Substituting $t = t/n$ in (7.2.5) yields

$$(e^{\frac{t}{n}\Delta} f)(x) = \left(\frac{n}{2\pi t}\right)^{m/2} \int e^{-\frac{n|x-y|^2}{2t}} f(y)\, dy, \tag{7.2.7}$$

and substituting $e^{\frac{t}{n}V(x)} f$ for f, we get the integral representation for K_n:

$$(K_n f)(x) = (e^{\frac{t}{n}\Delta} e^{\frac{t}{n}V(x)} f)(x) = \left(\frac{n}{2\pi t}\right)^{m/2} \int e^{-\frac{n|x-y|^2}{2t}} e^{\frac{t}{n}V(y)} f(y)\, dy,$$

or

$$\widehat{K_n}(x, y) = \left(\frac{n}{2\pi t}\right)^{m/2} e^{-\frac{n|x-y|^2}{2t}} e^{\frac{t}{n}V(y)}.$$

Using Proposition 7.2.2, part (2), we obtain the integral kernel of the product operator K_n^n:

$$\begin{aligned} \widehat{K_n^n}(x, y) &= \int \cdots \int \widehat{K}_n(x, u_1) \widehat{K}_n(u_1, u_2) \ldots \widehat{K}_n(u_{n-1}, y)\, du_{n-1} \ldots du_1 \\ &= \left(\frac{n}{2\pi t}\right)^{nm/2} \int \cdots \int e^{-\frac{n}{2t}\sum_{j=0}^{n-1}|u_{j+1}-u_j|^2} e^{\frac{t}{n}\sum_{j=0}^{n-1}V(u_{j+1})}\, du \\ &= \left(\frac{n}{2\pi t}\right)^{nm/2} \int \cdots \int e^{\sum_{j=0}^{n-1}\left(-\frac{n}{2t}|u_{j+1}-u_j|^2 + \frac{t}{n}V(u_{j+1})\right)}\, du \\ &= \left(\frac{n}{2\pi t}\right)^{nm/2} \int \cdots \int e^{-S_n}\, du, \end{aligned}$$

where $u_0 = x$, $u_n = y$, $du = du_{n-1} \cdots du_1$ and

$$S_n = \sum_{j=0}^{n-1} \left(\frac{n}{2t}|u_{j+1} - u_j|^2 - \frac{t}{n} V(u_{j+1}) \right). \tag{7.2.8}$$

Using the Trotter formula (7.2.6), the integral kernel of $K(t)$ is

$$\widehat{K(t)} = \lim_{n\to\infty} \widehat{K_n^n} = \lim_{n\to\infty} \int \cdots \int e^{-S_n} \left(\frac{n}{2\pi t}\right)^{nm/2} du. \tag{7.2.9}$$

We shall make sense of the above integral as a *Feynman path integral.*

The space of continuous, finite-energy paths from x to y in $\mathbb{R}^m$ within time t is defined by

$$\mathfrak{P}_{x,y;t} = \left\{\phi : [0,t] \to \mathbb{R}^n; \int_0^t |\dot{\phi}|^2 < \infty, \phi(0) = x, \phi(t) = y\right\}.$$

Let the points $u_0(=x), u_1, \ldots, u_{n-1}, u_n(=y) \in \mathbb{R}^m$ be fixed. Consider the piecewise linear function $\phi_n \in \mathfrak{P}_{x,y;t}$ such that $\phi_n(0) = x$, $\phi_n(\frac{kt}{n}) = u_k$, $\phi_n(t) = y$, for all $k = 0, \ldots, n$. Then (7.2.8) becomes

$$\begin{aligned} S_n &= \sum_{j=0}^{n-1} \left(\frac{n}{2t}\left|\phi_n\left(\frac{(j+1)t}{n}\right) - \phi_n\left(\frac{jt}{n}\right)\right|^2 - \frac{t}{n} V\left(\phi_n\left(\frac{(j+1)t}{n}\right)\right)\right) \\ &= \sum_{j=0}^{n-1} \left(\frac{1}{2}\frac{\left|\phi_n\left(\frac{(j+1)t}{n}\right) - \phi_n\left(\frac{jt}{n}\right)\right|^2}{t/n} - V\left(\phi_n\left(\frac{(j+1)t}{n}\right)\right)\right)\frac{t}{n}. \end{aligned}$$

This is a Riemannian sum for the integral action

$$S(\phi,t) = \int_0^t \left(\frac{1}{2}|\dot{\phi}(s)|^2 - V(\phi(s))\right) ds,$$

which is associated with the Lagrangian

$$L(\phi,\dot{\phi}) = \frac{1}{2}|\dot{\phi}(s)|^2 - V(\phi(s)).$$

Since the associated Hamiltonian is

$$H(p,q) = \frac{1}{2}|p|^2 + V(q),$$

we note that $S(\phi,t)$ satisfies the Hamilton–Jacobi equation

$$\frac{\partial S}{\partial t} + \frac{1}{2}|\nabla_\phi S|^2 + V(\phi(x)) = 0.$$

The measure element becomes

$$d\mathfrak{m}(\phi_n) = \left(\frac{n}{2\pi t}\right)^{nm/2} du = \left(\frac{n}{2\pi t}\right)^{nm/2} d\phi_n\left(\frac{t}{n}\right) \cdots d\phi_n\left(\frac{(n-1)t}{n}\right).$$

When taking the limit, we assume $\phi_n \to \phi$, $S_n \to S$, and $d\mathfrak{m}(\phi_n) \to d\mathfrak{m}(\phi)$, where $\phi \in \mathfrak{P}_{x,y;t}$, S is the classical action, and $d\mathfrak{m}$ is the Wiener measure defined on the space $\mathfrak{P}_{x,y;t}$. A rigorous proof of the existence of the Wiener measure is beyond the goal of this exposition. With this preparation we can formally treat the limit of (7.2.9) as an integral:

$$\lim_{n\to\infty} \widehat{K_n^n} = \lim_{n\to\infty} \int \cdots \int e^{-S_n} \left(\frac{n}{2\pi t}\right)^{nm/2} du = \int_{\mathfrak{P}_{x,y;t}} e^{-S(\phi,t)}\, d\mathfrak{m}(\phi).$$

This is called a *Feynman path integral* and is a formal expression that physicists use in their computations of propagators. We shall use it in this chapter to compute heat kernels.

With this introduction, the integral kernel of $K(t)$ can be written as a Feynman path integral:

$$\widehat{K(t)} = \int_{\mathfrak{P}_{x,y;t}} e^{-S(\phi,t)}\, d\mathfrak{m}(\phi). \tag{7.2.10}$$

Anzatz: The above limit does not depend on the sequence of piecewise functions ϕ_n chosen. Under this assumption, the path integral (7.2.10) makes sense.

The concept of "Feynman integral" stems from Feynman's 1948 paper [42], which contains a formula for the evolution of quantum systems. Making this idea rigorous has proven difficult. In this book we are using this concept just informally.

7.3 Formal Algorithm for Obtaining Heat Kernels

The following result states that the heat kernel can be represented as a path integral. This means that the heat kernel from point x to point y at time t depends on all continuous curves between x and y parameterized by $[0,t]$. Each curve is counted with a certain weight which is given as a solution of a transport equation. We shall just check the results informally, but the reader can elaborate a proof using the Trotter formula, in a similar way as we did in the case of the operator $\frac{1}{2}\Delta + V(x)$.

Theorem 7.3.1. *Let (a_{ij}) be a symmetric, nondegenerate, positive definite $m \times m$ matrix. The heat kernel for the operator*

$$\mathbb{L} = \frac{1}{2}\sum_{i,j=1}^{m} a_{ij}(x)\partial_{x_i}\partial_{x_j} + V(x)$$

can be obtained by the following algorithm:

1. Associate the Hamiltonian as the principal symbol of the operator $\mathbb{L}$:

$$H(p,x) = \frac{1}{2}\sum_{i,j=1}^{m} a_{ij}(x)p_i p_j + V(x).$$

2. *Using the Legendre transform, associate the Lagrangian*

$$L(x, \dot{x}) = \frac{1}{2} \sum_{i,j=1}^{m} a^{ij}(x) x_i x_j - V(x),$$

where $(a^{ij}) = (a_{ij})^{-1}$.

3. *Solve the Euler–Lagrange equation*

$$\frac{d}{ds} \frac{\partial L}{\partial \dot{x}} = \frac{\partial L}{\partial x}.$$

4. *Find the classical action* $S(x, t)$ *by integrating the Lagrangian along the solution of the Euler–Langrange equation:*

$$S(x, t) = \int_0^t L\big(x(s), \dot{x}(s)\big)\, ds.$$

5. *The action* $S(x, t)$ *also verifies the Hamilton–Jacobi equation*

$$\frac{\partial S}{\partial t} + H(\nabla_x S, x) = 0.$$

6. *Then the integral kernel of the operator* $e^{t\mathbb{L}}$ *is given by the path integral*

$$\widehat{K}(x, y; t) = \int_{\mathfrak{P}_{x,y;t}} e^{-S(\phi,t)}\, d\mathfrak{m}(\phi),$$

where the measure $d\mathfrak{m}(\phi)$ *is determined as in step 8, and* $S(\phi, t)$ *is the classical action* S *evaluated along the continuous curve* $\phi \in \mathfrak{P}_{x,y;t}$.

7. *Consider the solution* $v_n(t)$ *of the "transport equation"*

$$v_n'(t) + \frac{1}{2n}\Big(\sum_{i,j=1}^{m} a_{ij}(x) \partial_{x_i} \partial_{x_j} S_n \Big) v_n(t) = 0, \qquad n \geq 1,$$

satisfying the boundary condition

$$\lim_{t \searrow 0} \left(\frac{2\pi t}{n}\right)^{m/2} v_n(t) = 1,$$

where the S_n *are the partial Riemannian sums of the integral action* $S(\phi, t)$.

8. *The Wiener measure in the previous path integral is formally defined by*

$$dm(\phi) = \lim_{n \to \infty} dm(\phi_n),$$

where

$$d\mathfrak{m}(\phi_n) = v_n(t)^n d\phi_n\left(\frac{t}{n}\right)\cdots d\phi_n\left(\frac{(n-1)t}{n}\right).$$

Proof. We proceed by an informal computation, i.e., we shall assume, when necessary, that the conditions of the dominated convergence theorem always hold, so we can use the commutativity between the limit symbol and the integral. Using $S_n = S_n(x,t)$, we have

$$\begin{aligned}
\mathbb{L}(e^{-S_n}) &= \frac{1}{2}\sum_{i,j=1}^{m} a_{ij}\partial_{x_i}\partial_{x_j}e^{-S_n} + V(x)e^{-S_n} \\
&= \frac{1}{2}e^{-S_n}\left\{\sum_{i,j=1}^{m} a_{ij}(\partial_{x_i}S_n)(\partial_{x_j}S_n) - \sum_{i,j=1}^{m} a_{ij}\partial_{x_i}\partial_{x_j}S_n\right\} + V(x)e^{-S_n} \\
&= e^{-S_n}\left\{H(\nabla S_n) - \frac{1}{2}\sum_{i,j=1}^{m} a_{ij}\partial_{x_i}\partial_{x_j}S_n\right\}. \qquad (7.3.11)
\end{aligned}$$

Using (7.3.11) yields

$$\begin{aligned}
&\left(\partial_t - \mathbb{L}\right)\left(e^{-S_n}v_n^n(t)\right) \\
&= \partial_t\left(e^{-S_n}v_n^n(t)\right) - \left(\mathbb{L}e^{-S_n}\right)v_n^n(t) \\
&= -e^{-S_n}v_n^n\left(\partial_t S_n - n\frac{v_n'}{v_n}\right) - e^{-S_n}v_n^n\left\{H(\nabla S_n) - \frac{1}{2}\sum_{i,j=1}^{m} a_{ij}\partial_{x_i}\partial_{x_j}S_n\right\} \\
&= -e^{-S_n}v_n^n\left\{(\partial_t S_n + H(\nabla S_n)) - \frac{n}{v_n}\underbrace{\left(v_n' + \frac{v_n}{2n}\sum_{i,j=1}^{m} a_{ij}\partial_{x_i}\partial_{x_j}S_n\right)}_{=0 \text{ by } 7.}\right\} \\
&= -e^{-S_n}v_n^n\left(\partial_t S_n + H(\nabla S_n)\right). \qquad (7.3.12)
\end{aligned}$$

Assume there is a constant $M > 0$ such that $|e^{-S_n}v_n^n| < M$ for all $n \geq 1$. Applying the operator $\partial_t - \mathbb{L}$ to the path integral

$$\widehat{K}(x,y;t) = \int_{\mathfrak{P}_{x,y;t}} e^{-S(\phi,t)}\,d\mathfrak{m}(\phi) = \lim_{n\to\infty}\int\cdots\int e^{-S_n}v_n^n(t)\,du_{n-1}\cdots du_1$$

and using (7.3.12) yields

$$\begin{aligned}(\partial_t - \mathbb{L})\widehat{K}(x, y; t) &= \lim_{n\to\infty} \int \cdots \int (\partial_t - \mathbb{L})\big(e^{-S_n} v_n^n(t)\big)\, du_{n-1} \cdots du_1 \\ &= -\lim_{n\to\infty} \int \cdots \int e^{-S_n} v_n^n \Big(\partial_t S_n + H(\nabla S_n)\Big)\, du_{n-1} \cdots du_1 \\ &= 0,\end{aligned}$$

because

$$\lim_{n\to\infty} \Big(\partial_t S_n + H(\nabla S_n)\Big) = \partial_t S + H(\nabla S) = 0,$$

by the Hamilton–Jacobi equation. ■

The next theorem deals with second-order operators with a linear part.

Theorem 7.3.2. *Let (a_{ij}) be a symmetric, nondegenerate, positive definite $m \times m$ matrix. The heat kernel for the operator*

$$\mathbb{L} = \frac{1}{2} \sum_{i,j=1}^{m} a_{ij}(x)\partial_{x_i}\partial_{x_j} + \sum_j V_j(x)\partial_{x_j}$$

can be obtained following this algorithm:

1. *Associate the following Hamiltonian:*[2]

$$H(p, x) = \frac{1}{2} \sum_{i,j=1}^{m} a_{ij}(x) p_i p_j - \sum_j V_j(x) p_j.$$

2. *Using the Legendre transform, associate the Lagrangian*

$$L(x, \dot{x}) = \frac{1}{2} \sum_{i,j=1}^{m} a^{ij}(x)\big(x_i - V_i(x)\big)\big(x_j - V_j(x)\big),$$

where $(a^{ij}) = (a_{ij})^{-1}$.

3. *Solve the Euler–Lagrange equation*

$$\frac{d}{ds}\frac{\partial L}{\partial \dot{x}} = \frac{\partial L}{\partial x}.$$

[2] Note the minus sign in front of the second term.

4. *Find the classical action $S(x,t)$ by integrating the Lagrangian along the solution of the Euler–Lagrange equation:*

$$S(x,t) = \int_0^t L\big(x(s), \dot{x}(s)\big)\, ds.$$

5. *The action $S(x,t)$ also verifies the Hamilton–Jacobi equation*

$$\frac{\partial S}{\partial t} + H(\nabla_x S, x) = 0.$$

6. *Then the integral kernel of the operator $e^{t\mathbb{L}}$ is given by the path integral*

$$\widehat{K}(x,y;t) = \int_{\mathfrak{P}_{x,y;t}} e^{-S(\phi,t)}\, d\mathfrak{m}(\phi),$$

where the measure $d\mathfrak{m}(\phi)$ is determined as in step 8, and $S(\phi,t)$ is the classical action S evaluated along the continuous curve $\phi \in \mathfrak{P}_{x,y;t}$.

7. *Consider the solution $v_n(t)$ of the "transport equation"*

$$v_n'(t) + \frac{1}{2n}\left(\sum_{i,j=1}^{m} a_{ij}(x)\partial_{x_i}\partial_{x_j} S_n\right) v_n(t) = 0, \qquad n \ge 1,$$

satisfying the boundary condition

$$\lim_{t \searrow 0} \left(\frac{2\pi t}{n}\right)^{m/2} v_n(t) = 1,$$

where the S_n are the partial Riemannian sums of the integral action $S(\phi,t)$.

8. *The Wiener measure in the above path integral is formally defined by*

$$dm(\phi) = \lim_{n\to\infty} dm(\phi_n),$$

where

$$d\mathfrak{m}(\phi_n) = v_n(t)^n\, d\phi_n\Big(\frac{t}{n}\Big) \cdots d\phi_n\left(\frac{(n-1)t}{n}\right).$$

Proof. We shall proceed formally, providing a verification by a direct computation without worrying about the details. First we check the form of the Lagrangian given in step 2. Consider the vectors $x = (x_1, \ldots, x_n)$ and $V = (V_1, \ldots, V_n)$. The Hamiltonian can also be written as

$$H = \frac{1}{2}\langle ap, p\rangle - \langle V, p\rangle.$$

One of the Hamiltonian equations yields

$$\dot{x} = H_p = ap - V \Longrightarrow p = a^{-1}(\dot{x} + V).$$

The Legendre transform yields

$$\begin{aligned} L = p\dot{x} - H &= \langle p, ap - V\rangle - \frac{1}{2}\langle ap, p\rangle + \langle V, p\rangle \\ &= \frac{1}{2}\langle ap, p\rangle = \frac{1}{2}\langle aa^{-1}(\dot{x} + V), a^{-1}(\dot{x} + V)\rangle \\ &= \frac{1}{2}\langle(\dot{x} + V), a^{-1}(\dot{x} + V)\rangle = \frac{1}{2}\sum_{i,j} a^{ij}(x)\big(\dot{x}_i + V_i(x)\big)\big(\dot{x}_j + V_j(x)\big). \end{aligned}$$

By computation, we have

$$\begin{aligned} \mathbb{L}\Big(e^{-S_n}\Big) &= \frac{1}{2}\sum_{i,j=1}^{m} a_{ij}(x)\partial_{x_i}\partial_{x_j} e^{-S_n} + \sum_j V_j(x)\partial_{x_j} e^{-S_n} \\ &= \frac{1}{2}e^{-S_n}\left\{\sum_{i,j=1}^{m} a_{ij}\big(\partial_{x_i} S_n\big)\big(\partial_{x_j} S_n\big) - \sum_{i,j} a_{ij}\partial_{x_i}\partial_{x_j} S_n\right\} \\ &\quad - e^{-S_n}\sum_j V_j(x)\partial_{x_j} S_n \\ &= e^{-S_n}\left\{\frac{1}{2}\sum_{i,j} a_{ij}\big(\partial_{x_i} S_n\big)\big(\partial_{x_j} S_n\big) - \sum_j V_j(x)\partial_{x_j} S_n\right\} \\ &\quad - \frac{1}{2}e^{-S_n}\sum_{i,j} a_{ij}\partial_{x_i}\partial_{x_j} S_n \\ &= e^{-S_n}\left\{H(\nabla S_n) - \frac{1}{2}\sum_{i,j}^{m} a_{ij}\partial_{x_i}\partial_{x_j} S_n\right\}. \end{aligned} \tag{7.3.13}$$

Using (7.3.13) yields

$$\begin{aligned} &\big(\partial_t - \mathbb{L}\big)\Big(e^{-S_n} v_n^n(t)\Big) \\ &\quad = -e^{-S_n} v_n^n(t)\left(\partial_t S_n - n\frac{v_n'}{v_n}\right) - e^{-S_n} v_n^n(t)\left\{H(\nabla S_n) - \frac{1}{2}\sum_{i,j} a_{ij}\partial_{x_i}\partial_{x_j} S_n\right\} \end{aligned}$$

$$= -e^{-S_n} v_n^n(t) \left\{ \left(\partial_t S_n + H(\nabla S_n) \right) - \frac{n}{v_n(t)} \underbrace{\left(v_n' + \frac{v_n}{2n} \sum_{i,j} a_{ij} \partial_{x_i} \partial_{x_j} S_n \right)}_{=0} \right\}$$

$$= -e^{-S_n} v_n^n(t) \left(\partial_t S_n + H(\nabla S_n) \right).$$

The rest of the proof is the same as the last part of the proof of Theorem 7.3.1. ∎

When $\det(a_{ij}(x)) = 0$, the matrix a_{ij} cannot be inverted, and we might have difficulty in assigning a Lagrangian. In this case we may skip from step 1 to step 5. Since the main point is finding the action, we may always do this skip as long as we are able to solve the Hamilton–Jacobi equation.

We shall consider next several examples of operators for which we are able to compute the heat kernels using the previous theorems. We start the presentation with the easiest case. For more worked-out examples, the reader can consult, for instance [102]. We end this section by reminding the reader of two important integrals that will play a major role in computing path integrals in the sequel.

Let $M = (m)_{ij}$ be a $k \times k$ symmetric nonsingular matrix and $(\xi_1, \dots, \xi_k) \in \mathbb{R}^k$ be a given vector. Then

$$\int_{\mathbb{R}^k} e^{-\frac{1}{2}\sum_{i,j=1}^k m_{ij} y_i y_j} = (2\pi)^{k/2} (\det M)^{-1/2}, \tag{7.3.14}$$

$$\int_{\mathbb{R}^k} e^{-\frac{1}{2}\sum_{i,j=1}^k m_{ij} y_i y_j + \sum_{i=1}^k \xi_i y_i} = (2\pi)^{k/2} (\det M)^{-1/2} e^{\frac{1}{2}\sum_{i,j=1}^k \xi_i (M^{-1})_{ij} \xi_j}. \tag{7.3.15}$$

7.4 The Operator $\frac{1}{2}\partial_x^2$

We are interested in finding the heat kernel of the operator $\mathbb{L} = \frac{1}{2}\partial_x^2$ by using path integrals. The Hamiltonian is given by the principal symbol $H(p, x) = \frac{1}{2}p^2$ and the associated Lagrangian is given by $L(x, \dot{x}) = \frac{1}{2}\dot{x}^2$. The Euler–Lagrange equation is $\ddot{x} = 0$. The classical solution satisfying the boundary conditions $x(0) = x_0$ and $x(t) = x$ is

$$x_{cl}(s) = \frac{x - x_0}{t} s + x_0.$$

The classical action from x_0 to x within time t is

$$S_{cl}(x, x_0, t) = \int_0^t \frac{1}{2} \dot{x}_{cl}(s)^2 \, ds = \int_0^t \frac{\left(x(t) - x(0)\right)^2}{2t^2} \, ds = \frac{\left(x - x_0\right)^2}{2t}.$$

The heat kernel of the operator $\partial_t - \frac{1}{2}\partial_x^2$ is given by the path integral

$$K(x_0, x; t) = \int_{\mathfrak{P}_{x_0,x;t}} e^{-S(\phi,t)}\, d\mathfrak{m}(\phi).$$

In order to evaluate the above path integral, we shall change the integration variable as follows. For any $\phi \in \mathfrak{P}_{x_0,x;t}$, we consider the continuous curve $\psi \in \mathfrak{P}_{0,0;t}$ defined by

$$\phi(s) = x_{cl}(s) + \psi(s). \tag{7.4.16}$$

Since $\phi(0) = x_{cl}(0) = x_0$ and $\phi(t) = x_{cl}(t) = x$, it follows that $\psi(0) = \psi(t) = 0$. The reason for the change of variables (7.4.16) is to simplify a later computation from the more complicated integral (7.3.15) to the simpler one (7.3.14). This technique might not always work, in which case we need to use (7.3.15) directly.

The value of the action integral along the curve ϕ is denoted by

$$S(\phi, t) = S\big(\phi \mid \phi(0) = x_0, \phi(t) = x\big).$$

The action along ϕ can be written in terms of the action along ψ. One obtains

$$\begin{aligned} S(\phi,t) &= S\big(\phi|\phi(0)=x_0, \phi(t)=x\big) = \int_0^t \frac{1}{2}\dot{\phi}(s)^2\, ds = \int_0^t \frac{1}{2}\big(\dot{x}_{cl}(s)+\dot{\psi}(s)\big)^2\, ds \\ &= \int_0^t \frac{1}{2}\dot{x}_{cl}(s)^2\, ds + \int_0^t \dot{x}_{cl}(s)\dot{\psi}(s)\, ds + \int_0^t \frac{1}{2}\dot{\psi}(s)^2\, ds \\ &= S_{cl}(x, x_0, t) + \frac{x - x_0}{t}\int_0^t \dot{\psi}(s)\, ds + S\big(\psi \mid \psi(0) = 0, \psi(t) = 0\big). \end{aligned}$$

The second term in the sum vanishes because

$$\int_0^t \dot{\psi}(s)\, ds = \psi(t) - \psi(0) = 0.$$

Therefore, the action integral along ϕ is

$$S(\phi, t) = \frac{(x - x_0)^2}{2t} + S\big(\psi \mid \psi(0) = 0, \psi(t) = 0\big),$$

and substituting in our initial path integral yields

$$\int_{\mathfrak{P}_{x_0,x;t}} e^{-S(\phi,t)}\, d\mathfrak{m}(\phi) = e^{-\frac{(x-x_0)^2}{2t}} \int_{\mathfrak{P}_{0,0;t}} e^{-S(\psi,t)}\, d\mathfrak{m}(\psi). \tag{7.4.17}$$

This way, the path integral over $\mathfrak{P}_{x_0,x;t}$ was replaced by a path integral over the space of continuous loops $\mathfrak{P}_{0,0;t}$. We shall next compute the right-hand side of (7.4.17).

Let $u_j = \psi(jt/n)$, for $j = 0, \ldots, n$. Then the Riemannian sum S_n of the action integral along ψ is

$$S_n = \sum_{j=0}^{n-1} \frac{1}{2}\left(\frac{u_{j+1} - u_j}{t/n}\right)^2 \frac{t}{n} = \sum_{j=0}^{n-1} \frac{(u_{j+1} - u_j)^2}{2t/n}$$

$$= \frac{1}{2t/n}\left\{(u_1 - u_0)^2 + (u_2 - u_1)^2 + \cdots + (u_n - u_{n-1})^2\right\}. \quad (7.4.18)$$

Since $u_0 = u_n = 0$, we get

$$S_n = \frac{1}{2t/n}\{2u_1^2 - 2u_0u_1 + 2u_2^2 - 2u_1u_2 + \cdots + 2u_{n-1}^2\}$$

$$= \frac{1}{2t/n} \sum_{i,j=1}^{n-1} b_{ij}u_iu_j, \quad (7.4.19)$$

where

$$(b_{ij}) = \begin{pmatrix} 2 & -1 & 0 & \cdots & 0 & 0 \\ -1 & 2 & -1 & \cdots & 0 & 0 \\ 0 & -1 & 2 & \cdots & 0 & 0 \\ \vdots & \vdots & \vdots & \ddots & \vdots & \vdots \\ 0 & 0 & 0 & \cdots & 2 & -1 \\ 0 & 0 & 0 & \cdots & -1 & 2 \end{pmatrix} \quad (7.4.20)$$

is an $(n-1) \times (n-1)$ matrix.

In order to construct the measure $d\mathfrak{m}(\psi)$, we need to construct the volume elements v_n given by part 8 of Theorem 7.3.1. First we need to construct the transport equation. Since $x = u_n = \psi(t)$, differentiating in formula (7.4.18) by u_n yields

$$\partial_x^2 S_n = \frac{2}{2t/n} = \frac{n}{t}.$$

Hence the transport equation given by part 8 of Theorem 7.3.1 is

$$v_n'(t) + \frac{1}{2t}v_n(t) = 0.$$

Integrating yields

$$v_n(t) = \frac{c_n}{\sqrt{t}}.$$

The constant c_n is determined from the boundary condition

$$1 = \lim_{t \searrow 0}\left(\frac{2\pi t}{n}\right)^{1/2} v_n(t) = \lim_{t \searrow 0}\left(\frac{2\pi t}{n}\right)^{1/2} \frac{c_n}{\sqrt{t}}.$$

It follows that $c_n = \sqrt{\frac{n}{2\pi}}$. The volume element becomes

$$v_n(t) = \sqrt{\frac{n}{2\pi t}},$$

and the measure element is

$$dm(\psi) = \lim_{n\to\infty} v_n(t)^n \, du_1 \cdots du_{n-1} = \lim_{n\to\infty} \left(\sqrt{\frac{n}{2\pi t}}\right)^n du_1 \cdots du_{n-1}.$$

Now we are ready to evaluate the path integral on the right side of (7.4.17). Using (7.4.19), we have

$$\begin{aligned}
&\int_{\mathfrak{P}_{0,0;t}} e^{-S(\psi,t)} \, d\mathfrak{m}(\psi) \\
&= \lim_{n\to\infty} \int \cdots \int e^{-S_n} \left(\sqrt{\frac{n}{2\pi t}}\right)^n du_1 \cdots du_{n-1} \\
&= \lim_{n\to\infty} \left(\sqrt{\frac{n}{2\pi t}}\right)^n \int \cdots \int e^{-\frac{1}{2t/n}\sum_{i,j=0}^{n-1} b_{ij} u_i u_j} \, du_1 \cdots du_{n-1} \\
&= \lim_{n\to\infty} \left(\sqrt{\frac{n}{2\pi t}}\right)^n \int_{\mathbb{R}^{n-1}} e^{-\frac{1}{2}\sum_{i,j=0}^{n-1} m_{ij}\, u_i u_j} \, du_1 \cdots du_{n-1} \\
&= \lim_{n\to\infty} \left(\sqrt{\frac{n}{2\pi t}}\right)^n \sqrt{\frac{(2\pi)^{n-1}}{\det(m_{ij})}} \\
&= \lim_{n\to\infty} \frac{1}{\sqrt{2\pi}} \left(\frac{n}{t}\right)^{n/2} \frac{1}{\sqrt{\det(m_{ij})}}.
\end{aligned} \tag{7.4.21}$$

where we let $m_{ij} = b_{ij}/(t/n)$. Using a property of determinants,

$$\det(m_{ij}) = \left(\frac{n}{t}\right)^{n-1} \det(b_{ij}). \tag{7.4.22}$$

We shall now compute $\det(b_{ij})$. Consider the following $(n-1)\times(n-1)$ determinant:

$$B_{n-1} = \begin{vmatrix} 2 & -1 & 0 & \cdots & 0 & 0 \\ -1 & 2 & -1 & \cdots & 0 & 0 \\ 0 & -1 & 2 & \cdots & 0 & 0 \\ \vdots & \vdots & \vdots & \ddots & \vdots & \vdots \\ 0 & 0 & 0 & \cdots & 2 & -1 \\ 0 & 0 & 0 & \cdots & -1 & 2 \end{vmatrix},$$

Expanding after the last row yields the following linear recursion equation:

$$B_{n-1} = 2B_{n-2} - B_{n-3}.$$

The associated characteristic equation $r^2 = 2r - 1 = 0$ has the double root $r_1 = r_2 = 1$. Then the general term of the sequence B_n can be written as

$$B_n = r_1^n(C_1 + nC_2) = C_1 + nC_2, \tag{7.4.23}$$

where C_1 and C_2 are two real constants which will be determined from the initial conditions. Since

$$B_1 = |2| = 2, \qquad B_2 = \begin{vmatrix} 2 & -1 \\ -1 & 2 \end{vmatrix} = 3,$$

the constants C_1, C_2 satisfy the linear system

$$C_1 + C_2 = 2,$$
$$C_1 + 2C_2 = 3,$$

with solution $C_1 = C_2 = 1$. Substituting in (7.4.23) yields $B_n = n + 1$. Hence

$$\det(b_{ij}) = B_{n-1} = n,$$

and then (7.4.22) yields

$$\det(m_{ij}) = n\left(\frac{n}{t}\right)^{n-1}.$$

Substituting in (7.4.21), we obtain

$$\begin{aligned}\int_{\mathfrak{P}_{0,0;t}} e^{-S(\psi,t)}\, d\mathfrak{m}(\psi) &= \lim_{n\to\infty} \frac{1}{\sqrt{2\pi}}\left(\frac{n}{t}\right)^{n/2} \frac{1}{n^{1/2}\left(\frac{n}{t}\right)^{\frac{n-1}{2}}} \\ &= \frac{1}{\sqrt{2\pi t}}. \end{aligned} \tag{7.4.24}$$

Substituting in (7.4.17) yields the well-known heat kernel for $\frac{1}{2}\partial_x^2$:

$$K(x_0, x; t) = \int_{\mathfrak{P}_{x_0,x;t}} e^{-S(\phi,t)}\, d\mathfrak{m}(\phi) = \frac{1}{\sqrt{2\pi t}} e^{-\frac{(x-x_0)^2}{2t}}, \qquad t > 0. \tag{7.4.25}$$

7.5 The Operator $\frac{1}{2}(\partial_x^2 + \partial_x)$

Consider the operator $L = \frac{1}{2}(\partial_x^2 + \partial_x)$. Since the operator has a linear term, we shall use the method given by Theorem 7.3.2. The associated Hamiltonian is obtained by replacing ∂_x by the momentum p and changing the sign[3] in front of p:

$$H(x, p) = \frac{1}{2}(p^2 - p).$$

The Hamiltonian system

$$\dot{x} = H_p = p - \frac{1}{2} \Longrightarrow p = \frac{1}{2} + \dot{x},$$
$$\dot{p} = -H_x = 0 \Longrightarrow p \text{ constant},$$

with the boundary conditions

$$x(0) = x_0, \qquad x(t) = x,$$

has the solution

$$x(s) = \frac{x - x_0}{t} s + x_0. \tag{7.5.26}$$

The Legendre transform provides the Lagrangian

$$\begin{aligned} L(x, \dot{x}) = p\dot{x} - H &= \left(\frac{1}{2} + \dot{x}\right)\dot{x} - \frac{1}{2}\left(\frac{1}{2} + \dot{x}\right)^2 + \frac{1}{2}\left(\frac{1}{2} + \dot{x}\right) \\ &= \frac{\dot{x}^2}{2} + \frac{\dot{x}}{2} + \frac{1}{8}. \end{aligned}$$

The Euler–Lagrange equation is $\ddot{x} = 0$ and its solution is given by (7.5.26). The classical action is

$$\begin{aligned} S_{cl}(x_0, x; t) &= \int_0^t L\big(x_{cl}(s), \dot{x}_{cl}(s)\big)\, ds = \int_0^t \left(\frac{1}{2}\dot{x}(s)^2 + \frac{1}{2}\dot{x}(s) + \frac{1}{8}\right) ds \\ &= \int_0^t \left(\frac{(x - x_0)^2}{2t^2} + \frac{x - x_0}{2t} + \frac{1}{8}\right) ds \\ &= \frac{(x - x_0)^2}{2t} + \frac{1}{2}(x - x_0) + \frac{t}{8}. \end{aligned} \tag{7.5.27}$$

[3] This is the Hamiltonian of the conjugate operator.

Let ϕ be a curve satisfying $\phi(0) = x_0$ and $\phi(t) = x$. We make a "change of variables," introducing the curve ψ,

$$\phi(s) = x_{cl}(s) + \psi(s).$$

We note that $\psi(0) = \psi(t) = 0$. Next we shall express the action along ϕ in terms of the action along ψ:

$$\begin{aligned} S(\phi,t) &= S(\phi \mid \phi(0) = x_0, \phi(t) = x) = \int_0^t L(\phi(s), \dot{\phi}(s))\, ds \\ &= \int_0^t \left\{ \frac{1}{2}\dot{\phi}^2(s) + \frac{1}{2}\dot{\phi}(s) + \frac{1}{8} \right\} ds \\ &= \int_0^t \left\{ \frac{1}{2}\dot{x}_{cl}^2(s) + \frac{1}{2}\dot{x}_{cl}(s) + \frac{1}{8} \right\} ds + \int_0^t \dot{x}_{cl}(s)\dot{\psi}(s)\, ds \\ &\quad + \int_0^t \left\{ \frac{1}{2}\dot{\psi}^2(s) + \frac{1}{2}\dot{\psi}(s) + \frac{1}{8} - \frac{1}{8} \right\} ds \\ &= S_{cl}(x_0, x; t) + \frac{x - x_0}{t} \int_0^t \dot{\psi}(s)\, ds + S(\psi \mid \psi(0) = 0, \psi(t) = 0) - \frac{t}{8}. \end{aligned}$$

Since $\int_0^t \dot{\psi}(s)\, ds = \psi(t) - \psi(0) = 0$, using relation (7.5.27), the above equation becomes

$$S(\phi,t) = \frac{(x - x_0)^2}{2t} + \frac{1}{2}(x - x_0) + S(\psi \mid \psi(0) = 0, \psi(t) = 0). \qquad (7.5.28)$$

By Theorem 7.3.2, the heat kernel is

$$K(x_0, x; t) = \int_{\mathfrak{P}_{x_0,x;t}} e^{-S(\phi,t)}\, d\mathfrak{m}(\phi) = e^{-\frac{(x-x_0)^2}{2t} - \frac{1}{2}(x-x_0)} \int_{\mathfrak{P}_{0,0;t}} e^{-S(\psi,t)}\, d\mathfrak{m}(\psi). \qquad (7.5.29)$$

We still need to compute the path integral on the right side. We note that this is taken over the space of loops starting at the origin.

Consider the intermediate points $u_j = \psi(jt/n)$, $j = 0, \ldots, n$, with $u_0 = \psi(0) = 0$, $u_n = \psi(t) = 0$. The Riemannian sum of the action is given by

$$\begin{aligned} S_n &= \sum_{j=0}^{n-1} \left\{ \frac{1}{2} \frac{(u_{j+1} - u_j)^2}{(t/n)^2} \frac{t}{n} + \frac{1}{2} \frac{u_{j+1} - u_j}{t/n} + \frac{1}{8} \right\} \frac{t}{n} \\ &= \sum_{j=0}^{n-1} \frac{1}{2} \frac{(u_{j+1} - u_j)^2}{t/n} + \frac{1}{2} \underbrace{\sum_{j=0}^{n-1} (u_{j+1} - u_j)}_{=0} + \frac{t}{8} \\ &= \frac{1}{2t/n} \sum_{i,j=0}^{n-1} b_{ij}\, u_i u_j + \frac{t}{8}, \end{aligned} \qquad (7.5.30)$$

with (b_{ij}) given by (7.4.20). The same computation as before shows that the volume element is

$$v_n(t) = \left(\frac{n}{2\pi t}\right)^{1/2}$$

and the measure is

$$d\mathfrak{m}(\psi) = \left(\frac{n}{2\pi t}\right)^{n/2} du_1 \cdots du_{n-1}. \tag{7.5.31}$$

Using (7.5.30) and (7.5.31), the path integral on the right side of (7.5.29) can be evaluated as in the following:

$$\begin{aligned}
&\int_{\mathfrak{P}_{0,0;t}} e^{-S(\psi,t)}\, d\mathfrak{m}(\psi) \\
&= \lim_{n\to\infty} \int \cdots \int e^{-S_n} \left(\frac{n}{2\pi t}\right)^{n/2} du_1 \cdots du_{n-1} \\
&= e^{-\frac{t}{8}} \lim_{n\to\infty} \left(\frac{n}{2\pi t}\right)^{n/2} \int \cdots \int e^{-\frac{1}{2t/n}\sum_{i,j=0}^{n-1} b_{ij}\, u_i u_j}\, du_1 \cdots du_{n-1} \\
&= e^{-\frac{t}{8}} \frac{1}{(2\pi t)^{1/2}},
\end{aligned}$$

by a previous calculation; see (7.4.24). Substituting in (7.5.29), we obtain the heat kernel for $\frac{1}{2}(\partial_x^2 + \partial_x)$:

$$K(x_0, x; t) = \frac{1}{\sqrt{2\pi t}} e^{-\frac{(x-x_0)^2}{2t} - \frac{1}{2}(x-x_0) - \frac{t}{8}}, \qquad t > 0. \tag{7.5.32}$$

7.6 Heat Kernel for $L = \frac{1}{2}(\partial_x^2 - \partial_x)$

The Hamiltonian in this case is $H(p, x) = \frac{1}{2}(p^2 + p)$. A similar computation as in the previous section shows that the associated Lagrangian is given by

$$L(x, \dot{x}) = \frac{\dot{x}^2}{2} - \frac{\dot{x}}{2} + \frac{1}{8},$$

with the classical action

$$S_{cl}(x_0, x; t) = \frac{(x - x_0)^2}{2t} - \frac{1}{2}(x - x_0) + \frac{t}{8}.$$

The same computation as in the previous example yields the heat kernel for $\frac{1}{2}(\partial_x^2 - \partial_x)$:

$$K(x_0, x; t) = \frac{1}{\sqrt{2\pi t}} e^{-\frac{(x-x_0)^2}{2t} + \frac{1}{2}(x-x_0) - \frac{t}{8}}, \qquad t > 0, \tag{7.6.33}$$

7.7 Heat Kernel for $L = \frac{1}{2}x^2\partial_x^2 + x\partial_x$

The operator is elliptic away from the origin. We shall find the heat kernel for $x_0, x > 0$. We can obtain a similar formula for $x_0, x < 0$. Note that there is no heat transfer between the regions $x_0 > 0$ and $x < 0$. An efficient way to solve this problem is to reduce the operator to the previous operator. Changing the variable $x = e^u$, and using

$$x\frac{\partial}{\partial x} = \frac{\partial}{\partial u}, \qquad x^2\frac{\partial^2}{\partial x^2} = \frac{\partial^2}{\partial u^2} - \frac{\partial}{\partial u},$$

we obtain

$$\begin{aligned} e^{t(\frac{1}{2}x^2\partial_x^2 + x\partial_x)} &= e^{\frac{t}{2}(\partial_u^2 + \partial_u)} \\ &= \frac{1}{\sqrt{2\pi t}} e^{-\frac{(u-u_0)^2}{2t} - \frac{1}{2}(u-u_0) - \frac{t}{8}} \\ &= \frac{1}{\sqrt{2\pi t}} e^{-\frac{(\ln x - \ln x_0)^2}{2t} - \frac{1}{2}(\ln x - \ln x_0) - \frac{t}{8}} \\ &= \frac{1}{\sqrt{2\pi t}} \sqrt{\frac{x_0}{x}}\, e^{-\frac{1}{2t}(\ln x - \ln x_0)^2 - \frac{t}{8}}, \qquad t > 0 \end{aligned}$$

where we used formula (7.5.32).

7.8 The Hermite Operator $\frac{1}{2}(\partial_x^2 - a^2x^2)$

Consider the operator $\mathbb{L} = \frac{1}{2}(\partial_x^2 - a^2x^2)$, with a constant. The associated Hamiltonian is $H(p, x) = \frac{1}{2}p^2 - \frac{1}{2}a^2x^2$ and the Hamiltonian equations are

$$\begin{aligned} \dot{x} &= H_p = p, \\ \dot{p} &= -H_x = a^2x. \end{aligned}$$

Then the classical path between x_0 and x within time t satisfies

$$\begin{aligned} \ddot{x} &= a^2x, \\ x(0) &= x_0, \ x(t) = x. \end{aligned} \tag{7.8.34}$$

The solution is

$$x(s) = A\sinh(as) + B\cosh(as).$$

From the boundary conditions, we get

$$B = x_0, \qquad A = \frac{x - x_0\cosh(at)}{\sinh(at)}.$$

Therefore, the classical path is

$$x_{cl}(s) = \frac{x - x_0\cosh(at)}{\sinh(at)}\sinh(as) + x_0\cosh(as). \tag{7.8.35}$$

The associated Lagrangian is

$$L(x,\dot{x}) = p\dot{x} - H = \dot{x}^2 - \left(\frac{1}{2}\dot{x}^2 - \frac{1}{2}a^2x^2\right) = \frac{1}{2}\dot{x}^2 + \frac{1}{2}a^2x^2,$$

with the Euler–Lagrange equation (7.9.50).

The classical action from x_0 to x within time t is

$$S_{cl}(x,x_0,t) = \int_0^t \frac{1}{2}\left(\dot{x}_{cl}(s)^2 + a^2x_{cl}^2(s)\right)ds.$$

We shall compute the above integrand first. In order to simplify the calculations, we let

$$\sigma = \sinh(at), \qquad \kappa = \cosh(at).$$

Using (7.8.35) yields

$$\begin{aligned}
x_{cl}(s) &= \frac{x - x_0\kappa}{\sigma}\sinh(as) + x_0\cosh(as),\\
\frac{1}{2}\dot{x}_{cl}^2(s) &= \frac{1}{2}a^2\left(\frac{x - x_0\kappa}{\sigma}\cosh(as) + x_0\sinh(as)\right)^2\\
&= \frac{1}{2}a^2\bigg(\frac{(x - x_0\kappa)^2}{\sigma^2}\cosh^2(as) + 2\,\frac{x - x_0\kappa}{\sigma}\cosh(as)x_0\sinh(as)\\
&\qquad + x_0^2\sinh^2(as)\bigg),\\
\frac{1}{2}a^2x_{cl}^2(s) &= \frac{1}{2}a^2\bigg(\frac{(x - x_0\kappa)^2}{\sigma^2}\sinh^2(as) + 2\,\frac{x - x_0\kappa}{\sigma}\cosh(as)x_0\sinh(as)\\
&\qquad + x_0^2\cosh^2(as)\bigg).
\end{aligned}$$

Using

$$\cosh^2(as) = \frac{1 + \cosh(2as)}{2}, \qquad \sinh^2(as) = \frac{-1 + \cosh(2as)}{2},$$

adding the above relations yields

$$\begin{aligned}
\frac{1}{2}\dot{x}_{cl}^2(s) + \frac{1}{2}a^2x_{cl}^2(s) &= \frac{a^2}{2}\bigg(\frac{(x - x_0\kappa)^2}{\sigma^2}\cosh(2as) + 2x_0\frac{(x - x_0\kappa)}{\sigma}\sinh(2as)\\
&\qquad + x_0^2\cosh(2as)\bigg).
\end{aligned}$$

Integrating between 0 and t, we obtain the classical action

$$\begin{aligned}
S_{cl}(x,x_0,t) &= \frac{a^2}{2}\frac{1}{2a}\left(\frac{(x-x_0\kappa)^2}{\sigma^2}\sinh(2at)+2x_0\frac{(x-x_0\kappa)}{\sigma}[\cosh(2at)-1]\right.\\
&\qquad\left.+x_0^2\sinh(2at)\right)\\
&= \frac{a^2}{2}\frac{1}{2a}\left(\frac{(x-x_0\kappa)^2}{\sigma^2}2\sigma\kappa+2x_0\frac{(x-x_0\kappa)}{\sigma}2\sigma^2+x_0^2\cdot 2\sigma\kappa\right)\\
&= \frac{a}{2\sigma}\left((x-x_0\kappa)^2\kappa+2x_0(x-x_0\kappa)\sigma^2+x_0^2\sigma^2\kappa\right)\\
&= \frac{a}{2\sigma}\left(\left(x^2-2xx_0\kappa+x_0^2\kappa^2\right)\kappa+2xx_0\sigma^2-x_0^2\sigma^2\kappa\right)\\
&= \frac{a}{2\sigma}\left(x^2\kappa+x_0^2(\kappa^2-\sigma^2)\kappa+2xx_0(\sigma^2-\kappa^2)\right)\\
&= \frac{a}{2\sigma}\left(\left(x^2+x_0^2\right)\kappa-2xx_0\right),
\end{aligned}$$

where we used $\kappa^2-\sigma^2=1$ Hence

$$S_{cl}(x_0,x,t)=\frac{a}{2\sinh(at)}\left((x^2+x_0^2)\cosh(at)-2xx_0\right). \tag{7.8.36}$$

Let $x:[0,t]\to\mathbb{R}$ be a path satisfying $x(0)=x_0$ and $x(t)=x$. Consider the deviation $v(s)$ from the classical path

$$x(s)=x_{cl}(s)+v(s),$$

where $v(0)=v(0)=0$. The action along $x(s)$ becomes

$$\begin{aligned}
S\big(x\,|\,x(0)=x_0,x(t)=x\big) &= \int_0^t\frac{1}{2}\left\{\dot{x}^2(s)+\frac{a^2}{2}x(s)^2\right\}ds\\
&= \int_0^t\frac{1}{2}\left\{\left(\dot{x}_{cl}(s)+\dot{v}(s)\right)^2+\frac{a^2}{2}\left(x_{cl}(s)+v(s)\right)^2\right\}ds\\
&= \int_0^t\frac{1}{2}\left\{\dot{x}_{cl}^2(s)+\frac{a^2}{2}x_{cl}(s)^2\right\}ds+\int_0^t\frac{1}{2}\left\{\dot{v}^2(s)+\frac{a^2}{2}v(s)^2\right\}ds\\
&\quad+\int_0^t\left(\dot{x}_{cl}(s)\dot{v}(s)+a^2x_{cl}(s)v(s)\right)ds\\
&= S_{cl}(x_0,x,t)+S\big(v\,|\,v(0)=0,v(t)=0\big)\\
&\quad+\int_0^t\left(\dot{x}_{cl}(s)\dot{v}(s)+a^2x_{cl}(s)v(s)\right)ds.
\end{aligned}$$

We shall show that the last integral vanishes:

$$\begin{aligned}
&\int_0^t \big(\dot{x}_{cl}(s)\dot{v}(s) + a^2 x_{cl}(s)v(s)\big)\, ds \\
&\quad = \int_0^t \frac{d}{ds}(\dot{x}_{cl}(s)v(s))\, ds - \int_0^t \big(\ddot{x}_{cl}(s) - a^2 x_{cl}(s)\big)v(s)\, ds \\
&\quad = \dot{x}_{cl}(t)v(t) - \dot{x}_{cl}(0)v(0) = 0,
\end{aligned}$$

where we used $v(0) = v(t) = 0$ and the Euler–Lagrange equations (7.9.50).
Hence

$$\begin{aligned}
S\big(x(s)\big) &= S\big(x \mid x(0) = x_0, x(t) = x\big) = S_{cl}(x_0, x, t) + S\big(v \mid v(0) = 0, v(t) = 0\big) \\
&= S_{cl}(x_0, x, t) + S\big(v(s)\big).
\end{aligned}$$

Using (7.8.36), Theorem 7.3.1 yields the heat kernel

$$\begin{aligned}
K(x_0, x; t) &= \int_{\mathfrak{P}_{x_0,x;t}} e^{-S(x(s))}\, d\mathfrak{m}(x) = e^{-S_{cl}(x_0,x,t)} \int_{\mathfrak{P}_{0,0;t}} e^{-S(v(s))}\, d\mathfrak{m}(v) \\
&= e^{-\frac{a}{2\sinh(at)}\left((x^2+x_0^2)\cosh(at) - 2xx_0\right)} \int_{\mathfrak{P}_{0,0;t}} e^{-S(v(s))}\, d\mathfrak{m}(v). \quad (7.8.37)
\end{aligned}$$

In order to compute the path integral on the right side of (7.8.37), we need to consider the intermediate points $u_j = v(jt/n)$, $j = 0, \ldots, n$, with $u_0 = v(0) = 0$ and $u_n = v(t) = 0$. The Riemannian sum associated with the action

$$S = \int_0^t \left(\frac{1}{2}\dot{x}^2 + \frac{1}{2}a^2x^2\right) ds$$

is

$$\begin{aligned}
S_n &= \sum_{j=0}^{n-1} \frac{1}{2}\frac{(u_{j+1} - u_j)^2}{(t/n)^2} \cdot \frac{t}{n} + \sum_{j=0}^{n-1} \frac{1}{2}a^2 u_j^2 \frac{t}{n} \\
&= \frac{1}{2t/n} \sum_{i,j=0}^{n-1} b_{ij} u_i u_j + \frac{a^2}{2}\frac{t}{n} \sum_{j=0}^{n-1} u_j^2 \\
&= \frac{1}{2t/n} \sum_{i,j=0}^{n-1} a_{ij}(t)\, u_i u_j, \quad (7.8.38)
\end{aligned}$$

where

$$a_{ij}(t) = b_{ij} + \left(\frac{at}{n}\right)^2 \delta_{ij}$$

$$= \begin{pmatrix} 2 & -1 & 0 & \cdots & 0 & 0 \\ -1 & 2 & -1 & \cdots & 0 & 0 \\ 0 & -1 & 2 & \cdots & 0 & 0 \\ \vdots & \vdots & \vdots & \ddots & \vdots & \vdots \\ 0 & 0 & 0 & \cdots & 2 & -1 \\ 0 & 0 & 0 & \cdots & -1 & 2 \end{pmatrix} + \left(\frac{at}{n}\right)^2 \delta_{ij}$$

$$= \begin{pmatrix} 2+\left(\frac{at}{n}\right)^2 & -1 & 0 & \cdots & 0 & 0 \\ -1 & 2+\left(\frac{at}{n}\right)^2 & -1 & \cdots & 0 & 0 \\ 0 & -1 & 2+\left(\frac{at}{n}\right)^2 & \cdots & 0 & 0 \\ \vdots & \vdots & \vdots & \ddots & \vdots & \vdots \\ 0 & 0 & 0 & \cdots & 2+\left(\frac{at}{n}\right)^2 & -1 \\ 0 & 0 & 0 & \cdots & -1 & 2+\left(\frac{at}{n}\right)^2 \end{pmatrix}$$

is an $(n-1)\times(n-1)$ matrix. Let $f_n(t) = \frac{t}{n}\det(a_{ij}(t))$. Expanding the determinant over the last row yields the recursion

$$f_n = \left[2 + \left(\frac{at}{n}\right)^2\right] f_{n-1} - f_{n-2}.$$

This can be written as a difference equation:

$$\frac{f_n - 2f_{n-1} + f_{n-2}}{(t/n)^2} = a^2 f_{n-1}.$$

Assume the pointwise limit $f(t) = \lim_{n\to\infty} f_n(t)$ exists. Then the above difference equation becomes the ordinary differential equation

$$f''(t) = a^2 f(t),$$

with solution

$$f(t) = A\sinh(at) + B\cosh(at), \quad A, B \in \mathbb{R}.$$

We shall determine the value of the constants A and B from the initial conditions.

Note that $\det(a_{ij})(0) = n$, so

$$f_n(0) = \lim_{t \searrow 0} f_n(t) = \lim_{t \searrow 0} \frac{t}{n} \cdot \lim_{t \searrow 0} \det(a_{ij}(t)) = 0.$$

Hence, $f(0) = \lim_{n\to\infty} f_n(0) = 0$. It follows that $B = 0$ and

$$f(t) = A \sinh(at).$$

Differentiating at $t = 0$,

$$\begin{aligned} f_n'(0) &= \lim_{t \searrow 0} \left\{ \left(\frac{t}{n}\right)' \det a_{ij}(t) \right\} + \lim_{t \searrow 0} \left\{ \frac{t}{n} \frac{d}{dt} \det a_{ij}(t) \right\} \\ &= 1, \end{aligned}$$

since the second limit is zero. Therefore, $f'(0) = 1$ and hence $A = 1/a$. It follows that

$$f(t) = \lim_{n\to\infty} f_n(t) = \frac{1}{a} \sinh(at). \tag{7.8.39}$$

A similar computation as in a previous example shows that the volume elements are

$$v_n(t) = \sqrt{\frac{n}{2\pi t}},$$

and hence the Wiener measure becomes

$$d\mathfrak{m}(v) = \left(\frac{n}{2\pi t}\right)^{n/2} du_1 \cdots du_{n-1}.$$

Next we evaluate the path integral on the right side of (7.8.37):

$$\begin{aligned} &\int_{\mathfrak{P}_{0,0;t}} e^{-S(v(s))}\, d\mathfrak{m}(v) \\ &= \lim_{n\to\infty} \int \cdots \int e^{-S_n} \left(\frac{n}{2\pi t}\right)^{n/2} du_1 \cdots du_{n-1} \\ &= \lim_{n\to\infty} \left(\frac{n}{2\pi t}\right)^{n/2} \int \cdots \int e^{-\frac{1}{2t/n} \sum_{i,j=0}^{n-1} a_{ij} u_i u_j}\, du_1 \cdots du_{n-1} \\ &= \lim_{n\to\infty} \left(\frac{n}{2\pi t}\right)^{n/2} \int_{\mathbb{R}^{n-1}} e^{-\frac{1}{2} \sum_{i,j=0}^{n-1} m_{ij}\, u_i u_j}\, du_1 \cdots du_{n-1}, \end{aligned}$$

where $m_{ij} = \dfrac{a_{ij}}{t/n}$.

$$\begin{aligned}
\int_{\mathfrak{P}_{0,0;t}} e^{-S(v(s))}\, d\mathfrak{m}(v) &= \lim_{n\to\infty} \left(\frac{n}{2\pi t}\right)^{n/2} \sqrt{\frac{(2\pi)^{n-1}}{\det(m_{ij})}} \\
&= \lim_{n\to\infty} \frac{1}{\sqrt{2\pi}} \left(\frac{t}{n}\right)^{n/2} \frac{1}{\sqrt{\frac{\det a_{ij}(t)}{(t/n)^{n-1}}}} \\
&= \lim_{n\to\infty} \frac{1}{\sqrt{2\pi}} \frac{1}{\sqrt{\frac{t}{n}\det a_{ij}(t)}} = \lim_{n\to\infty} \frac{1}{\sqrt{2\pi}} \frac{1}{\sqrt{f_n(t)}} \\
&= \frac{1}{\sqrt{2\pi}} \sqrt{\frac{a}{\sinh(at)}},
\end{aligned}$$

by (7.8.39). Substituting in (7.8.37), we obtain the following heat kernel for the Hermite operator:

$$K(x_0, x; t) = \frac{1}{\sqrt{2\pi t}} \sqrt{\frac{at}{\sinh(at)}} e^{-\frac{a}{2\sinh(at)}\left((x^2+x_0^2)\cosh(at) - 2xx_0\right)}, \quad t > 0. \tag{7.8.40}$$

7.9 Evaluating Path Integrals

We have seen in the previous sections that the heat kernel can be expressed formally as a path integral. However, computing path integrals from the definition, even in the simplest cases, is a tedious job and reminds us of computing Riemann integrals starting from the Riemann sum. Since this is not always efficient, we shall next present a couple of methods that overpass the computation of the path integral.

7.9.1 Van Vleck's Formula

Given two points x, y in space and a time $t > 0$, the propagator from x to y within time t is given by the path integral

$$K(x_0, x; t) = \int_{\mathfrak{P}_{x,y;t}} e^{-S(\phi,t)}\, d\mathfrak{m}(\phi). \tag{7.9.41}$$

This means that K depends on all continuous paths joining x and y parameterized by $[0, t]$. Among all the possible paths between the aforementioned points, the *classical path* plays a distinguished role; see Fig. 7.3. This is the path on which a

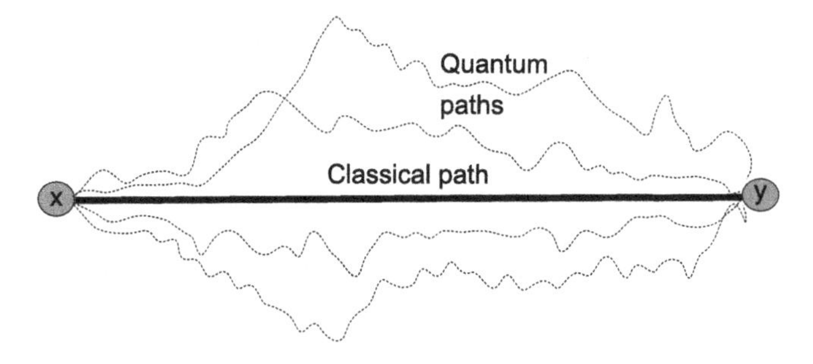

Fig. 7.3 Classical and quantum paths between x and y

classical particle would travel on and is given by the solution of the Euler–Lagrange system of equations. It is a remarkable fact that for a *classical Lagrangian*, i.e., a Lagrangian which is at most quadratic in $\dot{x}_i$ and x_i, the path integral (7.9.41) depends only on the classical action. This is the famous van Vleck formula which expresses the path integral in the following closed-form formula(see [109]):

$$\int_{\mathfrak{P}_{x,y;t}} e^{-S(\phi,t)}\, d\mathfrak{m}(\phi) = \sqrt{\det\left(-\frac{1}{2\pi}\frac{\partial^2 S_{cl}(x,y)}{\partial x\, \partial y}\right)}\, e^{-S_{cl}(x,y,t)}, \qquad (7.9.42)$$

where

$$S_{cl}(x,y,t) = \int_0^t L\big(x(s), \dot{x}(s)\big)\, ds$$

is the classical action obtained by integrating the Lagrangian along the solution $x(s)$ of the Euler–Lagrange equation. The factor

$$V(t) = \sqrt{\det\left(-\frac{1}{2\pi}\frac{\partial^2 S_{cl}(x,y)}{\partial x\, \partial y}\right)}$$

is called the *van Vleck determinant*, and it plays the role of volume element in the geometric method described in Chap. 3. The aforementioned formula can be applied successfully in the cases of second-order elliptic operators with constant, linear and quadratic potential.

7.9.2 Applications of van Vleck's Formula

In the following we shall present some applications. In all of the next examples the heat kernel will be calculated using the formula

$$K(x,y;t) = V(t)e^{-S_{cl}(x,y;t)},$$

where V is the van Vleck determinant.

(*a*) *One-dimensional case.* We have seen in Sect. 7.4 that the classical action associated with the operator $\mathbb{L} = \frac{1}{2}\partial_x^2$ is

$$S_{cl}(x_0, x, t) = \frac{(x - x_0)^2}{2t}.$$

Since $\frac{\partial^2 S_{cl}}{\partial x_0 \partial x} = -\frac{1}{t}$, formula (7.9.42) yields the following familiar formula of the heat kernel:

$$K(x_0, x, t) = \frac{1}{\sqrt{2\pi t}} e^{-\frac{1}{2t}(x-x_0)^2}, \qquad t > 0.$$

(*b*) *One-dimensional case with linear part.* The associated action with the operator $\mathbb{L} = \frac{1}{2}(\partial_x^2 + \partial_x)$ as provided by Sect. 7.5 is

$$S_{cl}(x_0, x, t) = \frac{(x - x_0)^2}{2t} + \frac{1}{2}(x - x_0) + \frac{t}{8}.$$

Since

$$\frac{\partial^2 S_{cl}}{\partial x_0 \partial x} = -\frac{1}{t},$$

van Vleck's formula (7.9.42) yields the following heat kernel:

$$K(x_0, x, t) = \frac{1}{\sqrt{2\pi t}} e^{-\frac{1}{2t}(x-x_0)^2 - \frac{1}{2}(x-x_0) - \frac{t}{8}}, \qquad t > 0.$$

(*c*) *The case of linear potential.* In the following we shall find the heat kernel of the operator $\mathbb{L} = \frac{1}{2}\partial_x^2 - ax$, with $a \in \mathbb{R}$.
The Hamiltonian function associated with the operator $\mathbb{L}$ is given by

$$H(p, x) = \frac{1}{2}p^2 - ax.$$

The Hamiltonian system becomes

$$\dot{x} = H_p = p,$$
$$\dot{p} = -H_x = a.$$

The associated Lagrangian function is $L(x, \dot{x}) = \frac{1}{2}\dot{x}^2 + ax$. The Euler–Lagrange equation $\ddot{x} = a$ with the boundary conditions $x(0) = x_0$, $x(t) = x$ has the unique solution

$$x(s) = \frac{1}{2}as^2 + bs + x_0, \tag{7.9.43}$$

with

$$b = \frac{x - x_0}{t} - \frac{at}{2}. \tag{7.9.44}$$

The Lagrangian along the solution (7.9.43) is

$$\begin{aligned} L\big(x(s), \dot{x}(s)\big) &= \frac{1}{2}\dot{x}(s)^2 + ax(s) \\ &= \frac{1}{2}(as + b)^2 + a\left(\frac{1}{2}as^2 + bs + x_0\right) \\ &= a^2s^2 + 2abs + ax_0 + \frac{1}{2}b^2. \end{aligned}$$

The classical action is obtained by integrating the Lagrangian along the solution:

$$\begin{aligned} S(x_0, x; t) &= \int_0^t L\big(x(s), \dot{x}(s)\big)\, ds \\ &= \frac{1}{3}a^2t^3 + abt^2 + ax_0t + \frac{1}{2}b^2t. \end{aligned}$$

Substituting b from (7.10.67) yields the classical action

$$\begin{aligned} S_{cl}(x_0, x; t) &= \frac{1}{3}a^2t^3 + at^2\Big(\frac{x - x_0}{t} - \frac{at}{2}\Big) + ax_0t + \frac{1}{2}t\Big(\frac{x - x_0}{t} - \frac{at}{2}\Big)^2 \\ &= \frac{(x - x_0)^2}{2t} + \frac{1}{2}a(x + x_0)t - \frac{1}{24}a^2t^3. \end{aligned} \tag{7.9.45}$$

The heat kernel is now given by van Vleck's formula:

$$\begin{aligned} K(x_0, x, t) &= \sqrt{\det\left(-\frac{1}{2\pi}\frac{\partial^2 S_{cl}}{\partial x\, \partial x_0}\right)}\, e^{-S_{cl}(x_0, x, t)} \\ &= \frac{1}{\sqrt{2\pi t}} e^{-\frac{(x-x_0)^2}{2t} - \frac{1}{2}a(x+x_0)t + \frac{1}{24}a^2t^3}, \quad t > 0. \end{aligned}$$

(*d*) *The n-dimensional linear potential.* Consider the n-dimensional operator

$$\mathbb{L} = \frac{1}{2}\sum_{i=1}^{n} \partial_{x_i}^2 - \sum_{i=1}^{n} a_i x_i, \tag{7.9.46}$$

with $a_i \in \mathbb{R}$. In this case the Hamiltonian function is

$$H(p, x) = \frac{1}{2}\sum_{i=1}^{n} p_i^2 - \sum_{i=1}^{n} a_i x_i,$$

and the Hamiltonian equations are

$$\begin{aligned} \dot{x}_j &= p_j, \\ \dot{p}_j &= -a_j, \qquad j = 1, \dots, n. \end{aligned}$$

The solution joining the points x_0 and x is

$$x^j(s) = \frac{1}{2} a_j s^2 + b_j s + x_0^j, \tag{7.9.47}$$

with

$$b_j = \frac{x^j - x_0^j}{t} - \frac{a_j t}{2}, \qquad j = 1, \dots, n. \tag{7.9.48}$$

The Lagrangian

$$L(x, \dot{x}) = \frac{1}{2} |\dot{x}_i|^2 + \langle a, x \rangle = \frac{1}{2} \sum_{i=1}^{n} \dot{x}_i^2 + \sum_{i=1}^{n} a_i x_i$$

evaluated along the above solution is

$$L\big(x(s), \dot{x}(s)\big) = |a|^2 s^2 + 2\langle a, b \rangle s + \langle a, x_0 \rangle + \frac{1}{2} |b|^2.$$

Integrating the Lagrangian along the solution yields the following classical action:

$$S(x_0, x; t) = \frac{|x - x_0|^2}{2t} + \frac{1}{2} \langle a, x + x_0 \rangle t - \frac{1}{24} |a|^2 t^3. \tag{7.9.49}$$

The van Vleck determinant is

$$\sqrt{\det\left(-\frac{1}{2\pi} \frac{\partial^2 S_{cl}(x_0, x)}{\partial x_0 \, \partial x} \right)} = (2\pi t)^{-n/2},$$

so we conclude that the heat kernel of the operator (7.9.46) is

$$K(x, x_0, t) = \frac{1}{(2\pi t)^{n/2}} e^{-\frac{|x - x_0|^2}{2t} - \frac{1}{2} \langle a, x + x_0 \rangle t + \frac{1}{24} |a|^2 t^3}, \quad t > 0.$$

(*e*) We shall next find the heat kernel for the Hermite operator

$$\mathbb{L} = \frac{1}{2} \left(\frac{d}{dx}^2 - a^2 x^2 \right),$$

with $a \in \mathbb{R}$. The associated Hamiltonian is given by

$$H(p, x) = \frac{1}{2}p^2 - \frac{1}{2}a^2 x^2$$

and the Hamiltonian system of equations is

$$\dot{x} = H_p = p,$$
$$\dot{p} = -H_x = a^2 x.$$

The classical path between x_0 and x at time t satisfies

$$\ddot{x} = a^2 x, \qquad x(0) = x_0, \ x(t) = x, \tag{7.9.50}$$

with the solution

$$x(s) = \frac{x - x_0 \cosh(at)}{\sinh(at)} \sinh(as) + x_0 \cosh(as). \tag{7.9.51}$$

The Lagrangian associated with the above Hamiltonian is given by the Legendre transform

$$L(x, \dot{x}) = p\dot{x} - H = \frac{1}{2}\dot{x}^2 + \frac{1}{2}a^2 x^2. \tag{7.9.52}$$

Integrating the solution (7.9.51) along the Lagrangian (7.9.52) yields the classical action

$$S_{cl}(x_0, x, t) = \frac{a}{2\sinh(at)}\Big((x^2 + x_0^2)\cosh(at) - 2xx_0\Big). \tag{7.9.53}$$

The van Vleck determinant is

$$V(t) = \sqrt{\det\left(-\frac{1}{2\pi}\frac{\partial^2 S_{cl}}{\partial x \, \partial x_0}\right)} = \frac{1}{\sqrt{2\pi t}}\sqrt{\frac{at}{\sinh(at)}}.$$

Applying formula (7.9.42) yields the following heat kernel:

$$K(x_0, x, t) = \frac{1}{\sqrt{2\pi t}}\sqrt{\frac{at}{\sinh(at)}}\, e^{-\frac{1}{2t}\frac{at}{\sinh(at)}[(x^2+x_0^2)\cosh(at) - 2xx_0]},$$

where $t > 0$.

In spite of the fact that van Vleck's formula is a fast way of obtaining function-type heat kernels, it is not useful in the case when the potential is more than quadratic or when the operator is sub-elliptic, since in these cases the operators might not have function-type heat kernels.

7.9.3 The Feynman–Kac Formula

This famous formula was first used by Kac in order to compute heat kernels for Laplace operators with potential; see [75]. Let $\Delta_m = \frac{1}{2}\sum_{i=1}^m \partial_{x_i}^2$ be the Laplace operator on $\mathbb{R}^m$ and V be a smooth potential function. Let C_0^t denote the space of $\mathbb{R}^n$-valued functions $\phi(s)$ continuous on $[0,t]$ such that $\phi(0) = 0$. Then the following formula holds (see [73, 74]):

$$(e^{t(\Delta_m - V)} f)(x) = \int_{C_0^t} e^{-\int_0^t V\left(\phi(s)+x\right) ds} f\left(\phi(t)+x\right) d\mathfrak{m}(x), \tag{7.9.54}$$

where $\mathfrak{m}$ is the Wiener measure, which is the product of m one-dimensional Wiener measures. The right-hand side is a Wiener integral, which is computed as a path integral. Even if in the general case the integral cannot be computed explicitly, we shall work out just a few simple examples of heat kernels for which the Feynman–Kac formula can be successfully applied.

1. The operator $\frac{1}{2}\partial x^2$. In this case $V = 0$ and $m = 1$ and hence formula (7.9.54) becomes

$$\begin{aligned}
(e^{\frac{t}{2}\partial_x^2} f)(x) &= \int_{C_0^t} f\left(\phi(t)+x\right) d\mathfrak{m}(\phi) \\
&= \lim_{n\to\infty} \left(\frac{n}{2\pi t}\right)^{n/2} \int_{\mathbb{R}^n} e^{-\sum_{j=1}^n \frac{(u_{j-1}-u_j)^2}{2(t/n)}} f(u_n)\, du_n \cdots du_1 \\
&= \lim_{n\to\infty} \left(\frac{n}{2\pi t}\right)^{n/2} \int_{\mathbb{R}} \left\{ \int_{\mathbb{R}^{n-1}} e^{-\sum_{j=1}^n \frac{(u_{j-1}-u_j)^2}{2(t/n)}} du_{n-1} \cdots du_1 \right\} \\
&\quad \times f(u_n)\, du_n \\
&= \lim_{n\to\infty} \left(\frac{n}{2\pi t}\right)^{n/2} \int_{\mathbb{R}} \left(\frac{2\pi t}{n}\right)^{\frac{n-1}{2}} n^{-1/2} e^{-\frac{(u_n-u_0)^2}{2t}} f(u_n)\, du_n \\
&= \frac{1}{(2\pi t)^{1/2}} \int_{\mathbb{R}} e^{-\frac{(v-x)^2}{2t}} f(v)\, dv,
\end{aligned} \tag{7.9.55}$$

where we used that $x = u_0$. Making $f(v) = \delta_{x_0}(v)$ yields the familiar formula of the heat kernel

$$K(x_0, x; t) = (e^{\frac{t}{2}\partial_x^2}\delta_{x_0})(x) = \frac{1}{(2\pi t)^{1/2}} \int_{\mathbb{R}} e^{-\frac{(v-x)^2}{2t}} \delta_{x_0}\, dv = \frac{1}{(2\pi t)^{1/2}} e^{-\frac{(x-x_0)^2}{2t}}.$$

Equation (7.9.55) can be written in the equivalent form

$$\int_{C_0^t} f\big(\phi(t)+x\big)\, d\mathfrak{m}(\phi) = E^x\big(f(W_t+x)\big),$$

where E^x is the expectation operator and W_t is a Brownian motion starting at zero. This result holds in general for the right-hand side of (7.9.54):

$$\int_{C_0^t} e^{-\int_0^t V(X_s)\,ds} f\big(\phi(t)+x\big)\, d\mathfrak{m}(x) = E^x\Big[e^{-\int_0^t V\big(\phi(s)+x\big)\,ds} f\big(X_t\big)\Big],$$

where E^x is the expectation operator given that X_t is a Brownian motion starting at x. This shows that Feynman–Kac's formula can be represented either as a path integral or as an expectation operator; see also formula (8.23.89).

2. *The Hermite operator in m dimensions.* In the case $V(u) = \frac{a^2}{2}|u|^2$, formula (7.9.54) becomes

$$\begin{aligned}
(e^{t(\Delta_m - \frac{a^2}{2}|x|^2)} f)(x) &= \int_{C_0^t} e^{-\frac{a^2}{2}\int_0^t |\phi(s)+x|^2\,ds} f\big(\phi(t)+x\big)\, d\mathfrak{m}(\phi) \\
&= \lim_{n\to\infty} \Big(\frac{2\pi t}{n}\Big)^{-nm} \int_{\mathbb{R}^{nm}} e^{-\sum_{j=1}^n \left\{\frac{|u_{j-1}-u_j|^2}{2(t/n)} + \frac{t}{n}\frac{a^2}{2}|u_j|^2\right\}} \\
&\quad \times f(u_n)\, du, \\
&= \lim_{n\to\infty} \Big(\frac{2\pi t}{n}\Big)^{-nm} \int_{\mathbb{R}^{nm}} e^{-S_{n+1}} f(u_n)\, du_1 \cdots du_n,
\end{aligned}$$

where S_{n+1} has the meaning given in formula (7.8.38). The rest of the path integral computation follows Sect. 7.8, with $u_j \in \mathbb{R}^m$ and $x = u_0$. The result will be the n-dimensional analog of formula (7.8.40):

$$\Big(e^{t(\Delta_m - \frac{a^2}{2}|x|^2)} f\Big)(x) = \int K(v, x; t) f(v)\, dv,$$

with

$$K(x_0, x; t) = \frac{1}{(2\pi t)^{n/2}} \Big(\frac{at}{\sinh(at)}\Big)^{n/2} e^{-\frac{a}{2\sinh(at)}\left[(|x|^2+|x_0|^2)\cosh(at) - 2\langle x, x_0\rangle\right]},$$

$$t > 0. \tag{7.9.56}$$

An application of the above formula for computing the heat kernel for the Heisenberg operator can be found in [112].

7.10 Non-Commutativity of Sums of Squares

Let X and Y be two vector fields on $\mathbb{R}^m$. We denote by $X^n = \underbrace{X \cdots X}_{n\ times}$ the n-fold iterated composition of X. It is easy to note that if the vector fields X and Y commute, then the operators X^n and Y^n also commute; i.e.,

$$[X, Y] = 0 \Longrightarrow [X^n, Y^n] = 0, \quad \forall n \geq 1.$$

Since the case of concern is the sum of squares of vector fields, we shall limit ourselves to the case $n = 2$. Next we introduce a special class of noncommutative vector fields.

Definition 7.10.1. Two vector fields X and Y satisfy condition $\mathfrak{N}$ at the point p if $[X^2, Y^2]_p \neq 0$. If the above condition holds at each point, the distribution $\mathcal{D}$ spanned by the vector fields X and Y is called an $\mathfrak{N}$-distribution, and the vector fields X and Y are called $\mathfrak{N}$-fields.

Example 7.10.1. The vector fields $X = \partial_x$ and $Y = \partial_y + x\partial_z$ satisfy condition $\mathfrak{N}$ everywhere on $\mathbb{R}^3_{(x,y,z)}$. We have

$$\begin{aligned}
[X, Y] &= \partial_z \neq 0, \\
X^2 &= \partial_z^2, \\
Y^2 &= \partial_y^2 + 2x\partial_y\partial_z + x^2\partial_z^2 \\
[X^2, Y^2] &= 4\partial_x\partial_y\partial_z + 2\partial_z^2 + 4x\partial_x\partial_z^2 \neq 0.
\end{aligned}$$

The computation is left to the reader.

In the following we shall provide a sufficient condition for a distribution to be of class $\mathfrak{N}$. First we need a few definitions.

Definition 7.10.2. A distribution $\mathcal{D} = span\{X, Y\}$ is called nilpotent if there is an integer $k \geq 1$ such that all the Lie brackets of X and Y iterated k times vanish. The smallest integer k with this property is called the nilpotence class of $\mathcal{D}$.

For instance, the vector fields of Example 7.10.1 span a nilpotent distribution with nilpotence class $k = 2$:

$$[X, [X, Y]] = 0, \qquad [Y, [X, Y]] = 0, \qquad \text{but} \quad [X, Y] \neq 0.$$

The *nilpotence class* of a distribution is not the same thing as the *step* of the distribution. The nilpotence class describes the functional nature of the distribution (i.e., of polynomial, exponential type), while the step describes the non-holonomy of the distribution (i.e., the degree of non-integrability). More precisely, the step at the point p is the number of brackets plus 1 needed to span $\mathbb{R}^n$ at p. The nilpotence class and the step are the same thing only in the case when the distribution is generated by left invariant vector fields on nilpotent Lie groups (for instance, on the Heisenberg group).

Example 7.10.2. For example, the distribution spanned by the vector fields $X = \partial_x + e^y \partial_z$ and $Y = \partial_y$ on $\mathbb{R}^3$ is bracket generating with step 2 everywhere, but it is not nilpotent.

On the other side, the distribution spanned by the vector fields $X = \partial_x$ and $Y = \partial_y + zx\partial_z$ is nilpotent with the nilpotence class 2. However, this distribution is not bracket generating on the plane $\{z = 0\}$; i.e., it does not have a finite step there.

The following result can be found in [23]:

Theorem 7.10.3. *Any distribution $\mathcal{D} = span\{X, Y\}$ with nilpotence class 2, is an $\mathfrak{N}$-distribution.*

Proof. Since the nilpotence class is 2, we have

$$[X, [X, Y]] = 0, \qquad [Y, [Y, X]] = 0, \qquad [X, Y] \neq 0. \tag{7.10.57}$$

The first two relations of (7.10.57) become

$$X^2 Y + Y X^2 = 2XYX, \tag{7.10.58}$$
$$Y^2 X + X Y^2 = 2YXY. \tag{7.10.59}$$

Multiplying on the right of (7.10.58) by Y and of (7.10.59) by X yields

$$\begin{aligned} X^2Y^2 + YX^2Y &= 2(XY)^2, \\ Y^2X^2 + XY^2X &= 2(YX)^2, \end{aligned}$$

and subtracting, we have

$$\begin{aligned} [X^2, Y^2] = X^2Y^2 - Y^2X^2 &= \left(2(XY)^2 - YX^2Y\right) - \left(2(YX)^2 - XY^2X\right) \\ &= 2\{(XY)^2 - (YX)^2\} + (XY^2X - YX^2Y). \end{aligned} \tag{7.10.60}$$

Multiplying on the left of (7.10.58) by Y and of (7.10.59) by X yields

$$\begin{aligned} YX^2Y + Y^2X^2 &= 2(YX)^2, \\ XY^2X + X^2Y^2 &= 2(XY)^2, \end{aligned}$$

and subtracting, we get

$$\begin{aligned} [X^2, Y^2] = X^2Y^2 - Y^2X^2 &= \left(2(XY)^2 - XY^2X\right) - \left(2(YX)^2 - YX^2Y\right) \\ &= 2\{(XY)^2 - (YX)^2\} + (YX^2Y - XY^2X). \end{aligned} \tag{7.10.61}$$

Comparing (7.10.60) and (7.10.61) yields

$$\begin{aligned} XY^2X - YX^2Y = YX^2Y - XY^2X \Longleftrightarrow \\ XY^2X = YX^2Y, \end{aligned} \tag{7.10.62}$$

and hence relations (7.10.60) and (7.10.61) become

$$[X^2, Y^2] = 2\{(XY)^2 - (YX)^2\}. \tag{7.10.63}$$

If we let $A = XY$ and $B = YX$, using (7.10.62), the operators A and B commute,

$$AB = (XY)(YX) = XY^2X = YX^2Y = (YX)(XY) = BA,$$

so $A^2 - B^2 = (A - B)(A + B)$.

Then (7.10.63) becomes

$$[X^2, Y^2] = 2(XY - YX)(XY + YX). \tag{7.10.64}$$

We need to show that X and Y are $\mathfrak{N}$-fields; i.e., if $[X, Y] \neq 0$, then $[X^2, Y^2] \neq 0$. By contradiction, assume $[X^2, Y^2] = 0$. Then by (7.10.64), we have

either: $XY - YX = 0$, i.e., $[X, Y] = 0$, which is a contradiction.
or: $XY + YX = 0$, i.e.,

$$XY = -YX. \tag{7.10.65}$$

The remainder of the proof deals with showing that (7.10.65) cannot hold. By contradiction, we assume that (7.10.65) holds. Then

$$[X, Y] = XY - YX = 2XY = -2YX.$$

Therefore, we have

$$[X, [X, Y]] = 0 \Longrightarrow [X, XY] = 0 \Longrightarrow X^2Y = XYX, \tag{7.10.66}$$
$$[Y, [Y, X]] = 0 \Longrightarrow [Y, YX] = 0 \Longrightarrow Y^2X = YXY. \tag{7.10.67}$$

Using (7.10.66) and (7.10.67), we have

$$\begin{aligned} X[X, Y] &= X(XY - YX) = X^2Y - XYX = 0, \\ Y[X, Y] &= Y(XY - YX) = YXY - Y^2X = 0. \end{aligned}$$

Combining the last two relations, we obtain

$$\begin{aligned} [X, Y]^2 &= [X, Y][X, Y] = (XY - YX)[X, Y] \\ &= X(Y[X, Y]) - Y(X, [X, Y]) \\ &= 0. \end{aligned}$$

Hence $[X, Y] = 0$, which is a contradiction. It turns out that (7.10.65) cannot hold.

It follows that X and Y are $\mathfrak{N}$-fields. ■

Next we present an example of a distribution with the nilpotence class 2.

Example 7.10.3 (Heisenberg-type distributions). Consider the vector fields X, Y, T on $\mathbb{R}^3$. If

$$[X,Y] = T, \quad [X,T] = 0, \quad [Y,T] = 0,$$

we say that $\mathcal{D} = span\{X, Y\}$ is a Heisenberg-type distribution. Since

$$[[X,Y],X] = [T,X] = 0, \quad [[X,Y],Y] = [T,Y] = 0,$$

it follows that $\mathcal{D}$ is a nilpotent distribution with the nilpotence class 2.

One of the classical examples of vector fields with the above properties is

$$X = \partial_x + 2y\partial_z, \quad Y = \partial_y - 2x\partial_z, \quad T = -4\partial_z.$$

It is left as an exercise for the reader to show that X^2 and Y^2 do not commute.

7.11 Path Integrals and Sub-Elliptic Operators

In the following we shall assume that X_1 and X_2 are $\mathfrak{N}$-vector fields on $\mathbb{R}^3$; i.e., $[X_1, X_2] \neq 0$ and $[X_1^2, X_2^2] \neq 0$. We also assume that the distribution $\mathcal{D} = span\{X_1, X_2\}$ is step 2; i.e., $\{X_1, X_2, [X_1, X_2]\}$ span $\mathbb{R}^3$ at each point.

In this section we shall describe the heat kernel of the sub-elliptic operator $\Delta_X = \frac{1}{2}(X_1^2 + X_2^2)$. We are led to the study of the heat semigroup $e^{t\Delta_X}$, with $t > 0$. Since the operators X_1^2 and X_2^2 do not commute, we need to invoke the Trotter formula. Hence we have

$$K(t) = e^{t\Delta_X} = e^{\frac{1}{2}tX_1^2 + \frac{1}{2}tX_2^2} = \lim_{n\to\infty} \left(e^{\frac{t}{2n}X_1^2} e^{\frac{t}{2n}X_2^2} \right)^n = \lim_{n\to\infty} K_n^n, \quad (7.11.68)$$

where

$$K_n = e^{\frac{t}{2n}X_1^2} e^{\frac{t}{2n}X_2^2} = K_n^1 K_n^2,$$

with $K_n^1 = e^{\frac{t}{2n}X_1^2}$, $K_n^2 = e^{\frac{t}{2n}X_2^2}$. Let $\widehat{K_n}$, $\widehat{K_n^1}$, $\widehat{K_n^2}$ be the integral kernels of the operators K_n, K_n^1, K_n^2, respectively. By Proposition 7.2.2, part (1), we have

$$\widehat{K_n}(x,y) = \int_{\mathbb{R}^3} \widehat{K_n^1}(x,u)\widehat{K_n^2}(u,y)\,du. \quad (7.11.69)$$

Hence in order to compute $\widehat{K_n}$, it suffices to know $\widehat{K_n^1}$ and $\widehat{K_n^2}$. Since the computations of the above integral kernels are virtually identical, it suffices to do it only for the first one.

$\widehat{K_n^1}$ is the integral kernel for the operator $\partial_t - \frac{t}{2n}X_1^2$. If $X_1 = \sum_{i=1}^3 a_i \partial_{x_i}$, then

$$X_1^2 = \sum_{i,j=1}^3 a_i a_j \partial_{x_i}\partial_{x_j} + \sum_{i,j=1}^3 a_i(\partial_{x_i} a_j)\partial_{x_j}$$

is a second-order operator with a linear term. Using Theorem 7.3.1 with $\mathbb{L} = \frac{t}{2n}X_1^2$ yields a path integral form for the kernel

$$\widehat{K_n^1}(x, y; t) = \int_{\mathfrak{P}_{x,y;t}} e^{-S(\phi,t)}\, d\mathfrak{m}(\phi), \tag{7.11.70}$$

where S satisfies the Hamilton–Jacobi equation (2.3.17).

Knowing $\widehat{K_n^1}$ and $\widehat{K_n^2}$ yields $\widehat{K_n}$ by formula (7.11.69). Applying (7.11.68) yields the integral kernel for $\partial_t - \Delta_X$:

$$\begin{aligned}
&\widehat{K(t)}(x, y) \\
&\quad = \lim_{n\to\infty} \widehat{K_n^n} = \lim_{n\to\infty} \int\cdots\int \widehat{K_n}(x, u_1)\widehat{K_n}(u_1, u_2)\cdots\widehat{K_n}(u_{n-1}, y)\, du_{n-1}\cdots du_1 \\
&\quad = \lim_{n\to\infty} \int_{\mathbb{R}^{6n}} \widehat{K_n^1}(x, v_1)\widehat{K_n^2}(v_1, u_1)\widehat{K_n^1}(u_1, v_2)\widehat{K_n^2}(v_2, u_2)\cdots\widehat{K_n^1}(u_{n-1}, v_n) \\
&\qquad \times \widehat{K_n^2}(v_n, y)\, dudv,
\end{aligned}$$

with $dudv = du_{n-1}dv_{n-1}\cdots du_1 dv_1$, and $\widehat{K_n^j}$ given by (7.11.70).

Even if the above computation is always theoretically possible, the computation complexity might make the integral almost impossible to evaluate, even in the case of relative simple vector fields such as $X_1 = \partial_{x_1}$ and $X_2 = x_1\partial_{x_2}$. The conclusion is that the method of path integrals is not well suited to this type of operator.

Chapter 8
The Stochastic Analysis Method

This chapter deals with probabilistic methods of obtaining the heat kernel. The main idea of this subject is that the heat kernel can be represented as a transition density of an associated stochastic process, as pointed out by Kolmogorov [80] in the early 1930s. The probabilistic methods were also useful in obtaining the heat kernel of the Heisenberg Laplacian, as shown by Hulanicki [68] and Gaveau [49] in the late 1970s.

8.1 Elements of Stochastic Processes

A *stochastic process* describes a random phenomenon that changes in time. For instance, the velocity of a small particle in a liquid, the price of a stock and the interest rates for bonds are random phenomena depending on time.

A stochastic process is a family of random variables X_t depending smoothly on a parameter $t \in T$. In our case, the parameter set T is usually the half-line $[0, \infty)$. For each t, the random variable X_t is defined on a *sample space* Ω, where we have a *Borel field* $\mathcal{F}$ of subsets of Ω, called *events*, and a *probability measure* P on $\mathcal{F}$. We shall assume the reader is already familiar with the basic notions of the *probability space* $(\Omega, \mathcal{F}, P)$ the and random variables defined on it.

For each *state of the world* $\omega \in \Omega$, the function $t \to X_t(\omega)$ is called a *realization* or *sample path* of the stochastic process. In Fig. 8.1 we have represented a few realizations of a process X_t.

Stochastic processes are also called *random processes* or *random functions*. For convenience, a stochastic process will be denoted by X_t or $X(t)$. The process can be real-valued or vectorial. In the latter case $X_t = (X_1(t), \dots, X_n(t)) \in \mathbb{R}^n$, for $t \in T$.

The *joint distribution function* of a real-valued stochastic process X_t is defined by

$$F(x_1, x_2, \dots, x_n; t_1, t_2, \dots, t_n) = P\big(X_{t_1} \le x_1, X_{t_2} \le x_2, \dots, X_{t_n} \le x_n\big),$$

O. Calin et al., *Heat Kernels for Elliptic and Sub-elliptic Operators*,
Applied and Numerical Harmonic Analysis, DOI 10.1007/978-0-8176-4995-1_8,

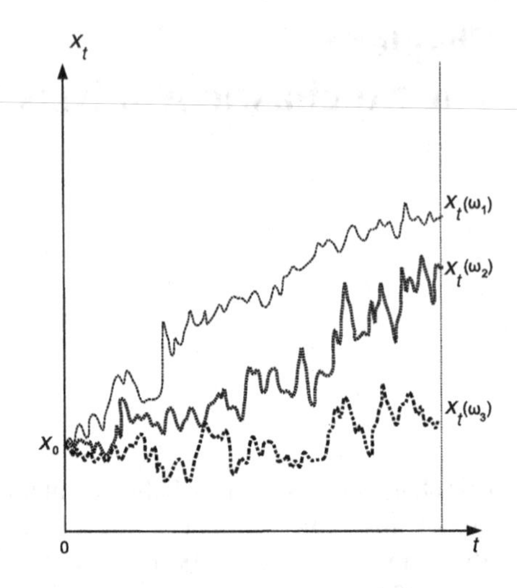

Fig. 8.1 Three realizations of the process X_t

for $t_1, t_2, \ldots, t_n \in T$ and $n \geq 1$ integer. The previous function satisfies the following two conditions:

(1) *Symmetry:* For any permutation $(i_1, i_2, \ldots, i_n)$ of $(1, 2, \ldots, n)$ we have

$$F(x_{i_1}, x_{i_2}, \ldots, x_{i_n}; t_{i_1}, t_{i_2}, \ldots, t_{i_n}) = F(x_1, x_2, \ldots, x_n; t_1, t_2, \ldots, t_n).$$

(2) *Compatibility:* For any $k < n$, we have

$$\begin{aligned} &F(x_1, x_2, \ldots, x_k, \infty, \ldots, \infty; t_1, t_2, \ldots, t_k, t_{k+1}, \ldots, t_n) \\ &\quad = F(x_1, x_2, \ldots, x_k; t_1, t_2, \ldots, t_k). \end{aligned}$$

In 1931 Kolmogorov proved that given any system of distribution functions F satisfying the previous conditions (1) and (2), there are a probability space $(\Omega, \mathcal{F}, P)$ and a stochastic process $X_t(\omega)$ such that $F(x_1, x_2, \ldots, x_n; t_1, t_2, \ldots, t_n)$ gives the joint distribution of the random variables

$$X_{t_1}(\omega), X_{t_2}(\omega), \ldots, X_{t_n}(\omega).$$

For a proof of Kolmogorov's extension theorem, the reader can consult reference [85]. Using Kolmogorov's result, one may define a stochastic process given the distribution function. We shall consider an important example below.

The Gaussian process. This is a process X_t such that for any $t_1, \ldots, t_n \in T$, the random variables

$$X_{t_1}, X_{t_2}, \ldots, X_{t_n}$$

have an n-variate Gaussian distribution[1] with

$$E(X_{t_j}) = m(t_j),$$
$$\text{Cov}(X_{t_j}, X_{t_k}) = E\big([X_{t_j} - m(t_j)]\,[X_{t_k} - m(t_k)]\big) = \Gamma(t_j, t_k) = \Gamma_{jk}.$$

Let $\Gamma = (\Gamma_{jk})$ denote the covariance matrix, which is assumed nonsingular. The joint distribution function in this case is

$$F(x_1, x_2, \ldots, x_n; t_1, t_2, \ldots, t_n)$$
$$= \int_{-\infty}^{x_1} \cdots \int_{-\infty}^{x_n} \frac{|G(u)|^{1/2}}{(2\pi)^{n/2}} \exp\left\{-\frac{1}{2}\sum_{i,j=1}^{n} G_{ij}(u)\big(u_i - m(t_i)\big)\big(u_j - m(t_j)\big)\right\} du_n \cdots du_1,$$

where $(G_{ij}) = G = \Gamma^{-1}$.

Sometimes the following equivalent characterization of Gaussian processes is useful in practice.

Proposition 8.1.1. *The process X_t is Gaussian if and only if the linear combination $\sum_{i=1}^{n} a_i X_{t_i}$ is normally distributed for all $n \geq 1$, all coefficients $a_i \in \mathbb{R}$, and all times $t_i \in T$.*

The stationary process. A stochastic process X_t is called *stationary in the strict sense* if

$$F(x_1, x_2, \ldots, x_n; t_1 + \tau, t_2 + \tau, \ldots, t_n + \tau) = F(x_1, x_2, \ldots, x_n; t_1, t_2, \ldots, t_n),$$

for any $t_1, \ldots, t_n, \tau \in T$. For these kinds of processes,

$$P(X(\tau) \leq x) = F(x, \tau) = F(x, 0) = F(x),$$

and the mean and the variance are constants:

$$E(X_t) = m, \qquad \text{Var}(X_t) = \sigma^2. \tag{8.1.1}$$

[1] This means that $X_{t_1}, X_{t_2}, \ldots, X_{t_n}$ have a joint normal distribution.

Furthermore, the covariance function depends only on the difference $|t-s|$:

$$\Gamma(s,t) = \int_{\mathbb{R}^2} xy\, dF(x,y;s,t) = \int_{\mathbb{R}^2} xy\, dF(x,y;0,t-s) \qquad (8.1.2)$$
$$= \Gamma(0,t-s).$$

A process for which relations (8.1.1)–(8.1.2) hold is said to be *stationary in the wide sense*.

Example. Let $X_t = A_1 \cos t + A_2 \sin t$, with A_1, A_2 uncorrelated random variables, having mean 0 and variance σ^2. Standard properties of random variables show that

$$\begin{aligned} E(X_t) &= E(A_1)\cos t + E(A_2)\sin t = 0, \\ \mathrm{Var}(X_t) &= \sigma^2 \cos^2 t + \sigma^2 \sin^2 t = \sigma^2, \\ \Gamma(s,t) &= E(X_s X_t) = \sigma^2(\cos s \cos t + \sin s \sin t) \\ &= \sigma^2 \cos(t-s). \end{aligned}$$

Hence X_t is a stationary process in the wide sense.

The stationary Gaussian process. Another important process is the Gaussian process with constant mean $m(t) = m$ and covariance function $\Gamma(t_j, t_k) = \phi(t_j - t_k)$. This process is stationary in the wide sense. Since a Gaussian distribution is completely determined by its mean and variance, it follows that the process is also stationary in the strict sense.

Independent increments process. This is a stochastic process X_t such that for any $t_1 < t_2 < \ldots < t_n$, the random variables

$$X_{t_2} - X_{t_1},\ X_{t_3} - X_{t_2}, \ldots, X_{t_n} - X_{t_{n-1}}$$

are mutually independent. The process is said to be *stationary* if the increment $X_{t_j} - X_{t_{j-1}}$ has a distribution depending on the difference $t_j - t_{j-1}$.

If the process X_t has independent increments and is stationary such that $X_t - X_s$ has a Gaussian distribution with mean

$$E(X_t - X_s) = 0$$

and variance

$$\mathrm{Var}(X_t - X_s) = \sigma^2 |t-s|,$$

then X_t is called a *Wiener–Einstein process.*

Brownian motion. A stochastic process W_t is called a *Brownian motion* if it satisfies the following properties:

(1) The process starts at zero: $W_0 = 0$
(2) For $s < t$, the random variable $W_t - W_s$ is normally distributed with mean 0 and variance $t - s$

(3) W_t has independent increments
(4) The sample paths $t \to W_t(\omega)$ are continuous

A Brownian motion is a stationary Gaussian process with independent increments. Its distribution function is given by

$$F(x,t) = \int_{-\infty}^{x} \frac{1}{\sqrt{2\pi t}} e^{-u^2/(2t)} \, du,$$

and

$$P\Big(\alpha \le W_t - W_s \le \beta\Big) = \frac{1}{\sqrt{2\pi(t-s)}} \int_{\alpha}^{\beta} e^{-x^2/2(t-s)} \, dx,$$

for $0 \le s < t$ and $\alpha < \beta$.

Since the process starts at zero, we note that W_t is normally distributed with mean 0 and variance t, see Fig. 8.2. An important fact is that the covariance $\mathrm{Cov}(W_t, W_s) = \min\{t, s\}$. This can be shown by using conditions (2) and (3). Assuming $s < t$, we have

$$\begin{aligned}\mathrm{Cov}(W_t, W_s) &= E(W_s W_t) = E\big((W_s - W_0)(W_t - W_s) + W_s^2\big)\\ &= E(W_s - W_0)E(W_t - W_s) + E(W_s^2)\\ &= 0 + \mathrm{Var}(W_s) = s = \min\{t, s\}.\end{aligned}$$

The infinitesimal notation. If $0 = t_0 < t_1 < \cdots < t_n$, then the increment $W_{t_j} - W_{t_{j-1}}$ is a random variable normally distributed with zero mean and variance $t_j - t_{j-1}$, and can be denoted by $(\Delta W)_{t_j}$. In the case when the difference $dt = t_j - t_{j-1}$ is infinitesimal, the random variable $(\Delta W)_{t_j}$ will be denoted by dW_t. The random variable dW_t has zero mean and variance dt. It is a well-known result of stochastic analysis that dW_t has the magnitude of $\sqrt{dt}$; more precisely, the random variable dW_t^2 is completely predictable and we have $dW_t^2 = dt$ in the mean sense.

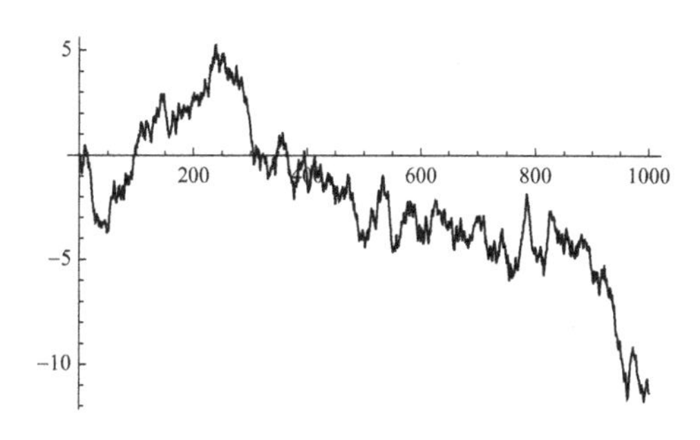

Fig. 8.2 A realization of a Brownian motion

Markov processes. Brownian motion belongs to a larger class of "memoryless" processes, called *Markov processes*. A stochastic process X_t is said to be a Markov process if

$$P(X_t \mid X_{t_1}, X_{t_2}, \dots, X_{t_k}) = P(X_t \mid X_{t_k}),$$

for any ascending sequence $t_1 < t_2 < \cdots < t_k < t$. This means that the probability that the particle has position X_t at time t, given that its position is X_{t_i} at time t_i, for $i = 1, \dots, k$, depends only on the last position X_{t_k} of the particle. If X_t is a real-valued process, consider the function

$$F(x_0, t_0; x, t) = P(X_t \le x \mid X_{t_0} = x_0),$$

called the *transition distribution function* of the process X_t. This function can be interpreted as the probability that the particle will be in a state included in $(-\infty, x]$, given that the particle starts at the point x_0 at time t_0. Using the properties of the probability, we can check that

$$F(x_0, t_0; x, t) \ge 0, \qquad F(x_0, t_0; -\infty, t) = 0, \qquad F(x_0, t_0; \infty, t) = 1.$$

Furthermore, the transition distribution function satisfies the *Chapman–Kolmogorov equation*

$$F(x_0, t_0; x, t) = \int_{-\infty}^{\infty} F(y, s; x, t)\, d_y F(x_0, t_0; y, s), \qquad t_0 < s < t. \tag{8.1.3}$$

The derivative of the transition distribution function

$$f(x_0, t_0; x, t) = \frac{\partial}{\partial x} F(x_0, t_0; x, t) \tag{8.1.4}$$

is called *the transition density function* of X_t. It satisfies the *Chapman–Kolmogorov equation*

$$f(x_0, t_0; x, t) = \int_{-\infty}^{\infty} f(y, s; x, t)\, f(x_0, t_0; y, s)\, dy, \qquad t_0 < s < t. \tag{8.1.5}$$

The transition density function will play an important role in finding the heat kernel of certain second-order partial differential operators which generate stochastic processes. In the case $t_0 = 0$, we shall denote $p_t(x_0, x) = f(x_0, 0; x, t)$. This makes sense in the case when the transition distribution function $F(x_0, t_0; x, t)$ depends on the difference $t - t_0$. These kinds of processes are called *homogeneous*. In this case

$$F(x_0; x, t) = P(X_t \le x \mid X_0 = x_0),$$

and $p_t(x_0, x) = f(x_0; x, t) = \frac{\partial}{\partial x} F(x_0; x, t)$.

Stochastic differential equations. In the case of ordinary differential equations we deal with equations of the form

$$\frac{dX_t}{dt} = aX_t + b\frac{dY_t}{dt}, \qquad t \geq 0,$$

which can also be written in the differential form

$$dX_t = aX_t\,dt + b\,dY_t,$$

where Y_t is a given function. In the case of stochastic differential equations the continuous process $t \to X_t$ is not differentiable, so one may not write dX_t/dt. The stochastic processes of concern in this chapter have the form

$$dX_t = a(t, X_t)dt + b(t, X_t)\,dW_t. \tag{8.1.6}$$

This can be interpreted as the infinitesimal notation for the finite difference equation

$$X_{t+h} - X_t = a(t, X_t)h + b(t, X_t)\,(W_{t+h} - W_t), \qquad h \searrow 0.$$

Another way of looking at (8.1.6) is by using integrals:

$$\int_0^t dX_s = \int_0^t a(s, X_s)\,ds + \int_0^t b(s, X_s)\,dW_s,$$

where the last term on the right-hand side is an Ito integral. More details about stochastic differential equations and stochastic integrals can be found in references [95] and [83].

Ito's formula. A process of the type (8.1.6) is called an *Ito process.* Consider the process $F_t = F(t, X_t)$ derived[2] from X_t, with F twice differentiable in both arguments. Ito's formula provides the expression of dF_t, which looks like a usual Taylor series up to the second derivative in X_t:

$$dF_t = \frac{\partial F}{\partial X_t}\,dX_t + \frac{\partial F}{\partial t}\,dt + \frac{1}{2}\frac{\partial^2 F}{\partial X_t^2}\,(dX_t)^2.$$

Using the stochastic relations

$$(dW_t)^2 = dt, \qquad dt\,dW_t = dW_t\,dt = (dt)^2 = 0,$$

[2] If X_t is the stock price at time t, a financial instrument with the price $F_t = F(t, X_t)$ derived from X_t is called a *financial derivative.*

the aforementioned formula becomes the *Ito formula*

$$dF_t = \left[\frac{\partial F}{\partial X_t} a(t, X_t) + \frac{\partial F}{\partial t} + \frac{1}{2}\frac{\partial^2 F}{\partial X_t^2} b(t, X_t)^2 \right] dt + \frac{\partial F}{\partial X_t} b(t, X_t)\, dW_t. \tag{8.1.7}$$

For instance, if $F_t = F(t, W_t) = W_t^2$ is the square of the one-dimensional Brownian motion, then

$$dF_t = 2W_t\, dW_t + \frac{1}{2}[2(dW_t)^2] = 2W_t\, dW_t + dt.$$

There is also a multidimensional version of Ito's formula. In this case $W_1(t), \dots, W_m(t)$ denote m independent Brownian motions and consider n Ito processes

$$\begin{aligned} dX_j(t) &= a_j(t, X_t)\, dt + b_{j1}(t, X_t)\, dW_1(t) + \cdots + b_{jm}(t, X_t)\, dW_m(t), \\ j &= 1, \dots, n. \end{aligned} \tag{8.1.8}$$

We have the following result.

Lemma 8.1.2 (Ito). *Let $F(t,x) = \big(F_1(t,x), \dots, F_p(t,x)\big)$ be a C^2 map from $[0,\infty) \times \mathbb{R}^n$ into $\mathbb{R}^p$, where $X_t = \big(X_1(t), \dots, X_n(t)\big)$ is given by (8.1.8). Then the process $Y_t = F(t, X_t)$ is also an Ito process satisfying*

$$dY_t = \sum_k \frac{\partial F}{\partial x_k}(t, X)\, dX_k(t) + \frac{\partial F}{\partial t}(t, X)\, dt + \frac{1}{2}\sum_{i,j} \frac{\partial^2 F}{\partial x_i \partial x_j}(t, X)\, dX_i(t)\, dX_j(t),$$

with the conventions $dW_i(t)\, dW_j(t) = \delta_{ij}\, dt$, $dW_i(t)\, dt = dt\, dW_i(t) = 0$.

8.2 Ito Diffusion

This section deals with a particular type of Ito process that is related to our study of heat kernels.

Definition 8.2.1. A time-homogeneous Ito diffusion is an n-dimensional stochastic process

$$dX_t = b(X_t)dt + \sigma(X_t)\, dW(t), \qquad t \geq 0,\ X_0 = x, \tag{8.2.9}$$

where $W(t) = \big(W_1(t), \dots W_m(t)\big)$ is an m-dimensional Brownian motion and $b : \mathbb{R}^n \to \mathbb{R}^n$ and $\sigma : \mathbb{R}^n \to \mathbb{R}^{n \times m}$ are measurable functions.

In general, the stochastic differential equation (8.2.9) does not have a unique solution. For instance, the one-dimensional equation

$$dX_t = 3X_t^{1/3}dt + 3X_t^{2/3}dW_t, \quad X_0 = 0, \tag{8.2.10}$$

has infinitely many solutions. If for any $a > 0$ we define

$$\varphi_a(x) = \begin{cases} (x-a)^3, & x \geq a, \\ 0, & x < a, \end{cases}$$

then a straightforward computation provides the derivatives $\varphi_a'(x) = 3\varphi_a^{2/3}(x)$ and $\varphi_a''(x) = 6\varphi_a^{1/3}(x)$. Denoting $F_a(t) = \varphi_a\big(W_t\big)$, an application of Ito's formula yields

$$\begin{aligned} dF_t &= \varphi_a'(W_t)dW_t + \frac{1}{2}\varphi_a''(W_t)dt, \\ &= 3F_a(t)^{2/3}dW_t + 3F_a(t)^{1/3}dt. \end{aligned}$$

Moreover, $F_a(0) = \varphi_a(W_0) = \varphi_a(0) = 0$. Hence the aforementioned equation has a solution $F_a(t) = \varphi_a(W_t)$ for each $a > 0$.

Sometimes the solution of (8.2.9) might explode in finite time. An application of Ito's formula shows that $X_t = 1/(1 - W_t)$ is a solution for the following stochastic differential equation:

$$dX_t = X_t^3 dt + X_t^2\, dW_t, \qquad t \geq 0,\ X_0 = 1, \tag{8.2.11}$$

which blows up when the Brownian motion W_t reaches 1.

As in the theory of ordinary differential equations, in order to ensure the existence and uniqueness of the solution of the stochastic differential equation (8.2.9), we need to require an additional Lipschitz condition on the functions $b(x)$ and $\sigma(x)$. Let

$$|\sigma|^2 = \sum \sigma_{ij}^2$$

denote the norm of the matrix function σ. The following result can be found, for instance, in reference [95], Theorem 5.2.1.

Theorem 8.2.2. *If there is a constant $K > 0$ such that*

$$|b(x) - b(y)| + |\sigma(x) - \sigma(y)| \leq K|x - y|, \qquad \forall x, y \in \mathbb{R}^n, \tag{8.2.12}$$

then there is a unique stochastic process $(X_t)_{t \geq 0}$ satisfying the stochastic differential equation (8.2.9).

Equation (8.2.10) does not satisfy condition (8.2.12). To show this, let's assume there is a positive constant K such that

$$|x^{1/3} - y^{1/3}| + |x^{2/3} - y^{2/3}| \le K|x - y| \qquad (8.2.13)$$

for all $x, y \in \mathbb{R}$. Let $u = x^{1/3}$ and $v = y^{1/3}$. Then (8.2.13) becomes

$$|u - v| + |u^2 - v^2| \le K|u^3 - v^3|, \qquad \forall u, v \in \mathbb{R}.$$

Assuming $u \neq v$ and dividing by $|u - v|$, the preceding equation becomes

$$1 + |u + v| < K|u^2 + uv + v^2|.$$

This implies $1 < K|u^2 + uv + v^2|$ or

$$\frac{1}{K} < |u^2 + uv + v^2|, \qquad u, v \in \mathbb{R},\ u \neq v. \qquad (8.2.14)$$

If $\epsilon > 0$ and we let $|u| < \epsilon$, $|v| < \epsilon$, then the right term is bounded by

$$|u^2 + uv + v^2| \le |u|^2 + |u|\,|v| + |v|^2 < 3\epsilon^2.$$

This contradicts the inequality of (8.2.14) if we choose $\epsilon < 1/\sqrt{3K}$. Hence the inequality (8.2.14) does not hold for x and y small enough, with $x \neq y$. The Lipschitz condition (8.2.12) does not hold for the stochastic equation (8.2.10), and hence the uniqueness result given by Theorem 8.2.2 does not hold.

8.3 The Generator of an Ito Diffusion

Let $(X_t)_{t\ge 0}$ be a stochastic process with $X_0 = x_0$. In this section we shall deal with the operator associated with X_t.

The *generator* of the stochastic process X_t is the second-order partial differential operator A defined by

$$Af(x) = \lim_{t \searrow 0} \frac{E[f(X_t)] - f(x)}{t},$$

for any smooth function (at least of class C^2) with compact support $f : \mathbb{R}^n \to \mathbb{R}$. Here E stands for the expectation operator taken at $t = 0$; i.e.,

$$E[f(X_t)] = \int_{\mathbb{R}^n} f(y) p_t(x, y)\, dy,$$

where $p_t(x, y)$ is the transition density of X_t.

In the following we shall find the generator associated with the Ito diffusion (8.2.9). The main tool used in deriving the formula for A is Ito's formula in several variables. If we let $F_t = f(X_t)$, then using Ito's formula, we have

$$dF_t = \sum_i \frac{\partial f}{\partial x_i}(X_t)\, dX_t^i + \frac{1}{2}\sum_{i,j} \frac{\partial^2 f}{\partial x_i \partial x_j}(X_t)\, dX_t^i\, dX_t^j, \tag{8.3.15}$$

where $X_t = (X_t^1, \ldots, X_t^n)$ satisfies the Ito diffusion (8.2.9) on components; i.e.,

$$\begin{aligned} dX_t^i &= b_i(X_t)dt + [\sigma(X_t)\, dW(t)]_i \\ &= b_i(X_t)dt + \sum_k \sigma_{ik}\, dW_k(t). \end{aligned} \tag{8.3.16}$$

Using the stochastic relations $dt^2 = dt\, dW_k(t) = 0$ and $dW_k(t)\, dW_r(t) = \delta_{kr} dt$, a computation provides

$$\begin{aligned} dX_t^i\, dX_t^j &= \Big(b_i dt + \sum_k \sigma_{ik} dW_k(t)\Big)\Big(b_j dt + \sum_k \sigma_{jk} dW_k(t)\Big) \\ &= \Big(\sum_k \sigma_{ik} dW_k(t)\Big)\Big(\sum_r \sigma_{jr} dW_r(t)\Big) \\ &= \sum_{k,r} \sigma_{ik}\sigma_{jr}\, dW_k(t) dW_r(t) = \sum_k \sigma_{ik}\sigma_{jk}\, dt \\ &= (\sigma\sigma^T)_{ij}\, dt. \end{aligned}$$

Therefore,

$$dX_t^i\, dX_t^j = (\sigma\sigma^T)_{ij}\, dt. \tag{8.3.17}$$

Substituting (8.3.16) and (8.3.17) into (8.3.15) yields

$$\begin{aligned} dF_t &= \left[\frac{1}{2}\sum_{i,j} \frac{\partial^2 f}{\partial x_i \partial x_j}(X_t)(\sigma\sigma^T)_{ij} + \sum_i b_i(X_t)\frac{\partial f}{\partial x_i}(X_t)\right] dt \\ &\quad + \sum_{i,k} \frac{\partial f}{\partial x_i}(X_t)\sigma_{ik}(X_t)\, dW_k(t). \end{aligned}$$

integrating, we obtain

$$\begin{aligned} F_t = F_0 &+ \int_0^t \left[\frac{1}{2}\sum_{i,j} \frac{\partial^2 f}{\partial x_i \partial x_j}(\sigma\sigma^T)_{ij} + \sum_i b_i \frac{\partial f}{\partial x_i}\right](X_s)\, ds \\ &+ \sum_k \int_0^t \sum_i \sigma_{ik} \frac{\partial f}{\partial x_i}(X_s)\, dW_k(s). \end{aligned}$$

Since $F_0 = f(X_0) = f(x)$ and $E(f(x)) = f(x)$, applying the ,expectation operator in the previous relation, we obtain

$$E(F_t) = f(x) + E\left[\int_0^t \left(\frac{1}{2}\sum_{i,j}(\sigma\sigma^T)_{ij}\frac{\partial^2 f}{\partial x_i \partial x_j} + \sum_i b_i \frac{\partial f}{\partial x_i}\right)(X_s)\, ds\right]. \tag{8.3.18}$$

Using the commutation between the operator E and the integral $\int_0^t$ yields

$$\lim_{t\searrow 0} \frac{E(F_t) - f(x)}{t} = \frac{1}{2}\sum_{i,j}(\sigma\sigma^T)_{ij}\frac{\partial^2 f(x)}{\partial x_i \partial x_j} + \sum_k b_k \frac{\partial f(x)}{\partial x_k}.$$

We conclude the previous computations with the following result.

Theorem 8.3.1. *The generator of the Ito diffusion (8.2.9) is given by*

$$A = \frac{1}{2}\sum_{i,j}(\sigma\sigma^T)_{ij}\frac{\partial^2}{\partial x_i \partial x_j} + \sum_k b_k \frac{\partial}{\partial x_k}. \tag{8.3.19}$$

Substituting (8.3.19) in (8.3.18), we obtain *Dynkin's formula:*

$$E[f(X_t)] = f(x) + E\left[\int_0^t Af(X_s)\, ds\right], \tag{8.3.20}$$

for any $f \in C_0^2(\mathbb{R}^n)$.

8.4 Kolmogorov's Backward Equation and Heat Kernel

For any function $f \in C_0^2(\mathbb{R}^n)$, let $v(t,x) = E[f(X_t)]$, given that $X_0 = x$. As usual, E denotes the expectation at time $t = 0$. Then $v(0,x) = f(x)$, and differentiating in Dynkin's formula (8.3.20),

$$v(t,x) = f(x) + \int_0^t E[Af(X_s)]\, ds,$$

yields

$$\frac{\partial v}{\partial t} = E[Af(X_t)] = AE[f(X_t)] = Av(t,x).$$

We have arrived at the following result.

Theorem 8.4.1 (Kolmogorov's backward equation). *For any $f \in C_0^2(\mathbb{R}^n)$, the function $v(t,x) = E[f(X_t)]$ satisfies the following Cauchy problem:*

$$\frac{\partial v}{\partial t} = Av, \qquad t > 0,$$
$$v(0,x) = f(x),$$

where A denotes the generator of the Ito diffusion (8.2.9).

Using the expression of the expectation operator, the previous solution can also be written in the familiar form

$$v(t,x) = \int_{\mathbb{R}^n} f(y)\, p_t(x,y)\, dy,$$

which implies that the transition density $p_t(x,y)$ of the process X_t is the heat kernel for the operator A. This observation is fundamental and will lead to a method of computing heat kernels for second-order partial differential operators.

The formalization of this fact will be done as in the following. Let $f(x) = \delta_x$ be the Dirac distribution centered at $x \in \mathbb{R}^n$; i.e.,

$$\delta_x(\varphi) = \int_{\mathbb{R}^n} \delta_x(y)\varphi(y)\, dy = \varphi(x).$$

Then for any $x_0 \in \mathbb{R}^n$,

$$v(t,x_0) = \int_{\mathbb{R}^n} \delta_x(y)\, p_t(x_0,y)\, dy = p_t(x_0,x)$$

satisfies

$$\frac{\partial v}{\partial t} = Av, \qquad t > 0,$$
$$\lim_{t \searrow 0} v(t,x) = \delta_x;$$

i.e., $p_t(x_0,x)$ is the heat kernel of the operator A. Equivalently, $p_t(x_0,x)$ is the transition density function between the points $x_0 = X_0$ and $x = X_t$ for the diffusion process (8.2.9).

8.5 Algorithm for Finding the Heat Kernel

We shall conclude the last few sections with the following algorithm for finding heat kernels for second-order differential operators:

1. Consider a second-order operator of the form

$$A = \frac{1}{2}\sum_{i,j=1}^{n} g^{ij}(x)\frac{\partial^2}{\partial x_i \partial x_j} + \sum_{k=1}^{n} b_k \frac{\partial}{\partial x_k},$$

   with $\det g^{ij}(x)$ eventually zero at some points.
2. Decompose the matrix $g(x) \in \mathbb{R}^{n\times n}$ as $g(x) = \sigma(x)\cdot\sigma^T(x)$, where $\sigma(x) \in \mathbb{R}^{n\times m}$.
3. Construct the homogeneous Ito diffusion process

$$dX_t = b(X_t)dt + \sigma(X_t)\,dW(t), \qquad t \geq 0,\ X_0 = x_0,$$

   where $W(t) = \big(W_1(t), \ldots, W_m(t)\big)$ are m independent Brownian motions.
4. Solve the stochastic differential equation for X_t and find the transition density $p_t(x_0, x)$ with $X_t = x$, $X_0 = x_0$.
5. The heat kernel (propagator) of the operator A is given by the transition density $K(x_0, x; t) = p_t(x_0, x)$.

All the previous steps are almost obvious, with the exception of steps 2 and 4. When solving for X_t, we need also to check the uniqueness of solution by Theorem 8.2.2. The delicate part is to figure out the transition density associated with the process X_t, especially in cases when X_t has a complicated expression. Several methods of finding the transition density will be presented in the next section.

We finalize this section with a simple application of the aforementioned algorithm for the operator $\frac{1}{2}\partial_x^2$. In this case $n = m = 1$ and $b = 0$, $g = 1$. Then $\sigma = 1$ and the associated Ito diffusion is

$$dX_t = dW_t, \qquad X_0 = x_0.$$

The solution is given by $X_t = x_0 + W_t$, $t \geq 0$. Since the Brownian motion W_t is normally distributed with mean 0 and variance t, it follows that X_t is normally distributed with mean x_0 and variance t, so the transition density for X_t is

$$p_t(x_0, x) = \frac{1}{\sqrt{2\pi t}} e^{\frac{-(x-x_0)^2}{2t}}, \qquad t > 0.$$

Hence the heat kernel for $\frac{1}{2}\partial_x^2$ is given by

$$K(x_0, x; t) = \frac{1}{\sqrt{2\pi t}} e^{\frac{-(x-x_0)^2}{2t}}, \qquad t > 0.$$

8.6 Finding the Transition Density

We concluded in the previous section that the problem of finding the heat kernel is now reduced to the problem of finding the transition density of an Ito diffusion process. Next we shall present a few techniques of obtaining the transition density that will be useful in the sequel when computing heat kernels.

1. The case of a well-known distribution. In this case X_t is one of the familiar stochastic processes for which the density is well known. We shall next consider a few examples.

- If $X_t = W_t$ is a Brownian motion, then $p_t(0,x) = \frac{1}{\sqrt{2\pi t}} e^{-x^2/(2t)}$; see Fig. 8.3(a).
- If $X_t = x_0 + W_t$, i.e., a Brownian motion starting at x_0, then the transition density is $p_t(x_0,x) = \frac{1}{\sqrt{2\pi t}} e^{-(x-x_0)^2/(2t)}$.
- If $X_t = e^{Y_t}$, with Y_t normally distributed with mean μ and variance σ^2, then X_t is lognormally distributed, with

$$p = \frac{1}{x\sqrt{2\pi}\sigma} e^{\frac{-(\ln x-\mu)^2}{2\sigma^2}}, \qquad x > 0.$$

For instance, in the particular case when $X_t = e^{W_t + c}$, with $c \in \mathbb{R}$ given by $c = \ln x_0$, the transition density is

$$p_t(x_0,x) = \frac{1}{x\sqrt{2\pi t}} e^{\frac{-(\ln x - \ln x_0)^2}{2t}}, \qquad x > 0; \tag{8.6.21}$$

see Fig. 8.3b.

2. Reduction to a well-known distribution. Here we shall distinguish between the cases of one and more variables.

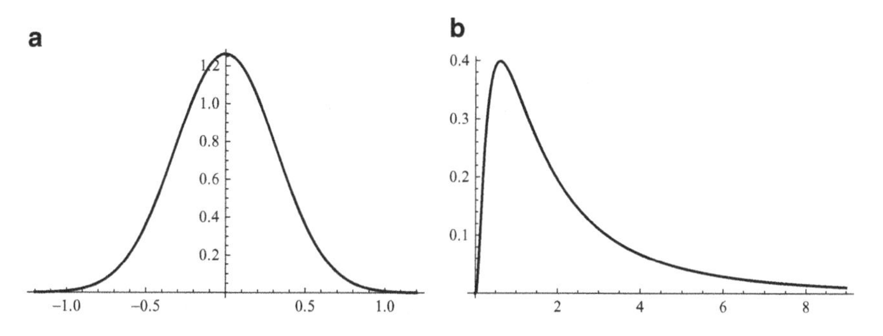

Fig. 8.3 (**a**) The standard normal distribution; (**b**) the lognormal distribution with $\sigma = 1, \mu = 0.5$

The case of one variable. Let $X_t = h(W_t + c)$, with h an increasing, smooth function with the inverse $h^{-1} = g$. The process starts at $x_0 = h(c)$, so $c = g(x_0)$. If F_{X_t} and F_{W_t} denote the distribution functions of X_t and W_t, respectively, then we have

$$\begin{aligned} F_{X_t}(x) &= P(X_t \le x) = P(h(W_t + c) \le x) = P(W_t \le g(x) - c) \\ &= F_{W_t}(g(x) - g(x_0)). \end{aligned}$$

Differentiating, we obtain the transition probability of X_t:

$$\begin{aligned} p_t(x_0, x) &= \frac{d}{dx} F_{X_t}(x) = F'_{W_t}(g(x) - g(x_0))\, g'(x) \\ &= \frac{1}{\sqrt{2\pi t}} e^{\frac{-(g(x)-g(x_0))^2}{2t}} g'(x). \end{aligned}$$

We note that we recover formula (8.6.21) in the case $h(x) = \ln x$.

The case of several variables. We shall next deal with the case of two variables. The general case of n variables can be treated in a similar way. Consider a continuous differentiable transformation of the region $(x_1, x_2) \in \mathcal{R}$ into the region $(y_1, y_2) \in \mathcal{S}$ given by

$$\begin{aligned} y_1 &= g(x_1, x_2), \\ y_2 &= h(x_1, x_2), \end{aligned}$$

with the nonzero Jacobian

$$\frac{\partial(y_1, y_2)}{\partial(x_1, x_2)} = \det\left(\frac{\partial y_i}{\partial x_j}\right)_{i,j} \neq 0, \qquad \forall (x_1, x_2) \in \mathcal{R}.$$

By the inverse function theorem, we can write

$$\begin{aligned} x_1 &= G(y_1, y_2), \\ x_2 &= H(y_1, y_2), \qquad \forall (y_1, y_2) \in \mathcal{S}. \end{aligned}$$

Consider the two-dimensional stochastic process $X_t = (X_t^1, X_t^2)$ such that the vector (x_1, x_2) is a possible value of X_t. Consider another stochastic process $Y_t = (Y_t^1, Y_t^2)$ defined by

$$\begin{aligned} Y_t^1 &= g(X_t^1, X_t^2), \\ Y_t^2 &= h(X_t^1, X_t^2). \end{aligned}$$

Let $p_X(x)$ and $p_Y(y)$ be the probability density functions of X_t and Y_t, respectively. Since we have

$$P(X_t \in \mathcal{R}) = P(Y_t \in \mathcal{S}),$$

and the left side is given by

$$P(X_t \in \mathcal{R}) = \iint_{\mathcal{R}} p_X(x)\,dx = \iint_{\mathcal{S}} p_X(G(y), H(y)) \det\left(\frac{\partial x_i}{\partial y_j}\right) dy,$$

while the right side is

$$P(Y_t \in \mathcal{S}) = \iint_{\mathcal{S}} p_Y(y)\,dy,$$

equating yields

$$\iint_{\mathcal{S}} p_X(G(y), H(y)) \det\left(\frac{\partial x_i}{\partial y_j}\right) dy = \iint_{\mathcal{S}} p_Y(y)\,dy.$$

Since this relation holds for any set $\mathcal{S}$, it follows that

$$p_Y(y) = p_X(G(y), H(y)) \det\left(\frac{\partial x_i}{\partial y_j}\right).$$

Inverting, we get a formula that provides the density function of X_t in terms of the density function of Y_t:

$$p_X(x) = p_Y(g(x), h(x)) \det\left(\frac{\partial y_i}{\partial x_j}\right). \tag{8.6.22}$$

A similar relation holds for the n-dimensional case.

If the process X_t can be transformed into the process Y_t, and the joint density function for (Y_t^1, Y_t^2) is known, then formula (8.6.22) provides a formula for the joint density function of (X_t^1, X_t^2). In the particular case when Y_t^1 and Y_t^2 are independent processes, then $p_Y(y) = p_{Y_1}(y_1)\,p_{Y_2}(y_2)$ and the aforementioned formula becomes

$$p_X(x) = p_{Y_1}(g(x)) p_{Y_2}(h(x)) \det\left(\frac{\partial y_i}{\partial x_j}\right). \tag{8.6.23}$$

3. Reduction to a Wiener integral. Sometimes we deal with stochastic differential equations of the form

$$dX_t = f(t) dW_t, \qquad X_0 = x_0,$$

where W_t is a one-dimensional Brownian motion. The solution can be written in terms of a *Wiener integral*,

$$X_t = x_0 + \int_0^t f(s)\,dW_s. \tag{8.6.24}$$

The following result can be found in [83], p. 11.

Theorem 8.6.1. *Let $f \in L^2[a,b]$. Then the Wiener integral $\int_a^b f(s)\,dW_s$ is a Gaussian random variable with mean 0 and variance $\|f\|^2 = \int_a^b f(s)^2\,ds$.*

The idea of the proof is to show that the result holds for the step functions first. Then approximating any function $f \in L^2[a,b]$ by step functions, and using that the limit of a Gaussian random variable is Gaussian, one obtains the desired result. The proof details can be found in [83].

Using Theorem 8.6.1, the process X_t given by (8.6.24) is Gaussian, with mean x_0 and variance $\sigma^2 = \int_0^t f(s)^2\,ds$. Then the transition density of X_t is

$$p_t(x_0,x) = \left(2\pi \int_0^t f^2(s)\,ds\right)^{-1/2} e^{-\frac{1}{2}\frac{(x-x_0)^2}{\int_0^t f^2(s)\,ds}}, \qquad t > 0. \tag{8.6.25}$$

4. The case of an integrated Brownian motion. The solution of the stochastic differential equation

$$dX_t = W_t\,dt, \qquad X_0 = x_0 \qquad t \geq 0,$$

is given by

$$X_t = x_0 + \int_0^t W_s\,ds \tag{8.6.26}$$

and is called *an integrated Brownian motion*. A realization of an integrated Brownian motion starting at $x_0 = 0$ is given in Fig. 8.4. We shall find next the transition density of X_t.

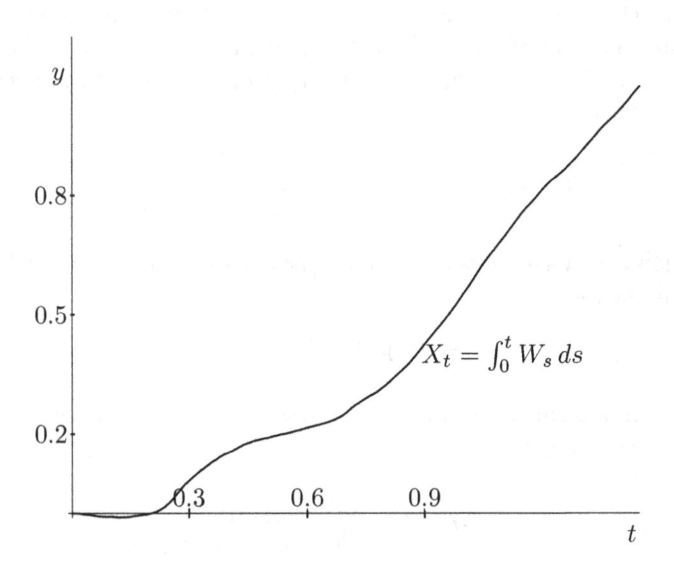

Fig. 8.4 A realization of an integrated Brownian motion $X_t = \int_0^t W_s\,ds$

For any smooth function f, the integration by parts for a Riemann–Stieltjes integral yields

$$\int_a^b f(s)\,dW_s = f(b)W_b - f(a)W_a - \int_a^b W_s\,df(s)$$
$$= f(b)W_b - f(a)W_a - \int_a^b W_s\,f'(s)\,ds.$$

Making $f(s) = s$, $a = 0$, $b = t$, and using that $W_0 = 0$, the previous relation implies

$$\int_0^t W_s\,ds = tW_t - \int_0^t s\,dW_s.$$

Then (8.6.26) becomes

$$X_t = (x_0 + tW_t) - \int_0^t s\,dW_s, \qquad t > 0. \tag{8.6.27}$$

Then X_t is a difference of two Gaussian random variables (see Theorem 8.6.1), and hence it will be Gaussian. Its mean is equal to

$$E(X_t) = E(x_0 + tW_t) - E\left(\int_0^t f(s)\,dW_s\right) = x_0.$$

A direct computation (see [99], p. 370) shows that the variance is

$$\mathrm{Var}(X_t) = \mathrm{Var}\left(\int_0^t W_s\,ds\right) = \frac{t^3}{3}.$$

Hence the transition probability of X_t is given by

$$p_t(x_0, x) = \sqrt{\frac{3}{2\pi t^3}}\,e^{-\frac{3(x-x_0)^2}{2t^3}}, \qquad t > 0. \tag{8.6.28}$$

5. Using the transition distribution function. If X_t is a one-dimensional continuous stochastic process with $X_0 = x_0$, we recall that the *transition distribution function* is defined by

$$F(x_0; x, t) = P(X_t \le x \mid X_0 = x_0),$$

where P denotes the probability. Then the transition density is obtained by differentiating the distribution function

$$p_t(x_0, x) = \frac{d}{dx}F(x_0; x, t).$$

A similar relation works in the case of several variables. For instance, consider a two-dimensional continuous process $X_t = (X_1(t), X_2(t))$; then define

$$F(x_1^0, x_2^0; x_1, x_2; t) = P(X_1(t) \le x_1,\ X_2(t) \le x_2 \mid X_1(0) = x_1^0,\ X_2(0) = x_2^0).$$

Then the transition probability is obtained as the following mixed derivative:

$$p_t(x_0, x) = \frac{\partial^2}{\partial x_1 \partial x_2} F(x_1^0, x_2^0; x_1, x_2; t).$$

6. Using the expectation operator. Assume again that X_t is a one-dimensional continuous stochastic process with $X_0 = x_0$, and transition density $p_t(x_0, x)$. Then the expectation operator is given by

$$E[f(X_t)] = \int f(x) p_t(x_0, x)\, dx, \tag{8.6.29}$$

for any function f. This will lead to the following formula for computing the transition probability:

$$p_t(x_0, x) = E[\delta(X_t - x)], \tag{8.6.30}$$

where δ is the Dirac distribution and X_t is the associated stochastic process with $X_0 = x_0$ and $X_t = x$. The reason behind the aforementioned formula is the simple computation

$$E[\delta(X_t - x)] = \int \delta(y - x) p_t(x_0, y)\, dy = p_t(x_0, x);$$

see formula (8.6.29). We shall exemplify this in the case of the operator $\frac{1}{2}\partial_x^2$. The associated Ito diffusion satisfies $dX_t = dW_t$ with the solution $X_t = x_0 + W_t$. Consequently,

$$\begin{aligned} p_t(x_0, x) &= E[\delta(x_0 + W_t - x)] = \int \delta(y - (x - x_0)) \frac{1}{\sqrt{2\pi t}} e^{-\frac{y^2}{2t}}\, dy \\ &= \frac{1}{\sqrt{2\pi t}} e^{-\frac{(x-x_0)^2}{2t}}. \end{aligned}$$

However, for sub-elliptic operators, this easy computation does not work. Then we shall proceed as in the following. The trick is to use the expression of the Dirac distribution as the inverse Fourier transform of 1:

$$\delta(y) = \frac{1}{2\pi} \int e^{i\xi y}\, d\xi.$$

Then we have

$$\begin{aligned} p_t(x_0, x) &= \int_{\mathbb{R}_y} \delta(y - x + x_0) \frac{1}{\sqrt{2\pi t}} e^{-\frac{y^2}{2t}}\, dy \\ &= \frac{1}{\sqrt{2\pi t}} \int_{\mathbb{R}_y} \frac{1}{2\pi} \int_{\mathbb{R}_\xi} e^{i\xi(y-x+x_0)}\, d\xi\; e^{-\frac{y^2}{2t}}\, dy \\ &= \frac{1}{\sqrt{2\pi t}} \frac{1}{2\pi} \int_{\mathbb{R}} \left[\int_{\mathbb{R}} e^{i\xi(y-x+x_0)-\frac{y^2}{2t}}\, dy \right] d\xi \\ &= \frac{1}{\sqrt{2\pi t}} \frac{1}{2\pi} \int_{\mathbb{R}} e^{-i\xi(x-x_0)} \left[\int_{\mathbb{R}} e^{i\xi y-\frac{y^2}{2t}}\, dy \right] d\xi. \end{aligned} \tag{8.6.31}$$

From relation (1.6.6), we have

$$\int_{\mathbb{R}} e^{i\xi y - \frac{y^2}{2t}}\, dy = \sqrt{2\pi t} e^{-\xi^2 t/2}.$$

Substituting in (8.6.31) and using (1.6.6) yields

$$\begin{aligned} p_t(x_0, x) &= \frac{1}{2\pi} \int_{\mathbb{R}} e^{-i\xi(x-x_0)} e^{-\frac{t}{2}\xi^2}\, d\xi \\ &= \frac{1}{2\pi} \sqrt{\frac{2\pi}{t}} e^{-\frac{(x-x_0)^2}{2t}} = \frac{1}{\sqrt{2\pi t}} e^{-\frac{(x-x_0)^2}{2t}}. \end{aligned} \tag{8.6.32}$$

The aforementioned computation can also be applied in the case of several variables. For instance, in the case of the operator $\frac{1}{2}(\partial_x^2 + \partial_y^2)$, the associated stochastic process is $(X_t, Y_t) = (x_0 + W_t, y_0 + W_t)$ and the probability density can be computed as

$$\begin{aligned} p_t(x_0, y_0; x, y) &= E[\delta(X_t - x)\delta(Y_t - y)] \\ &= E[\delta(x_0 + W_t - x)] E[\delta(y_0 + W_t - y)] \\ &= \frac{1}{\sqrt{2\pi t}} e^{-\frac{(x-x_0)^2}{2t}} \frac{1}{\sqrt{2\pi t}} e^{-\frac{(y-y_0)^2}{2t}} = \frac{1}{2\pi t} e^{-\frac{1}{2t}[(x-x_0)^2 + (y-y_0)^2]}. \end{aligned}$$

The main ingredient of the computation was $E[XY] = E[X]E[Y]$ for X and Y independent random variables. However, in general, X_t and Y_t are not independent processes and the computation becomes more complicated. We shall deal with this case in Sect. 8.17.

7. Exponential martingale. Let X_t be a one-dimensional stochastic process satisfying $E[\int_0^T X_t^2\, dt] < \infty$, with $T \le \infty$. Define the process

$$Z_t = e^{\int_0^t X_s\, dW_s - \frac{1}{2}\int_0^t X_s^2\, ds}.$$

A sufficient condition that Z_t be a martingale is Novikov condition (see [95], p. 55):

$$E\big[e^{\frac{1}{2}\int_0^T X_s^2\,ds}\big] < \infty.$$

This property has a useful consequence. Since $Z_0 = 1$, from the martingale condition $E[Z_t] = Z_0 = 1$, so

$$E\Big[e^{\int_0^t X_s\,dW_s}/e^{\frac{1}{2}\int_0^t X_s^2\,ds}\Big] = 1, \quad \forall t > 0.$$

If the process X_t is independent of W_t, then

$$E\Big[e^{\int_0^t X_s\,dW_s}\Big] = e^{-\frac{1}{2}\int_0^t X_s^2\,ds}. \tag{8.6.33}$$

In particular, if $X_t = \alpha$ is a constant process, the aforementioned relation becomes $E[e^{\alpha W_t}] = e^{-\frac{1}{2}\alpha^2 t}$.

8.7 Kolmogorov's Forward Equation

The *adjoint operator* of the operator

$$Af(x) = \sum_{i,j} a_{ij}(x)\frac{\partial^2 f}{\partial x_i\,\partial x_j} + \sum_k b_k(x)\frac{\partial f}{\partial x_k}$$

is given by

$$A^*\phi(x) = \sum_{i,j}\frac{\partial^2}{\partial x_i\,\partial x_j}(a_{ij}\phi) - \sum_k \frac{\partial}{\partial x_k}(b_k\phi),$$

and satisfies the property $\langle Af, \phi\rangle_{L^2} = \langle f, A^*\phi\rangle_{L^2}$, for any $f, \phi \in C_0^2$. Let $p_t(x, y)$ be a smooth transition density of the process X_t which satisfies the stochastic differential equation (8.2.9). Denote by A the generator of the Ito diffusion X_t. (x_0 and x have been replaced in this section by x and y, respectively.)

Theorem 8.7.1 (Kolmogorov's forward equation.). *The transition density p_t (x, y) satisfies the equation*[3]

$$\frac{d}{dt}p_t(x, y) = A_y^*\,p_t(x, y), \qquad \forall x, y \in \mathbb{R}^n.$$

[3] This equation is also called the *Fokker–Plank equation.*

Proof. Substituting the expectation operator $E[f(X_t)] = \int_{\mathbb{R}^n} f(y)p_t(x,y)\,dy$ into Dynkin's formula (8.3.20) yields

$$\int_{\mathbb{R}^n} f(y)p_t(x,y)\,dy = f(x) + \int_0^t \left(\int_{\mathbb{R}^n} A_y f(y)\, p_s(x,y)\,dy \right) ds.$$

Differentiating with respect to t and using the definition of the adjoint, we obtain

$$\begin{aligned}\int_{\mathbb{R}^n} f(y)\,\frac{d}{dt}p_t(x,y)\,dy &= \int_{\mathbb{R}^n} A_y f(y)\, p_t(x,y)\,dy \\ &= \int_{\mathbb{R}^n} f(y)\, A_y^* p_t(x,y)\,dy, \qquad \forall f \in C_0^2(\mathbb{R}^n),\end{aligned}$$

and hence $\frac{d}{dt}p_t(x,y) = A_y^*\, p_t(x,y)$. ■

In the following sections we shall consider several examples of second-order partial differential operators associated with an Ito diffusion for which we shall calculate the heat kernels following the algorithm presented in Sect. 8.5.

8.8 The Operator $\frac{1}{2}\partial_x^2 - x\partial_x$

In this section we shall study an operator associated with *Langevin's equation* in one dimension,

$$dX_t = -X_t dt + dW_t, \qquad t \geq 0, \tag{8.8.34}$$

with $X_0 = x_0$. The aforementioned equation is a homogeneous Ito diffusion with coefficients $\sigma(x) = 1$ and $b(x) = -x$. Using formula (8.3.19), we obtain the generator for Langevin's equation:

$$A = \frac{1}{2}\frac{\partial^2}{\partial x^2} - x\frac{\partial}{\partial x}. \tag{8.8.35}$$

Next we shall solve (8.8.34) by multiplying by the integrating factor e^t:

$$\begin{aligned} e^t dX_t + X_t e^t dt = e^t dW_t &\iff \\ d(e^t X_t) &= e^t dW_t. \end{aligned}$$

Integrating yields

$$\begin{aligned} e^t X_t &= x_0 + \int_0^t e^s\, dW_s \iff \\ X_t &= e^{-t}x_0 + e^{-t}\int_0^t e^s\, dW_s. \end{aligned}$$

Since X_t can be expressed in terms of Wiener integral, using Theorem 8.6.1 it follows that X_t is a Gaussian process, with mean $E(X_t) = e^{-t}x_0$ and variance

$$\operatorname{Var}(X_t) = e^{-2t}\int_0^t e^{2s}\,ds = \frac{1-e^{-2t}}{2}.$$

If we denote the Gaussian density with mean μ and variance t by

$$G_t(\mu, x) = \frac{1}{\sqrt{2\pi t}}e^{-\frac{(x-\mu)^2}{2t}}, \qquad t > 0,$$

then X_t has density $G_{(1-e^{-2t})/2}(e^{-t}x_0, \cdot)$.

Following the algorithm of Sect. 8.5, the heat kernel of the operator (8.8.35) is given by

$$\begin{aligned}
K(x_0, x; t) &= p_t(x_0, x) = G_{(1-e^{-2t})/2}(e^{-t}x_0, x) \\
&= \frac{1}{\sqrt{\pi(1-e^{-2t})}}e^{-\frac{(x-x_0e^{-t})^2}{1-e^{-2t}}} \\
&= \frac{1}{\sqrt{\pi(1-e^{-2t})}}e^{-\frac{x_0^2+x^2}{1-e^{-2t}}+x_0^2+\frac{x_0x}{\sinh t}}, \qquad t > 0.
\end{aligned}$$

8.9 Generalized Brownian Motion

Consider the stochastic differential equation

$$dX_t = a\,dt + b\,dW_t, \qquad X_0 = x_0, \tag{8.9.36}$$

with $a, b \in \mathbb{R}$ constants, called the *drift* and *diffusion* coefficients. Sometimes this is also called *Brownian motion with drift*.

Let $X_0 = x_0$. Integrating in (8.9.36) yields

$$\begin{aligned}
X_t &= x_0 + at + b\int_0^t dW_s \\
&= x_0 + at + bW_t,
\end{aligned}$$

which is Gaussian distributed with mean $E(X_t) = x_0 + at$ and variance $\operatorname{Var}(X_t) = b^2t$, so the transition density is

$$G_{b^2t}(x_0 + at, x) = \frac{1}{|b|\sqrt{2\pi t}}e^{-\frac{1}{2tb^2}(x-x_0-at)^2}. \tag{8.9.37}$$

Using formula (8.3.19), the generator of the generalized Brownian motion (8.9.36) is given by

$$A = \frac{1}{2}b^2 \frac{\partial^2}{\partial x^2} + a\frac{\partial}{\partial x}. \tag{8.9.38}$$

Using (8.9.37), it follows that the heat kernel of the operator (8.9.38) is given by

$$p_t(x_0, x) = \frac{1}{b\sqrt{2\pi t}} e^{-\frac{1}{2tb^2}(x-x_0-at)^2}, \qquad t > 0, \tag{8.9.39}$$

where we assumed $b > 0$.

We can easily extend this result to the case of n variables considering the operator

$$L = \frac{1}{2}\sum_{j=1}^{n} b_j^2 \frac{\partial^2}{\partial x_j^2} + \sum_{j=1}^{n} a_j \frac{\partial}{\partial x_j}. \tag{8.9.40}$$

The aforementioned operator is the generator of the n-dimensional generalized Brownian motion

$$dX_t^j = a_j\, dt + b_j\, dW_j(t), \qquad j = 1, \ldots, n. \tag{8.9.41}$$

Since $W_1(t), \ldots, W_n(t)$ are independent Brownian motions, the processes $X_t^1, \ldots, X_t^n$ are also independent, and hence the joint density function for $(X_t^1, \ldots, X_t^n)$ is the product of the probability density functions associated with each X_t^j:

$$p_t(x_0, x) = \prod_{j=1}^{n} p_t^j(x_0^j, x^j),$$

where

$$p_t^j(x_0^j, x^j) = \frac{1}{|b_j|\sqrt{2\pi t}} e^{-\frac{1}{2tb_j^2}(x_j - x_0^j - a_j t)^2}, \qquad t > 0.$$

This is equivalent to the relation

$$e^{Lt} = e^{L_1 t + \cdots + L_n t} = e^{L_1 t} \cdots e^{L_n t},$$

where

$$L_j = \frac{1}{2}b_j^2 \frac{\partial^2}{\partial x_j^2} + a_j \frac{\partial}{\partial x_j}.$$

Proposition 8.9.1. *The heat kernel of the operator (8.9.40) is*

$$K(x_0, x; t) = \frac{1}{|b_1 \cdots b_n|(2\pi t)^{n/2}} e^{-\frac{1}{2t}\sum_{j=1}^n \left(\frac{x_j - x_0^j - a_j t}{b_j}\right)^2}, \qquad t > 0. \quad (8.9.42)$$

Taking $b_1 = \cdots = b_n$, we obtain the following result.

Corollary 8.9.2. *The heat kernel of* $\frac{1}{2}\sum_{j=1}^n \frac{\partial^2}{\partial x_j^2} + \sum_{j=1}^n a_j \frac{\partial}{\partial x_j}$ *is*

$$K(x_0, x; t) = \frac{1}{(2\pi t)^{n/2}} e^{-\frac{\|x-x_0\|^2}{2t} - \frac{\|a\|^2}{2}t + \langle a, x-x_0\rangle}, \qquad t > 0,$$

where $x_0 = (x_0^1, \ldots, x_0^n)$, $x = (x_1, \ldots, x_n)$, $a = (a_1, \cdots a_n)$ *and* $\| \ \|$, $\langle\ ,\ \rangle$ *denote the length and the inner product of vectors in* $\mathbb{R}^n$.

Taking $a = -\frac{1}{2}$, we obtain the next result.

Corollary 8.9.3. *The heat kernel of* $\frac{1}{2}\frac{\partial^2}{\partial x^2} - \frac{1}{2}\frac{\partial}{\partial x}$ *is*

$$K(x_0, x; t) = \frac{1}{\sqrt{2\pi t}} e^{-\frac{(x-x_0)^2}{2t} - \frac{1}{2}(x-x_0) - \frac{1}{8}t}, \qquad t > 0.$$

8.10 Linear Noise

The following Ito diffusion

$$dX_t = X_t \, dW_t, \qquad X_0 = x_0 > 0, \qquad (8.10.43)$$

is modeling *linear noise*. According to formula (8.3.19), the generator associated with the previous equation is

$$A = \frac{1}{2}x^2 \frac{\partial^2}{\partial x^2}, \qquad x > 0. \qquad (8.10.44)$$

Following the algorithm of Sect. 8.5, we next need to solve (8.10.43). In order to do this, we shall consider the process $F_t = \ln X_t$. Applying Ito's lemma, we obtain

$$\begin{aligned} dF_t &= \frac{1}{X_t} dX_t - \frac{1}{2}\frac{1}{X_t^2} dX_t^2 \\ &= \frac{1}{X_t} dX_t - \frac{1}{2}\frac{1}{X_t^2} X_t^2 \, dt \\ &= dW_t - \frac{1}{2} dt, \end{aligned}$$

where we used (8.10.43). Integrating yields

$$
\begin{aligned}
F_t &= F_0 + \int_0^t dW_s - \frac{1}{2}\int_0^t ds \Longleftrightarrow \\
\ln X_t &= \ln x_0 + W_t - \frac{t}{2} \Longleftrightarrow \\
X_t &= e^{\ln x_0 + W_t - \frac{t}{2}} = x_0 e^{W_t - \frac{t}{2}}.
\end{aligned}
$$

Since $Y_t = \ln x_0 + W_t - t/2$ is normally distributed with mean $\mu = \ln x_0 - t/2$ and variance $\sigma^2 = t$, then following item 1 of Sect. 8.6, the process X_t has a lognormal distribution with the transition density

$$
\begin{aligned}
p_t(x_0, x) &= \frac{1}{x\sqrt{2\pi\sigma^2}} e^{-\frac{(\ln x - \mu)^2}{2\sigma^2}} = \frac{1}{x\sqrt{2\pi t}} e^{-\frac{(\ln x - \ln x_0 + \frac{t}{2})^2}{2t}} \\
&= \frac{1}{x\sqrt{2\pi t}} e^{-\frac{1}{2t}(\ln x - \ln x_0)^2 - \frac{1}{2}(\ln x - \ln x_0) - \frac{t}{8}} \\
&= \frac{x_0}{x}\frac{1}{\sqrt{2\pi t}} e^{-\frac{1}{2t}(\ln x - \ln x_0)^2 - \frac{1}{2}(\ln x + \ln x_0) - \frac{t}{8}}, \quad t > 0, \qquad (8.10.45)
\end{aligned}
$$

with $x, x_0 > 0$. To conclude, the operator (8.10.44) has the heat kernel given by (8.10.45).

8.11 Geometric Brownian Motion

The following stochastic differential equation

$$
dX_t = \alpha X_t\, dt + \beta X_t\, dW_t, \qquad X_0 = x_0 > 0, \qquad (8.11.46)
$$

is called *geometric Brownian motion* and is famous for its applications in the stock market.[4] Both the drift and diffusion coefficients are linear, $b(x) = \alpha x$, $\sigma(x) = \beta x$, and using (8.3.19) the generator associated with (8.11.46) becomes

$$
A = A_{\alpha,\beta} = \frac{1}{2}\beta^2 x^2 \frac{\partial^2}{\partial x^2} + \alpha x \frac{\partial}{\partial x}. \qquad (8.11.47)
$$

[4] In mathematical finance X_t represents the stock price with return α and volatility β.

We shall first solve the stochastic equation (8.11.46). Let $F_t = \ln X_t$, with $F_0 = \ln x_0$. An application of Ito's lemma yields

$$\begin{aligned} dF_t &= \frac{1}{X_t} dX_t - \frac{1}{2}\frac{1}{X_t^2}(dX_t)^2 \\ &= \frac{1}{X_t}(\alpha X_t\, dt + \beta X_t\, dW_t) - \frac{1}{2}\frac{1}{X_t^2}\beta^2 X_t^2\, dt \\ &= (\alpha - \frac{1}{2}\beta^2)dt + \beta\, dW_t. \end{aligned}$$

Integrate and get

$$\begin{aligned} F_t &= F_0 + \int_0^t \left(\alpha - \frac{1}{2}\beta^2\right) ds + \beta \int_0^t dW_s \Longleftrightarrow \\ \ln X_t &= \ln x_0 + \left(\alpha - \frac{1}{2}\beta^2\right) t + \beta W_t, \end{aligned}$$

and hence the solution is

$$X_t = e^{\ln x_0 + (\alpha - \frac{1}{2}\beta^2)t + \beta W_t}. \tag{8.11.48}$$

Let $U_t = \ln x_0 + (\alpha - \frac{1}{2}\beta^2)t + \beta W_t$. The process U_t is normally distributed with mean $\mu = \ln x_0 + (\alpha - \frac{1}{2}\beta^2)t$ and variance $\sigma^2 = \beta t$. Using the item 1 of Sect. 8.6, we obtain that $X_t = e^{U_t}$ is lognormally distributed with the transition density

$$\begin{aligned} p_t(x_0, x) &= \frac{1}{x}\frac{1}{\sqrt{2\pi\sigma^2}} e^{-\frac{(\ln x - \mu)^2}{2\sigma^2}} \\ &= \frac{1}{x}\frac{1}{\sqrt{2\pi\beta t}} e^{-\frac{1}{2\beta t}[\ln x - \ln x_0 - (\alpha - \frac{\beta^2}{2})]^2}, \qquad t > 0, \end{aligned} \tag{8.11.49}$$

with $x, x_0 > 0$. To conclude, the heat kernel of the operator (8.11.47) is given by the relation (8.11.49).

In the particular case when $\beta = 1$ and $\alpha = \frac{1}{2}$, the Ito diffusion (8.11.48) becomes $X_t = x_0 e^{W_t}$; see (Fig. 8.5). The generator operator

$$A_{\frac{1}{2},1} = \frac{1}{2}x^2\frac{\partial^2}{\partial x^2} + \frac{1}{2}x\frac{\partial}{\partial x}, \quad x > 0,$$

has the heat kernel

$$K(x_0, x; t) = \frac{1}{x}\frac{1}{\sqrt{2\pi t}} e^{-\frac{1}{2t}(\ln x - \ln x_0)^2}, \qquad t > 0.$$

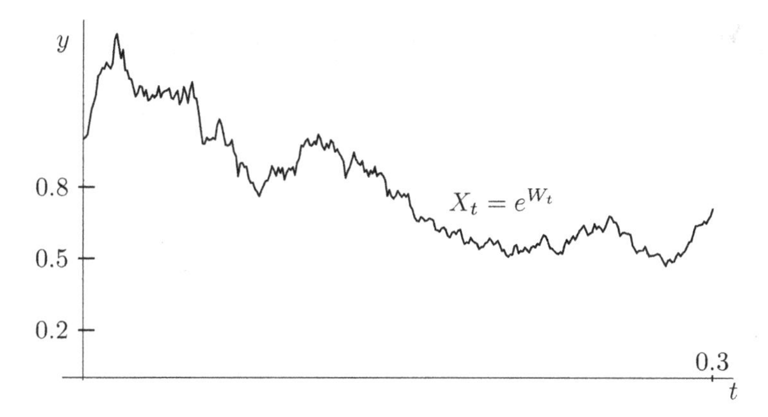

Fig. 8.5 A realization of the geometric Brownian motion $X_t = e^{W_t}$

This result can be generalized to n dimensions in a way similar to that presented at the end of Sect. 8.9. It follows that the operator

$$\frac{1}{2}\sum_{j=1}^{n} x_j^2 \frac{\partial^2}{\partial x_j^2} + \frac{1}{2}\sum_{j=1}^{n} x_j \frac{\partial}{\partial x_j}, \qquad x_j > 0,$$

has the heat kernel

$$K(x_0, x; t) = \frac{1}{x_1 \cdots x_n \, (2\pi t)^{n/2}} \, e^{-\frac{1}{2t}\sum_j (\ln x_j - \ln x_0^j)^2}, \qquad t > 0.$$

Remark 8.11.1. The generator (8.11.47) is an Euler operator, which, after the change of variable $y = \ln x$, becomes a second-order operator with constant coefficients

$$\begin{aligned} A &= \frac{1}{2}\beta^2 x^2 \frac{\partial^2}{\partial x^2} + \alpha x \frac{\partial}{\partial x} \\ &= \frac{1}{2}\beta^2 \frac{\partial^2}{\partial y^2} + \left(\alpha - \beta^2/2\right) \frac{\partial}{\partial y}, \end{aligned}$$

which is the generator of the Brownian motion with drift $\alpha - \beta^2/2$ and diffusion coefficient β given by

$$dY_t = (\alpha - \beta^2/2)\, dt + \beta\, dW_t.$$

8.12 Mean Reverting Ornstein–Uhlenbeck Process

In this section we shall study the heat kernel associated with the generator of the stochastic differential equation

$$dX_t = (m - X_t)dt + \sigma dW_t, \qquad X_0 = x_0, \quad m, \sigma \in \mathbb{R}. \tag{8.12.50}$$

Equation (8.12.50) has been used to model stochastic interest rates in finance that exhibit mean reverting properties. Since the drift is $b(x) = m - x$, the associated generator is given by

$$A = \frac{1}{2}\sigma^2 \frac{\partial^2}{\partial x^2} + (m - x)\frac{\partial}{\partial x}. \tag{8.12.51}$$

We may assume $\sigma > 0$. The next step in the algorithm of Sect. 8.5 is to solve the aforementioned stochastic equation. We proceed by writing the relation (8.12.50) in the form

$$dX_t + X_t dt = m\, dt + \sigma\, dW_t.$$

Multiplying by e^t, we obtain the exact equation:

$$d(e^t X_t) = me^t\, dt + \sigma e^t\, dW_t.$$

Integrating yields

$$\begin{aligned} e^t X_t &= x_0 + m\int_0^t e^s\, ds + \sigma \int_0^t e^s\, dW_s \Longleftrightarrow \\ X_t &= e^{-t}x_0 + m(1 - e^{-t}) + \sigma e^{-t}\int_0^t e^s\, dW_s, \end{aligned} \tag{8.12.52}$$

which is the *mean reverting Ornstein–Uhlenbeck process*.[5] We can see from formula (8.12.52) that X_t is the sum of a deterministic function of t and a Wiener integral. Using Theorem 8.6.1, we obtain that X_t is Gaussian with mean

$$E(X_t) = e^{-t}x_0 + m(1 - e^{-t})$$

and variance

$$\mathrm{Var}(X_t) = \sigma^2 e^{-2t}\int_0^t e^{2s}\, ds = \sigma^2 \frac{1 - e^{-2t}}{2}.$$

[5] The name *mean reverting* comes from the fact that in the long run the mean $E(X_t)$ tends to m.

Using the notation for the Gaussian distribution, the probability density of X_t is

$$\begin{aligned} p_t(x_0, x) &= G_{\sigma^2(1-e^{-2t})/2}\left(e^{-t}x_0 + m(1-e^{-t}), x\right) \\ &= \frac{1}{\sigma\sqrt{\pi(1-e^{-2t})}} e^{-\frac{1}{\sigma^2(1-e^{-2t})}[x-e^{-t}x_0-m(1-e^{-t})]^2} \\ &= \frac{1}{\sigma\sqrt{\pi(1-e^{-2t})}} e^{-\frac{1}{\sigma^2(1-e^{-2t})}[(x-x_0)+(x_0-m)(1-e^{-t})]^2}, \quad t > 0. \quad (8.12.53) \end{aligned}$$

It follows by the last step of the algorithm in Sect. 8.5 that the operator (8.12.51) has the heat kernel given by (8.12.53).

It is interesting to note that when $t \to \infty$, the heat kernel tends to

$$p_\infty(x_0, x) = \frac{1}{\sigma\sqrt{\pi}} e^{-\frac{(x-x_0)^2}{\sigma^2}},$$

which resembles a normal distribution. This shows that the distribution of heat does not go to zero in the long range, but it tends to a steady-state distribution.

Particular cases: If we let $\sigma = m = 1$, we obtain that the heat kernel of the operator

$$\frac{1}{2}\frac{\partial^2}{\partial x^2} + (1-x)\frac{\partial}{\partial x}$$

is given by

$$K(x_0, x; t) = \frac{1}{\sqrt{\pi(1-e^{-2t})}} e^{-\frac{1}{1-e^{-2t}}[(x-x_0)+(x_0-1)(1-e^{-t})]^2}, \qquad t > 0.$$

If we take $m = 0$, we obtain that the operator

$$A = \frac{1}{2}\sigma^2\frac{\partial^2}{\partial x^2} - x\frac{\partial}{\partial x},$$

has the heat kernel given by

$$K(x_0, x; t) = \frac{1}{\sigma\sqrt{\pi(1-e^{-2t})}} e^{-\frac{(x-e^{-t}x_0)^2}{\sigma^2(1-e^{-2t})}}, \qquad t > 0.$$

8.13 Bessel Operator and Bessel Process

In this section we shall deal with the process satisfied by the Euclidean distance from the origin to a particle following a Brownian motion in $\mathbb{R}^n$. More precisely, let $W(t) = (W_1(t), \dots, W_n(t))$ be a Brownian motion in $\mathbb{R}^n$, $n \geq 2$. Let

$$R(t) = \text{dist}(O, W(t)) = \sqrt{W_1(t)^2 + \cdots + W_n(t)^2}.$$

The process $R(t)$ is not smooth at $t = 0$, but since $W(t)$ never hits the origin almost surely for $n \geq 2$, we may still apply Ito's formula and obtain

$$dR(t) = \sum_{j=1}^{n} \frac{W_j(t)}{R(t)} dW_j(t) + \frac{n-1}{2R(t)} dt, \tag{8.13.54}$$

called the n-dimensional *Bessel process*. In this case

$$b(R) = \frac{n-1}{2R}, \qquad \sigma = \left[\frac{W_1}{R}, \dots, \frac{W_n}{R}\right], \qquad \sigma\sigma^T = \frac{\sum W_i^2}{R^2} = 1.$$

Formula (8.3.19) provides the generator

$$A = \frac{1}{2}\frac{\partial^2}{\partial\rho^2} + \frac{n-1}{2\rho}\frac{\partial}{\partial\rho}, \qquad \rho > 0, \tag{8.13.55}$$

called the *Bessel operator*.

The rest of this section deals with finding the transition density $p_t(\rho_0, \rho)$ of the Bessel process (8.13.54). The computation will be different for the cases $\rho_0 = 0$ and $\rho_0 \neq 0$.

In the case $\rho_0 = 0$ we have the following result. The next computation can be found in reference [83], p. 133.

Proposition 8.13.1. *If $\rho_0 = 0$, the transition density function of $R(t)$, $t > 0$, is given by*

$$p_t(0, \rho) = \begin{cases} \dfrac{2}{(2t)^{n/2}\,\Gamma(n/2)}\, \rho^{n-1} e^{-\frac{\rho^2}{2t}}, & \rho \geq 0, \\ 0, & \rho < 0, \end{cases}$$

with

$$\Gamma\left(\frac{n}{2}\right) = \begin{cases} (\frac{n}{2} - 1)! & \text{for } n \text{ even}, \\ (\frac{n}{2} - 1)(\frac{n}{2} - 2)\cdots\frac{3}{2}\frac{1}{2}\sqrt{\pi}, & \text{for } n \text{ odd}. \end{cases}$$

Proof. Since the Brownian motions $W_1(t), \ldots, W_n(t)$ are independent, their joint density function is

$$\begin{aligned} f_{W_1 \cdots W_n}(t) &= f_{W_1}(t) \cdots f_{W_n}(t) \\ &= \frac{1}{(2\pi t)^{n/2}} e^{-(x_1^2 + \cdots + x_n^2)/(2t)}, \qquad t > 0. \end{aligned}$$

In the next computation we shall use the following formula of integration which follows from the use of polar coordinates:

$$\int_{\{|x| \le \rho\}} f(x)\, dx = \sigma(\mathbb{S}^{n-1}) \int_0^\rho r^{n-1} g(r)\, dr, \tag{8.13.56}$$

where $f(x) = g(|x|)$ is a function on $\mathbb{R}^n$ with spherical symmetry, and where

$$\sigma(\mathbb{S}^{n-1}) = \frac{2\pi^{n/2}}{\Gamma(n/2)}$$

is the area of the $(n-1)$-dimensional sphere in $\mathbb{R}^n$.

Let $\rho \ge 0$. The distribution function of $R(t)$ is

$$\begin{aligned} F_R(\rho) = P(R(t) \le \rho) &= \int_{\{R(t) \le \rho\}} f_{W_1 \cdots W_n}(t)\, dx_1 \cdots dx_n \\ &= \int_{x_1^2 + \cdots + x_n^2 \le \rho^2} \frac{1}{(2\pi t)^{n/2}} e^{-(x_1^2 + \cdots + x_n^2)/(2t)}\, dx_1 \cdots dx_n \\ &= \int_0^\rho r^{n-1} \left(\int_{\mathbb{S}(0,1)} \frac{1}{(2\pi t)^{n/2}} e^{-(x_1^2 + \cdots + x_n^2)/(2t)}\, d\sigma \right) dr \\ &= \frac{\sigma(\mathbb{S}^{n-1})}{(2\pi t)^{n/2}} \int_0^\rho r^{n-1} e^{-r^2/(2t)}\, dr. \end{aligned}$$

Differentiating yields

$$\begin{aligned} p_t(0, \rho) = \frac{d}{d\rho} F_R(\rho) &= \frac{\sigma(\mathbb{S}^{n-1})}{(2\pi t)^{n/2}} \rho^{n-1} e^{-\frac{\rho^2}{2t}} \\ &= \frac{2}{(2t)^{n/2}\Gamma(n/2)} \rho^{n-1} e^{-\frac{\rho^2}{2t}}, \qquad \rho > 0, t > 0. \end{aligned}$$

■

In the two-dimensional case the aforementioned density becomes *Wald's distribution*, (see Fig. 8.6a)

$$f_t(x) = \frac{1}{t} x e^{-\frac{x^2}{2t}}, \qquad x > 0, t > 0.$$

The case $\rho_0 > 0$ is more complex and requires a few preliminary definitions.

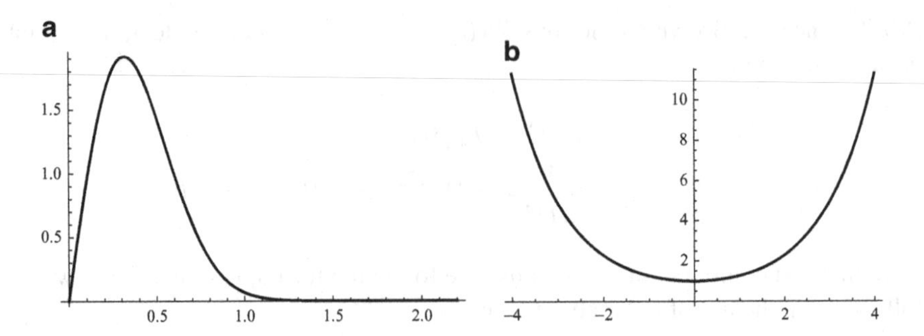

Fig. 8.6 (**a**) Wald's distribution; (**b**) the Bessel function $I_0(x)$

The second-order differential equation

$$z^2 \frac{d^2}{dz^2} + z \frac{dw}{dz} - (z^2 + \nu^2)w = 0$$

is called the *modified Bessel equation*. If $\nu = n$ is an integer, a fundamental system of solutions is given by the solutions $I_n(z)$ and $K_n(z)$. The modified Bessel function of the first kind has the expansion

$$I_n(z) = \sum_{m \geq 0} \frac{(z/2)^{2m+n}}{m!(m+n)!}; \tag{8.13.57}$$

see reference [41], p. 9. The case $\nu = 0$ serves our purposes, when the modified Bessel function of order zero of the first kind is given by

$$I_0(z) = \sum_{m \geq 0} \frac{(z/2)^{2m}}{(m!)^2}; \tag{8.13.58}$$

see Fig. 8.6b.

Lemma 8.13.2. *For any* $z \in \mathbb{C}$ *we have*

$$\int_0^{2\pi} e^{z \cos \theta} \, d\theta = 2\pi I_0(z), \tag{8.13.59}$$

where I_0 *is the modified Bessel function of order zero of the first kind.*

Proof. Integrating in the Taylor expansion

$$e^{z \cos \theta} = \sum_{n \geq 0} \frac{z^n \cos^n \theta}{n!}$$

yields

$$\int_0^{2\pi} e^{z\cos\theta}\,d\theta = \sum_{n\geq 0}\frac{z^n}{n!}\int_0^{2\pi}\cos^n\theta\,d\theta = \sum_{m\geq 0}\frac{z^{2m}}{(2m)!}\int_0^{2\pi}(\cos\theta)^{2m}\,d\theta$$

$$= 4\sum_{m\geq 0}\frac{z^{2m}}{(2m)!}\int_0^{\pi/2}(\cos\theta)^{2m}d\theta = 4\sum_{m\geq 0}\frac{z^{2m}}{(2m)!}\int_0^{\pi/2}(\sin\theta)^{2m}d\theta$$

$$= 4\sum_{m\geq 0}\frac{z^{2m}}{(2m)!}\frac{1\cdot 3\cdot 5\cdot\dots\cdot(2m-1)}{2\cdot 4\cdot\dots\cdot(2m)}\frac{\pi}{2}$$

$$= 2\pi\sum_{m\geq 0}\frac{z^{2m}}{(2m)!}\frac{(2m)!}{[2\cdot 4\cdot\dots\cdot(2m)]^2} = 2\pi\sum_{m\geq 0}\frac{z^{2m}}{2^{2m}}\frac{1}{(m!)^2} = 2\pi I_0(z),$$

by virtue of (8.13.57). In the previous proof we also used the following relations, which are left as an exercise for the reader:

$$\int_0^{2\pi}(\cos\theta)^{2m+1}\,d\theta = 0,\ \int_0^{2\pi}(\cos\theta)^{2m}\,d\theta = 4\int_0^{\pi/2}(\cos\theta)^{2m}\,d\theta,$$

$$\int_0^{\pi/2} f(\cos\theta)\,d\theta = \int_0^{\pi/2} f(\sin\theta)\,d\theta,$$

$$\int_0^{\pi/2}(\sin\theta)^{2m}\,d\theta = \frac{1\cdot 3\cdot 5\cdot\dots\cdot(2m-1)}{2\cdot 4\cdot\dots\cdot(2m)}\frac{\pi}{2}.$$

■

Now we go back to the problem of finding the transition density $p_t(\rho_0,\rho)$, with $t,\rho_0,\rho > 0$. The Brownian motion starts at the point $x_0 = (x_1^0,\dots,x_n^0)$ at $t = 0$ and we let $\rho_0 = |x_0|$. The conditional distribution function of $R(t)$ given $R(0) = \rho_0$ is given by

$$F_{R(t)|R(0)}(\rho|\rho_0) = P\big(R(t)\leq\rho|R(0)=\rho_0\big)$$

$$= \int_{\{R(t)\leq\rho\}} f_{W_1\cdots W_n|W_1(0)=x_1^0\cdots W_n(0)=x_n^0}(t)\,dx_1\cdots dx_n$$

$$= \int_{\{x_1^2+\dots+x_n^2\leq\rho^2\}}\frac{1}{(2\pi t)^{n/2}}e^{-\frac{1}{2t}\sum_{j=1}^n\left(x_j-x_j^0\right)^2}\,dx_1\cdots dx_n$$

$$= \frac{1}{(2\pi t)^{n/2}}\int_{\mathbf{B}(0,\rho)} e^{-\frac{1}{2t}|x-x_0|^2}\,dx$$

$$= \frac{1}{(2\pi t)^{n/2}}e^{-\frac{1}{2t}|x_0|^2}\int_{\mathbf{B}(0,\rho)} e^{-\frac{1}{2t}|x|^2}\,e^{\frac{1}{t}\langle x,x_0\rangle}\,dx. \qquad (8.13.60)$$

The integral over the ball $\mathbf{B}(0,\rho)$ can be easily computed in the particular case $n = 2$. The general case is more complicated and we shall omit it. Using polar coordinates, the area element changes as $dx_1 dx_2 = r\, d\rho\, d\theta$, where $r = |x|$. Since the inner product is invariant under orthogonal transformations, if $M \in \mathcal{O}_2$, $\det M = 1$, and then

$$\int_{\mathbf{B}(0,\rho)} e^{-\frac{1}{2t}|x|^2} e^{\frac{1}{t}\langle x, x_0\rangle} dx = \int_{\mathbf{B}(0,\rho)} e^{-\frac{1}{2t}|Mx|^2} e^{\frac{1}{t}\langle Mx, Mx_0\rangle} dx. \tag{8.13.61}$$

Choosing the orthogonal matrix M such that $Mx_0 = |x_0|e_1 = \rho_0 e_1$, by using polar coordinates and Fubini's theorem, the integral (8.13.61) becomes

$$\begin{aligned}\int_0^\rho \int_0^{2\pi} e^{-\frac{r^2}{2t}} e^{\frac{1}{t} r\rho_0 \cos\theta}\, r\, dr\, d\theta &= \int_0^\rho re^{-\frac{r^2}{2t}} \left[\int_0^{2\pi} e^{\frac{1}{t} r\rho_0 \cos\theta} d\theta\right] dr \\ &= \int_0^\rho re^{-\frac{r^2}{2t}} I_0\Big(\frac{r\rho_0}{t}\Big)\, r,\end{aligned}$$

by Lemma 8.13.2. Substituting back in (8.13.60) with $n = 2$ yields the following conditional distribution function:

$$F_{R(t)|R(0)}(\rho|\rho_0) = \frac{1}{2\pi t} e^{-\frac{1}{2t}\rho_0^2} \int_0^\rho re^{-\frac{r^2}{2t}} I_0\Big(\frac{r\rho_0}{t}\Big)\, dr.$$

Differentiating with respect to ρ yields the transition density function

$$p_t(\rho_0, \rho) = \frac{d}{d\rho} F_{R(t)|R(0)}(\rho|\rho_0) = \frac{1}{2\pi t} e^{-\frac{1}{2t}\rho_0^2} \rho e^{-\frac{\rho^2}{2t}} I_0\Big(\frac{\rho\rho_0}{t}\Big).$$

We have obtained the following result.

Proposition 8.13.3. *The heat kernel of the Bessel operator*

$$A = \frac{1}{2}\frac{\partial^2}{\partial \rho^2} + \frac{1}{2\rho}\frac{\partial}{\partial \rho}, \qquad \rho > 0, \tag{8.13.62}$$

is given by

$$p_t(\rho_0, \rho) = \frac{\rho}{2\pi t} e^{-\frac{1}{2t}(\rho_0^2 + \rho^2)} I_0\Big(\frac{\rho\rho_0}{t}\Big), \qquad t > 0, \rho > 0. \tag{8.13.63}$$

Since the heat kernel of a sum of commuting operators is the product of their kernels, we obtain the following consequence.

Corollary 8.13.4. *The heat kernel of the operator*

$$\frac{1}{2}\sum_{i=1^n}\frac{\partial^2}{\partial x_i^2}+\frac{1}{2}\sum_{i=1}^{n}\frac{1}{x_i}\frac{\partial}{\partial x_i}$$

is

$$K(x_0,x;t)=\frac{x_1\cdots x_n}{(2\pi t)^n}\,e^{-\frac{1}{2t}(|x_0|^2+|x|^2)}\prod_{i=1}^{n}I_0\left(\frac{x_ix_i^0}{t}\right),\qquad x_i,x_i^0,t>0.$$

8.14 Brownian Motion on a Circle

Let $x_0 = (\cos\phi_0,\sin\phi_0)\in S^1$ be a fixed point on the unit circle. If W_t is the one-dimensional Brownian motion, consider the following system of stochastic differential equations:

$$dX_1(t) = -\frac{1}{2}X_1(t)\,dt - X_2(t)\,dW_t, \tag{8.14.64}$$

$$dX_2(t) = -\frac{1}{2}X_2(t)\,dt + X_1(t)\,dW_t, \tag{8.14.65}$$

with the initial condition $X_0 = x_0$. According to Theorem 8.2.2, the aforementioned system has a unique solution $X(t) = \big(X_1(t),X_2(t)\big)$. If we let

$$X_t = \big(X_1(t),X_2(t)\big) = \big(\cos(\phi_0+W_t),\sin(\phi_0+W_t)\big), \tag{8.14.66}$$

a direct application of Ito's formula yields

$$dX_1(t) = -\frac{1}{2}\cos(\phi_0+W_t)\,dt - \sin(\phi_0+W_t)\,dW_t,$$

$$dX_2(t) = -\frac{1}{2}\sin(\phi_0+W_t)\,dt + \cos(\phi_0+W_t)\,dW_t,$$

which shows that (8.14.66) is the desired solution for the system (8.14.64)–(8.14.65). The process (8.14.66) is called the *Brownian motion on the circle* S^1 starting at x_0. The stochastic differential equation can also be written as

$$dX_t = b(X_t)\,dt + \sigma(X_t)\,dW_t,$$

with the drift and diffusion given by

$$b(x)=\begin{pmatrix}-x_1/2\\-x_2/2\end{pmatrix},\qquad \sigma(x_1,x_2)=\begin{pmatrix}-x_2\\x_1\end{pmatrix}.$$

Then

$$\sigma\sigma^T = \begin{pmatrix} x_2^2 & -x_1x_2 \\ -x_1x_2 & x_1^2 \end{pmatrix},$$

and hence the generator of the process X_t is

$$\begin{aligned} A &= \frac{1}{2}\sum_{i,j}(\sigma\sigma^T)_{ij}\frac{\partial^2}{\partial_{x_i}\partial_{x_j}} + \sum_j b_j(x)\frac{\partial}{\partial x_j} \\ &= \frac{1}{2}\Big[x_2^2\partial_{x_1}^2 + x_1^2\partial_{x_2}^2 - 2x_1x_2\partial_{x_1x_2}^2 - x_1\partial_{x_1} - x_2\partial_{x_2}\Big] \\ &= \frac{1}{2}(x_1\partial_{x_2} - x_2\partial_{x_1})^2. \end{aligned}$$

In polar coordinates $x_1 = r\cos\theta$, $x_2 = r\sin\theta$, we have $A = \frac{1}{2}\partial_\theta^2$, which is the Laplace operator on S^1.

The heat kernel of A is given by the transition density of the process

$$X_t = \big(\cos(\phi_0 + W_t), \sin(\phi_0 + W_t)\big).$$

The distribution function given that the Brownian motion on the circle starts from ϕ_0 at $t = 0$ is

$$\begin{aligned} F(\phi \mid \phi_0) &= P(\phi_0 + W_t + 2n\pi < \phi,\ \forall n \in \mathbb{Z}) \\ &= \sum_{n\in\mathbb{Z}} P(W_t \le 2n\pi + \phi - \phi_0) = \sum_{n\in\mathbb{Z}} F_{W_t}(2n\pi + \phi - \phi_0), \end{aligned}$$

and then the transition density is

$$\begin{aligned} p_t(\phi_0, \phi) &= \frac{d}{d\phi}F(\phi|\phi_0) = \sum_{n\in\mathbb{Z}} \frac{1}{\sqrt{2\pi t}}e^{-\frac{(2n\pi+\phi-\phi_0)^2}{2t}} \\ &= \frac{1}{\sqrt{2\pi t}}\sum_{n\in\mathbb{Z}} e^{-\frac{1}{2t}\left(4n^2\pi^2+(\phi-\phi_0)^2+4n\pi(\phi-\phi_0)\right)} \\ &= \frac{1}{\sqrt{2\pi t}}e^{-\frac{(\phi-\phi_0)^2}{2t}}\sum_{n\in\mathbb{Z}} e^{-\frac{2n^2\pi^2}{t}}e^{-\frac{2n\pi(\phi-\phi_0)}{t}} \\ &= \frac{1}{\sqrt{2\pi t}}e^{-\frac{(\phi-\phi_0)^2}{2t}}\left(1 + 2\sum_{n\ge1} e^{-\frac{2n^2\pi^2}{t}}\cosh\frac{2n\pi(\phi-\phi_0)}{t}\right). \end{aligned}$$

This series can also be expressed as a theta-function, see (3.11.35).

8.15 An Example of a Heat Kernel

The following example follows [25]. Consider the stochastic differential equation

$$dX_t = \left(\sqrt{1+X_t^2} + \frac{1}{2}X_t\right) dt + \sqrt{1+X_t^2}\, dW_t, \qquad X_0 = x_0. \qquad (8.15.67)$$

The drift and diffusion coefficients are given by $b(x) = \sqrt{1+x^2} + \frac{1}{2}x$, $\sigma(x) = \sqrt{1+x^2}$. The following estimations hold:

$$\begin{aligned}|\sigma(x) - \sigma(y)| &= |\sqrt{1+x^2} - \sqrt{1+y^2}| = \frac{|x^2-y^2|}{\sqrt{1+x^2}+\sqrt{1+y^2}} \\ &= \frac{|x+y|}{\sqrt{1+x^2}+\sqrt{1+y^2}} \cdot |x-y| < 2|x-y|,\end{aligned}$$

since

$$\begin{aligned}\frac{|x+y|}{\sqrt{1+x^2}+\sqrt{1+y^2}} &\le \frac{|x|}{\sqrt{1+x^2}+\sqrt{1+y^2}} + \frac{|y|}{\sqrt{1+x^2}+\sqrt{1+y^2}} \\ &< \frac{|x|}{\sqrt{1+x^2}} + \frac{|y|}{\sqrt{1+x^2}} < 1+1, \\ |b(x) - b(y)| &\le |\sqrt{1+x^2} - \sqrt{1+y^2}| + \frac{1}{2}|x-y| \\ &< 2|x-y| + \frac{1}{2}|x-y| = \frac{5}{2}|x-y|.\end{aligned}$$

Hence

$$|b(x) - b(y)| + |\sigma(x) - \sigma(y)| \le \frac{9}{2}|x-y|, \qquad \forall x, y \in \mathbb{R},$$

and by Theorem 8.2.2 it follows that the stochastic equation (8.15.67) has a unique solution. Therefore, instead of solving the equation, it suffices to guess a solution.

Let $U_t = c + t + W_t$, where $c = \sinh^{-1}(x_0)$. Then $X_t = \sinh(U_t)$, with $X_0 = x_0$, and by Ito's formula we have

$$\begin{aligned}dX_t &= \cosh(U_t)\, dU_t + \frac{1}{2}\sinh U_t\, (dU_t)^2 \\ &= \cosh(U_t)\,(dt + dW_t) + \frac{1}{2}\sinh(U_t)\, dt \\ &= \left(\cosh(U_t) + \frac{1}{2}\sinh(U_t)\right) dt + \cosh(U_t)\, dW_t\end{aligned}$$

$$= \left(\sqrt{1+\sinh^2(U_t)} + \frac{1}{2}\sinh(U_t)\right) dt + \sqrt{1+\sinh^2(U_t)}\, dW_t$$
$$= \left(\sqrt{1+X_t^2} + \frac{1}{2}X_t\right) dt + \sqrt{1+X_t^2}\, dW_t.$$

Hence

$$X_t = \sinh(c+t+W_t), \qquad c = \sinh^{-1}(x_0) \tag{8.15.68}$$

is the solution of (8.15.67). The generator of the aforementioned process X_t is the operator

$$A = \frac{1}{2}(1+x^2)\frac{d^2}{dx^2} + \left(\sqrt{1+x^2} + \frac{x}{2}\right)\frac{d}{dx}. \tag{8.15.69}$$

The conditional distribution function of X_t is

$$\begin{aligned} F_{X|X_0}(x|x_0) &= P(X_t \le x \mid X_0 = x_0) = P(\sinh(U_t) \le x \mid U_0 = \sinh^{-1} x_0) \\ &= P(c+t+W_t \le \sinh^{-1} x \mid x = \sinh^{-1} x_0) \\ &= P(W_t \le \sinh^{-1} x - c - t \mid c = \sinh^{-1} x_0) \\ &= P(W_t \le \sinh^{-1} x - \sinh^{-1} x_0 - t) \\ &= \int_{-\infty}^{\sinh^{-1} x - \sinh^{-1} x_0 - t} \frac{1}{\sqrt{2\pi t}} e^{-\frac{u^2}{2t}}\, du, \end{aligned}$$

so the transition density of X_t is

$$p_t(x_0, x) = \frac{d}{dx}F_{X|X_0}(x|x_0) = \frac{1}{\sqrt{2\pi t}} e^{-\frac{(\sinh^{-1} x - \sinh^{-1} x_0 - t)^2}{2t}}, \quad t > 0. \tag{8.15.70}$$

We may further transform this formula by using

$$\begin{aligned} \sinh^{-1} x - \sinh^{-1} x_0 &= \ln\left(x + \sqrt{1+x^2}\right) - \ln\left(x_0 + \sqrt{1+x_0^2}\right) \\ &= \ln \frac{x+\sqrt{1+x^2}}{x_0 + \sqrt{1+x_0^2}}, \end{aligned}$$

$$(\sinh^{-1} x - \sinh^{-1} x_0 - t)^2 = \ln^2 \frac{x+\sqrt{1+x^2}}{x_0+\sqrt{1+x_0^2}} - 2t \ln \frac{x+\sqrt{1+x^2}}{x_0+\sqrt{1+x_0^2}} + t^2.$$

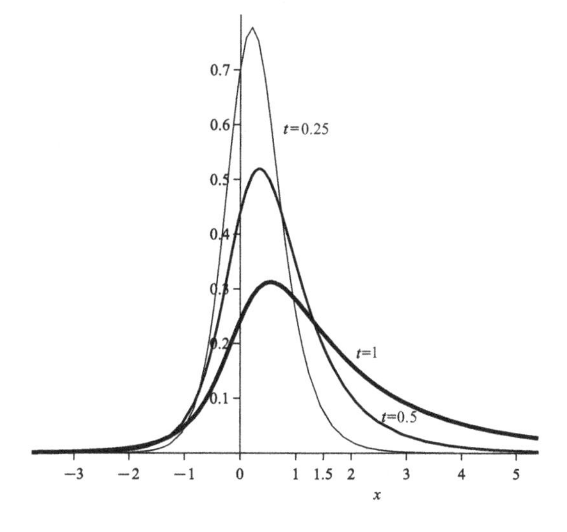

Fig. 8.7 The transition density as a function of x, with $x_0 = 0$ in the cases: $t = 1$, $t = 0.5$ and $t = 0.25$. For t small, the graph tends to the Dirac distribution centered at $x_0 = 0$

Then the transition density (8.15.70) becomes

$$p_t(x_0, x) = \frac{1}{\sqrt{2\pi t}} \frac{x + \sqrt{1+x^2}}{x_0 + \sqrt{1+x_0^2}} e^{-\frac{1}{2t}\ln^2 \frac{x+\sqrt{1+x^2}}{x_0+\sqrt{1+x_0^2}} - \frac{t}{2}}, \qquad t > 0. \quad (8.15.71)$$

To conclude, the heat kernel of the operator (8.15.69) is given by (8.15.71). Its graph is given by Fig. 8.7.

8.16 A Two-Dimensional Case

Let $W_1(t)$ and $W_2(t)$ be two independent Brownian motions and consider the random process $X_t = (X_1(t), X_2(t))$ given by

$$\begin{aligned} X_1(t) &= W_1(t) + W_2(t) + x_1^0, \\ X_2(t) &= \sinh\big(W_2(t) + c\big), \quad c = \sinh^{-1}(x_2^0). \end{aligned}$$

The process starts at $X_0 = (x_1^0, x_2^0) = x_0$. The associated stochastic differential equation is

$$\begin{aligned} dX_1(t) &= dW_1(t) + dW_2(t), \\ dX_2(t) &= \cosh\big(W_2(t) + c\big)\, dW_2(t) + \frac{1}{2} X_2(t)\, dt \\ &= \sqrt{1 + X_2(t)^2}\, dW_2(t) + \frac{1}{2} X_2(t)\, dt. \end{aligned}$$

In matrix form this becomes

$$\begin{pmatrix} dX_1(t) \\ dX_2(t) \end{pmatrix} = \begin{pmatrix} 0 \\ \frac{1}{2}X_2(t) \end{pmatrix} dt + \begin{pmatrix} 1 & 1 \\ 0 & \sqrt{1+X_2(t)^2} \end{pmatrix} \begin{pmatrix} dW_1(t) \\ dW_2(t) \end{pmatrix},$$

with the coefficients

$$b(x) = \begin{pmatrix} 0 \\ \frac{1}{2}x_2 \end{pmatrix}, \qquad \sigma = \begin{pmatrix} 1 & 1 \\ 0 & \sqrt{1+x_2^2} \end{pmatrix}, \qquad \sigma\sigma^T = \begin{pmatrix} 1 & \sqrt{1+x_2^2} \\ \sqrt{1+x_2^2} & 1+x_2^2 \end{pmatrix}.$$

Following relation (8.3.19), the generator of X_t is given by

$$\begin{aligned} A &= \frac{1}{2}\sum_{i,j}(\sigma\sigma^T)_{ij}\frac{\partial^2}{\partial x_i \partial x_j} + \sum_j b_j(x)\frac{\partial}{\partial x_j} \\ &= \frac{1}{2}\begin{pmatrix} 1 & \sqrt{1+x_2^2} \\ \sqrt{1+x_2^2} & 1+x_2^2 \end{pmatrix}\frac{\partial^2}{\partial x_i \partial x_j} + \begin{pmatrix} 0 \\ \frac{1}{2}x_2 \end{pmatrix} \cdot (\partial_{x_1}, \partial_{x_2}) \\ &= \frac{1}{2}\left(\frac{\partial^2}{\partial x_1^2} + 2\sqrt{1+x_2^2}\frac{\partial^2}{\partial x_1 \partial x_2} + (1+x_2^2)\frac{\partial^2}{\partial x_2^2}\right) + \frac{1}{2}x_2\partial_{x_2}. \end{aligned}$$

In order to find the heat kernel of the operator A, we need to find the transition density of the process $X_t = (X_1(t), X_2(t))$. This can be obtained from the transition density of the Brownian motion $\big(W_1(t), W_2(t)\big)$ using formula (8.6.23) and the technique presented in Sect. 8.6, item 2. The transformation

$$\begin{aligned} x_1 &= u_1 + u_2 + x_1^0, \\ x_2 &= \sinh(u_2 + c), \quad c = \sinh^{-1}(x_2^0), \end{aligned}$$

has the inverse

$$u_1 = (x_1 - x_1^0) - \big(\sinh^{-1}(x_2) - \sinh^{-1}(x_2^0)\big), \tag{8.16.72}$$

$$u_2 = \sinh^{-1}(x_2) - \sinh^{-1}(x_2^0), \tag{8.16.73}$$

with the Jacobian

$$\det\frac{\partial(u_1, u_2)}{\partial(x_1, x_2)} = \frac{1}{\sqrt{1+x_2^2}} \neq 0.$$

The density function of X_t is

$$\begin{aligned}
p_t(x_0, x) &= f_{X|X(0)=x_0}(x|x_0) = f_{W_1,W_2}\big(u_1(x_0,x), u_2(x_0,x)\big) \det \frac{\partial(u_1,u_2)}{\partial(x_1,x_2)} \\
&= f_{W_1}\big(u_1(x_0,x)\big)\, f_{W_2}\big(u_2(x_0,x)\big) \det \frac{\partial(u_1,u_2)}{\partial(x_1,x_2)} \\
&= \frac{1}{\sqrt{2\pi t}} e^{-\frac{u_1^2}{2t}} \frac{1}{\sqrt{2\pi t}} e^{-\frac{u_2^2}{2t}} \frac{1}{\sqrt{1+x_2^2}} \\
&= \frac{1}{2\pi t} \frac{1}{\sqrt{1+x_2^2}} e^{-\frac{|u|^2}{2t}},
\end{aligned}$$

with u given by (8.16.72)-(8.16.73). A computation provides

$$\begin{aligned}
|u|^2 &= u_1^2 + u_2^2 \\
&= (x_1 - x_1^0)^2 - 2(x_1 - x_1^0)\big(\sinh^{-1}(x_2) - \sinh^{-1}(x_2^0)\big) \\
&\quad + 2\big(\sinh^{-1}(x_2) - \sinh^{-1}(x_2^0)\big)^2 \\
&= (x_1 - x_1^0)^2 - 2(x_1 - x_1^0) \ln \frac{x_2 + \sqrt{1+x_2^2}}{x_2^0 + \sqrt{1+(x_2^0)^2}} + 2\big(\sinh^{-1}(x_2) - \sinh^{-1}(x_2^0)\big)^2
\end{aligned}$$

and hence

$$e^{-\frac{|u|^2}{2t}} = \left[\frac{x_2 + \sqrt{1+x_2^2}}{x_2^0 + \sqrt{1+(x_2^0)^2}}\right]^{\frac{x_1 - x_1^0}{t}} e^{-\frac{1}{2t}(x_1 - x_1^0)^2 - \frac{1}{t}\left(\sinh^{-1}(x_2) - \sinh^{-1}(x_2^0)\right)^2}.$$

We have arrived at the following result.

Proposition 8.16.1. *The heat kernel of the operator*

$$A = \frac{1}{2}\left(\frac{\partial^2}{\partial x_1^2} + 2\sqrt{1+x_2^2}\,\frac{\partial^2}{\partial x_1 \partial x_2} + (1+x_2^2)\frac{\partial^2}{\partial x_2^2}\right) + \frac{1}{2} x_2 \partial_{x_2}$$

is given by

$$\begin{aligned}
p_t(x_0, x) = {} & \frac{1}{2\pi t} \frac{1}{\sqrt{1+x_2^2}} \left[\frac{x_2 + \sqrt{1+x_2^2}}{x_2^0 + \sqrt{1+(x_2^0)^2}}\right]^{\frac{x_1 - x_1^0}{t}} \\
& e^{-\frac{1}{2t}(x_1 - x_1^0)^2 - \frac{1}{t}\left(\sinh^{-1}(x_2) - \sinh^{-1}(x_2^0)\right)^2}, \quad t > 0.
\end{aligned}$$

8.17 Kolmogorov's Operator

One of the classical examples of heat kernels obtained by using the transition density of an associated stochastic process is the degenerate operator of Kolmogorov on $\mathbb{R}^2$ (see [80, 81]):

$$L = \frac{1}{2}\partial_{x_1}^2 - x_1\partial_{x_2}. \tag{8.17.74}$$

Since

$$\sigma\sigma^T = \begin{pmatrix} 1 & 0 \\ 0 & 0 \end{pmatrix}, \qquad b = \begin{pmatrix} 0 \\ -x_1 \end{pmatrix}, \qquad \sigma = \begin{pmatrix} 1 & 0 \\ 0 & 0 \end{pmatrix},$$

the operator L is the generator of the following two-dimensional Ito diffusion:

$$\begin{aligned} dX_t &= b\,dt + \sigma\,dW \\ &= \begin{pmatrix} 0 \\ -X_1(t) \end{pmatrix} dt + \begin{pmatrix} 1 & 0 \\ 0 & 0 \end{pmatrix}\begin{pmatrix} dW_1(t) \\ dW_2(t) \end{pmatrix}, \end{aligned}$$

where $W_1(t)$ and $W_2(t)$ are independent Brownian motions. If $X_t = (X_1(t), X_2(t))$, then

$$\begin{aligned} dX_1(t) &= dW_1(t), \\ dX_2(t) &= -X_1(t)\,dt, \end{aligned}$$

with the solution

$$\begin{aligned} X_1(t) &= x_1^0 + W_1(t), \\ X_2(t) &= x_2^0 - \int_0^t \big(x_1^0 + W_1(s)\big)\,ds = x_2^0 - x_1^0 t - \int_0^t W_1(s)\,ds, \end{aligned}$$

where $X_0 = (x_1^0, x_2^0)$. We shall use the expected value method presented in item 6 of Sect. 8.6. Denote $W(t) = W_1(t)$:

$$\begin{aligned} p_t(x_0, x) &= E[\delta(X_1(t) - x_1)\delta(X_2(t) - x_2)] \\ &= E\left[\delta(x_1 - x_1^0 - W(t))\delta(x_2^0 - x_2 - x_1^0 t - \int_0^t W(s)\,ds)\right] \\ &= E\left[\delta(x_1^0 - x_1 + W(t))\frac{1}{2\pi}\int e^{i\eta(x_2^0 - x_2 - x_1^0 t - \int_0^t W(s)\,ds)}\,d\eta\right] \\ &= \frac{1}{2\pi}\int e^{i\eta(x_2^0 - x_2 - x_1^0 t)}\,E[\delta(x_1^0 - x_1 + W(t))e^{-i\eta\int_0^t W(s)\,ds}]\,d\eta. \end{aligned} \tag{8.17.75}$$

Let

$$v = v(x_1^0, x_1, t; \eta) = E[\delta(x_1^0 - x_1 + W(t))e^{-i\eta\int_0^t W(s)\,ds}].$$

According to the Feynman–Kac formula (see Theorem 8.23.1) with the linear potential $V(x) = ax$ and $a = i\eta$, the element v verifies the initial boundary problem

$$\partial_t v = \Big(\frac{1}{2}\partial_{x_1}^2 - ax_1\Big)v,$$

$$v_{|t=0} = \delta_{x_1^0}(x).$$

Theorem 3.15.1 yields

$$v = \frac{1}{\sqrt{2\pi t}} e^{-\frac{(x_1-x_1^0)^2}{2t} - \frac{1}{2}a(x_1+x_1^0)t + \frac{1}{24}a^2t^3} = \frac{1}{\sqrt{2\pi t}} e^{-\frac{(x_1-x_1^0)^2}{2t} - \frac{i\eta}{2}(x_1+x_1^0)t - \frac{t^3}{24}\eta^2}.$$

Substituting back in (8.17.75) yields

$$\begin{aligned} p_t(x_0, x) &= \frac{1}{2\pi}\frac{1}{\sqrt{2\pi t}} e^{-\frac{(x_1-x_1^0)^2}{2t}} \int e^{-\frac{t^3}{24}\eta^2 + i\eta[x_2 - x_2^0 - \frac{1}{2}(x_1+x_1^0)t]}\,d\eta \\ &= \frac{1}{2\pi}\frac{1}{\sqrt{2\pi t}} e^{-\frac{(x_1-x_1^0)^2}{2t}} \sqrt{\frac{24\pi}{t^3}} e^{-\frac{24}{4t^3}[x_2 - x_2^0 - \frac{1}{2}(x_1+x_1^0)t]^2} \\ &= \frac{\sqrt{3}}{\pi t^2} e^{-\frac{(x_1-x_1^0)^2}{2t} - \frac{6}{t^3}[x_2 - x_2^0 - \frac{1}{2}(x_1+x_1^0)t]^2}, \qquad t > 0, \end{aligned}$$

which is the heat kernel of the operator (8.17.74).

8.18 The Operator $\frac{1}{2}\mathbf{x}_1^2\partial_{\mathbf{x}_2}^2$

The operator $A = \frac{1}{2}x_1^2\partial_{x_2}^2$ is the generator of the two-dimensional Ito diffusion

$$dX_1(t) = 0,$$

$$dX_2(t) = X_1(t)\,dW_t, \qquad t \geq 0,$$

where W_t is a one-dimensional Brownian motion. Solving yields

$$X_1(t) = x_1^0,$$

$$X_2(t) = x_2^0 + x_1^0\,W_t.$$

Assume $x_1^0 \neq 0$. Then the transition distribution function is

$$\begin{aligned} F_{X|X_0}(x_1, x_2 \mid x_1^0, x_2^0) &= P\big(X_1(t) \le x_1,\, X_2(t) \le x_2 \mid X_1(0) = x_1^0, X_2(0) = x_2^0\big) \\ &= P(x_1^0 \le x_1,\, x_2^0 + x_1^0 W_t \le x_2) \\ &= P(x_1^0 \le x_1) P\left(W_t \le \frac{x_2 - x_2^0}{x_1^0}\right) \\ &= H_{x_1^0}(x_1) \int_{-\infty}^{\frac{x_2 - x_2^0}{x_1^0}} \frac{1}{\sqrt{2\pi t}} e^{-\frac{u^2}{2t}}\, du, \end{aligned}$$

where

$$H_{x_1^0}(x_1) = \begin{cases} 1, & x_1 \ge x_1^0, \\ 0, & x_1 < x_1^0, \end{cases}$$

is the Heaviside function. The transition density is obtained by differentiating the previous transition density:

$$\begin{aligned} p_t(x_0, x) &= \frac{\partial^2}{\partial x_1 \partial x_2} F_t(x_1, x_2 \mid x_1^0, x_2^0) \\ &= \frac{\partial}{\partial x_1} H_{x_1^0}(x_1) \frac{\partial}{\partial x_2} \int_{-\infty}^{\frac{x_2 - x_2^0}{x_1^0}} \frac{1}{\sqrt{2\pi t}} e^{-\frac{u^2}{2t}}\, du \\ &= \delta_{x_1^0}(x_1) \frac{1}{x_1^0 \sqrt{2\pi t}} e^{-\frac{1}{2t}\left(\frac{x_2 - x_2^0}{x_1^0}\right)^2}. \end{aligned}$$

Hence the heat kernel of $A = \frac{1}{2} x_1^2 \partial_{x_2}^2$ is given by

$$K(x_0, x; t) = \delta_{x_1^0}(x_1) \otimes \frac{1}{x_1^0 \sqrt{2\pi t}} e^{-\frac{1}{2t}\left(\frac{x_2 - x_2^0}{x_1^0}\right)^2},$$

for any initial point (x_1^0, x_2^0) with $x_1^0 \neq 0$.

If $x_1^0 = 0$, the transition distribution is

$$\begin{aligned} F_{X|X_0}(x_1, x_2 \mid 0, x_2^0) &= P\big(X_1(t) \le x_1,\, X_2(t) \le x_2 \mid X_1(0) = 0, X_2(0) = x_2^0\big) \\ &= P(0 \le x_1,\, x_2^0 \le x_2) = H_0(x_1)\, H_{x_2^0}(x_2), \end{aligned}$$

so by taking the derivative, we get

$$p_t(0, x) = \delta_0(x_1) \delta_{x_2^0}(x_2).$$

8.19 Grushin's Operator

The Grushin operator

$$\Delta_G = \frac{1}{2}(\partial^2_{x_1} + x_1^2 \partial^2_{x_2})$$

is a sub-elliptic operator which is the generator for the Ito diffusion $\big(X_1(t), X_2(t)\big)$ given by

$$dX_1(t) = dW_1(t),$$

$$dX_2(t) = X_1(t)\, dW_2(t),$$

with $X_1(0) = x_1^0$, $X_2(0) = x_2^0$ and $W_1(t)$, $W_2(t)$ one-dimensional independent Brownian motions starting at zero. Solving yields

$$X_1(t) = x_1^0 + W_1(t),$$

$$X_2(t) = x_2^0 + \int_0^t X_1(s)\, dW_2(s),$$

and the heat kernel of Δ_G is given by the following transition probability (see item 6 of Sect. 8.6):

$$\begin{aligned} p_t(x_0, x) &= E[\delta(x_1 - X_1(t))\delta(x_2 - X_2(t))] \\ &= E\left[\delta(x_1 - X_1(t))\frac{1}{2\pi}\int e^{i\eta(x_2 - X_2(t))}\, d\eta\right] \\ &= \frac{1}{2\pi}\int e^{i\eta(x_2 - x_2^0)} E\left[\delta(x_1 - X_1(t)) e^{-i\eta\int_0^t X_1(s) dW_2(s)}\right] d\eta. \quad (8.19.76) \end{aligned}$$

The following idea of computing comes from [31]. The expectation operator E is the Wiener integral over $W_1(t)$ and $W_2(t)$. Since $X_1(t)$ is independent of $W_2(t)$, we may integrate first with respect to $W_2(t)$ and get

$$E_{W_2}\left[e^{i\eta\int_0^t X_1(s)\, dW_2(s)}\right] = e^{-\frac{1}{2}\eta^2\int_0^t X_1^2(s)\, ds},$$

see (8.6.33). Substituting back in (8.19.76) yields

$$\begin{aligned} p_t(x_0, x) &= \frac{1}{2\pi}\int e^{i\eta(x_2 - x_2^0)}\, E[\delta(x_1 - X_1(t)) e^{-\frac{1}{2}\eta^2\int_0^t X_1^2(s); ds}]\, d\eta \\ &= \frac{1}{2\pi}\int e^{i\eta(x_2 - x_2^0)}\, v(x_1^0, x_1, t; \eta)\, d\eta, \quad (8.19.77) \end{aligned}$$

where here E denotes the expectation with respect to $W_1(t)$ and

$$v = v(x_1^0, x_1, t; \eta) = E\left[\delta(x_1 - X_1(t))e^{-\frac{1}{2}\eta^2 \int_0^t X_1^2(s)\,ds}\right]$$
$$= E\left[\delta(x_1 - (x_1^0 + W_1(t)))e^{-\frac{1}{2}\eta^2 \int_0^t (x_1^0 + W_1(s))^2\,ds}\right].$$

From the Feynman–Kac formula(see Theorem 8.23.1), with the quadratic potential $V(x) = \frac{1}{2}\eta^2 x^2$, the element v verifies the initial boundary problem

$$\partial_t v = \frac{1}{2}\partial_{x_1^2} - \frac{1}{2}\eta^2 x_1^2 v,$$
$$v_{|t=0} = \delta_{x_1^0}(x_1).$$

From Theorem 3.16.1, we get

$$v = \frac{1}{\sqrt{2\pi t}}\sqrt{\frac{\eta t}{\sinh(\eta t)}}\, e^{-\frac{1}{2t}\frac{\eta t}{\sinh(\eta t)}\left[\left(x_1^2 + (x_1^0)^2\right)\cosh(\eta t) - 2x_1 x_1^0\right]}, \quad t > 0.$$

Substituting back (8.19.77) yields

$$p_t(x_0, x) = \frac{1}{2\pi}\frac{1}{\sqrt{2\pi t}}\int \sqrt{\frac{\eta t}{\sinh(\eta t)}}\, e^{i\eta(x_2 - x_2^0) - \frac{1}{2t}\frac{\eta t}{\sinh(\eta t)}[\left(x_1^2 + (x_1^0)^2\right)\cosh(\eta t) - 2x_1 x_1^0]}\,d\eta,$$

which, after the substitution $\tau = \eta t$, becomes

$$p_t(x_0, x) = \frac{1}{(2\pi t)^{3/2}}\int \sqrt{\frac{\tau}{\sinh \tau}}\, e^{\frac{1}{t} f(x_0, x, t; \tau)},$$

with

$$f(x_0, x, t; \tau) = i\tau(x_2 - x_2^0) - \frac{\tau}{2}\left[\left(x_1^2 + (x_1^0)^2\right)\coth\tau - 2x_1 x_1^0 \operatorname{sech}\tau\right].$$

8.20 Squared Bessel Process

The operator

$$\frac{1}{2}x\partial_x^2 + \lambda\partial_x, \qquad x > 0, \tag{8.20.78}$$

is the generator of the following *squared Bessel process*:

$$dX_t = \lambda dt + \sqrt{X_t/2}\,dW_t, \qquad t \geq 0,$$

with $\lambda > 1/8$. The heat kernel of (8.20.78) with zero boundaries at $x = 0, \infty$ is given by the transition density of the aforementioned process in terms of the modified Bessel function(see [29]):

$$K(x_0, x; t) = \frac{e^{-(x+x_0)/(t/4)}}{t/4}\left(\frac{x}{x_0}\right)^{2\lambda-1/2} I_{4\lambda-1}(8\sqrt{xx_0}/t), \quad x, x_0 > 0, t > 0. \tag{8.20.79}$$

8.21 CIR Processes

The operator

$$\frac{1}{2}x\partial_x^2 + (\lambda_0 - \lambda_1 x)\partial_x, \qquad x > 0, \tag{8.21.80}$$

is the generator of the CIR process[6]

$$dX_t = (\lambda_0 - \lambda_1 x)dt + \sqrt{X_t/2}\,dW_t, \qquad t > 0,$$

with $\lambda_0, \lambda_1 > 0$, constants. These types of processes have been used to model stochastic interest rates in finance: see [37]. The heat kernel of the operator (8.21.80), given as the transition density of the CIR process, is given for instance in [29]:

$$K(x_0, x; t) = \frac{4\lambda_1 e^{\lambda_1 t}}{e^{\lambda_1 t} - 1}\left(\frac{x}{x_0}e^{\lambda_1 t}\right)^{2\lambda_0 - \frac{1}{2}} e^{-4\lambda_1 \frac{xe^{\lambda_1 t}+x_0}{e^{\lambda_1 t}-1}}$$
$$I_{4\lambda_0-1}\left(\frac{8\lambda_1}{e^{\lambda_1 t} - 1}\sqrt{xx_0e^{\lambda_1 t}}\right),$$

$x_0, x > 0, t > 0$.

8.22 Limitations of the Method

The *stochastic method* presented in this chapter works for a large class of operators. However, it does not apply to all operators. There are several limitations of the method that will be discussed in the following. These limitations can be overcome by using some of the methods presented in the other chapters.

---

[6] CIR stands for Cox, Ingersoll and Ross.

1. Explicit solution but the transition density is hard to get. Consider the case of the one-dimensional *linear noise with drift* given by

$$dX_t = r\,dt + \alpha X_t\,dW_t, \qquad r, \alpha \in \mathbb{R}. \tag{8.22.81}$$

Given $x_0 \in \mathbb{R}$, Theorem 8.2.2 ensures the existence and uniqueness of a process X_t satisfying (8.22.81) with the initial condition $X_0 = x_0$. According to Theorem 8.3.1, the generator associated with the process X_t is given by

$$A = \frac{1}{2}\alpha^2 x^2 \frac{\partial^2}{\partial x^2} + r\frac{\partial}{\partial x}. \tag{8.22.82}$$

Without loss of generality, we may assume $\alpha = 1$. Equation (8.22.81) becomes

$$dX_t - X_t\,dW_t = r\,dt.$$

Multiplying by the integrating factor $\mu_t = e^{-W_t + \frac{1}{2}t}$ yields

$$\mu_t\,dX_t - \mu_t X_t\,dW_t = r\mu_t\,dt. \tag{8.22.83}$$

We shall show that the left side is exact. Using the product rule and Ito's formula yields

$$\begin{aligned} d\mu_t &= -\mu_t\,dW_t + \mu_t\,dt, \\ d\mu_t\,dX_t &= -\mu_t X_t\,dt, \end{aligned}$$

since $dt^2 = dt\,dW_t = dW_t\,dt = 0$ and $dW_t\,dW_t = dt$. Then

$$\begin{aligned} d(\mu_t X_t) &= d\mu_t\,X_t + \mu_t\,dX_t + d\mu_t\,dX_t \\ &= -\mu_t X_t\,dW_t + \mu_t X_t\,dt + \mu_t\,dX_t - \mu_t X_t\,dt \\ &= \mu_t\,dX_t - \mu_t X_t\,dW_t, \end{aligned}$$

which is the left side of (8.22.83). Integrating in

$$d(\mu_t X_t) = r\mu_t\,dt$$

yields

$$\begin{aligned} \mu_t X_t &= \mu_0 X_0 + r\int_0^t \mu_s\,ds \iff \\ X_t &= \mu_t^{-1} x_0 + r\mu_t^{-1}\int_0^t \mu_s\,ds \iff \\ X_t &= e^{W_t - \frac{1}{2}t} x_0 + re^{W_t - \frac{1}{2}t}\int_0^t e^{-W_s + \frac{1}{2}s}\,ds, \end{aligned} \tag{8.22.84}$$

where $\mu_0 = 1$ and $X_0 = x_0$. If $r = 0$, then X_t has a lognormal distribution and this case was treated in Sect. 8.10. However, in the case $r \neq 0$, the transition density of the process (8.22.84) is not easy to obtain.

2. The stochastic differential equation has a unique solution, but it cannot be solved explicitly. Consider the following equation:

$$dX_t = dt + (\cos X_t)\, dW_t,$$
$$X_0 = x_0.$$

It is obvious that the existence and uniqueness condition of Theorem 8.2.2 applies and the equation has a unique solution X_t. However, a closed-form solution is hard to obtain. Hence the stochastic method is not suitable for finding the heat kernel of the operator

$$A = \frac{1}{2}(\cos x)^2 \frac{d^2}{dx^2} + \frac{d}{dx}.$$

3. The stochastic differential equation might have multiple solutions. This is the case of equation

$$dX_t = 3X_t^{1/3}\, dt + 3X_t^{2/3}\, dW_t, \tag{8.22.85}$$
$$X_0 = 0,$$

considered in Sect. 8.2. A direct application of the algorithm in Sect. 8.5 will not produce the heat kernel for

$$A = \frac{9}{2} x^{4/3} \frac{d^2}{dx^2} + 3x \frac{d}{dx}.$$

The aforementioned operator is singular at $x = 0$ and each solution of (8.22.85) has a contribution to the kernel that starts at the origin.

8.23 Operators with Potential and the Feynman–Kac Formula

In the previous sections we computed the heat kernel for a number of operators of the form

$$A = \frac{1}{2} \sum_{i,j} (\sigma\sigma^T)_{ij} \frac{\partial^2}{\partial x_i \partial x_j} + \sum_k b_k \frac{\partial}{\partial x_k} \tag{8.23.86}$$

associated with the n-dimensional Ito diffusion process

$$dX_t = b(X_t)\,dt + \sigma(X_t)\,dW(t), \tag{8.23.87}$$

which is supposed to satisfy the existence and uniqueness conditions of Theorem 8.2.2.

In this section we deal with an extension of Kolmogorov's backward equation (see Theorem 8.4.1), for operators of the type $A - V(x)$, where $V(x)$ is a potential function.

The first formula of this type deals with the n-dimensional Laplacian; it was found by Kac [75] in 1949. The result states that the equation

$$\begin{aligned} \frac{\partial v}{\partial t} &= \frac{1}{2}\Delta v - V(x)v, \qquad t > 0, \\ v(0,x) &= f(x), \end{aligned}$$

has the solution given by

$$v(t,x) = E^x\Big[f(X_t)\,e^{-\int_0^t V(X_s)\,ds}\Big], \tag{8.23.88}$$

where E^x is the conditional expectation given that the process X_t is a Brownian motion starting at x; i.e., $X_t = W_t + x$ with $W_0 = 0$.

The extension of the aforementioned result to any operator of the type (8.23.86) is given below. For a proof the reader can consult the references [83, 95].

Theorem 8.23.1 (Feynman–Kac formula). *Let X_t be the continuous solution of the stochastic differential equation (8.23.87) and let A be its generator. Let $f \in C_0^2(\mathbb{R}^n)$ and $V(x)$ be lower-bounded continuous functions on $\mathbb{R}^n$. Then the function*

$$v(t,x) = E^x[f(X_t)\,e^{-\int_0^t V(X_s)\,ds}], \qquad t \ge 0, x \in \mathbb{R}^n, \tag{8.23.89}$$

satisfies the Cauchy problem

$$\begin{aligned} \frac{\partial v}{\partial t} &= Av - Vv, \qquad t > 0, \\ v(0,x) &= f(x). \end{aligned}$$

Here E^x denotes the conditional expectation given that $X_0 = x$.

We note that if $V = 0$, then the Feynman–Kac formula becomes the Kolmogorov forward equation result; see Theorem 8.4.1.

In spite of its simplicity, formula (8.23.89) cannot be worked out explicitly unless the potential $V(x)$ is of a very particular form. Its importance is more of a theoretical value, since it provides an integral representation of the heat equation even in cases that do not have closed-form solutions. For example, in the case of quartic potential, the solution of the equation

$$\begin{aligned}\frac{\partial v}{\partial t} &= \frac{1}{2}\partial_x^2 v - x^4 v, \qquad t > 0,\\ v(0,x) &= f(x)\end{aligned}$$

is

$$v(t,x) = E^x[f(W_t + x)\, e^{-\int_0^t (W_s+x)^4\, ds}], \qquad t \geq 0, x \in \mathbb{R}.$$

It is worth noting that the Feynman–Kac formula can also be expressed in terms of path integrals; see Chap. 7.

Part II
Heat Kernel on Nilpotent Lie Groups and Nilmanifolds

Chapter 9
Laplacians and Sub-Laplacians

9.1 Sub-Riemannian Structure and Heat Kernels

We have seen in the previous chapters how an elliptic operator can be associated in a natural way with a geometric Riemannian structure. In a similar way sub-elliptic operators arise from similar structures, called sub-Riemannian structures, which will be discussed next. References for sub-Riemannian manifolds are [27] and [92].

9.1.1 Sub-Riemannian Structure

Let M be a smooth manifold. We shall assume M without boundary throughout this chapter. A sub-bundle $\mathcal{H}$ of the tangent bundle $T(M)$ is called *nonholonomic* if the vector fields of the sub-bundle $\mathcal{H}$ and a finite number of iterations of their Lie brackets span the whole tangent space at each point. The sub-bundle $\mathcal{H}$ is sometimes called the *horizontal distribution*.

Definition 9.1.1. A connected manifold M is called a sub-Riemannian manifold if it has a nonholonomic sub-bundle $\mathcal{H}$ with a positive definite, nondegenerate metric on $\mathcal{H}$.

The aforementioned nonholonomic property of the sub-bundle $\mathcal{H}$ is also called the *bracket generating property* or *Hörmander's condition* [65]. Sometimes it is also called *Chow's condition* [35].

Next we provide the physical interpretation for the previous definition. The coordinates of a physical system can be represented as a point on a manifold, called the *configuration space*. The states of the system, which consist of pairs of coordinates and momenta, form the cotangent bundle of a manifold, called the *phase space*.

The physical system is Riemannian if the states "can move in any direction"; i.e., there are curves in the configuration space starting at a given point pointing in all directions. These systems have the freedom to move in all directions.

However, there are examples of mechanical systems with nonholonomic constraints which are not of the type just described. A rolling ball on a surface or a

O. Calin et al., *Heat Kernels for Elliptic and Sub-elliptic Operators*, Applied and Numerical Harmonic Analysis, DOI 10.1007/978-0-8176-4995-1_9,

penny rolling on a plane are two examples of these kind of systems. In this case the system is sub-Riemannian, which means that the structure moves only along horizontal directions, i.e., directions tangent to the sub-bundle $\mathcal{H}$.

Given an initial configuration and a final configuration, it is natural to ask whether it is always possible for a sub-Riemannian system to move from the initial position to the final position in the configuration space. Everyone has faced this problem while parking a car. Starting from a given position, the car has to move into a given final parking position. Since this motion cannot be done by only one smooth maneuver, the horizontal curve on the associated sub-Riemannian manifold will be just piecewise smooth. The following global connectedness result holds for sub-Riemannian manifolds:

Theorem 9.1.2 (Chow's theorem, [35]). *Any two points of a sub-Riemannian manifold can be joined by a piecewise smooth horizontal curve, i.e., a curve tangent to the horizontal distribution $\mathcal{H}$.*

Gromov showed that in some particular cases the "piecewise" attribute can be omitted, but in general it cannot.

There are examples of physical systems which have states that cannot be connected by a horizontal curve. This is the case with thermodynamical systems where the horizontal curves correspond to adiabatic processes, i.e., processes along which there is no heat exchange. J. Charney was the first to note that there are always states which cannot be joined by an adiabatic curve, i.e., there are states which cannot be connected by a horizontal curve; see, for instance, [27], Chap. 3.

We shall list below several basic definitions related to a sub-Riemannian structure.

Definition 9.1.3.

1. The dimension $\dim T(M)/\mathcal{H}$ is called the *codimension* of the sub-Riemannian structure $\mathcal{H}$.
2. Let $\Gamma(M, T(M))$ be the space of all smooth vector fields on M, and set $\Gamma^0(M,\mathcal{H}) = \Gamma(M,\mathcal{H})$, the space of smooth vector fields on M taking values in $\mathcal{H}$. For any $k \in \mathbb{N}$, let

$$\Gamma^k(M,\mathcal{H}) = \Gamma^0(M,\mathcal{H}) + [\Gamma^0(M,\mathcal{H}), \Gamma^{k-1}(M,\mathcal{H})].$$

   For any point $x \in M$, we denote by $\Gamma_x^\ell(\mathcal{H})$ the values of the tangent vectors in $\Gamma^\ell(M,\mathcal{H})$ at the point $x \in M$. If for some $k > 0$, $\Gamma_x^k(\mathcal{H}) = T_x(M)$ but $\Gamma_x^{k-1}(\mathcal{H}) \neq T_x(M)$, then we say the sub-Riemannian structure is of *step* $k+1$ at the point x. The sub-Riemannian structure is step k on the domain U if it is step k at each point $x \in U$.
3. If there is an ascending sequence of sub-bundles $\{\mathcal{H}_\ell\}$ of $T(M)$

$$\mathcal{H} = \mathcal{H}_0 \subset \mathcal{H}_1 \subset \mathcal{H}_1 \subset \cdots \subset \mathcal{H}_k = T(M)$$

such that $\Gamma^{\ell}(M, \mathcal{H}) = \Gamma(M, \mathcal{H}_{\ell})$, then the sub-Riemannian structure is called *regular*. We have denoted by $\Gamma(M, \mathcal{H}_{\ell})$ the space of smooth vector fields taking values in $\mathcal{H}_{\ell}$. Note that in this case all sub-bundles $\mathcal{H}_{\ell}$ are nonholonomic.
4. A sub-Riemannian structure $\mathcal{H}$ is called *minimal* if there is no sub-bundle in $\mathcal{H}$ which satisfies the bracket generating property.

Two additional conditions on the nonholonomic sub-bundle $\mathcal{H}$ shall be required throughout the book.

(Sub-1) *The horizontal sub-bundle $\mathcal{H}$ is trivial.*

This means that we can consider globally defined, nowhere-vanishing smooth vector fields $\{X_i\}$, $i = 1, \ldots, \dim \mathcal{H}$, satisfying Hörmander's condition, which guarantees the hypoellipticity of the following second-order operator:

$$\sum_{i=1}^{\dim \mathcal{H}} X_i^{*} X_i .$$

The second assumption is

(Sub-2) *There is a volume form d_{VM} with respect to which the vector fields $\{X_i\}$ are all skew-symmetric.*

If the aforementioned conditions hold, we say that the sub-Riemannian structure is "trivializable," or is of the strong sense. It is worth noting that even if the typical examples of sub-Riemannian manifolds are contact manifolds, they are not always trivializable structures.

Furthermore, the associated sub-elliptic operator

$$\Delta_{\text{sub}} = -\sum_{i=1}^{\dim \mathcal{H}} X_i^{2}$$

is called the *sub-Laplacian*. Our next main purpose is to construct heat kernels for sub-Laplacians on sub-Riemannian manifolds. A special role will be played by the Grushin-type operators that will be treated in Sect. 10.3. Most of the operators in this chapter are sub-Laplacian operators on nilpotent Lie groups and their quotient spaces.

It is a rather convenient assumption to consider the metric on the sub-bundle $\mathcal{H}$ in such a way that the vector fields $\{X_i\}$ are orthonormal.

9.1.2 Heat Kernel of the Sub-Laplacian and Laplacian

Under the aforementioned conditions **(Sub-1)** and **(Sub-2)**, the sub-Laplacian Δ_{sub} is symmetric and positive with respect to the volume form $d_M V$, whose existence is assumed by the latter condition. If the metric is extended to the entire tangent space, then the Laplacian is symmetric and positive with respect to the Riemannian volume form. The volume element that we assumed to exist for the sub-Riemannian manifold will coincide in this case with the Riemannian volume form, up to a multiplicative constant, as we shall see in the following examples.

Moreover, we have

Theorem 9.1.4 ([105]). *If a sub-Riemannian metric on a noncompact manifold can be extended to a complete Riemannian metric, then the sub-Laplacian is essentially self-adjoint on the space of smooth functions with compact support.*

In all the cases treated in the sequel, we encounter left invariant operators on nilpotent Lie groups, for which the volume form $d_V M$ postulated by **(Sub-2)** is a Haar measure, and the conditions **(Sub-1)** and **(Sub-2)** are satisfied with respect to this measure. In this case, a left invariant sub-Riemannian metric defined on a left invariant nonholonomic sub-bundle can be extended to a left invariant complete metric on the entire space.

This is the reason why in the following cases we shall not distinguish between the sub-Laplacian Δ_{sub}, respectively the Laplacian Δ, both defined on $C_0^\infty(M)$, and their unique self-adjoint realizations on the space $L_2(M,\ d_M V)$.

Invoking the spectral decomposition theorem

$$\Delta_{\text{sub}} = \int_0^\infty \lambda dE_\lambda,$$

with $\{E_\lambda\}$ the spectral measure, we know that the heat kernel $K(t;\tilde{g},\ g)$ is the kernel distribution of the bounded operator

$$e^{-t\Delta_{\text{sub}}} = \int_0^\infty e^{-\lambda t} dE_\lambda : L_2(M) \to L_2(M), \qquad t > 0.$$

This kernel is smooth because

1. Δ_{sub} is hypoelliptic
2. For any integer k, the operator $\Delta_{\text{sub}}^{\ k} \circ e^{-t\Delta_{\text{sub}}} = e^{-t\Delta_{\text{sub}}} \circ \Delta_{\text{sub}}^{\ k}$ is defined on $L_2(M)$ and $e^{-t\Delta_{\text{sub}}}$ maps continuously $L_2(M)$ to $\bigcap_{k=1}^{\infty}\left(\text{domain of } \Delta_{\text{sub}}^{\ k}\right) = C^\infty(M)$
3. $e^{-t\Delta_{\text{sub}}}$ can be extended to the entire space of distributions $\mathcal{D}'(M)$ on M from $L_2(M)$. We have $e^{-t\Delta_{\text{sub}}} = e^{-t/2\Delta_{\text{sub}}} \circ e^{-t/2\Delta_{\text{sub}}}$ as a map from $\mathcal{D}'(M)$ to $C^\infty(M)$

Of course, the previous arguments are also valid for the case of Laplacians and the heat kernel for Laplacians on $C^\infty(\mathbb{R}_+ \times M \times M)$.

We shall next review the purpose of our future exposition.

(1) We are interested in the spectral decomposition of the sub-Laplacian and Laplacian in a *multiplication operator form.* This means finding a measure space $(\mathbf{X},\ \mathbf{dm})$, a (positive) function φ on $\mathbf{X}$, and a unitary transformation

$$\mathbf{U} : L_2(M) \xrightarrow{\sim} L_2(\mathbf{X}, \mathbf{dm}) \text{ such that}$$
$$\Delta_{\text{sub}} = \mathbf{U}^{-1} \circ \mathbf{M}_{\varphi} \circ \mathbf{U},$$

where $\mathbf{M}_{\varphi} : L_2(\mathbf{X}, \mathbf{dm}) \ni f \mapsto \varphi f \in L_2(\mathbf{X}, \mathbf{dm})$ is the multiplication operator by the function φ.

(2) We shall provide an *explicit* construction for the heat kernel $K(t; x, y)$ of the sub-Laplacian Δ_{sub} (and also Laplacian). This means finding a solution $K(t; x, y)$ for

$$\left(\Delta_{\text{sub}} + \frac{\partial}{\partial t}\right) K(t; x, y) = 0,$$
$$\lim_{t \to 0} \int_M K(t; x, y) f(y) d_M V(y) = f(x).$$

In the case of nilpotent Lie groups, using the (left) invariance of the heat kernel

$$K(t; g \cdot x, g \cdot y) = K(t; x, y), \qquad \forall g, x, y \in M,$$

it follows that the heat kernel $K(t; x, y)$ is of the form

$$K(t; x, y) = k_t(y^{-1} * x),$$

with $k_t(x)$ a smooth function on $\mathbb{R}_+ \times M$.

(3) We shall find *explicit* expressions for the Green functions and fundamental solutions of the operators Δ_{sub} and Δ.

(4) We aim to obtain *explicit* expressions for the heat kernel of the Grushin-type operators defined in Sect. 10.3.

There are many papers in the literature treating the hypoelliptic operators on nilpotent Lie groups (for instance, [15, 100] and papers cited therein). However, our concern here is to construct the heat kernel in terms of special functions. We shall use Hermite functions and hyperbolic functions for the two-step cases, while for the three-step cases the computation might be achieved by using elliptic functions and other special functions; see [21,22,58]. However, until now there are no explicit formulas for Laplacians and sub-Laplacians on nilpotent Lie groups with step greater than three (including the Engel group). Therefore, we are limited in our endeavor of finding- closed form formulas only for the two-step nilpotent Lie groups.

Remark 9.1.5. If we have a spectral decomposition of Δ (or Δ_{sub}) in the multiplication operator form, then the heat kernel is expressed as

$$e^{-t\Delta} = \mathbf{U} \circ e^{-t\mathbf{M}_{\varphi}} \circ \mathbf{U}^{-1}.$$

This is similar to the case of Euclidean spaces where the operator $\mathbf{U}$ is the Fourier transform. So **(1)** leads to **(2)**, but not the other way around.

Integrating the heat kernels, we obtain Green functions. If the heat kernel is

$$K(t; x, y) \in C^\infty(\mathbb{R}^+ \times M \times M),$$

then the Green function $G(x, y)$ can be expressed as

$$\int_M G(x, y) f(y) d_M V(y) = \int_0^\infty \int_G K(t; x, y) f(y) d_M V(y) \wedge dt.$$

Even if we have an *explicit* expression for the heat kernel, this does not provide the spectral decomposition of the (sub-)Laplacian in a multiplication operator form directly. For this purpose we are still required to find a measure space $(\mathbf{X}, \mathbf{dm})$, a unitary transformation between $L_2(G)$ and $L_2(\mathbf{X}, \mathbf{dm})$, and a positive function on $\mathbf{X}$, with respect to which the Laplacian (sub-Laplacian) can be expressed as a multiplication operator.

9.1.3 The Volume Form

In the definition of the sub-Riemannian manifold we have assumed that there exists a volume form $d_M V(x)$ such that each vector field X_k is skew-symmetric with respect to the inner product induced by this volume form:

$$\int_M X_k(f)(x) \overline{g(x)} d_M V(x) = -\int_M f(x) \overline{X_k(g)(x)} d_M V(x), \quad f \text{ or } g \in C_0^\infty(M).$$

For regular sub-Riemannian manifolds, we can construct a volume form in an intrinsic way (see [92] for Popp's measure) based on the inner product defined on the nonholonomic sub-bundle, even if the vector fields taking values in the nonholonomic sub-bundle are not necessarily skew-symmetric.

For the sake of simplicity, in this section we shall explain the construction of the volume form just for the case of two-step regular sub-Riemannian manifolds.

Let $X \in \mathcal{H}_x$ and $Y \in \mathcal{H}_x$ be tangent vectors which are extended to the vector fields $\tilde{X}$ and $\tilde{Y}$ on M taking values in $\mathcal{H}$. Then the value of the bracket $[\tilde{X}, \tilde{Y}] \in T_x(M)$ (mod $\mathcal{H}_x$) does not depend on the extensions $\tilde{X}$ and $\tilde{Y}$. Thus we have a well-defined bundle map

$$\rho : \mathcal{H} \otimes \mathcal{H} \to T(M)/\mathcal{H}.$$

The map ρ is surjective by the bracket generating assumption

$$\Gamma(M, \mathcal{H}) + [\Gamma(M, \mathcal{H}), \Gamma(M, \mathcal{H})] = \Gamma(T(M)).$$

Now we can introduce a metric on $T(M)/\mathcal{H}$ such that it is isometric with the orthogonal complement of the kernel of the map ρ. Note that $\mathcal{H} \otimes \mathcal{H}$ has a tensorial product metric from $\mathcal{H}$. Then we have an exact sequence of vector bundles with metrics except total space $T(M)$:

$$0 \to \mathcal{H} \to T(M) \to T(M)/\mathcal{H} \to 0. \tag{9.1.1}$$

Since the isomorphism

$$\bigwedge^{\dim \mathcal{H}} \mathcal{H}^* \otimes \bigwedge^{\dim M - \dim \mathcal{H}} \left(T(M)/\mathcal{H}\right)^* \cong \bigwedge^{\dim M} T^*(M)$$

does not depend on the choice of the splitting of the previous exact sequence, with the help of the metrics on $\mathcal{H}$ and $T(M)/\mathcal{H}$, we can trivialize $\bigwedge^{\dim M} T^*(M)$ by choosing dual orthonormal bases of these two bundles at each point. This provides a volume form on M.

Remark 9.1.6. In the following examples the vector fields trivializing a nonholonomic sub-bundle are skew-symmetric with respect to the volume form constructed by this way (up to a multiplicative constant).

9.2 Laplacian and Sub-Laplacian on Nilpotent Lie Groups

In this section we shall explain the basic properties of nilpotent Lie groups and sub-Riemannian structures on such Lie groups. We shall also deal with several examples of nilpotent Lie groups and sub-Laplacians on them for which we shall construct heat kernels in later sections.

9.2.1 Nilpotent Lie Groups

Let G be a nilpotent Lie group with the Lie algebra $\mathfrak{g}$. This means that if we put $\mathfrak{g}^1 = [\mathfrak{g}, \mathfrak{g}]$, $\mathfrak{g}^2 = [\mathfrak{g}, \mathfrak{g}^1], \ldots, \mathfrak{g}^k = [\mathfrak{g}, \mathfrak{g}^{k-1}], \ldots$, then there exists $N \in \mathbb{N}$ such that $\mathfrak{g}^N = \{0\}$. For an elementary treatment of nilpotent Lie groups, we refer the reader to [110].

We begin by reminding the reader of one of the most fundamental results.

Theorem 9.2.1. *Let G be a connected and simply connected nilpotent Lie group. Then the exponential map* $\exp : \mathfrak{g} \longrightarrow G$ *is a diffeomorphism.*

Because of this fact, we can work on the linear space $\mathfrak{g}$ using the Campbell–Hausdorff formula

$$\exp X \cdot \exp Y = \exp(X + Y + 1/2[X, Y] + 1/12[[X, Y], Y - X] + \ldots);$$
$$X,\ Y \in \mathfrak{g},$$

which says that we can introduce the group multiplication in the Lie algebra $\mathfrak{g}$ through the identification $\exp : \mathfrak{g} \xrightarrow{\sim} G$. The group law is defined on $\mathfrak{g}$ as in the following:

If $\exp X \cdot \exp Y = \exp Z$, then we define $X * Y = Z$, with Z given by
$Z = X + Y + 1/2[X, Y] + 1/12[[X, Y], Y - X] + \dots$ (finite sum).

We shall use the notations $g, h, \dots$ (or $\tilde{g}, \tilde{h}, \dots,$) $\in G \cong \mathfrak{g}$ if we consider them as elements of the Lie group $\mathfrak{g}$, and $X, Y, \dots \in \mathfrak{g}$ if we consider them as elements of the Lie algebra $\mathfrak{g}$. This way we eliminate the confusion of whether we are dealing with $\mathfrak{g}$ as a group or as a Lie algebra.

Let $X \in \mathfrak{g}$, and denote by $\widetilde{X}$ the left invariant vector field on the Lie group G defined by

$$\widetilde{X}(f)(g) = \frac{df(g \cdot \exp tX)}{dt}_{|t=0},$$

where $f \in C^\infty(\mathfrak{g})$. By the correspondence $\mathfrak{g} \ni X \to \widetilde{X}$, we have the following trivialization of the tangent bundle $T(G)$ of the group G:

$$G \times \mathfrak{g} \longrightarrow T(G),$$

$$(g, X) \longmapsto \widetilde{X}_g \in T_g(G).$$

We also have a trivialization of the cotangent bundle $T^*(G)$:

$$G \times \mathfrak{g}^* \longrightarrow T^*(G),$$

$$(g, \theta) \longmapsto \widetilde{\theta}_g \in T_g(G),$$

where the left invariant one-form $\widetilde{\theta}$ for $\theta \in \mathfrak{g}^*$ is defined by the formula

$$\widetilde{\theta}_g(\widetilde{X}_g) = \theta(X), \ X \in \mathfrak{g}.$$

Let $\{X_i\}_{i=1}^{\dim \mathfrak{g}}$ be a basis of the Lie algebra $\mathfrak{g}$. We consider the coordinates on the Lie group $G = \mathfrak{g}$ introduced by the correspondence

$$(x_1, \dots, x_{\dim \mathfrak{g}}) \leftrightarrow g = \sum_{i=1}^{\dim \mathfrak{g}} x_i X_i.$$

Then we can take the differential form

$$dg = dx_1 \wedge \dots \wedge dx_{\dim \mathfrak{g}}$$

as a Haar measure.

Proposition 9.2.2. *Let $X \in \mathfrak{g}$. If we consider the left invariant vector field $\widetilde{X}$ as a first-order differential operator on the group G,*

$$\widetilde{X} : C^\infty(G) \to C^\infty(G),\ f \mapsto \widetilde{X}(f),$$

then it is skew-symmetric with respect to the inner product defined by a Haar measure.

Proof. Let $\varphi,\ \psi \in C_0^\infty(G)$. We have

$$\begin{aligned}\int_G \widetilde{X}(\varphi)(g)\cdot\psi(g)dg &= \lim_{t\to 0}\int_G \frac{d\varphi\big(g\cdot\exp(tX)\big)}{dt}_{|t=0}\cdot\psi(g)dg\\ &= \lim_{t\to 0}\int_G \varphi(g)\cdot\frac{d\psi\big(g\cdot\exp(-tX)\big)}{dt}_{|t=0}dg\\ &= -\int_G \varphi(g)\cdot\widetilde{X}(\psi)(g)dg,\end{aligned}$$

since the Haar measure is left and right invariant in the case of the nilpotent Lie groups. ■

Let $\rho_1 : \mathfrak{g} \to \mathfrak{g}/[\mathfrak{g},\mathfrak{g}]$ be the projection map and let $\{X_i\}_{i=1}^{\ell}$ be linearly independent elements such that $\{\rho_1(X_i)\}_{i=1}^{\ell}$ spans the quotient space $\mathfrak{g}/[\mathfrak{g},\mathfrak{g}]$. Then the subspace $\{\tilde{X}_i\}_{i=1}^{\ell}$ defines a regular sub-Riemannian structure on the group G. If the group is a k-step nilpotent Lie group, then this structure is minimal, nonholonomic, and of step k.

9.2.2 The Heisenberg Group

Let H_3 be the three-dimensional Heisenberg group identified by $\mathbb{R}^3$. The product law is given by the formula

$$\begin{aligned}\mathbb{R}^3\times\mathbb{R}^3 \ni \Big((x,y,z),(\tilde{x},\tilde{y},\tilde{z})\Big) &\longmapsto (x,y,z)*(\tilde{x},\tilde{y},\tilde{z})\\ &= \big(x+\tilde{x},y+\tilde{y},z+\tilde{z}+1/2(x\tilde{y}-\tilde{y}x)\big)\in\mathbb{R}^3.\end{aligned}$$

Then its Lie algebra $\mathfrak{h}_3$ is identified by itself; i.e., the exponential map is the identity.

H_3 can also be realized as a real 3×3 matrix space

$$H_3 \cong \left\{\begin{pmatrix}1 & x & z\\ 0 & 1 & y\\ 0 & 0 & 1\end{pmatrix};\ x,y,z\in\mathbb{R}\right\}.$$

The identification is given by the map

$$H_3 \ni (x,y,z) \longmapsto \begin{pmatrix}1 & x & z+\frac{xy}{2}\\ 0 & 1 & y\\ 0 & 0 & 1\end{pmatrix}.$$

Remark 9.2.3. There are also other ways to realize the group H_3 as a matrix space. For example, if we let

$$X = \begin{pmatrix} 0 & 1 & 0 & 0 \\ 0 & 0 & 0 & 0 \\ 0 & 0 & 0 & -1 \\ 0 & 0 & 0 & 0 \end{pmatrix}, \quad Y = \begin{pmatrix} 0 & 0 & 1 & 0 \\ 0 & 0 & 0 & 1 \\ 0 & 0 & 0 & 0 \\ 0 & 0 & 0 & 0 \end{pmatrix} \quad \text{and } Z = \begin{pmatrix} 0 & 0 & 0 & 2 \\ 0 & 0 & 0 & 0 \\ 0 & 0 & 0 & 0 \\ 0 & 0 & 0 & 0 \end{pmatrix},$$

we have $[X, Y] = Z$, and the space of matrices

$$\left\{ \begin{pmatrix} 0 & x & y & z \\ 0 & 0 & 0 & y \\ 0 & 0 & 0 & -x \\ 0 & 0 & 0 & 0 \end{pmatrix} ; x, y, z \in \mathbb{R} \right\}$$

is isomorphic to $\mathfrak{h}_3$. So the three-dimensional Heisenberg group H_3 is identified by a subgroup in 4×4 matrix space as

$$H_3 = \left\{ \begin{pmatrix} 1 & x & y & z \\ 0 & 1 & 0 & y \\ 0 & 0 & 1 & -x \\ 0 & 0 & 0 & 1 \end{pmatrix} ; x, y, z \in \mathbb{R} \right\}.$$

The exponential map is given by

$$\exp : \mathfrak{h}_3 \ni \begin{pmatrix} 0 & x & y & z \\ 0 & 0 & 0 & y \\ 0 & 0 & 0 & -x \\ 0 & 0 & 0 & 0 \end{pmatrix} \mapsto \mathbf{Id} + \begin{pmatrix} 0 & x & y & z \\ 0 & 0 & 0 & y \\ 0 & 0 & 0 & -x \\ 0 & 0 & 0 & 0 \end{pmatrix} = \begin{pmatrix} 1 & x & y & z \\ 0 & 1 & 0 & y \\ 0 & 0 & 1 & -x \\ 0 & 0 & 0 & 1 \end{pmatrix} \in H_3.$$

In the following we shall consider the three-dimensional Heisenberg group as $\mathbb{R}^3$. Let $X = (1, 0, 0)$, $Y = (0, 1, 0)$ and $Z = (0, 0, 1)$ be the basis of the Lie algebra $\mathfrak{h}_3$. The left invariant vector fields $\widetilde{X}$, $\widetilde{Y}$ and $\widetilde{Z}$ are given by

$$\widetilde{X} = \frac{\partial}{\partial x} - \frac{y}{2}\frac{\partial}{\partial z},$$

$$\widetilde{Y} = \frac{\partial}{\partial y} + \frac{x}{2}\frac{\partial}{\partial z},$$

$$\widetilde{Z} = \frac{\partial}{\partial z}.$$

A sub-Riemannian structure $\mathcal{H}$ on H_3 in the strong sense is defined by a sub-bundle spanned by $\{\widetilde{X}, \widetilde{Y}\}$. We consider a left invariant metric on $\mathcal{H}$ such that the vector fields $\widetilde{X}$ and $\widetilde{Y}$ are orthonormal at each point. In this case the sub-Laplacian Δ_{sub} is

$$\Delta_{\mathrm{sub}} = -(\widetilde{X}^2 + \widetilde{Y}^2). \tag{9.2.2}$$

Furthermore, if we consider the left invariant Riemannian metric on H_3 such that $\widetilde{X}, \widetilde{Y}$ and $\widetilde{Z}$ are orthonormal at each point, then the metric tensor is given by

$$\begin{pmatrix} 1+\frac{y^2}{4} & -\frac{xy}{4} & \frac{y}{2} \\ -\frac{xy}{4} & 1+\frac{x^2}{4} & -\frac{x}{2} \\ \frac{y}{2} & -\frac{x}{2} & 1 \end{pmatrix},$$

with the inverse matrix

$$\begin{pmatrix} 1 & 0 & -\frac{y}{2} \\ 0 & 1 & \frac{x}{2} \\ -\frac{y}{2} & \frac{x}{2} & 1+\frac{x^2+y^2}{4} \end{pmatrix}.$$

Hence the Laplacian Δ_{H_3} is

$$\begin{aligned} \Delta_{H_3} &= -\left(\frac{\partial^2}{\partial x^2} + \frac{\partial^2}{\partial y^2} - y\frac{\partial^2}{\partial x \partial z} + x\frac{\partial^2}{\partial y \partial z} + \left(1 + \frac{x^2+y^2}{4}\right)\frac{\partial^2}{\partial z^2}\right) \\ &= -\widetilde{X}^2 - \widetilde{Y}^2 - \widetilde{Z}^2 = \Delta_{\mathrm{sub}} - \widetilde{Z}^2. \end{aligned}$$

9.2.3 *Higher-Dimensional Heisenberg Groups*

The higher-dimensional Heisenberg group is given by $H_{2n+1} = \mathbb{R}^{2n+1} = \mathbb{R}^{2n} \times \mathbb{R} \cong \mathbb{R}^n \times \mathbb{R}^n \times \mathbb{R}$ together with the following group law:

$$\big((x,y;z),(\tilde{x},\tilde{y};\tilde{z})\big) \longmapsto \left(x+\tilde{x}, y+\tilde{y}; z+\tilde{z}+\frac{<x,\tilde{y}> - <y,\tilde{x}>}{2}\right),$$

for any $(x,y;z),(\tilde{x},\tilde{y};\tilde{z}) \in \mathbb{R}^n \times \mathbb{R}^n \times \mathbb{R}$. In the aforementioned formula, $< \cdot, \cdot >$ denotes the standard inner product on $\mathbb{R}^n$. H_{2n+1} is realized in the matrix space

$$H_{2n+1} \cong \left\{ \begin{pmatrix} 1 & x_1 & x_2 & \cdots & x_n & z \\ 0 & 1 & 0 & \cdots & 0 & y_1 \\ 0 & 0 & 1 & \cdots & 0 & y_2 \\ 0 & 0 & 0 & \cdots & 0 & \vdots \\ \vdots & & & & 1 & y_n \\ 0 & \cdots & \cdots & \cdots & 0 & 1 \end{pmatrix} ; \; x, \; y \in \mathbb{R}^n, \; z \in \mathbb{R} \right\}$$

by the map

$$H_{2n+1} = \mathbb{R}^n \times \mathbb{R}^n \times \mathbb{R} \ni (x, y; z) \mapsto \begin{pmatrix} 1 & x_1 & x_2 & \cdots & x_n & z + \frac{<x,y>}{2} \\ 0 & 1 & 0 & \cdots & 0 & y_1 \\ 0 & 0 & 1 & \cdots & 0 & y_2 \\ 0 & 0 & 0 & \cdots & 0 & \vdots \\ \vdots & & & & 1 & y_n \\ 0 & \cdots & \cdots & \cdots & 0 & 1 \end{pmatrix}.$$

The Lie algebra of H_{2n+1}, which can also be regarded as $\mathbb{R}^{2n+1}$, is denoted by $\mathfrak{h}_{2n+1}$. For $1 \leq i \leq n$, the vectors

$$\begin{aligned} X_i &= (\underbrace{0, \ldots, 0}_{i-1 \text{ zero}}, 1, 0, 0, \ldots, 0, ; 0), \\ Y_i &= (\underbrace{0, \ldots, 0, 0, \ldots, 0}_{n+i-1 \text{ zero}}, 1, 0, \ldots; 0), \\ Z &= (\underbrace{0, \ldots, 0, 0, \ldots, 0}_{2n \text{ zero}}; 1) \end{aligned}$$

form a basis in the Heisenberg algebra $\mathfrak{h}_{2n+1}$. The corresponding left invariant vector fields are expressed as

$$\widetilde{X}_i = \frac{\partial}{\partial x_i} - \frac{y_i}{2}\frac{\partial}{\partial z}, \tag{9.2.3}$$

$$\widetilde{Y}_i = \frac{\partial}{\partial y_i} + \frac{x_i}{2}\frac{\partial}{\partial z}, \tag{9.2.4}$$

$$\widetilde{Z}_i = \frac{\partial}{\partial z}. \tag{9.2.5}$$

Now we have a left invariant sub-Riemannian structure $\mathcal{H}$ spanned by the left invariant vector fields $\{\widetilde{X}_i, \widetilde{Y}_i;\ i = 1, \ldots, n\}$. We shall consider the left invariant metric on $\mathcal{H}$ such that $\{\widetilde{X}_i, \widetilde{Y}_i;\ i = 1, \ldots, n\}$ are orthonormal at each point of H_{2n+1}. The sub-Laplacian in this case is given by

$$\Delta_{\text{sub}} = -\sum_{i=1}^{n} \widetilde{X}_i^{\,2} + \widetilde{Y}_i^{\,2}. \tag{9.2.6}$$

We may also consider the left invariant Riemannian metric g on H_{2n+1} such that the left invariant vectors fields $\{\widetilde{X}_i, \widetilde{Y}_i;\ i = 1, \ldots, n\}$ and $\widetilde{Z}$ are orthonormal. The metric tensor in this case is given by

$$g_{ij} = \begin{pmatrix} g\left(\frac{\partial}{\partial x_i}, \frac{\partial}{\partial x_j}\right) & g\left(\frac{\partial}{\partial x_i}, \frac{\partial}{\partial y_j}\right) & g\left(\frac{\partial}{\partial x_i}, \frac{\partial}{\partial z}\right) \\ g\left(\frac{\partial}{\partial y_i}, \frac{\partial}{\partial x_j}\right) & g\left(\frac{\partial}{\partial y_i}, \frac{\partial}{\partial y_j}\right) & g\left(\frac{\partial}{\partial y_i}, \frac{\partial}{\partial z}\right) \\ g\left(\frac{\partial}{\partial z}, \frac{\partial}{\partial x_j}\right) & g\left(\frac{\partial}{\partial z}, \frac{\partial}{\partial y_j}\right) & g\left(\frac{\partial}{\partial z}, \frac{\partial}{\partial z}\right) \end{pmatrix}$$

$$= \begin{pmatrix} 1+\frac{{y_1}^2}{4} & \cdots & \cdots & \frac{y_1 y_n}{4} & \cdots & \cdots & \cdots & & \\ \vdots & \ddots & \frac{y_i y_j}{4} & & \cdots & \ddots & -\frac{x_i y_j}{4} & \cdots & \frac{y_i}{2} \\ \vdots & \frac{y_j y_i}{4} & \ddots & & \cdots & -\frac{y_j x_i}{4} & & & \\ \frac{y_n y_1}{4} & \cdots & \cdots & 1+\frac{y_n^2}{4} & \cdots & \cdots & \vdots & & \\ \cdots & \cdots & \cdots & \cdots & 1+\frac{x_1^2}{4} & \cdots & & & \frac{x_1}{2} \\ \cdots & \cdots & \cdots & -\frac{x_i y_j}{4} & \cdots & \ddots & \frac{x_i x_j}{4} & & -\frac{x_i}{2} \\ \cdots & \cdots & -\frac{x_i y_j}{4} & & \cdots & \frac{x_j x_i}{4} & \ddots & & \\ \cdots & \cdots & & \cdots & \cdots & \cdots & \cdots & 1+\frac{x_n^2}{4} & \vdots \\ \cdots & \frac{y_i}{2} & \cdots & \cdots & \cdots & -\frac{x_i}{2} & \cdots & \cdots & 1 \end{pmatrix}.$$

Then the Laplacian $\Delta_{H_{2n+1}}$ is

$$\Delta_{H_{2n+1}} = \Delta_{\text{sub}} - \widetilde{Z}^2.$$

In terms of coordinates $(x_1, \ldots, x_n, y_1, \ldots, y_n, z)$, the Laplacian and sub-Laplacian can be expressed by

$$\begin{aligned} \Delta_{\text{sub}} = &- \sum \left(\frac{\partial^2}{\partial {x_i}^2} + \frac{\partial^2}{\partial {y_i}^2} \right) + \sum y_i \frac{\partial^2}{\partial x_i \partial z} - \sum x_i \frac{\partial^2}{\partial y_i \partial z} \\ &- \left(\frac{\sum ({x_i}^2 + {y_i}^2)}{4} \right) \frac{\partial^2}{\partial z^2}, \end{aligned} \tag{9.2.7}$$

$$\begin{aligned} \Delta_{H_{2n+1}} = &- \sum \left(\frac{\partial^2}{\partial {x_i}^2} + \frac{\partial^2}{\partial {y_i}^2} \right) + \sum y_i \frac{\partial^2}{\partial x_i \partial z} - \sum x_i \frac{\partial^2}{\partial y_i \partial z} \\ &- \left(1 + \frac{\sum ({x_i}^2 + {y_i}^2)}{4} \right) \frac{\partial^2}{\partial z^2}. \end{aligned} \tag{9.2.8}$$

9.2.4 Quaternionic Heisenberg Group

Let $\mathbb{H}$ be the field of quaternions over the real numbers with the basis $\{\mathbf{1}, \mathbf{i}, \mathbf{j}, \mathbf{k}\}$. Their products are given by the following usual relations:

$$\mathbf{1i} = \mathbf{i1} = \mathbf{i}, \quad \mathbf{1j} = \mathbf{j1} = \mathbf{j}, \quad \mathbf{1k} = \mathbf{k1} = \mathbf{k}, \quad \mathbf{i}^2 = \mathbf{j}^2 = \mathbf{k}^2 = -\mathbf{1},$$
$$\mathbf{ij} = -\mathbf{ji} = \mathbf{k}, \quad \mathbf{jk} = -\mathbf{kj} = \mathbf{i}, \quad \mathbf{ki} = -\mathbf{ik} = \mathbf{j},$$

which obviously satisfy the associativity but not the commutativity. The elements of $\mathbb{H}$ are denoted by $h = x_0\mathbf{1} + x_1\mathbf{i} + x_2\mathbf{j} + x_3\mathbf{k}$, with real coefficients $x_0, x_1, x_2, x_3 \in \mathbb{R}$. The "conjugate"of h is defined by $\overline{h} = x_0\mathbf{1} - x_1\mathbf{i} - x_2\mathbf{j} - x_3\mathbf{k}$. Then we have

$$\overline{hk} = \overline{k}\,\overline{h}, \ \forall h, k \in \mathbb{H}.$$

The norm, or the absolute value, of the element $h \in \mathbb{H}$ is denoted by $|h|$ and is given by

$$|h| = \sqrt{h\overline{h}} = \sqrt{\sum x_i{}^2}.$$

We shall now consider the antisymmetric bilinear form

$$\mathfrak{Im}(h,k) = \frac{1}{2}\left(h\overline{k} - k\overline{h}\right),$$

which introduces a Lie bracket on the vector space $\mathbb{H} \oplus \mathbb{R}^3$ by

$$\begin{aligned} &\mathbb{H} \times \mathbb{H} \to \mathbb{R}^3 \\ &(h, \tilde{h}) \mapsto [h, \tilde{h}] = \mathfrak{Im}(h, \tilde{h}). \end{aligned}$$

Then the space

$$\begin{aligned} q\mathfrak{h}_7 = \mathbb{H} \oplus \mathbb{R}^3 =&\big\{\big(x_0\mathbf{1} + x_1\mathbf{i} + x_2\mathbf{j} + x_3\mathbf{k}; z_1, z_2, z_3\big) \\ &= (x_0, x_1, x_2, x_3; z_1, z_2, z_3) \mid x_i,\ z_i \in \mathbb{R}\big\} \end{aligned}$$

becomes a two-step nilpotent Lie algebra with respect to the aforementioned bracket. We shall introduce a group multiplication on $q\mathfrak{h}_7$ by

$$\begin{aligned} &\mathbb{H} \oplus \mathbb{R}^3 \times \mathbb{H} \oplus \mathbb{R}^3 \ni \big((h,z),(k,\tilde{z})\big) \mapsto (h,z) * (k,\tilde{z}), \\ &(h,z) * (k,\tilde{z}) = \big(h + k\big) \oplus (z + \tilde{z} + \mathfrak{Im}(h,k)/2). \end{aligned}$$

We shall denote by $q\mathbb{H}_7$ the Lie group with this multiplication law. Let

$$\begin{aligned} X_0 &= (1,0,0,0;0,0,0) = \mathbf{1} \oplus 0 \in \mathbb{H} \oplus \mathbb{R}^3, \\ X_1 &= (0,1,0,0;0,0,0) = \mathbf{i} \oplus 0 \in \mathbb{H} \oplus \mathbb{R}^3, \\ X_2 &= (0,0,1,0;0,0,0) = \mathbf{j} \oplus 0 \in \mathbb{H} \oplus \mathbb{R}^3, \\ X_3 &= (0,0,0,1;0,0,0) = \mathbf{k} \oplus 0 \in \mathbb{H} \oplus \mathbb{R}^3 \end{aligned}$$

be elements in $q\mathfrak{h}_7$ which, together with their bracket generate the entire space $q\mathfrak{h}_7$. The expressions of the corresponding left invariant vector fields are given by

$$\widetilde{X}_0 = \frac{\partial}{\partial x_0} + \frac{1}{2}\sum_{i=1}^{3} x_i \frac{\partial}{\partial z_i}, \tag{9.2.9}$$

$$\widetilde{X}_1 = \frac{\partial}{\partial x_1} + \frac{x_0}{2}\frac{\partial}{\partial z_1} + \frac{x_3}{2}\frac{\partial}{\partial z_2} - \frac{x_2}{2}\frac{\partial}{\partial z_3}, \tag{9.2.10}$$

$$\widetilde{X}_2 = \frac{\partial}{\partial x_2} - \frac{x_3}{2}\frac{\partial}{\partial z_1} + \frac{x_0}{2}\frac{\partial}{\partial z_2} + \frac{x_1}{2}\frac{\partial}{\partial z_3}, \tag{9.2.11}$$

$$\widetilde{X}_3 = \frac{\partial}{\partial x_3} + \frac{x_2}{2}\frac{\partial}{\partial z_1} - \frac{x_1}{2}\frac{\partial}{\partial z_2} + \frac{x_0}{2}\frac{\partial}{\partial z_3}. \tag{9.2.12}$$

We shall introduce on $q\mathfrak{H}_7$ a sub-Riemannian structure $\mathcal{H}$ in the strong sense, i.e., a sub-bundle spanned by the left invariant vector fields $\{\tilde{X}_i\}_{i=0}^3$. The sub-Laplacian in this case is

$$\begin{aligned}
\Delta_{\text{sub}} &= -\sum_{i=0}^{3} \widetilde{X}_i^{\,2} \\
&= -\sum_{i=0}^{3}\left(\frac{\partial^2}{\partial x_i^2} + x_i \frac{\partial^2}{\partial x_0 \partial z_i}\right) - x_0 \frac{\partial^2}{\partial x_1 \partial z_1} - x_3 \frac{\partial^2}{\partial x_1 \partial z_2} + x_2 \frac{\partial^2}{\partial x_1 \partial z_3} \\
&\quad + x_3 \frac{\partial^2}{\partial x_2 \partial z_1} - x_0 \frac{\partial^2}{\partial x_2 \partial z_2} - x_1 \frac{\partial^2}{\partial x_2 \partial z_3} - x_2 \frac{\partial^2}{\partial x_3 \partial z_1} + x_1 \frac{\partial^2}{\partial x_3 \partial z_2} \\
&\quad - x_0 \frac{\partial^2}{\partial x_3 \partial z_3} - \frac{1}{4}\sum_{i=1}^{3}\left(\sum_{i=0}^{3} x_i^2\right)\frac{\partial^2}{\partial z_i^2}.
\end{aligned} \tag{9.2.13}$$

When considering a Riemannian metric on the entire space such that the vector fields $\{\widetilde{X}_i\}$ and $\{\widetilde{Z}_j\}$ are orthonormal at each point, this metric can be regarded as an extension of the metric defined on the sub-Riemannian structure. Consequently, the Laplacian in this case is given by

$$\Delta = \Delta_{\text{sub}} - \frac{\partial^2}{\partial z_1^2} - \frac{\partial^2}{\partial z_2^2} - \frac{\partial^2}{\partial z_3^2}.$$

Higher-dimensional quaternionic Heisenberg algebras $q\mathfrak{h}_{4n+3} \cong \mathbb{H}^n \oplus \mathbb{R}^3$ are defined by

$$\begin{aligned}
&\big((h_1, \dots, h_n; z), (\tilde{h}_1, \dots, \tilde{h}_n; \tilde{z})\big) \\
&\quad \mapsto \left(h_1 + \tilde{h}_1, \dots, h_n + \tilde{h}_n; z + \tilde{z} + \frac{1}{2}\sum_{i=1}^{n} \mathfrak{Im}(h_i, \tilde{h}_i)\right),
\end{aligned}$$

for all $(h_1, \ldots, h_n; z), (\tilde{h}_1, \ldots, \tilde{h}_n; \tilde{z}) \in \mathbb{H}^n \oplus \mathbb{R}^3$, and with notation $\mathfrak{Im}\left(h_i, \tilde{h}_i\right) = \frac{1}{2}(h_i \overline{\tilde{h}_i} - \widetilde{h_i}\ \bar{h}_i)$.

The Lie group associated with the Lie algebra $q\mathfrak{h}_{4n+3}$ will be denoted by $q\mathbb{H}_{4n+3}$, and it will be called the $(4n+3)$-dimensional *quaternionic Heisenberg group*.

9.2.5 Heisenberg-Type Lie Algebra

Let $\mathfrak{n}$ and $\mathfrak{z}$ be two real vector spaces endowed with the inner products $< \cdot, \cdot >_{\mathfrak{n}}$ and $< \cdot, \cdot >_{\mathfrak{z}}$, respectively. We shall assume that there is a bilinear form

$$J : \mathfrak{z} \times \mathfrak{n} \longrightarrow \mathfrak{n}$$

such that

$$\textbf{(HT-1)}: \quad J(Z, J(Z, X)) \quad = - < Z, Z >_{\mathfrak{z}} X, \tag{9.2.14}$$

$$\textbf{(HT-2)}: \quad < J(Z, X), Y >_{\mathfrak{n}} = - < X, J(Z, Y) >_{\mathfrak{n}}, \qquad \forall X \in \mathfrak{n},\ Z \in \mathfrak{z}. \tag{9.2.15}$$

The bilinear form J is sometimes called *Clifford multiplication*. Now we will define the skew-symmetric bilinear map $B : \mathfrak{n} \times \mathfrak{n} \to \mathfrak{z}^*$ given by

$$B(X, Y) : \mathfrak{z} \to \mathbb{R}, \qquad B(X, Y)(Z) =< J(Z, X), Y >_{\mathfrak{n}}.$$

Proposition 9.2.4. *If Z_1 and $Z_2 \in \mathfrak{z}$ are orthogonal, then*

$$J(Z_1, J(Z_2, Y)) + J(Z_2, J(Z_1, X)) = 0.$$

Proof. Using the bilinearity of the map J, we have

$$\begin{aligned} &J(Z_1 + Z_2, J(Z_1 + Z_2, X)) \\ &\quad = J(Z_1, J(Z_1, X)) + J(Z_1, J(Z_2, X)) + J(Z_2, J(Z_1, X)) + J(Z_2, J(Z_2, X)) \\ &\quad = - < Z_1, Z_1 >_{\mathfrak{z}} X + J(Z_1, J(Z_2, X)) + J(Z_2, J(Z_1, X)) - < Z_2, Z_2 >_{\mathfrak{z}} X \\ &\quad = - < Z_1 + Z_2, Z_1 + Z_2, >_{\mathfrak{z}} X = - < Z_1, Z_1 >_{\mathfrak{z}} X - < Z_2, Z_2 >_{\mathfrak{z}} X, \end{aligned}$$

and hence

$$J(Z_1, J(Z_2, X)) + J(Z_2, J(Z_1, X)) = 0.$$

■

Consider the inner product $< \cdot, \cdot >_{\mathfrak{z}^*}$ on $\mathfrak{z}^*$ induced naturally by the inner product on $\mathfrak{z}$. We have the following result.

Proposition 9.2.5. *Let $\varphi \in \mathfrak{z}^*$ and assume that for any fixed $X \neq 0$,*

$$< B(X,Y), \varphi >_{\mathfrak{z}^*} = 0, \qquad \forall\, Y \in \mathfrak{n}.$$

Then $\varphi = 0$.

Proof. Let $\{Z_k\}$ be an orthonormal basis on $\mathfrak{z}$ and denote by $\{Z_k{}^*\}$ the dual basis on $\mathfrak{z}^*$. Then

$$< B(X,Y), \varphi >_{\mathfrak{z}^*} = \sum < J(Z_k, X), Y >_{\mathfrak{n}} a_k = 0,$$

where we expressed $\varphi = \sum a_k Z_k{}^*$. If we take $Y = J(Z_j, X)$, then by property (9.2.15) and Proposition 9.2.4, we have

$$\begin{aligned} < B(X,Y), \varphi >_{\mathfrak{z}^*} &= \sum < J(Z_k, X), Y >_{\mathfrak{n}} a_k \\ &= \sum_{k \neq j} < J(Z_k, X), J(Z_j, X) >_{\mathfrak{n}} - < Z_j, Z_j >_{\mathfrak{z}} a_j \\ &= - \sum_{k \neq j} < J(Z_j, X), J(Z_k, X) >_{\mathfrak{n}} - < Z_j, Z_j >_{\mathfrak{z}} a_j = 0. \end{aligned}$$

This implies that

$$a_j = 0.$$

Hence $\varphi = 0$. ■

Corollary 9.2.6. *For any fixed nonzero $X \in \mathfrak{n}$, the mapping*

$$\mathfrak{n} \to \mathfrak{z}^*, \; Y \mapsto B(X,Y)$$

is surjective.

We now define a two-step nilpotent Lie algebra structure on $V = \mathfrak{n} \oplus \mathfrak{z}^*$ (or on $\mathfrak{n} \oplus \mathfrak{z}$ by identifying $\mathfrak{z}^* \cong \mathfrak{z}$) in such a way that

$$\text{for } X,\; Y \in \mathfrak{n},\; [X,Y] = B(X,Y),$$

and all other brackets are zero. By Proposition 9.2.5, we have

Proposition 9.2.7.

$$[V,V] = \mathfrak{z}^* = \textit{ center}.$$

This induces a two-step nilpotent Lie group structure on V given by

$$\begin{aligned} &V \times V \to V, \\ &\big((x,w),(\tilde{x},\tilde{w})\big) \mapsto (x,w) * (\tilde{x},\tilde{w}) = \left(x + \tilde{x}, w + \tilde{w} + \frac{1}{2} B(X,Y)\right). \end{aligned}$$

Remark 9.2.8.

1. Let $C\ell_Q(\mathfrak{z})$ denote the Clifford algebra with respect to the positive bilinear form $Q(z) =< z, z >_{\mathfrak{z}}$. This means that if $\mathcal{T}(\mathfrak{z})$ is the tensor algebra

$$\mathcal{T}(\mathfrak{z}) = \mathbb{R} \oplus \mathfrak{z} \oplus \mathfrak{z} \otimes \mathfrak{z} \oplus \overset{3}{\otimes} \mathfrak{z} \oplus \cdots,$$

and I_Q is a two-sided ideal in $\mathcal{T}(\mathfrak{z})$ generated by the elements of the form

$$z \otimes z + Q(z),$$

then $C\ell_Q(\mathfrak{z})$ is the quotient algebra $\mathcal{T}(\mathfrak{z})/I_Q$. Conditions (9.2.14) and (9.2.15) required on the bilinear map J say that the vector space $\mathfrak{n}$ has a Clifford module structure of the Clifford algebra $C\ell_Q(\mathfrak{z})$.

2. The Heisenberg-type Lie algebra was first defined in [77], where the Lie algebra $\mathfrak{g}$ is two-step and is equipped with an inner product $< \cdot, \cdot >$ such that for any $Z \in [\mathfrak{g}, \mathfrak{g}] \equiv \mathfrak{z} =$ center, the map $J(Z) : \mathfrak{z}^{\perp} \to \mathfrak{z}^{\perp}$ defined by

$$\textbf{(Ka-1)}: < J(Z)(X), Y >=< Z, [X, Y] >=< ad(X)^*(Z), Y > \qquad (9.2.16)$$

satisfies

$$\textbf{(Ka-2)}: J(Z)^2 = - < Z, Z > Id. \qquad (9.2.17)$$

We have constructed a Heisenberg-type Lie algebra from the two vector spaces $\mathfrak{n}$ and $\mathfrak{z}$ satisfying conditions **(HT-1)** and **(HT-2)**. Conversely, let's assume that $\mathfrak{g}$ satisfies conditions **(Ka-1)** and **(Ka-2)**. Set $\mathfrak{n} = \mathfrak{z}^{\perp} = [\mathfrak{g}, \mathfrak{g}]^{\perp}$, and define the bilinear map

$$J : \mathfrak{z} \times \mathfrak{z}^{\perp} \to \mathfrak{z}^{\perp}$$

by $J(Z, X) = J(Z)(X)$. Then it can easily be seen that $\mathfrak{z}$, $\mathfrak{z}^{\perp}$ and the bilinear map J satisfy conditions **(HT-1)** and **(HT-2)**.

3. Heisenberg algebras are particular types of Heisenberg-type Lie algebras. Also, quaternionic Heisenberg algebras are Heisenberg-type Lie algebras. The latter case can be shown as in the following.

   Let $\mathbb{H}_0 = \{h \in \mathbb{H}; z = -\overline{z}\}$ and consider the map $\mathbb{H}_0 \times \mathbb{H} \to \mathbb{H}$

$$(z, h) \longmapsto z \cdot h.$$

Properties (9.2.14) and (9.2.15) are satisfied. The Heisenberg-type Lie algebra defined by this bilinear map is the quaternionic Heisenberg Lie algebra. Higher-dimensional quaternionic Heisenberg Lie algebras can be constructed in a similar way.

4. All the Heisenberg-type Lie algebras are realized in terms of the Clifford module following periodicity of Clifford algebras and modules [5].

We consider next a sub-Riemanian structure on the Heisenberg-type Lie group $V = \mathfrak{n} \oplus \mathfrak{z}^*$.

For that purpose we choose arbitrary orthonormal bases $\{X_i\}_{i=1}^N$ and $\{Z_k\}_{k=1}^d$ of $\mathfrak{n}$ and $\mathfrak{z}^*$, respectively, where $N = \dim \mathfrak{n}$ and $d = \dim \mathfrak{z}$. These bases induce coordinates on the Lie group $V = \mathfrak{n} \oplus \mathfrak{z}^*$ by

$$\mathbb{R}^{N+d} \ni (x_1, \ldots, x_N; z_1, \ldots, z_d) \leftrightarrow \sum x_i X_i + \sum z_j Z_j \in \mathfrak{n} \oplus \mathfrak{z}^*.$$

Then under the identification $T(V) \cong V \times V$, we consider a sub-Riemannian structure $\mathcal{H}$ spanned by the left invariant vector fields $\{\widetilde{X}_i\}_{i=1}^N$ given by

$$\mathcal{H} = V \times \mathfrak{n} \subset V \times V \cong T(V).$$

Considering the constants of structure given by $[X_i, X_j] = B(X_i, X_j) = \sum C_{i\,j}^k Z_k$, the left invariant vector fields are

$$\widetilde{X}_i = \frac{\partial}{\partial x_i} - \frac{1}{2} \sum_{j\,k} C_{i,j}^k x_j \frac{\partial}{\partial z_k},$$

and hence the sub-Laplacian Δ_{sub} is given by

$$\Delta_{\text{sub}} = -\sum \widetilde{X}_i^{\,2}.$$

Remark 9.2.9. According to a theorem in [55], we can choose an orthonormal basis in the Heisenberg-type Lie algebra $V = \mathfrak{n} \oplus \mathfrak{z}$ with respect to which the structure constants are one of $\{-1, 0, 1\}$. In this case, the expressions for the Laplacian and sub-Laplacian become very simple.

9.2.6 Free Two-Step Nilpotent Lie Algebra

Let $\{X_i\}_{i=1}^N$ be a basis of $\mathbb{R}^N$. Consider also a basis $\{Z_{i\,j}\}$ of $\mathbb{R}^{\frac{1}{2}N(N-1)}$, $1 \le i < j \le N$. Define the Lie bracket relations by

$$[X_i, X_j] = -[X_j, X_i] = \begin{cases} Z_{ij}, & \text{for } 1 \le i < j \le N, \\ 0, & \text{otherwise.} \end{cases}$$

Then we can introduce a Lie group structure on $\mathbb{R}^{N(N+1)/2} = \mathbb{R}^N \oplus \mathbb{R}^{{}_N C_2}$ by

$$\begin{aligned}(\mathbb{R}^N \oplus \mathbb{R}^{{}_N C_2}) \times (\mathbb{R}^N \oplus \mathbb{R}^{{}_N C_2}) &\ni (x \oplus z, \tilde{x} \oplus \tilde{z}) \\ &\mapsto (x + \tilde{x}) \oplus \left(z + \tilde{z} + \frac{1}{2} \sum_{1 \le i < j \le N} (x_i \tilde{x}_j - x_j \tilde{x}_i) Z_{ij} \right).\end{aligned}$$

This group is called a *free two-step nilpotent Lie group*.

Let $\mathfrak{g}$ be a two-step nilpotent Lie algebra. Consider the bases $\{X_i\}_{i=1}^N$ and $\{Z_j\}_{j=1}^d$ for the complement of the derived algebra $[\mathfrak{g}, \mathfrak{g}]$ and the derived algebra $[\mathfrak{g}, \mathfrak{g}]$, respectively. Let the structure constants $\{C_{i\,j}^{\,k}\}$ be given by

$$[X_i, X_j] = \sum_k C_{i\,j}^{\,k} Z_k.$$

Now let $\mathfrak{fg}_{N+C_N^2}$ $(C_N^2 = N(N-1)/2)$ be the free two-step nilpotent Lie algebra with the basis $\{X_i, Z_{i\,j}\}_{1 \le i < j \le N}$ such that

$$[X_i, X_j] = Z_{i\,j}.$$

We define a Lie algebra homomorphism ρ by

$$\rho : \mathfrak{fg}_{N+C_N^2} \to \mathfrak{g},$$

$$\rho\left(\sum x_i X_i + \sum z_{i\,j} Z_{i\,j}\right) = \sum x_i X_i + \sum z_{i\,j} C_{i\,j}^{\,k} Z_k.$$

Then we have the Lie group homomorphism between corresponding simply connected nilpotent Lie groups. In this sense, any two-step nilpotent Lie group (Lie algebra) is covered by a free two-step nilpotent Lie group (Lie algebra).

9.2.7 The Engel Group

As we explained in Sect. 9.1.2, no one has been successful yet in expressing the heat kernel of a sub-Laplacian (or Laplacian) on nilpotent Lie groups of step greater than or equal to 3 in an explicit (integral) form. The explicit formula is missing even in the minimal dimensional case (dimension 4) nilpotent Lie group of step 3, called the *Engel group*. This group appears in various contexts and we shall also introduce it here, although we shall not give its heat kernel (for the Laplacian or sub-Laplacian) in any form. The crucial reason for which an explicit form of the heat kernel is missing is that we cannot solve the quartic oscillator in terms of any special functions, such as is done with the harmonic oscillator.

Since the Engel group has codimension two and a three-step sub-Riemannian structure, we shall present here two sub-Laplacian operators and the Laplacian and note the difference from the two-step cases. Later, at the end of Sect. 10.3.3, we shall also express what the higher-step Grushin operators look like.

Let $\mathfrak{e}_4$ be a four-dimensional Lie algebra with the basis $\{X, Y, W, Z\}$ with the following nonzero bracket relations:

$$[X, Y] = W, \qquad [X, W] = Z. \tag{9.2.18}$$

Let $\mathfrak{E}_4$ (identified with $\mathbb{R}^4$) be the Lie group whose Lie algebra is $\mathfrak{e}_4$. Then the group multiplication is given by the formula

$$(x, y, w, z) * (\tilde{x}, \tilde{y}, \tilde{w}, \tilde{z}) = \Big(x + \tilde{x}, y + \tilde{y}, w + \tilde{w} + \frac{1}{2}(x\tilde{y} - y\tilde{x}), z + \tilde{z} + \frac{1}{2}(x\tilde{w} - w\tilde{x}) + \frac{1}{12}(x - \tilde{x})(x\tilde{y} - y\tilde{x})\Big).$$

This group is realized in 4×4 real matrix space as

$$\mathfrak{E}_4 \cong \left\{ \begin{pmatrix} 1 & x & \frac{x^2}{2} & z \\ 0 & 1 & x & w \\ 0 & 0 & 1 & y \\ 0 & 0 & 0 & 1 \end{pmatrix} ; x, y, z, w \in \mathbb{R} \right\}.$$

The identification is given by the map

$$(x, y, w, z) \mapsto \begin{pmatrix} 1 & x & \frac{x^2}{2} & z + \frac{xw}{2} + \frac{xy^2}{3!} \\ 0 & 1 & x & w + \frac{xy}{2} \\ 0 & 0 & 1 & y \\ 0 & 0 & 0 & 1 \end{pmatrix}.$$

If $A \in \mathfrak{e}_4$ is an element of Engel's algebra, denote by $\tilde{A}$ the left invariant vector field on the group $\mathfrak{E}_4$ defined by

$$\tilde{A}(f)(g) = \frac{d}{dt}\big(f(g \cdot \exp tA)\big)_{|t=0}, \qquad f \in C^\infty(\mathfrak{E}_4).$$

Let $\mathcal{H}_2 = \operatorname{span}\{\tilde{X}, \tilde{Y}\}$. By the bracket relations (9.2.18), we obtain that $\mathcal{H}_2$ is a three-step, regular, minimal, codimension-2 sub-Riemannian structure on $\mathfrak{E}_4$ in the strong sense. If we let $\mathcal{H}_1 = \operatorname{span}\{\tilde{X}, \tilde{Y}, \tilde{W}\}$, we obtain a codimension-1, two-step sub-Riemannian structure on $\mathfrak{E}_4$ in the strong sense.

The vector fields $\tilde{X}$, $\tilde{Y}$, $\tilde{W}$, $\tilde{Z}$ are expressed by

$$\tilde{X} = \frac{\partial}{\partial x} - y\frac{\partial}{\partial w} - w\frac{\partial}{\partial z},$$

$$\tilde{Y} = \frac{\partial}{\partial y} + x\frac{\partial}{\partial w},$$

$$\tilde{W} = \frac{\partial}{\partial w} + x\frac{\partial}{\partial z},$$

$$\tilde{Z} = \frac{\partial}{\partial z},$$

while the sub-Laplacians associated with the sub-Riemannian structures $\mathcal{H}_1$ and $\mathcal{H}_2$ are given by

$$\begin{aligned}
\Delta_{\text{sub}}^{(2)} &= -\tilde{X}^2 - \tilde{Y}^2 \\
&= -\frac{\partial^2}{\partial x^2} - \frac{\partial^2}{\partial y^2} - (x^2+y^2)\frac{\partial^2}{\partial w^2} - w^2\frac{\partial^2}{\partial z^2} + 2y\frac{\partial^2}{\partial x \partial w} \\
&\quad + 2w\frac{\partial^2}{\partial x \partial z} - 2yw\frac{\partial^2}{\partial w \partial z} - 2x\frac{\partial^2}{\partial y \partial w}, \\
\Delta_{\text{sub}}^{(1)} &= -\Delta_{\text{sub}}^{(2)} - \tilde{W}^2 \\
&= -\frac{\partial^2}{\partial x^2} - \frac{\partial^2}{\partial y^2} - (x^2+y^2+1)\frac{\partial^2}{\partial w^2} - (w^2+x^2)\frac{\partial^2}{\partial z^2} + 2y\frac{\partial^2}{\partial x \partial w} \\
&\quad + 2w\frac{\partial^2}{\partial x \partial z} - 2(yw+x)\frac{\partial^2}{\partial w \partial z} - 2x\frac{\partial^2}{\partial y \partial w}.
\end{aligned}$$

If we fix a left invariant Riemannian metric on the Engel group $\mathfrak{E}_4$ in such a way that $\{\widetilde{X}, \widetilde{Y}, \widetilde{W}, \widetilde{Z}\}$ are orthonormal at each point, then the Laplacian Δ is given by the formula

$$\begin{aligned}
\Delta &= \Delta_{\text{sub}}^{(2)} - \tilde{W}^2 - \tilde{Z}^2 = \Delta_{\text{sub}}^{(1)} - \widetilde{Z}^2 \\
&= -\frac{\partial^2}{\partial x^2} - \frac{\partial^2}{\partial y^2} - (x^2+y^2+1)\frac{\partial^2}{\partial w^2} - (w^2+x^2+1)\frac{\partial^2}{\partial z^2} + 2y\frac{\partial^2}{\partial x \partial w} \\
&\quad + 2w\frac{\partial^2}{\partial x \partial z} - 2(yw+x)\frac{\partial^2}{\partial w \partial z} - 2x\frac{\partial^2}{\partial y \partial w}.
\end{aligned}$$

When the aforementioned vector fields are considered on the group realized as a subgroup of the 4×4 matrix space, we have the following expressions:

$$\begin{aligned}
\tilde{X} &= \frac{\partial}{\partial x}, \\
\tilde{Y} &= \frac{\partial}{\partial y} + x\frac{\partial}{\partial w} + \frac{x^2}{2}\frac{\partial}{\partial z}, \\
\tilde{W} &= \frac{\partial}{\partial w} + x\frac{\partial}{\partial z}, \\
\tilde{Z} &= \frac{\partial}{\partial z},
\end{aligned}$$

and

$$\begin{aligned}
\Delta_{\mathrm{sub}}^{(2)} &= -\frac{\partial^2}{\partial x^2} - \frac{\partial^2}{\partial y^2} - x^2\frac{\partial^2}{\partial w^2} - \frac{x^4}{4}\frac{\partial^2}{\partial z^2} \\
&\quad - 2x\frac{\partial^2}{\partial y \partial w} - x^2\frac{\partial^2}{\partial y \partial z} - x^3\frac{\partial^2}{\partial w \partial z}, \\
\Delta_{\mathrm{sub}}^{(1)} &= -\frac{\partial^2}{\partial x^2} - \frac{\partial^2}{\partial y^2} - (x^2+1)\frac{\partial^2}{\partial w^2} - \left(x^2 + \frac{x^4}{4}\right)\frac{\partial^2}{\partial z^2} \\
&\quad - 2x\frac{\partial^2}{\partial y \partial w} - x^2\frac{\partial^2}{\partial y \partial z} - (x^3+2x)\frac{\partial^2}{\partial w \partial z}, \\
\Delta &= -\frac{\partial^2}{\partial x^2} - \frac{\partial^2}{\partial y^2} - (x^2+1)\frac{\partial^2}{\partial w^2} - \left(1 + x^2 + \frac{x^4}{4}\right)\frac{\partial^2}{\partial z^2} \\
&\quad - 2x\frac{\partial^2}{\partial y \partial w} - x^2\frac{\partial^2}{\partial y \partial z} - (x^3+2x)\frac{\partial^2}{\partial w \partial z}.
\end{aligned}$$

More details about Engel's group and its sub-Riemannian structure and a description of bicharacteristic curves of three-step Grushin operators can be found in Chap. 12 of the book [27] and in the paper [47].

Chapter 10
Heat Kernels for Laplacians and Step-2 Sub-Laplacians

10.1 Spectral Decomposition and Heat Kernel

In this section we shall provide an explicit multiplication form of the Laplacian for a certain class of two-step nilpotent Lie groups including Heisenberg groups; see [46]. The result presented in this section is more precise than the explicit formula of the heat kernel given for the first time in the paper of Hulanicki [68], as explained in Remark 9.1.5.

Let $\mathfrak{g}$ be a two-step nilpotent Lie algebra such that

$$\begin{aligned} &\mathfrak{g} = \mathfrak{g}_+ \oplus \mathfrak{g}_- \oplus \mathfrak{z}, \ [\mathfrak{g}_+, \mathfrak{g}_-] = \mathfrak{z}, \ [\mathfrak{g}_\pm, \ \mathfrak{g}_\pm] = 0 \\ &\mathfrak{z} \text{ is the center and} \\ &n = \dim \mathfrak{g}_+ = \dim \mathfrak{g}_-, \ \dim \mathfrak{z} = d. \end{aligned} \tag{10.1.1}$$

Let $\{X_i\}_{i=1}^n$, $\{Y_i\}_{i=1}^n$ and $\{Z_k\}_{k=1}^d$ be a basis of $\mathfrak{g}_+$, $\mathfrak{g}_-$ and $\mathfrak{z}$, respectively, with the structure constants $C_{i,j}^k$:

$$[X_i, \ Y_j] = \sum_{k=1}^d C_{i,j}^k \ Z_k$$

and all other brackets zero. For each $\lambda \in \mathfrak{z}^*$, associate the matrix $C(\lambda)$,

$$C(\lambda)_{i,j} = \sum_{k=1}^d C_{i,j}^k \ \lambda(Z_k). \tag{10.1.2}$$

Assume the matrix $C(\lambda)^t C(\lambda)$ is diagonal:

$$C(\lambda)^t C(\lambda) = \begin{pmatrix} c_1(\lambda) & & & \cdots & 0 \\ 0 & c_2(\lambda) & & \cdots & \\ & & \cdots & & \\ \cdots & & & \cdots & \\ 0 & \cdots & & & c_n(\lambda) \end{pmatrix}.$$

O. Calin et al., *Heat Kernels for Elliptic and Sub-elliptic Operators*,
Applied and Numerical Harmonic Analysis, DOI 10.1007/978-0-8176-4995-1_10,

with all the diagonal elements $c_i(\lambda)$ nondegenerate positive bilinear forms on $\mathfrak{z}$. We identify the Lie algebra

$$\mathfrak{g} = \mathfrak{g}_+ \oplus \mathfrak{g}_- \oplus \mathfrak{z}$$

and the Lie group

$$\exp : \mathfrak{g}_+ \times \mathfrak{g}_- \times \mathfrak{z} \cong G$$

through the exponential map. We denote an element g in G by

$$g = \sum x_i X_i + \sum y_i Y_i + \sum z_k Z_k = (x, y, z).$$

The multiplication group law is defined as

$$g * \tilde{g} = g + \tilde{g} + \frac{1}{2}[g, \tilde{g}].$$

Consider the left invariant vector fields $\{\tilde{X}_i\}$, $\{\tilde{Y}_j\}$ and $\{\tilde{Z}_k\}$ expressed by

$$\begin{aligned}
\widetilde{X}_i &= \frac{\partial}{\partial x_i} - \frac{1}{2}\sum C_{i\,j}^{k} y_j \frac{\partial}{\partial z_k},\\
\widetilde{Y}_j &= \frac{\partial}{\partial y_j} + \frac{1}{2}\sum C_{i\,j}^{k} x_i \frac{\partial}{\partial z_k},\\
\widetilde{Z}_k &= \frac{\partial}{\partial z_k}.
\end{aligned}$$

The sub-Riemannian structure $\mathcal{H}$ on G is defined by the sub-bundle spanned by the vector fields $\{\widetilde{X}_i, \widetilde{Y}_j\}_{i,j=1}^{n}$. We choose the metric on $\mathcal{H}$ such that the aforementioned left invariant vector fields are orthonormal at each point. This left invariant metric on $\mathcal{H}$ can be extended to the entire tangent space as a left invariant Riemannian metric by assuming that $\{\widetilde{X}_i, \widetilde{Y}_j, \widetilde{Z}_k\}$ are orthonormal at each point. The sub-Laplacian Δ_{sub} and the Laplacian Δ are given by the formulas

$$\begin{aligned}
-\Delta_{\text{sub}} &= \sum \widetilde{X}_i^{\,2} + \widetilde{Y}_i^{\,2},\\
-\Delta &= \sum \frac{\partial^2}{\partial x_i^2} + \frac{\partial^2}{\partial y_i^2} - \sum C_{i,j}^{k} y_j \frac{\partial^2}{\partial x_i \partial z_k} + \sum C_{i,j}^{k} x_i \frac{\partial^2}{\partial y_j \partial z_k}\\
&\quad + \frac{1}{2}\sum_i \sum_{k_1<k_2} \sum_{j_1,j_2} C_{i,j_1}^{k_1} C_{i,j_2}^{k_2} y_{j_1} y_{j_2} \frac{\partial^2}{\partial z_{k_1} \partial z_{k_2}}\\
&\quad + \frac{1}{2}\sum_j \sum_{k_1<k_2} \sum_{i_1,i_2} C_{i_1,j}^{k_1} C_{i_2,j}^{k_2} x_{i_1} x_{i_2} \frac{\partial^2}{\partial z_{k_1} \partial z_{k_2}}\\
&\quad + \sum_k \left(1 + \frac{1}{4}\left\{\sum_{i=1}^{n}\left(\sum_{j=1}^{n} y_j C_{i,j}^{k}\right)^2 + \sum_{j=1}^{n}\left(\sum_{i=1}^{n} x_i C_{i,j}^{k}\right)^2\right\}\right)\frac{\partial^2}{\partial z_k^2}.
\end{aligned}$$

We shall give two examples next of the Lie algebras satisfying the above assumptions.

Example 10.1.1. The Heisenberg algebra of any dimension.

Example 10.1.2. The Heisenberg-type algebra with the dimension of the center

$$d = \dim \mathfrak{z} = \dim[\mathfrak{g}, \mathfrak{g}] = 0 \ (\mathrm{mod}\ 4).$$

Let $\{Z_k\}_{k=1}^d$ be an orthonormal basis of the center $\mathfrak{z}$ and consider

$$T = J(Z_1) \circ J(Z_2) \circ \cdots \circ J(Z_d).$$

Using (9.2.4), we have

$$T^2 = Id,$$
$$T \circ J(Z) = -J(Z) \circ T, \qquad \forall Z \neq 0 \in \mathfrak{z}.$$

Now consider the vector spaces

$$V_+ = \{X \in \mathfrak{n};\ T(X) = X\}, \qquad V_- = \{X \in \mathfrak{n};\ T(X) = -X\}.$$

Then for any nonzero $Z \in \mathfrak{z}$, $J(Z)$ maps V_+ into V_-, and viceversa. We also have $[V_+, V_-] = 0$.

For the Laplacian associated with this group, we have an explicit spectral decomposition in a multiplication form; see [46].

Theorem 10.1.3. *There exist a measure space* $(\mathbf{X}, \mathbf{dm})$, *a positive function* φ *on* $\mathbf{X}$ *and a unitary operator* $\mathbf{U} : L_2(G) \cong L_2(\mathbf{X}, \mathbf{dm})$ *such that*

$$\mathbf{U}^{-1} \circ \mathbf{M}_\varphi \circ \mathbf{U} = \Delta.$$

Using the explicit form of the function φ and of the measure $\mathbf{dm}$ given by (10.1.3) and (10.1.6), we obtain the following consequence.

Corollary 10.1.4. *The spectrum of the Laplacian* Δ *is* $[0, \infty)$, *and it consists of all continuous spectrum.*

Even if we do not intend to provide a proof for this theorem, we shall describe the measure space $(\mathbf{X},\ \mathbf{dm})$, the unitary transformation $\mathbf{U}$ and the positive function φ on $\mathbf{X}$ that appear in this case. Let $\mathbf{k} = (k_1, \ldots, k_n)$, $k_i \in \mathbb{N}$, $k_i \geq 0$, be a multi-index and define the measure space

$$(\mathbf{X}_\mathbf{k}, \mathbf{dm}_\mathbf{k}) = \left(\mathfrak{g}_+ \times \mathfrak{z}^* \backslash \{0\}, \prod_{i=1}^n \sqrt{c_i(\lambda)} dv d\lambda\right) \tag{10.1.3}$$

and consider the direct union

$$(\mathbf{X}, \mathbf{dm}) = \coprod_{\mathbf{k}\in\mathbb{N}^n} (\mathbf{X_k}, \mathbf{dm_k}). \tag{10.1.4}$$

Let $\mathcal{F}$ be the partial Fourier transform

$$\mathcal{F} : C_0^\infty(\mathfrak{g}_+ \times \mathfrak{g}_- \times \mathfrak{z}) \to C^\infty(\mathfrak{g}_+ \times \mathfrak{g}_-^* \times \mathfrak{z}^*),$$

$$(\mathcal{F}f)(x, \xi, \eta) = (2\pi)^{-(n+d)/2} \int e^{i(<\xi,y>+<\eta,z>)} f(x, y, z)dydz, \qquad i = \sqrt{-1}$$

and consider the restriction map

$$\mathbf{R} : L_2(\mathfrak{g}_+ \times \mathfrak{g}_-^* \times \mathfrak{z}^*) \longrightarrow L_2(\mathfrak{g}_+ \times \mathfrak{g}_-^* \times (\mathfrak{z}^*\backslash\{0\})).$$

We also let

$$\Phi : \mathfrak{g}_+ \times \mathfrak{g}_+ \times (\mathfrak{z}^*\backslash\{0\}) \longrightarrow \mathfrak{g}_+ \times \mathfrak{g}_-^* \times (\mathfrak{z}^*\backslash\{0\})$$

be a diffeomorphism defined by

$$\begin{aligned}
\Phi(v, w, \lambda) &= (x, \xi, \eta), \\
x &= v - w, \\
< \xi, y > &= -\frac{1}{2} < \lambda, [v + w, y] > \quad [= \mathbf{T}_\lambda(v + w)(y),\ y \in \mathfrak{g}_-], \\
\eta &= \lambda,
\end{aligned}$$

where $\mathbf{T}_\lambda : \mathfrak{g}_+ \to \mathfrak{g}_-^*$ is a isomorphism for any $\lambda \neq 0$. Then the composition $\mathbf{K} = \Phi^* \circ \mathbf{R} \circ \mathcal{F}$ is a unitary transformation given by

$$\begin{aligned}
\mathbf{K} = \Phi^* \circ \mathbf{R} \circ \mathcal{F} &: L_2(\mathfrak{g}_+ \times \mathfrak{g}_- \times \mathfrak{z})\ [=\ L_2(G)] \\
&\longrightarrow L_2\left(\mathfrak{g}_+ \times \mathfrak{g}_+ \times (\mathfrak{z}^*\backslash\{0\}),\ \prod_{i=1}^n \sqrt{c_i(\lambda)}dv\, dw\, d\lambda\right).
\end{aligned}$$

Consider next the following first-order differential operators:

$$\mathbf{S}_i = \frac{\partial}{\partial w_i} - \sqrt{c_i(\lambda)}w_i,\ i = 1, \ldots, n,$$

and let h_i be the functions given by

$$h_i(w, \lambda) = e^{-\frac{1}{2}\sqrt{c_i(\lambda)}w_i^2}.$$

For any fixed $\lambda \neq 0$ and each multi-index $\mathbf{k} = (k_1, \ldots, k_n)$, let $h(w, \lambda, \mathbf{k})$ denote the Hermite function of the variables $w \in \mathfrak{g}_+$:

$$h(w, \lambda, \mathbf{k}) = (\mathbf{S}_1^{k_1} h_1)(w, \lambda) \cdots (\mathbf{S}_n^{k_n} h_n)(w, \lambda).$$

Now for a function $f \in C_0^\infty(\mathfrak{g}_+ \times \mathfrak{g}_+ \times (\mathfrak{z}^*\backslash\{0\}))$, we define

$$\mathbf{E}(f)(v,\lambda,\mathbf{k}) = \frac{1}{N_{\mathbf{k}}(\lambda)} \int_{\mathfrak{g}_+} f(v,w,\lambda)h(w,\lambda,\mathbf{k})dw \in C^\infty(\mathfrak{g}_+ \times (\mathfrak{z}^*\backslash\{0\})),$$

where $N_{\mathbf{k}}(\lambda)$ is the L_2-norm of the function $h(w,\lambda,\mathbf{k})$:

$$N_{\mathbf{k}}(\lambda)^2 = \int_{\mathfrak{g}_+} |h(w,\lambda,\mathbf{k})|^2\, dw_1 \cdots dw_n = 2^{|\mathbf{k}|} \cdot \mathbf{k}! \cdot \pi^{n/2} \cdot \prod_{i=1}^n c_i(\lambda)^{\frac{k_i - 1}{2}}.$$

Then the operator $\mathbf{E}$ is extended to a unitary transformation from $L_2(\mathfrak{g}_+ \times \mathfrak{g}_+ \times (\mathfrak{z}^*\backslash\{0\}))$ to $L_2(\mathbf{X}, \mathbf{dm})$, and the unitary operator $\mathbf{U}$ is defined by

$$\mathbf{U} = \mathbf{E} \circ \Phi^* \circ \mathbf{R} \circ \mathcal{F}. \tag{10.1.5}$$

Finally, let the function φ on $\mathbf{X}$ be

$$\varphi(v,\lambda,\mathbf{k}) = |\lambda|^2 + \sum_{i=1}^n (2k_i + 1)\sqrt{c_i(\lambda)}, \quad (v,\lambda) \in \mathfrak{g}_+ \times (\mathfrak{z}^*\backslash\{0\}). \tag{10.1.6}$$

With this introduction, we get the explicit integral expression for the heat kernel by calculating the kernel distribution of the composition operator

$$\mathbf{U}^{-1} \circ e^{-t\mathbf{M}_\varphi} \circ \mathbf{U}.$$

The heat kernel is given by the following result.

Theorem 10.1.5.

$$\begin{aligned} K(t;g,\tilde{g}) &= K(t;x,y,z,\tilde{x},\tilde{y},\tilde{z}) \qquad (10.1.7)\\ &= (2\pi)^{-(n+d)/2} \int_{\mathfrak{z}^*} e^{\sqrt{-1}<\eta,\tilde{z}-z+1/2[\tilde{x},y]-1/2[x,\tilde{y}]>} \\ &\quad \times e^{-t|\eta|^2} \cdot \prod_{i=1}^n \frac{\sqrt{c_i(\eta)}}{2\sinh t\sqrt{c_i(\eta)}} \cdot e^{-\frac{\sqrt{c_i(\eta)}}{4} \cdot \frac{\cosh t\sqrt{c_i(\eta)}}{\sinh t\sqrt{c_i(\eta)}} \cdot \{(x_i-\tilde{x}_i)^2 + (y_i - \tilde{y}_i)^2\}}\, d\eta. \end{aligned}$$

10.2 Complex Hamilton–Jacobi Theory

Complex Hamilton–Jacobi theory is different from the classical theory by the fact that the boundary values of the Hamiltonian system allow complex values, and hence the bicharacteristics and the action are complex. The fact that complex Hamilton–Jacobi theory is more suited to the study of sub-elliptic operators rather than classical theory was first noticed by Beals, Gaveau and Greiner in the 1990s; see [13, 14, 16]. The interested reader can also consult Chap. 5 of the book [28].

10.2.1 Path Integrals and Integral Expression of a Heat Kernel

It is known from statistical mechanics that the heat kernel $K_t(x,y)$ must be expressed as a path integral

$$K(t,x,y) = \int_{P_t(x,y)} e^{-S_t(\gamma)} d\mu(\gamma),$$

hopefully with a suitable *infinite-dimensional measure* $d\mu(\gamma)$, where $P_t(x,y)$ denotes the path space connecting x to y at a time t. The function $S_t(\gamma)$ is called the *classical action* and is given by

$$S_t(\gamma) = \frac{1}{2}\int_0^t |\dot{\gamma}(s)|^2 ds.$$

By normalizing the time parameter with $t = 1$, the aforementioned integral can also be written as

$$\frac{1}{t^N}\int_{P_1(x,y)} e^{-\frac{S_1(\gamma_t)}{2t}} d\mu(\gamma_t), \qquad \text{with} \quad \gamma_t(\sigma) = \gamma(\sigma t).$$

In the case of the Laplacian the aforementioned integral has the following asymptotic expansion:

$$K(t,x,y) \sim \frac{1}{(2\pi t)^{n/2}} e^{-\frac{d(x,y)^2}{2t}} u_0(x,y)(1 + O(t)).$$

Here $d(x,y)$ denotes the Riemannian distance between the points x and y. However, for sub-Laplacians the small time asymptotic expansion is more complicated: see [16].

In particular, when the Riemannian metric is Euclidean, there is only one geodesic (a line segment) that connects the points x and y, and hence the path integral reduces to the function

$$\frac{1}{(2\pi t)^{n/2}} e^{-\frac{|x-y|^2}{2t}}, \tag{10.2.8}$$

which is just the heat kernel of the Laplacian $\Delta = -\frac{1}{2}\sum_{i=1}^n \frac{\partial^2}{\partial x_i^2}$.

There are also similar arguments in the Heisenberg group case that reduce the path integral formula to a certain formula which makes full mathematical sense, see [78].

The heat kernel of the sub-Laplacian on the three-dimensional Heisenberg group was first constructed by Hulanicki in the mid-1970s, who provided an explicit integral formula using a probabilistic argument. Several papers published later in the 1990s deal with similar problems involving the heat kernel for Laplacians and sub-Laplacians on nilpotent Lie groups; see [14–16, 48, 98]. There are also quite a few recent papers dealing with a similar subject and calculations; see [32, 33, 78].

Unlike in Riemannian geometry, in sub-Riemannian geometry there are many geodesics connecting two points even locally; see [27, 28]. Because of this specific behavior, the heat kernel for sub-Laplacians will necessarily have an integral expression.

There are several ways to attack the problem of heat kernels for sub-Laplacians. Some methods involve Mehler's formula (see [107]), which is an important tool in the construction of the heat kernel on Heisenberg group, as done by Hulanicki [68]. This formula was also used in the previous section of this chapter to calculate the integrals included in the formula $\mathbf{U}^{-1} \circ e^{-t\mathcal{M}_\varphi} \circ \mathbf{U}$. This formula provides the generating function of Hermite polynomials explicitly.

Another method is the method of complex Hamilton–Jacobi theory originated by Beals, Gaveau and Greiner [14–16], where there is no need for the generating function formula of Hermite polynomials. The aforementioned authors obtained the same formula directly, using that the heat kernel itself can be seen as a generating function. Their method starts by assuming that the heat kernel $k_t(g)$ has a certain integral expression [see (10.2.10)]. The physical significance of the formula is that the heat flows *mostly* along the *geodesics* starting from the identity element at time $t = 0$, where the density of heat is the Dirac's δ-function. The total amount of heat at a point g at a time t, denoted by $k_t(g)$, should be equal to the sum (integral) over a certain class of geodesics arriving at point g at a time t from *somewhere*.

This class of geodesics is determined by solving the Hamiltonian system (bicharacteristics system) under an initial-boundary condition; i.e., we assume that the coordinates in $\mathfrak{g}/[\mathfrak{g}, \mathfrak{g}]$ are zero at $t = 0$ and the endpoint g must be arbitrarily given in the space G. In the Euclidean case there is only one such geodesic arriving at the point g under this condition, so no integration is needed and we have the well-known formula (10.2.8).

However, in the nilpotent (non-abelian) cases we must consider geodesics whose initial points are not the identity element. These will be parameterized by the dual space $[\mathfrak{g}, \mathfrak{g}]^*$ for both the Laplacian and sub-Laplacian cases. The reason for which we need to consider such geodesics is that on our curved space (which is otherwise topologically Euclidean) the wavefront set of the δ-function produces influence into the direction $[\mathfrak{g}, \mathfrak{g}]$. However, it is not clear whether this interpretation suffices to study the construction of the heat kernel on nilpotent Lie groups of step 3 or higher step, under the assumption that it has a prescribed integral form.

Inspired by the heat kernel formula provided in Theorem 10.1.5, we shall consider the following ansatz as one of the possibilities of constructing the heat kernel:

Theorem 10.2.1 (Meta-theorem). *We assume that the heat kernel of a general nilpotent Lie group has the following integral form, where f denotes the action function and W denotes the volume element:*

$$K(t; (x, z), (\tilde{x}, \tilde{z})) = k_t((\tilde{x}, \tilde{z})^{-1} * (x, z)), \tag{10.2.9}$$

$$k_t(x, z) = \frac{1}{t^N} \int_\tau e^{-\frac{f(x,z,\tau)}{t}} W(x, z, \tau) d\tau, \ (x, z) = g \in \big(\mathfrak{g}/[\mathfrak{g}, \mathfrak{g}] \times [\mathfrak{g}, \mathfrak{g}]\big), \tag{10.2.10}$$

with a specific order N. For two-step nilpotent groups, $N = \frac{1}{2} \dim \mathfrak{g}/[\mathfrak{g}, \mathfrak{g}] + \dim[\mathfrak{g}, \mathfrak{g}]$.

Consequently, the function $f(g,\tau)$ will include all the information of the *real* geodesics when $t \searrow 0$, and the volume element $W(g,\tau)$ will reflect the weight of the energy of the geodesics arriving at the point g.

We shall briefly present the content of the following few sections. In Sect. 10.2.2 we shall explain a method to construct a solution of a class of Hamilton–Jacobi equations that gives the action function f assumed to exist by the *Meta-theorem*. In Sect. 10.2.3 we shall assume that the action function f is a solution of the following generalized Hamilton–Jacobi equation:

$$\sum \tau_i \frac{\partial f(g,\tau)}{\partial \tau_i} + H(g;\nabla f) = f(g,\tau),$$

and we shall find an equation satisfied by the volume element $W(g,\tau)$, called the *generalized transport equation*. In Sects. 10.2.4 and 10.2.5 we shall show that this general mechanism works well in the case of two-step nilpotent Lie groups. In this case we have heat kernels for both sub-Laplacians and Laplacians in the integral form stated by the *Meta-theorem*.

10.2.2 A Solution of a Hamilton–Jacobi Equation

In this section we shall consider a class of Hamilton–Jacobi equations and construct solutions under a certain assumption. This assumption is satisfied in the case of the Hamiltonian being the principal symbol of (left or right invariant) sub-Laplacians or Laplacians on two-step nilpotent Lie groups.

Let $H(x,y;\xi,\eta)$ be a polynomial of the variables $(x,y;\xi,\eta) \in \mathbb{R}^m \times \mathbb{R}^d \times \mathbb{R}^m \times \mathbb{R}^d$ and total degree 2 with respect to the variables $\xi \in \mathbb{R}^m$ and $\eta \in \mathbb{R}^d$:

$$H(x,y;\xi,\eta) = \sum a_{i\,j}(x,y)\xi_i\xi_j + \sum b_{i\,k}(x,y)\xi_i\eta_k + \sum c_{k\,\ell}(x,y)\eta_k\eta_\ell, \tag{10.2.11}$$

where $a_{i\,j}(x,y)$, $b_{i\,k}(x,y)$ and $c_{k\,\ell}(x,y)$ are polynomials of the variables $x \in \mathbb{R}^m$ and $y \in \mathbb{R}^d$ with real coefficients. We may also allow the functions $a_{i\,j}$, $b_{i\,k}$ and $c_{k\,\ell}$ to be entire functions of the variables $x + \sqrt{-1}x' \in \mathbb{C}^m$ and $y + \sqrt{-1}y' \in \mathbb{C}^d$. This condition for the Hamiltonian is satisfied by the principal symbol of the (left)invariant sub-Laplacians and Laplacians on nilpotent Lie groups.

Consider the Hamiltonian system

$$\dot{x} = H_\xi = \frac{\partial H(x,y;\xi,\eta)}{\partial \xi}, \qquad \dot{y} = H_\eta = \frac{\partial H(x,y;\xi,\eta)}{\partial \eta},$$

$$\dot{\xi} = -H_x = -\frac{\partial H(x,y;\xi,\eta)}{\partial x}, \quad \dot{\eta} = -H_y = -\frac{\partial H(x,y;\xi,\eta)}{\partial y},$$

with the initial-boundary conditions

$$\begin{cases} x(0) = 0, \\ x(s) = x, \quad y(s) = y, \\ \eta(0) = \tau. \end{cases}$$

The following assumption will play an important role in the sequel.

Assumption 10.2.2 *We assume that there exists an open conic domain $\mathcal{D}$ in $\mathbb{C}^d$ such that for any $s \in \mathbb{R}$, $(x, y) \in \mathbb{R}^m \times \mathbb{R}^d$ and $\tau \in \mathcal{D}$, there exists a unique global solution of the above system*

$$\begin{aligned} X(t) &= X(t; s, x, y, \tau), & Y(t) &= Y(t; s, x, y, \tau), \\ \xi(t) &= \xi(t; s, x, y, \tau), & \eta(t) &= \eta(t; s, x, y, \tau), \end{aligned}$$

which is smooth with respect to the variables $(s, x, y, \tau) \in \mathbb{R} \times \mathbb{R}^m \times \mathbb{R}^d \times \mathcal{D}$.

Under this assumption, let g be a function defined by the formula

$$\begin{aligned} g(x, y; s, \tau) = &\sum_{j=1}^{d} \tau_j Y_j(0; s, x, y, \tau) \\ &+ \int_0^s \sum \xi_i(t)\dot{X}_i(t) + \sum \eta_j(t)\dot{Y}_j(t) - H(X(t), Y(t); \xi(t), \eta(t))dt. \end{aligned}$$

Then the function $g = g(x, y; s, \tau)$, where the variable τ is considered a parameter, is smooth and satisfies the following Hamilton–Jacobi equation:

Proposition 10.2.3. *We have*

$$\frac{\partial g}{\partial s} + H(x, y; \nabla g) = 0, \tag{10.2.12}$$

where $\nabla g = \left(\frac{\partial g}{\partial x}, \frac{\partial g}{\partial y}\right)$.

Proof. This is proved by an explicit computation of the derivatives

$$\frac{\partial g}{\partial s}(x, y; s, \tau), \qquad \frac{\partial g}{\partial x}(x, y; s, \tau), \qquad \frac{\partial g}{\partial y}(x, y; s, \tau).$$

First we shall show that

$$\frac{\partial g}{\partial s}(x, y; s, \tau) + H(x, y; \xi(s; s, x, y, \tau), \eta(s; s, x, y, \tau)) = 0. \tag{10.2.13}$$

We warn the reader that this will be a tedious computation:

$$
\begin{aligned}
&\frac{\partial g}{\partial s}(x,y;s,\tau)\\
&= \sum_{j=1}^{d} \tau_j \frac{\partial Y_j}{\partial s}(0;s,x,y,\tau)\\
&\quad + \sum \xi_i(s)\dot{X}_i(s) + \sum \eta(s)_j \dot{Y}_j(s) - H(x,y;\xi(s;s,x,y,\tau),\eta(s;s,x,y,\tau))\\
&\quad + \int_0^s \Bigg(\sum \frac{\partial \xi_i}{\partial s}(t;s,x,y,\tau)\dot{X}_i(t;s,x,y,\tau)\\
&\quad + \sum \xi_i(t;s,x,y,\tau)\frac{\partial \dot{X}_i}{\partial s}(t;s,x,y,\tau)\\
&\quad + \sum \frac{\partial \eta_j}{\partial s}(t;s,x,y,\tau)\dot{Y}_j(t;s,x,y,\tau) + \sum \eta_j(t;s,x,y,\tau)\frac{\partial \dot{Y}_j}{\partial s}(t;s,x,y,\tau)\\
&\quad - \sum \frac{\partial H}{\partial x_i}(X(t),Y(t);\xi(t),\eta(t))\frac{\partial X_i}{\partial s}(t;s,x,y,\tau)\\
&\quad - \sum \frac{\partial H}{\partial y_j}(X(t),Y(t);\xi(t),\eta(t))\frac{\partial Y_j}{\partial s}(t;s,x,y,\tau)\\
&\quad - \sum \frac{\partial H}{\partial \xi_i}(X(t),Y(t);\xi(t),\eta(t))\frac{\partial \xi_i}{\partial s}(t;s,x,y,\tau)\\
&\quad - \sum \frac{\partial H}{\partial \eta_j}(X(t),Y(t);\xi(t),\eta(t))\frac{\partial \eta_j}{\partial s}(t;s,x,y,\tau)\Bigg)\,dt\\
&= \sum_{j=1}^{d} \tau_j \frac{\partial Y_j}{\partial s}(0;s,x,y,\tau)\\
&\quad + \sum \xi_i(s)\dot{X}_i(s) + \sum \eta(s)_j \dot{Y}_j(s) - H(x,y;\xi(s;s,x,y,\tau),\eta(s;s,x,y,\tau))\\
&\quad + \int_0^s \Bigg\{\sum \left(\frac{\partial \xi_i}{\partial s}(t;s,x,y,\tau)\dot{X}_i(t;s,x,y,\tau) + \xi_i(t;s,x,y,\tau)\frac{\partial \dot{X}_i}{\partial s}(t;s,x,y,\tau)\right)\\
&\quad + \sum \left(\frac{\partial \eta_j}{\partial s}(t;s,x,y,\tau)\dot{Y}_j(t;s,x,y,\tau) + \eta_j(t;s,x,y,\tau)\frac{\partial \dot{Y}_j}{\partial s}(t;s,x,y,\tau)\right)\\
&\quad + \sum \dot{\xi}_i(t;s,x,y,\tau)\frac{\partial X_i}{\partial s}(t;s,x,y,\tau) + \sum \dot{\eta}_j(t;s,x,y,\tau)\frac{\partial Y_j}{\partial s}(t;s,x,y,\tau)\\
&\quad - \sum \dot{X}_i(t;s,x,y,\tau)\frac{\partial \xi_i}{\partial s}(t;s,x,y,\tau) - \sum \dot{Y}_j(t;s,x,y,\tau)\frac{\partial \eta_j}{\partial s}(t;s,x,y,\tau)\Bigg\}\,dt\\
&= \sum_{j=1}^{d} \tau_j \frac{\partial Y_j}{\partial s}(0;s,x,y,\tau)\\
&\quad + \sum \xi_i(s)\dot{X}_i(s) + \sum \eta(s)_j \dot{Y}_j(s) - H(x,y;\xi(s;s,x,y,\tau),\eta(s;s,x,y,\tau))\\
&\quad + \sum \xi_i(t)\frac{\partial X_i}{\partial s}(t;s,x,y,\tau)\Big|_0^s + \sum \eta_j(t)\frac{\partial Y_j}{\partial s}(t;s,x,y,\tau)\Big|_0^s.
\end{aligned}
$$

Now from the initial-boundary conditions

$$X(0; s, x, y, \tau) = 0, \qquad X(s; s, x, y, \tau) = x, \qquad Y(s; s, x, y, \eta) = y,$$

we have

$$\begin{aligned}
\dot{X}_i(s; s, x, y, \tau) + \frac{\partial X_i}{\partial s}(s; s, x, y, \tau) &= 0, \\
\dot{Y}_j(s; s, x, y, \tau) + \frac{\partial Y_j}{\partial s}(s; s, x, y, \tau) &= 0, \\
\frac{\partial X_i}{\partial s}(0; s, x, y, \tau) &= 0,
\end{aligned}$$

which finally imply (10.2.13).

Next we shall show that

$$\begin{aligned}
\frac{\partial g}{\partial x_k}(x, y; s, \tau) &= \xi_i(s; s, x, y, \tau), \\
\frac{\partial g}{\partial y_l}(x, y; s, \tau) &= \eta_l(s; s, x, y, \tau).
\end{aligned}$$

Another technical computation follows:

$$\begin{aligned}
&\frac{\partial g}{\partial x_k}(x, y; s, \tau) \\
&= \sum_{j=1}^{d} \tau_j \frac{\partial Y_j}{\partial x_k}(0; s, x, y, \tau) \\
&\quad + \int_0^s \Bigg(\sum \frac{\partial \xi_i}{\partial x_k}(t; s, x, y, \tau)\dot{X}_i(t; s, x, y, \tau) + \sum \xi_i(t; s, x, y, \tau)\frac{\partial \dot{X}_i}{\partial x_k}(t; s, x, y, \tau) \\
&\quad + \sum \frac{\partial \eta_j}{\partial x_k}(t; s, x, y, \tau)\dot{Y}_j(t; s, x, y, \tau) + \sum \eta_j(t; s, x, y, \tau)\frac{\partial \dot{Y}_j}{\partial x_k}(t; s, x, y, \tau) \\
&\quad - \sum \frac{\partial H}{\partial x_i}(X(t), Y(t); \xi(t), \eta(t))\frac{\partial X_i}{\partial x_k}(t; s, x, y, \tau) \\
&\quad - \sum \frac{\partial H}{\partial y_j}(X(t), Y(t); \xi(t), \eta(t))\frac{\partial Y_j}{\partial x_k}(t; s, x, y, \tau) \\
&\quad - \sum \frac{\partial H}{\partial \xi_i}(X(t), Y(t); \xi(t), \eta(t))\frac{\partial \xi_i}{\partial x_k}(t; s, x, y, \tau) \\
&\quad - \sum \frac{\partial H}{\partial \eta_j}(X(t), Y(t); \xi(t), \eta(t))\frac{\partial \eta_j}{\partial x_k}(t; s, x, y, \tau) \Bigg) dt \\
&= \sum_{j=1}^{d} \tau_j \frac{\partial Y_j}{\partial x_k}(0; s, x, y, \tau)
\end{aligned}$$

$$
\begin{aligned}
&+\int_0^s \Bigg(\sum \frac{\partial \xi_i}{\partial x_k}(t;s,x,y,\tau)\dot{X}_i(t;s,x,y,\tau)+\sum \xi_i(t;s,x,y,\tau)\frac{\partial \dot{X}_i}{\partial x_k}(t;s,x,y,\tau)\\
&+\sum \frac{\partial \eta_j}{\partial x_k}(t;s,x,y,\tau)\dot{Y}_j(t;s,x,y,\tau)+\sum \eta_j(t;s,x,y,\tau)\frac{\partial \dot{Y}_j}{\partial x_k}(t;s,x,y,\tau)\\
&+\sum \dot{\xi}_i(t;s,x,y,\tau)\frac{\partial X_i}{\partial x_k}(t;s,x,y,\tau)+\sum \dot{\eta}_j(t;s,x,y,\tau)\frac{\partial Y_j}{\partial x_k}(t;s,x,y,\tau)\\
&-\sum \dot{X}_i(t;s,x,y,\tau)\frac{\partial \xi_i}{\partial x_k}(t;s,x,y,\tau)-\sum \dot{Y}_j(t;s,x,y,\tau)\frac{\partial \eta_j}{\partial x_k}(t;s,x,y,\tau)\Bigg)dt\\
&=\sum_{j=1}^d \tau_j\frac{\partial Y_j}{\partial x_k}(0;s,x,y,\tau)\sum \xi_i(t;s,x,y,\tau)\frac{\partial X_i}{\partial x_k}\Big|_0^s+\sum \eta_j(t;s,x,y,\tau)\frac{\partial Y_j}{\partial x_k}\Big|_0^s\\
&=\xi_k(s;s,x,y,\tau),
\end{aligned}
$$

where we made use of the initial-boundary conditions

$$
\begin{aligned}
&\frac{\partial X_i}{\partial x_k}(s;s,x,y,\tau)=\frac{\partial x_i}{\partial x_k}=\delta_{i,k}, \quad \frac{\partial X_i}{\partial x_k}(0;s,x,y,\tau)=0,\\
&\frac{\partial Y_j}{\partial x_k}(s;s,x,y,\tau)=\frac{\partial y_j}{\partial x_k}=0.
\end{aligned}
$$

Using the similar boundary conditions

$$
\begin{aligned}
&\frac{\partial Y_j}{\partial y_l}(s;s,x,y,\tau)=\frac{\partial y_j}{\partial y_l}=\delta_{j,l}, \quad \frac{\partial X_i}{\partial y_l}(0;s,x,y,\tau)=0,\\
&\frac{\partial X_i}{\partial y_l}(s;s,x,y,\tau)=\frac{\partial x_i}{\partial y_l}=0,
\end{aligned}
$$

we have the following computation for the derivative of g:

$$
\begin{aligned}
&\frac{\partial g}{\partial y_l}(x,y;s,\tau)\\
&=\sum_{j=1}^d \tau_j\frac{\partial Y_j}{\partial y_l}(0;s,x,y,\tau)\\
&+\int_0^s \Bigg(\sum \frac{\partial \xi_i}{\partial y_l}(t;s,x,y,\tau)\dot{X}_i(t;s,x,y,\tau)+\sum \xi_i(t;s,x,y,\tau)\frac{\partial \dot{X}_i}{\partial y_l}(t;s,x,y,\tau)\\
&+\sum \frac{\partial \eta_j}{\partial y_l}(t;s,x,y,\tau)\dot{Y}_j(t;s,x,y,\tau)+\sum \eta_j(t;s,x,y,\tau)\frac{\partial \dot{Y}_j}{\partial y_l}(t;s,x,y,\tau)\\
&-\sum \frac{\partial H}{\partial x_i}(X(t),Y(t);\xi(t),\eta(t))\frac{\partial X_i}{\partial y_l}(t;s,x,y,\tau)\\
&-\sum \frac{\partial H}{\partial y_j}(X(t),Y(t);\xi(t),\eta(t))\frac{\partial Y_j}{\partial y_l}(t;s,x,y,\tau)
\end{aligned}
$$

$$
\begin{aligned}
&- \sum \frac{\partial H}{\partial \xi_i}(X(t), Y(t); \xi(t), \eta(t)) \frac{\partial \xi_i}{\partial y_l}(t; s, x, y, \tau) \\
&- \sum \frac{\partial H}{\partial \eta_j}(X(t), Y(t); \xi(t), \eta(t)) \frac{\partial \eta_j}{\partial y_l}(t; s, x, y, \tau) \Bigg) dt \\
&= \sum_{j=1}^{d} \tau_j \frac{\partial Y_j}{\partial y_l}(0; s, x, y, \tau) \\
&+ \int_0^s \Bigg(\sum \frac{\partial \xi_i}{\partial y_l}(t; s, x, y, \tau) \dot{X}_i(t; s, x, y, \tau) + \sum \xi_i(t; s, x, y, \tau) \frac{\partial \dot{X}_i}{\partial y_l}(t; s, x, y, \tau) \\
&+ \sum \frac{\partial \eta_j}{\partial y_l}(t; s, x, y, \tau) \dot{Y}_j(t; s, x, y, \tau) + \sum \eta_j(t; s, x, y, \tau) \frac{\partial \dot{Y}_j}{\partial y_l}(t; s, x, y, \tau) \\
&+ \sum \dot{\xi}_i(t; s, x, y, \tau) \frac{\partial X_i}{\partial y_l}(t; s, x, y, \tau) + \sum \dot{\eta}_j(t; s, x, y, \tau) \frac{\partial Y_j}{\partial y_l}(t; s, x, y, \tau) \\
&- \sum \frac{\partial H}{\partial \xi_i}(X(t), Y(t); \xi(t), \eta(t)) \frac{\partial \xi_i}{\partial y_l}(t; s, x, y, \tau) \\
&- \sum \frac{\partial H}{\partial \eta_j}(X(t), Y(t); \xi(t), \eta(t)) \frac{\partial \eta_j}{\partial y_l}(t; s, x, y, \tau) \Bigg) dt \\
&= \sum_{j=1}^{d} \tau_j \frac{\partial Y_j}{\partial y_l}(0; s, x, y, \tau) \\
&+ \sum \xi_i(t; s, x, y, \tau) \frac{\partial X_i}{\partial y_l}(t; s, x, y, \tau) \Big|_0^s + \sum \eta_j(t; s, x, y, \tau) \frac{\partial Y_j}{\partial y_l}(t; s, x, y, \tau) \Big|_0^s \\
&= \eta_l(s; s, x, y, \tau).
\end{aligned}
$$

Hence we conclude that $g = g(x, y; s, \tau)$ satisfies the equation

$$\frac{\partial g}{\partial s} + H(x, y; \nabla g) = 0.$$

■

10.2.3 The Generalized Transport Equation

In this section we shall deduce an equation, called the *generalized transport equation*, which is satisfied by the volume element present in the integral expression of the heat kernel for the sub-Laplacian Δ_{sub} on a nilpotent Lie group given by the *Meta-theorem* 10.2.1.

Let G be an n-dimensional, connected, simply connected nilpotent Lie group with the Lie algebra $\mathfrak{g}$, and let $\{X_i\}_{i=1}^m$ be a basis of the complement of the first derived ideal $[\mathfrak{g}, \mathfrak{g}]$, with $m = \dim \mathfrak{g}/[\mathfrak{g}, \mathfrak{g}]$.

Let

$$\Delta_{\text{sub}} = -\sum_{i=1}^{m} \tilde{X}_i^2 \tag{10.2.14}$$

be the sum of the left invariant vector fields $\tilde{X}_i$ on the group G. Then Δ_{sub} is a sub-Laplacian satisfying the Hörmander condition of hypoellipticity.

To apply the result of Sect. 10.2.2, we fix a basis $\{Z_j\}_{j=1}^d$ of the derived ideal $[\mathfrak{g}, \mathfrak{g}]$, and we work with the coordinates

$$G \ni \sum x_i \widetilde{X}_i + \sum y_j \widetilde{Z}_j \leftrightarrow (x, y) \in \mathbb{R}^m \times \mathbb{R}^d,$$

$$T^*(G) \ni \left(\sum x_i \widetilde{X}_i + \sum y_j \widetilde{Z}_j, \sum \xi_i \tilde{X}_i^* + \sum \eta_j \tilde{Z}_j^*\right)$$

$$\leftrightarrow (x, y; \xi, \eta) \in \mathbb{R}^{m+d} \times \mathbb{R}^{m+d}$$

where we have denoted by $\{\widetilde{X}_i, \widetilde{Z}_j\}$ the dual basis of $\{X_i, Z_j\}$. The corresponding left invariant 1-forms are given by $\{\xi_i, \eta_j\}$. This splitting (x, y) of the coordinates corresponds to the splitting of the variables in the preceding section.

Let H be the Hamiltonian associated with the sub-Laplacian (10.2.14):

$$H : T^*(G) \cong \mathbb{R}^m \times \mathbb{R}^d \times \mathbb{R}^m \times \mathbb{R}^d \ni (x, y; \xi, \eta) \mapsto H(x, y; \xi, \eta) \in \mathbb{R},$$

$$H(x, y; \xi, \eta) = \frac{1}{2} \sum_{i=1}^m \xi_i (\tilde{X}_i)^2.$$

Assume the function $f = f(x, y; \tau)$ is a solution of the generalized Hamilton–Jacobi equation

$$H(x, y; \nabla f) + \sum_{i=1}^d \tau_i \frac{\partial f}{\partial \tau_i} = f(x, y; \tau_1, \dots, \tau_d), \tag{10.2.15}$$

obtained by setting $s = 1$ in the solution g of the Hamilton–Jacobi equation (10.2.3) given in Sect. 10.2.2 under assumption (10.2.2).

To make our assumption clear, we state once again that the heat kernel $K(t; x, y, \tilde{x}, \tilde{y})$ is supposed to be of the form

$$K(t; x, y, \tilde{x}, \tilde{y}) = k_t\big((-\tilde{x}, -\tilde{y}) * (x, y)\big),$$

$$k_t(x, y) = \frac{1}{t^N} \int_{\mathbb{R}^d} e^{-\frac{f(x,y;\tau)}{t}} W(x, y, \tau) d\tau, \tag{10.2.16}$$

for some positive integer $N > 0$. In the case of a two-step nilpotent Lie group N is fixed and is given by $N = \frac{1}{2} \dim \mathfrak{g}/[\mathfrak{g}, \mathfrak{g}] + \dim[\mathfrak{g}, \mathfrak{g}] = m/2 + d$. However, the following calculations are valid for any $d > 0$ and $N > 0$.

Let the characteristic variety of Δ_{sub} be

$$Ch = \{(x, y, \xi, \eta);\ H(x, y, \xi, \eta) = 0\}.$$

This is a sub-bundle in T^*G and is trivialized as $G \times [\mathfrak{g},\mathfrak{g}]^*$, which is embedded in $T^*(G)$ according to the splitting $\mathfrak{g} = [\mathfrak{g},\mathfrak{g}] \oplus [\{X_i\}]$:

$$\mathcal{C}h = G \times [\mathfrak{g},\mathfrak{g}]^* \subset T^*(G);$$

see also Remark 10.2.12. The dimension of the integration variable τ is $d = \dim\, \mathcal{C}h - \dim\, G = \dim\, [\mathfrak{g},\mathfrak{g}]$.

In the following we shall accept the reasonable assumption that the integrand $e^{-\frac{f(x,y;\tau)}{t}} W(x,y;\tau)$ decreases fast enough such that the partial integrations can be performed in a convenient way when $|\tau| \to \infty$; see also Theorem 10.2.7. Then we have

$$\begin{aligned}
&\Delta_{\text{sub}}(e^{-\frac{f(x,y;\tau)}{t}} W)\\
&\quad = \frac{-1}{t^2} H(x,y;\nabla f)\cdot W \cdot e^{-\frac{f}{t}} - \frac{1}{t}\Big(\Delta_{\text{sub}}(f)\cdot W - \sum_{i=1}^{m} X_i(f)X_i(W)\Big)e^{-\frac{f}{t}}\\
&\qquad + \Delta_{\text{sub}}(W)e^{-\frac{f}{t}},
\end{aligned}$$

and

$$\begin{aligned}
&\frac{\partial}{\partial t}\left(\frac{1}{t^N}\int_{\mathbb{R}^d} e^{-\frac{f}{t}}\cdot W d\tau_1\cdots d\tau_d\right)\\
&\quad = -\frac{1}{t^N}\int_{\mathbb{R}^d}\frac{N}{t}\cdot e^{-\frac{f}{t}}\cdot W d\tau_1\cdots d\tau_d + \frac{1}{t^N}\int_{\mathbb{R}^d}\frac{f}{t^2}\cdot e^{-\frac{f}{t}}\cdot W d\tau_1\cdots d\tau_d.
\end{aligned}$$

Combining the last two relations, we obtain

$$\begin{aligned}
&-\left(\Delta_{\text{sub}} + \frac{\partial}{\partial t}\right)\Big(k_t(x,y)\Big)\\
&\quad = \frac{1}{t^N}\left\{\frac{1}{t^2}\int_{\mathbb{R}^d}(H(x,y;\nabla f) - f)\cdot e^{-\frac{f}{t}}\cdot W d\tau\right.\\
&\qquad - \frac{1}{t}\int_{\mathbb{R}^d}\left(\sum_i \tilde{X}_i(f)\tilde{X}_i(W) + (-\Delta_{\text{sub}}(f) - N)\cdot W\right)e^{-\frac{f}{t}} d\tau\\
&\qquad \left. - \int_{\mathbb{R}^d}\Delta_{\text{sub}}(W)e^{-\frac{f}{t}} d\tau\right\}\\
&\quad = \frac{1}{t^N}\left\{-\frac{1}{t^2}\int_{\mathbb{R}^d}\sum \tau_i\frac{\partial f}{\partial \tau_i}\cdot e^{-\frac{f}{t}}\cdot W d\tau\right.\\
&\qquad - \frac{1}{t}\int_{\mathbb{R}^d}\left(\sum_i \tilde{X}_i(f)\tilde{X}_i(W) - (\Delta_{\text{sub}}(f) + N)\cdot W\right)e^{-\frac{f}{t}} d\tau\\
&\qquad \left. - \int_{\mathbb{R}^d}\Delta_{\text{sub}}(W)e^{-\frac{f}{t}} d\tau\right\}\\
&\quad = \frac{1}{t^N}\left\{\frac{1}{t^2}\cdot A_2 - \frac{1}{t}\cdot A_1 - A_0\right\}.
\end{aligned}$$

Now we assume

$$W d\tau = df \wedge V \tag{10.2.17}$$

with the $(d-1)$-form[1]

$$V = \sum_{\alpha=1}^{d} (-1)^{\alpha-1} V_\alpha d\tau_1 \wedge \cdots \wedge \widehat{d\tau_\alpha} \wedge \cdots \wedge d\tau_d.$$

In general, the coefficients V_i are functions of both space variables $\{x_i, y_j\}$ and the characteristic variety variables $\{\tau_j\}$. Therefore, the operation $\Delta_{\text{sub}}(V)$ is defined as

$$\Delta_{\text{sub}}(V) = \sum_{\alpha=1}^{d} (-1)^{\alpha-1} \Delta_{\text{sub}}(V_\alpha) d\tau_1 \wedge \cdots \wedge \widehat{d\tau_\alpha} \wedge \cdots \wedge d\tau_d.$$

The terms A_1 and A_2 will be computed under assumption (10.2.17) on the function W in the following way:

$$\begin{aligned}
&\frac{1}{t^2} \cdot A_2 \\
&= -\frac{1}{t^2} \int_{\mathbb{R}^d} \left(\sum \tau_i \frac{\partial f}{\partial \tau_i} \right) \left(e^{-\frac{f}{t}} \right) df \wedge V \\
&= \frac{1}{t} \int_{\mathbb{R}^d} \left(\sum \tau_i \frac{\partial f}{\partial \tau_i} \right) d\left(e^{-\frac{f}{t}} \right) \wedge V \\
&= -\frac{1}{t} \int_{\mathbb{R}^d} e^{-\frac{f}{t}} d\left(\sum \tau_i \frac{\partial f}{\partial \tau_i} \cdot V \right) \\
&= -\frac{1}{t} \int_{\mathbb{R}^d} e^{-\frac{f}{t}} \left(df \wedge V + \left(\sum_j \sum_i \tau_i \frac{\partial^2 f}{\partial \tau_i \partial \tau_j} V_j \right) d\tau + \left(\sum_i \tau_i \frac{\partial f}{\partial \tau_i} \right) dV \right) \\
&= \int_{\mathbb{R}^d} d\left(e^{-f/t} \right) \wedge V - \frac{1}{t} \int_{\mathbb{R}^d} e^{-f/t} \left(\sum_j \sum_i \tau_i \frac{\partial^2 f}{\partial \tau_i \partial \tau_j} V_j \right) d\tau \\
&\quad + \int_{\mathbb{R}^d} \left(\sum \tau_i \frac{\partial e^{-f/t}}{\partial \tau_i} \right) \cdot \left(\sum_j \frac{\partial V_j}{\partial \tau_j} \right) d\tau \\
&= -\frac{1}{t} \int_{\mathbb{R}^d} e^{-f/t} \left(\sum_j \sum_i \tau_i \frac{\partial^2 f}{\partial \tau_i \partial \tau_j} V_j \right) d\tau \\
&\quad - \int_{\mathbb{R}^d} e^{-f/t} \left(\sum_i \sum_j \tau_i \frac{\partial^2 V_j}{\partial \tau_i \partial \tau_j} \right) d\tau - (d+1) \int_{\mathbb{R}^d} e^{-f/t} dV.
\end{aligned}$$

---

[1] P. Greiner worked out this form for the special case $d = 1$. See also Remark 10.2.4.

On the other hand, we have

$$\begin{aligned}
\frac{1}{t}A_1 &= \frac{1}{t}\int_{\mathbb{R}^d}\sum_i \tilde{X}_i(f)\tilde{X}_i\left(\sum_j \frac{\partial f}{\partial \tau_j}V_j\right)e^{-f/t}d\tau \\
&\quad -\frac{1}{t}\int_{\mathbb{R}^d}(\Delta_{\text{sub}}(f)+N)e^{-f/t}df\wedge V \\
&= \frac{1}{t}\int_{\mathbb{R}^d}\sum_j\left(\frac{\partial}{\partial \tau_j}H(x,y;\nabla f)\right)\cdot V_j\cdot e^{-f/t}d\tau \\
&\quad -\int_{\mathbb{R}^d}\sum_j\sum_i \tilde{X}_i(f)\tilde{X}_i(V_j)\frac{\partial e^{-f/t}}{\partial \tau_j}d\tau+\int_{\mathbb{R}^d}(\Delta_{\text{sub}}(f)+N)d\left(e^{-f/t}\right)\wedge V \\
&= \frac{1}{t}\int_{\mathbb{R}^d}e^{-f/t}\sum_j\frac{\partial H(x,y;\nabla f)}{\partial \tau_j}V_j d\tau \\
&\quad +\int_{\mathbb{R}^d}e^{-f/t}\sum_j\sum_i\frac{\partial}{\partial \tau_j}\left(\tilde{X}_i(f)\tilde{X}_i(V_j)\right)d\tau \\
&\quad -\int_{\mathbb{R}^d}e^{-f/t}d\Big((\Delta_{\text{sub}}(f)+N)V\Big).
\end{aligned}$$

Next we shall compute $-A_0+\frac{1}{t^2}A_2-\frac{1}{t}A_1$. Using that for any j

$$\frac{\partial}{\partial \tau_j}H(x,y;\nabla f)+\sum_i \tau_i\frac{\partial^2 f}{\partial \tau_j\partial \tau_i}=0,$$

we have

$$\begin{aligned}
&-A_0+\frac{1}{t^2}A_2-\frac{1}{t}A_1 \\
&\quad= \int -\Delta_{\text{sub}}\left(\sum\frac{\partial f}{\partial \tau_j}V_j\right)e^{-f/t}dt-(d+1)\int e^{-f/t}dV \\
&\qquad -\int e^{-f/t}\left(\sum\sum \tau_i\frac{\partial^2 V_j}{\partial \tau_i\partial \tau_j}\right)d\tau \\
&\qquad -\int e^{-f/t}\sum\sum\frac{\partial}{\partial \tau_j}\left(\tilde{X}_i(f)\tilde{X}_i(V_j)\right)d\tau \\
&\qquad +\int e^{-f/t}d\left((\Delta_{\text{sub}}(f)+N)V\right) \\
&\quad= -\int e^{-f/t}\Delta_{\text{sub}}\left(\sum\frac{\partial f}{\partial \tau_j}V_j\right)d\tau+(N-d-1)\int e^{-f/t}dV \\
&\qquad +\int e^{-f/t}d(\Delta_{\text{sub}}(f)V)-\int e^{-f/t}\sum\sum \tau_i\frac{\partial^2 V_j}{\partial \tau_i\partial \tau_j}d\tau \\
&\qquad -\int e^{-f/t}\left(\sum\sum\frac{\partial}{\partial \tau_j}\left(\tilde{X}_i(f)\tilde{X}_i(V_j)\right)\right)d\tau
\end{aligned}$$

$$= -\int e^{-f/t} \sum \left(\Delta_{\text{sub}}\Big(\frac{\partial f}{\partial \tau_j}\Big) V_j + \frac{\partial f}{\partial \tau_j} \Delta_{\text{sub}}(V_j) \right) d\tau$$

$$+ (N - d - 1) \int e^{-f/t} dV + \int e^{-f/t} d(\Delta_{\text{sub}}(f) V)$$

$$- \int e^{-f/t} \sum \sum \tau_i \frac{\partial^2 V_j}{\partial \tau_i \partial \tau_j} d\tau - \int e^{-f/t} \sum \tilde{X}_i(f) \tilde{X}_i \left(\sum \frac{\partial V_j}{\partial \tau_j} \right) d\tau$$

$$= -\int e^{-f/t} df \wedge \Delta_{\text{sub}}(V) - \int e^{-f/t} \left(-\Delta_{\text{sub}}(f) - N + d + 1\right) dV$$

$$- \int e^{-f/t} \sum \sum \tau_i \frac{\partial^2 V_j}{\partial \tau_i \partial \tau_j} d\tau - \int e^{-f/t} \sum \tilde{X}_i(f) \tilde{X}_i \left(\sum \frac{\partial V_j}{\partial \tau_j} \right) d\tau$$

$$= -\int e^{-f/t} df \wedge \Delta_{\text{sub}}(V) + \int e^{-f/t} \Delta_{\text{sub}}(f) dV + (N - d - 1) \int e^{-f/t} dV$$

$$- \int e^{-f/t} \mathfrak{D}(dV) - \int e^{-f/t} \sum \tilde{X}_i(f) \tilde{X}_i(dV).$$

Equating $-A_0 + \frac{1}{t^2} A_2 - \frac{1}{t} A_1 = 0$, it suffices for V to satisfy the following *generalized transport equation*:

$$df \wedge \Delta_{\text{sub}}(V) + \sum \tilde{X}_i(f) \cdot \tilde{X}_i(dV)$$
$$+ \mathfrak{D}(dV) - (\Delta_{\text{sub}}(f) + N - d - 1)\, dV = 0, \qquad (10.2.18)$$

where $\mathfrak{D}(V)$ is defined by

$$\mathfrak{D}(V) = \sum \tau_i \frac{\partial}{\partial \tau_i}(V) = \sum \sum (-1)^{j-1} \tau_i \frac{\partial V_j}{\partial \tau_i} d\tau_1 \wedge \cdots \wedge \widehat{d\tau_j} \wedge \cdots \wedge \tau_d$$

and we have

$$\mathfrak{D}(dV) = d\mathfrak{D}(V) - dV.$$

If we assume that all the coefficients V_i depend only on the variables τ_j, $j = 1, \ldots, d$, then the equation reduces to the transport equation for the two-step case.

Remark 10.2.4. Fundamental solutions have volume elements which do not depend on the number of missing directions (see [13]). The result here is similar for the heat kernel volume element W, under the assumption that the volume element is of the form $W d\tau = df \wedge V$; see (10.2.17). When there is only one missing direction, which, in the case of a nilpotent Lie group, means $\dim[\mathfrak{g}, \mathfrak{g}] = 1$, we recover an unpublished result of P. Greiner.

Next we shall work out the *first-order transport equation*. We consider

$$\left(\Delta_{\text{sub}} + \frac{\partial}{\partial t}\right) \left(k_t(g)\right)$$

$$= \frac{1}{t^N} \left\{ \frac{-1}{t^2} \int_{\mathbb{R}^d} (H(g; \nabla f) - f) \cdot e^{-\frac{f}{t}} \cdot W d\tau \right.$$

$$+\frac{1}{t}\int_{\mathbb{R}^d}\left(\sum_i \tilde{X}_i(f)\tilde{X}_i(W)-(\Delta_{\mathrm{sub}}(f)+N)\cdot W\right)e^{-\frac{f}{t}}d\tau$$

$$\left.+\int_{\mathbb{R}^d}\Delta_{\mathrm{sub}}(W)e^{-\frac{f}{t}}d\tau\right\}$$

$$=\frac{1}{t^N}\left\{\frac{1}{t^2}\int_{\mathbb{R}^d}\sum\tau_i\frac{\partial f}{\partial\tau_i}\cdot e^{-\frac{f}{t}}\cdot W d\tau\right.$$

$$+\frac{1}{t}\int_{\mathbb{R}^d}\left(\sum_i \tilde{X}_i(f)\tilde{X}_i(W)-(\Delta_{\mathrm{sub}}(f)+N)\cdot W\right)e^{-\frac{f}{t}}d\tau$$

$$\left.+\int_{\mathbb{R}^d}\Delta_{\mathrm{sub}}(W)e^{-\frac{f}{t}}d\tau\right\}$$

$$=\frac{-1}{t^N}\left\{\frac{1}{t}\int_{\mathbb{R}^d}\sum\tau_i\frac{\partial}{\partial\tau_i}\left(e^{-\frac{f}{t}}\right)\cdot W d\tau\right.$$

$$+\frac{1}{t}\int_{\mathbb{R}^d}\left(\sum_i \tilde{X}_i(f)\tilde{X}_i(W)-(\Delta_{\mathrm{sub}}(f)+N)\cdot W\right)e^{-\frac{f}{t}}d\tau$$

$$\left.+\int_{\mathbb{R}^d}\Delta_{\mathrm{sub}}(W)e^{-\frac{f}{t}}d\tau\right\}$$

$$=\frac{1}{t^N}\left\{\frac{1}{t}\int_{\mathbb{R}^d}e^{-\frac{f}{t}}\cdot\sum\frac{\partial}{\partial\tau_i}(\tau_i W)d\tau\right.$$

$$+\frac{1}{t}\int_{\mathbb{R}^{d\ell}}\left(\sum_i \tilde{X}_i(f)\tilde{X}_i(W)-(\Delta_{\mathrm{sub}}(f)+N)\cdot W\right)e^{-\frac{f}{t}}d\tau$$

$$\left.+\int_{\mathbb{R}^d}\Delta_{\mathrm{sub}}(W)e^{-\frac{f}{t}}d\tau\right\}$$

$$=\frac{1}{t^N}\left\{\frac{1}{t}\int_{\mathbb{R}^d}e^{-\frac{f}{t}}\cdot\sum\tau_i\frac{\partial W}{\partial\tau_i}d\tau\right.$$

$$+\frac{1}{t}\int_{\mathbb{R}^d}\left(\sum_i \tilde{X}_i(f)\tilde{X}_i(W)-(\Delta_{\mathrm{sub}}(f)+N-d)\cdot W\right)e^{-\frac{f}{t}}d\tau$$

$$\left.+\int_{\mathbb{R}^d}\Delta_{\mathrm{sub}}(W)e^{-\frac{f}{t}}d\tau\right\}.$$

If the function W does not depend on the space variables, then W satisfies the first-order transport equation

$$\sum\tau_i\frac{\partial W}{\partial\tau_i}-(\Delta_{\mathrm{sub}}(f)+N-d)\,W=0. \tag{10.2.19}$$

10.2.4 Heat Kernel for the Sub-Laplacian

This section deals with the complex Hamilton–Jacobi theory for two-step nilpotent Lie groups following [15]. In the sequel of this chapter we shall regard the heat equation as

$$\left(\frac{\partial}{\partial t} + \frac{1}{2}\Delta_{\text{sub}}\right)K = 0,$$

to avoid unnecessary constants in the description of the heat kernel.

Let G be an $(n+d)$-dimensional connected, simply connected two-step nilpotent Lie group with center $\mathfrak{z} = [\mathfrak{g}, \mathfrak{g}]$ of dimension $\dim \mathfrak{z} = d$. We identify T^*G with $\mathfrak{g} \times \mathfrak{g}^*$. Let $\{X_i\}_{i=1}^n$ be a basis of the complement of the derived algebra $[\mathfrak{g}, \mathfrak{g}]$. Denote the coordinates on $\mathfrak{g} \times \mathfrak{g}^*$ by $(x, z; \xi, \theta)$ by fixing a basis $\{Z_j\}_{j=1}^d$ in the center $\mathfrak{z} = [\mathfrak{g}, \mathfrak{g}]$.

Let $[X_i, X_j] = \sum_{k=1}^{d} 2a_{ij}^k Z_k$, with $a_{ij} = -a_{ji}$, and let $\Omega(\theta)$ be a $d \times d$ matrix with the entries

$$\Omega(\theta)_{i\,j} = \sum_{k=1}^{d} a_{i\,j}^k \theta_k. \tag{10.2.20}$$

For each X_j we denote by $\tilde{X}_j = \frac{\partial}{\partial x_j} + \sum_{i=1}^{n}\sum_{k=1}^{d} a_{ij}^k x_i \frac{\partial}{\partial z_k}$ the corresponding left invariant vector field on the group G. Then the sum

$$-\Delta_{\text{sub}} = \sum_{i=1}^{n} \tilde{X}_i^2$$

is a sub-Laplacian which satisfies the “Hörmander condition” of hypoellipticity as before. The Hamiltonian associated with the aforementioned sub-Laplacian is given by

$$H(x, z; \xi, \theta) = \frac{1}{2}\sum_{j=1}^{n}\left(\xi_j + \sum_{i=1}^{n}\sum_{k=1}^{d} a_{ij}^k x_i \theta_k\right)^2 = \frac{1}{2}\sum_{j}\left(\xi_j - \sum_{i} \Omega(\theta)_{j\,i} \cdot x_i\right)^2.$$

We consider the Hamiltonian system

$$\begin{cases} \dot{x}_j = H_{\xi_j} = \xi_j - \sum_i \Omega(\theta)_{j,i} x_i, \\ \dot{z}_k = H_{\theta_k}, \\ \dot{\xi}_j = -H_{x_j}, \\ \dot{\theta}_k = -H_{z_k} \equiv 0, \end{cases} \tag{10.2.21}$$

with the following *initial-boundary conditions*:

$$\begin{cases} x(0) = 0, \\ x(s) = x = (x_1, \dots, x_n) \in \mathbb{R}^n, \\ z(s) = z = (z_1, \dots, z_d) \in \mathbb{R}^d, \\ \theta(0) = \sqrt{-1}\tau, \\ \quad \tau = (\tau_1, \dots, \tau_d) \in \mathbb{R}^d, \end{cases} \tag{10.2.22}$$

where $s \in \mathbb{R}$, and x and z are arbitrarily given.

Since

$$\frac{d^2 x_j(t)}{dt^2} = -2 \sum_i \Omega(\theta)_{ji} \frac{dx_i(t)}{dt},$$

we have $\dot{x}(t) = e^{-2t\Omega(\theta)}\xi(0)$. Then by integrating the equation

$$\Omega(\theta)\dot{x}(t) = \Omega(\theta)e^{-2t\Omega(\theta)}\xi(0),$$

we have

$$\Omega(\theta)x(t) = -1/2\left(e^{-2t\Omega(\theta)} - Id\right)\xi(0).$$

Now by the condition that the value $\theta \equiv \theta(0) = \sqrt{-1}\tau$ is pure imaginary, the matrix $\sqrt{-1}\Omega(\tau)$ is self-adjoint. Hence the matrix

$$\frac{\sqrt{-1}s\Omega(\tau)}{\sinh\sqrt{-1}s\Omega(\tau)} = \frac{1}{2\pi\sqrt{-1}} \int_\Gamma \frac{\lambda}{\sinh\lambda}\left(\lambda - \sqrt{-1}s\Omega(\tau)\right)^{-1} d\lambda$$

is well defined and invertible for any $s \in \mathbb{R}$ and $\tau \in \mathbb{R}^d$, so that we have a one-to-one correspondence: between $\xi(0)$ and x:

$$\xi(0) = e^{s\sqrt{-1}\Omega(\tau)} \cdot \frac{\sqrt{-1}\Omega(\tau)}{\sinh s\sqrt{-1}\Omega(\tau)} \cdot x(s), \; s \neq 0. \tag{10.2.23}$$

The contour Γ is taken to be suitably surrounding the spectrum of the matrix $s\sqrt{-1}\Omega(\tau)$.

Now we shall solve the following initial value problem:

$$\begin{cases} \dot{x}_j(t) = H_{\xi_j} = \xi_j + \sqrt{-1}\sum_{i,k} a_{ij}^k x_i \tau_k = \xi_j - \sqrt{-1}\sum_i \Omega(\tau)_{ji} x_i, \\ \dot{\xi}_i(t) \; = -H_{x_i} = -\sqrt{-1}\sum_j \left(\xi_j - \sqrt{-1}\sum_\ell \Omega(\tau)_{j\ell} x_\ell\right) \cdot \Omega(\tau)_{ij}, \end{cases} \tag{10.2.24}$$

with the initial conditions

$$\begin{cases} x(0) = 0, \\ \xi(0) = e^{s\sqrt{-1}\Omega(\tau)} \cdot \frac{\sqrt{-1}\Omega(\tau)}{\sinh s\sqrt{-1}\Omega(\tau)} x. \end{cases} \tag{10.2.25}$$

The solutions are given by

$$
\begin{aligned}
x(t) &= x(t; s, x, \tau) = e^{(s-t)\sqrt{-1}\Omega(\tau)} \frac{\sinh t\sqrt{-1}\Omega(\tau)}{\sinh s\sqrt{-1}\Omega(\tau)} \cdot x, \\
\xi(t) &= \xi(t, s, x, \tau) \\
&= \frac{\sqrt{-1}\Omega(\tau)}{\sinh s\sqrt{-1}\Omega(\tau)} \cdot e^{s\sqrt{-1}\Omega(\tau)} \Big(Id - e^{-t\sqrt{-1}\Omega(\tau)} \sinh t\sqrt{-1}\Omega(\tau) \Big) \cdot x \\
&= \left(e^{-t\sqrt{-1}\Omega(\tau)} \cosh t\sqrt{-1}\Omega(\tau) \right) \cdot \left(e^{s\sqrt{-1}\Omega(\tau)} \frac{\sqrt{-1}\Omega(\tau)}{\sinh s\sqrt{-1}\Omega(\tau)} \right) x \\
&= \left(e^{-t\sqrt{-1}\Omega(\tau)} \cosh t\sqrt{-1}\Omega(\tau) \right) \xi(0).
\end{aligned}
$$

The solution $x(t)$ satisfies the boundary condition

$$
x(s) = x,
$$

and this leads to the solutions of the initial-boundary value problem (10.2.21) under the condition (10.2.22), together with the solutions

$$
\begin{aligned}
z_k(t) &= z_k + \int_s^t \sum_j \left(\left(e^{-2u\sqrt{-1}\Omega(\tau)} \xi(0) \right)_j \cdot \sum_i a_{i\,j}^{k} x_i(u) \right) du, \qquad k = 1, \ldots, d, \\
\theta(t) &\equiv \sqrt{-1}\tau, \qquad \tau = (\tau_1, \ldots, \tau_d) \in \mathbb{R}^d.
\end{aligned}
$$

It can be inferred from the previous expression that the functions $z_k(t)$, with $k = 1, \ldots, d$, are uniquely determined, but since we do not need their explicit form in the following calculations, their final form won't be provided.

Let $g = g(s; x, z, \tau) \in C^\infty(\mathbb{R} \times \mathbb{R}^n \times \mathbb{R}^d \times \mathbb{R}^d)$ be the *complex action integral* given by the formula

$$
\begin{aligned}
g(s; x, z, \tau) =& \sqrt{-1} \sum_{i=1}^{d} \tau \cdot z_i(0; s, x, z, \tau) \qquad (10.2.26) \\
&+ \int_0^s < \xi(t), \dot{x}(t) > + < \theta(t), \dot{z}(t) > - H(x(t), z(t); \xi(t), \theta(t)) dt.
\end{aligned}
$$

By Proposition 10.2.3, the action g satisfies the usual Hamilton–Jacobi equation

$$
\frac{\partial g}{\partial s} + H(x, z; \nabla g) = 0.
$$

The function g also satisfies the relation

$$
g(s; x, z, \ell \cdot \tau) = \frac{1}{\ell} \cdot g(1; x, z, \tau).
$$

Next we shall provide an explicit determination of the function $g(s; x, z, \tau)$. If we let

$$\zeta(t) = \dot{x}(t) = \xi(t) - \sqrt{-1}\Omega(\tau)x,$$

then

$$\dot{\zeta}(t) = -2\sqrt{-1}\Omega(\tau)\zeta(t),$$

and we have

$$\begin{aligned}
g(s; x, z, \tau) &= \sqrt{-1}\sum_{i=1}^{d} \tau_i \cdot z_i(0; s, x, z, \tau) \\
&\quad + \int_0^s < \xi(t),\ \dot{x}(t) > + < \theta(t),\ \dot{z}(t) > -H(x(t), z(t); \xi(t), \theta(t))dt \\
&= \sqrt{-1}\sum_{i=1}^{d} \tau_i \cdot z_i + \int_0^s < \xi(t),\ \dot{x}(t) > -\frac{1}{2} < \zeta(t),\ \zeta(t) > dt \\
&= \sqrt{-1}\sum_{i=1}^{d} \tau_i \cdot z_i + \int_0^s < \zeta(t) + \sqrt{-1}\Omega(\tau)x, \\
&\quad \zeta(t) > -\frac{1}{2} < \zeta(t),\ \zeta(t) > dt \\
&= \sqrt{-1}\sum_{i=1}^{d} \tau_i \cdot z_i + \int_0^s \frac{1}{2} < \zeta(t),\ \zeta(t) > - < x,\ \sqrt{-1}\Omega(\tau)\zeta(t) > dt \\
&= \sqrt{-1}\sum_{i=1}^{d} \tau_i \cdot z_i + \int_0^s 1/2 < \zeta(t),\ \zeta(t) > +\frac{1}{2} < x,\ \dot{\zeta}(t) > dt \\
&= \sqrt{-1}\sum_{i=1}^{d} \tau_i \cdot z_i + \frac{1}{2} < x, \zeta > \Big|_0^s \\
&= \sqrt{-1}\sum_{i=1}^{d} \tau_i \cdot z_i + \frac{1}{2}\Big\langle \sqrt{-1}\Omega(\tau)\coth\big(\sqrt{-1}s\Omega(\tau)\big)\cdot x,\ x\Big\rangle.
\end{aligned}$$

Let $f = f(x, z, \tau)$ be given by

$$\begin{aligned}
f(x, z, \tau) &= g(1; x, z, \tau) \\
&= \sqrt{-1}\sum_{i=1}^{d} \tau_i \cdot z_i + \frac{1}{2}\Big\langle \sqrt{-1}\Omega(\tau)\coth(\sqrt{-1}\Omega(\tau))\cdot x,\ x\Big\rangle. \quad (10.2.27)
\end{aligned}$$

Then f satisfies the identity

$$\frac{f(x, z, s \cdot \tau)}{s} = g(s; x, z, \tau)$$

and is a solution of the equation

$$H(x,z;\nabla f)+\sum_{i=1}^{d}\tau_i\frac{\partial f}{\partial \tau_i}=f(x,z,\tau_1,\dots,\tau_d),$$

called the *generalized Hamilton–Jacobi equation*. If $*$ denotes the group law on the group G, then the heat kernel $K(t;(x,z),(\tilde{x},\tilde{z}))=k_t((\tilde{x},\tilde{z})^{-1}*(x,z))$ is given by a function $k_t(x,z)$ of the form

$$k_t(x,z)=\frac{1}{t^{n/2+d}}\int_{\mathbb{R}^d}e^{-\frac{f(x,z,\tau)}{t}}\cdot W(x,z;\tau)d\tau,\qquad d=\dim\,[\mathfrak{g},\mathfrak{g}],$$

provided the function $W(x,z,\tau)$ satisfies the *first-order transport equation* (10.2.19):

$$\sum\tau_i\frac{\partial W}{\partial \tau_i}+\sum_j\tilde{X}_j(f)\tilde{X}_j(W)-\left(\Delta_{\text{sub}}(f)+\frac{d}{2}\right)\cdot W=0. \tag{10.2.28}$$

By noting that

$$\begin{aligned}-\Delta_{\text{sub}}(f)&=\frac{1}{2}\text{tr}\Big(\sqrt{-1}\Omega(\tau)\coth\big(\sqrt{-1}\Omega(\tau)\big)\Big)\\&=\frac{1}{2}\cdot\text{tr}\Big(\int_\Gamma\lambda\cdot\frac{\cosh\lambda}{\sinh\lambda}\cdot\big(\lambda-\sqrt{-1}\Omega(\tau)\big)^{-1}d\lambda\Big)\end{aligned}\tag{10.2.29}$$

does not depend on the space variables (x,z), we look for a solution of the transport equation (10.2.28) of the form $W(x,z,\tau)=W(\tau)$. In fact, the square root of the Jacobian of the correspondence (10.2.23) is a solution of the transport equation (10.2.28).

Proposition 10.2.5. *Let*

$$W(\tau)=\left(\det e^{\sqrt{-1}\Omega(\tau)}\cdot\frac{\sqrt{-1}\Omega(\tau)}{\sinh\sqrt{-1}\Omega(\tau)}\right)^{\frac{1}{2}}=\left(\det\frac{\sqrt{-1}\Omega(\tau)}{\sinh\sqrt{-1}\Omega(\tau)}\right)^{\frac{1}{2}},$$

where the branch is taken to be $W(0)=1$. Then the function $W(\tau)$ is a solution of the transport equation (10.2.28).

Proof. Let $\sigma(t)=W(t\tau)^2=\det\Big(\frac{\sqrt{-1}t\Omega(\tau)}{\sinh\sqrt{-1}t\Omega(\tau)}\Big)$. Then

$$\begin{aligned}\frac{d}{dt}\sigma(t)&=2W(t\tau)\sum_{k=1}^{d}\tau_k\left(\frac{\partial W}{\partial\tau_k}\right)(t\tau)\\&=\text{tr}\left(\frac{d}{dt}\left(\frac{\sqrt{-1}t\Omega(\tau)}{\sinh\sqrt{-1}t\Omega(\tau)}\right)\cdot\left(\frac{\sqrt{-1}t\Omega(\tau)}{\sinh\sqrt{-1}t\Omega(\tau)}\right)^{-1}\right)\cdot\sigma(t).\end{aligned}$$

By making use of the resolvent equation, we have

$$\sigma(t)^{-1}\cdot\frac{d}{dt}\sigma(t) = \mathrm{tr}\left(\int_\Gamma \lambda\cdot\frac{d}{d\lambda}\left(\frac{\lambda}{\sinh\lambda}\right)\cdot\frac{\sinh\lambda}{\lambda}\cdot\left(\lambda-\sqrt{-1}t\Omega(\tau)\right)^{-1}d\lambda\right)$$
$$= \mathrm{tr}\left(\int_\Gamma \lambda\cdot\frac{\sinh\lambda-\lambda\cosh\lambda}{\sinh^2\lambda}\cdot\frac{\sinh\lambda}{\lambda}\cdot\left(\lambda-\sqrt{-1}t\Omega(\tau)\right)^{-1}d\lambda\right),$$

and then

$$\sum_{k=1}^{d}\tau_k\frac{\partial W}{\partial\tau_k}(\tau)$$
$$= \frac{1}{2}\mathrm{tr}\left(\int_\Gamma \lambda\cdot\frac{\sinh\lambda-\lambda\cosh\lambda}{\sinh^2\lambda}\cdot\frac{\sinh\lambda}{\lambda}\cdot\left(\lambda-\sqrt{-1}\Omega(\tau)\right)^{-1}d\lambda\right)\cdot W(\tau).$$

Hence we have

$$\sum\tau_k\frac{\partial W}{\partial\tau_k}(\tau) - \Delta_{\mathrm{sub}}(f)\cdot W(\tau)$$
$$= \frac{1}{2}\mathrm{tr}\left(\int_\Gamma \lambda\cdot\frac{\sinh\lambda-\lambda\cosh\lambda}{\sinh^2\lambda}\cdot\frac{\sinh\lambda}{\lambda}\cdot\left(\lambda-\sqrt{-1}\Omega(\tau)\right)^{-1}d\lambda\right)\cdot W(\tau),$$
$$+\frac{1}{2}\mathrm{tr}\left(\int_\Gamma \lambda\cdot\frac{\cosh\lambda}{\sinh\lambda}\cdot\left(\lambda-\sqrt{-1}\Omega(\tau)\right)^{-1}d\lambda\right)\cdot W(\tau)$$
$$= \frac{d}{2}\cdot W(\tau),$$

which shows that $W(\tau)$ is a solution of the transport equation (10.2.28). ∎

Remark 10.2.6. The function $W(\tau)$ is similar to the van Vleck determinant (see [109] and [38]).

We arrive at the following result.

Theorem 10.2.7. *The function $k_t(x,z)$ is given by the following integral:*

$$k_t(x,z) = \frac{1}{(2\pi t)^{n/2+d}}\int_{\mathbb{R}^d} e^{-\frac{f(x,z,\tau)}{t}}\cdot\left(\det\frac{\sqrt{-1}\Omega(\tau)}{\sinh\sqrt{-1}\Omega(\tau)}\right)^{1/2}d\tau,$$

where the action function f is given by (10.2.27):

$$f(x,z,\tau) = \sqrt{-1}\sum_{i=1}^{d}\tau_i\cdot z_i + \frac{1}{2}\left\langle\sqrt{-1}\Omega(\tau)\coth(\sqrt{-1}\Omega(\tau))\cdot x,\ x\right\rangle.$$

Proof. By the arguments of the last two sections, we know that

$$\left(\frac{1}{2}\Delta_{\mathrm{sub}} + \frac{\partial}{\partial t}\right)k_t(x,z) = 0,$$

since we have constructed the action function f and the volume form W in such a way that the function $k_t(x,z)$ satisfies the heat equation.

Using the following asymptotic behaviors,

- $W(\tau) = O(|\tau|^{-j}),\ j > 0$ is arbitrary, (10.2.30)
- The bilinear form $\left\langle \left(\sqrt{-1}\Omega(\tau)\coth(\sqrt{-1}\Omega(t\tau))\right)\cdot x,\ x\right\rangle$ is *(strictly)* positive definite and
$$\left\langle \left(\sqrt{-1}\Omega(\tau)\coth(\sqrt{-1}\Omega(t\tau))\right)\cdot x,\ x\right\rangle = O(|\tau||x|^2) \tag{10.2.31}$$
(since nonzero eigenvalues of $\sqrt{-1}\Omega(\tau)$ are proportional to $|\tau|$),

we shall show that the Fourier inversion formula implies that

$$\lim_{t\to 0}\frac{1}{(2\pi t)^{n/2+d}}\int_{\mathfrak{g}}\int_{\mathbb{R}^d} e^{-\frac{f(x,z,\tau)}{t}}\cdot\left(\det\frac{\sqrt{-1}\Omega(\tau)}{\sinh\sqrt{-1}\Omega(\tau)}\right)^{1/2} d\tau\varphi(x,z)dxdz \tag{10.2.32}$$
$$= \varphi(0,0)$$

for $\forall\varphi\in C_0^\infty(\mathfrak{g})$. The computation will be provided in the following. Let

$$\mathcal{F}_{(z)}(\varphi)(x,\theta) = (2\pi)^{-d/2}\int_{\mathbb{R}^d} e^{-\sqrt{-1}<\theta,z>}\varphi(x,z)dz$$

be the partial Fourier transformation of functions $\varphi(x,z)$ with respect to the variable z. Then

$$\begin{aligned}
&\frac{1}{(2\pi t)^{n/2+d}}\int_{\mathfrak{g}}\int_{\mathbb{R}^d} e^{-\frac{f(x,z,\tau)}{t}}\cdot\left(\det\frac{\sqrt{-1}\Omega(\tau)}{\sinh\sqrt{-1}\Omega(\tau)}\right)^{1/2} d\tau\varphi(x,z)dxdz\\
&= \frac{1}{(2\pi t)^{n/2+d}}(2\pi)^{d/2}\int\int \mathcal{F}_{(z)}(\varphi)(x,\tau/t)\\
&\quad\times e^{-1/(2t)<\sqrt{-1}\Omega(\tau)\coth\sqrt{-1}\Omega(\tau)\cdot x,x>}\left(\det\frac{\sqrt{-1}\Omega(\tau)}{\sinh\sqrt{-1}\Omega(\tau)}\right)^{1/2} d\tau dx\\
&= \frac{1}{(2\pi t)^{n/2+d}}(2\pi)^{d/2}\int\int \mathcal{F}_{(z)}(\varphi)(\sqrt{t}x,\tau)\\
&\quad\times e^{-1/2<\sqrt{-1}\Omega(t\tau)\coth\sqrt{-1}\Omega(t\tau)\cdot x,x>}\left(\det\frac{\sqrt{-1}\Omega(t\tau)}{\sinh\sqrt{-1}\Omega(t\tau)}\right)^{1/2} t^d\,d\tau\, t^{n/2}dx\\
&\xrightarrow{t\to 0} \frac{1}{(2\pi)^{n/2+d/2}}\int\int \mathcal{F}_{(z)}(\varphi)(0,\tau)e^{-\frac{<x,x>}{2}}d\tau dx = \varphi(0,0).
\end{aligned}$$

■

10.2.5 Heat Kernel for Laplacian I

In this section we shall provide the heat kernel for the Laplacian on the general two-step nilpotent Lie groups, based on our arguments about the results of the last section.

Let Δ be a Laplacian on G

$$\Delta = \Delta_{\mathrm{sub}} - \sum_{k=1}^{d} \tilde{Z}_k^2,$$

with $\tilde{Z}_k = \frac{\partial}{\partial z_k}$. This means we are assuming that G is equipped with the left invariant metric with respect to which $\{X_i\}$ and $\{Z_k\}$ form an orthonormal basis at the identity element. The aforementioned operator becomes the Laplacian with respect to this Riemannian metric.

Since we have

$$[\Delta_{\mathrm{sub}}, \tilde{Z}] = 0 \quad (Z \text{ belongs to the center of } \mathfrak{g}),$$

the heat kernel $K_\Delta(t; (x, z), (\tilde{x}, \tilde{z}))$ for the Laplacian Δ is the kernel distribution [which belongs to $C^\infty(\mathbb{R}_+ \times G \times G)$] of the composed operator

$$e^{-t\Delta_{\mathrm{sub}}} \circ \Big(\mathrm{Id} \otimes e^{-t \sum \tilde{Z}_k^2} \Big).$$

Here Id denotes the identity operator on the space $L_2(\mathfrak{g}/[\mathfrak{g}, \mathfrak{g}])$. We regard

$$L_2(G) \cong L_2(\mathfrak{g}/[\mathfrak{g}, \mathfrak{g}]) \hat{\otimes} L_2([\mathfrak{g}, \mathfrak{g}])$$

by the identification

$$\mathrm{span}\{X_i\} \cong \mathfrak{g}/[\mathfrak{g}, \mathfrak{g}].$$

Theorem 10.2.8. *The heat kernel for the Laplacian on general two-step nilpotent Lie groups is given by*

$$\begin{aligned}
&K_\Delta(t; (x, z), (\tilde{x}, \tilde{z})) \\
&\quad = \int_{\mathrm{center}} K(t; (x, z), (\tilde{x}, y)) \cdot \frac{1}{(2\pi t)^{d/2}} e^{-\frac{|y-\tilde{z}|^2}{2t}} dy \\
&\quad = \frac{1}{(2\pi t)^{n/2+d}} \int_{\mathbb{R}^d} e^{-\frac{f^{\Delta}((\tilde{x}, \tilde{z})^{-1} * (x, z), \tau)}{t}} W(\tau) d\tau \\
&\quad = \frac{1}{(2\pi t)^{n/2+d}} \int_{\mathbb{R}^d} e^{-\frac{f^{\Delta}(x-\tilde{x}, z-\tilde{z}-1/2[\tilde{x}, x], \tau)}{t}} W(\tau) d\tau,
\end{aligned}$$

where

$$f^\Delta(x,z,\tau) = \sqrt{-1} < \tau, z > +1/2 < \sqrt{-1}\Omega(\tau)\big(\coth\sqrt{-1}\Omega(\tau)\big) \cdot x,\ x > +1/2|\tau|^2.$$

Remark 10.2.9. We shall show in the next section that the function f^Δ is the complex action integral for the Laplacian [see also (10.2.35)].

Proof. Since $K_\Delta(t;(x,z),(\tilde{x},\tilde{z}))$ is of the form

$$K_\Delta(t;(x,z),(\tilde{x},\tilde{z})) = k_t^\Delta((\tilde{x},\tilde{z})^{-1} * (x,z))$$

with function $k_t^\Delta(x,z) \in C^\infty(\mathbb{R}_+ \times \mathfrak{g})$, it will suffice to express this function as

$$\begin{aligned}
k_t^\Delta(x,z) &= \int_{\text{center}} K(t;(x,z),(0,y)) \cdot \frac{1}{(2\pi t)^{d/2}} e^{-\frac{|y|^2}{2t}} dy \\
&= \frac{1}{(2\pi t)^{n/2+d}} \int_{\mathbb{R}^d} \int_{\text{center}} e^{-\frac{\sqrt{-1}<\tau,z-y>+1/2<\sqrt{-1}\Omega(\tau)\coth\sqrt{-1}\Omega(\tau)\cdot x,x>}{t}} W(\tau) \\
&\quad \times \frac{1}{(2\pi t)^{d/2}} e^{-\frac{|y|^2}{2t}} d\tau dy \\
&= \frac{1}{(2\pi t)^{n/2+d}} \cdot \frac{1}{(2\pi t)^{d/2}} \int_{\mathbb{R}^d} \int_{\text{center}} e^{-\frac{|y|^2}{2t}} e^{\frac{\sqrt{-1}}{t}<\tau,y>} dy \\
&\quad \times e^{-\frac{\sqrt{-1}<\tau,z>+1/2<\sqrt{-1}\Omega(\tau)\coth\sqrt{-1}\Omega(\tau)\cdot x,x>}{t}} W(\tau) d\tau \\
&= \frac{1}{(2\pi t)^{n/2+d}} \int_{\mathbb{R}^d} e^{-\frac{|\tau|^2}{2t}} \cdot e^{-\frac{\sqrt{-1}<\tau,z>+1/2<\sqrt{-1}\Omega(\tau)\coth\sqrt{-1}\Omega(\tau)\cdot x,x>}{t}} W(\tau) d\tau \\
&= \frac{1}{(2\pi t)^{n/2+d}} \int_{\mathbb{R}^d} e^{-\frac{f^\Delta(x,z,\tau)}{t}} W(\tau) d\tau.
\end{aligned}$$

■

The aforementioned formula coincides with the earlier particular case of Theorem 10.1.5, and is the heat kernel we want to construct.

10.2.6 Heat Kernel for Laplacian II

In this section we shall construct the heat kernel for the Laplacian using the complex Hamilton–Jacobi method as in Sect. 10.2.4. We shall describe the Hamiltonian system and the related quantities associated with the sub-Laplacian.

1. *The Hamiltonian H^Δ of the Laplacian*:

$$\begin{aligned}
&H^\Delta(x,z;\xi,\theta) \\
&= \frac{1}{2}\left(\sum_{j=1}^n \left(\xi_j + \sum_{i=1}^n \sum_{k=1}^d \frac{1}{2} a_{ij}^k x_i \theta_k\right)^2 + \sum_{k=1}^d \theta_k^2\right)
\end{aligned}$$

$$= \frac{1}{2}\left(\sum_j \left(\xi_j - \sum_i \Omega(\theta)_{ji} \cdot x_i\right)^2 + \sum_{k=1}^{d} \theta_k^2\right)$$

$$= H(x, z; \xi, \theta) + \frac{1}{2}\sum_{k=1}^{d} \theta_k^2.$$

2. *The Hamiltonian system*:

$$\begin{cases} \dot{x}_i &= H^{\Delta}_{\xi_i} = H_{\xi_i} = \xi - \Omega(\theta)x, \\ \dot{z}_k &= H^{\Delta}_{\theta_k} = H_{\theta_k} + \theta_k, \\ \dot{\xi}_j &= -H^{\Delta}_{x_j} = -H_{x_j}, \\ \dot{\theta}_k &= -H^{\Delta}_{z} \equiv 0. \end{cases} \tag{10.2.33}$$

3. *The initial-boundary conditions*:

$$\begin{cases} x(0) = 0, \\ x(s) = x = (x_1, \ldots, x_n) \in \mathbb{R}^n, \\ z(s) = z = (z_1, \ldots, z_d) \in \mathbb{R}^d, \\ \theta(0) = \sqrt{-1}\tau, \\ \quad \tau = (\tau_1, \ldots, \tau_d) \in \mathbb{R}^d, \end{cases} \tag{10.2.34}$$

where $s \in \mathbb{R}$ and x, z are arbitrarily given.

In the Hamiltonian system (10.2.33), with the exception of the second equation, the other equations coincide with the corresponding equations in (10.2.21). Hence we have the same solutions $x^{\Delta}(t) = x(t)$ and $\xi^{\Delta}(t) = \xi(t)$ given by

$$\begin{aligned} x^{\Delta}(t) &= x^{\Delta}(t; s, x, \tau) = x(t; s, x, \tau) \\ &= e^{(s-t)\sqrt{-1}\Omega(\tau)} \frac{\sinh t\sqrt{-1}\Omega(\tau)}{\sinh s\sqrt{-1}\Omega(\tau)} \cdot x, \end{aligned}$$

$$\begin{aligned} \xi^{\Delta}(t) &= \xi(t) = \xi(t; s, x, \tau) \\ &= \frac{\sqrt{-1}\Omega(\tau)}{\sinh s\sqrt{-1}\Omega(\tau)} \cdot e^{s\sqrt{-1}\Omega(\tau)}\left(Id - e^{-t\sqrt{-1}\Omega(\tau)} \sinh t\sqrt{-1}\Omega(\tau)\right) \cdot x \\ &= \left(e^{-t\sqrt{-1}\Omega(\tau)} \cosh t\sqrt{-1}\Omega(\tau)\right) \cdot \left(e^{s\sqrt{-1}\Omega(\tau)} \frac{\sqrt{-1}\Omega(\tau)}{\sinh s\sqrt{-1}\Omega(\tau)}\right) x \\ &= \left(e^{-t\sqrt{-1}\Omega(\tau)} \cosh t\sqrt{-1}\Omega(\tau)\right) \xi(0), \end{aligned}$$

and satisfy the system (10.2.24) under the initial conditions (10.2.25). The solution $z^{\Delta}(t)$ is given by

$$z^{\Delta}(t) = z(t) + \sqrt{-1}(t-s)\tau_k.$$

Now the complex action integral $g^{\Delta} \in C^{\infty}(\mathbb{R} \times \mathbb{R}^n \times \mathbb{R}^d \times \mathbb{R}^d)$ is given by the formula

$$\begin{aligned} g^{\Delta}(s;x,z,\tau) &= \sqrt{-1}\sum_{i=1}^{d} \tau_i \cdot z_i^{\Delta}(0;s,x,z,\tau) \\ &\quad + \int_0^s < \xi^{\Delta}(t), \dot{x}^{\Delta}(t) > + < \theta(t), \dot{z}^{\Delta}(t) > \\ &\quad - H^{\Delta}(x^{\Delta}(t), z^{\Delta}(t); \xi^{\Delta}(t), \theta(t))dt \\ &= \sqrt{-1}\sum_{i=1}^{d} \tau_i \cdot z_i + \int_0^s < \xi(t), \dot{x}(t) > \qquad (10.2.35) \\ &\quad - \left(H(x(t), z(t); \xi(t), \theta(t)) + 1/2 \sum_{k=1}^{d} \theta_k(t)^2 \right) dt \\ &= g(s;x,z,\tau) + \frac{s}{2}\sum_{k=1}^{d} \tau_k^2, \qquad (10.2.36) \end{aligned}$$

where $g = g(s;x,z\tau)$ is the complex action function constructed using (10.2.26). We conclude with the following result.

Proposition 10.2.10. *The following hold:*

1. *g^{Δ} satisfies the usual Hamilton–Jacobi equation*

$$\frac{\partial g^{\Delta}}{\partial s} + H^{\Delta}(x,z;\nabla g^{\Delta}) = 0.$$

2. *The function g^{Δ} satisfies the relation*

$$g^{\Delta}(s;x,z,\ell \cdot \tau) = \frac{1}{\ell} \cdot g^{\Delta}(1;x,z,\tau).$$

We can see that the function $g^{\Delta}(1;x,z,\tau)$ coincides with the function which appears in the integrand of the heat kernel given in the last section. By the same reason as in the case of the sub-Laplacian, the function $f^{\Delta}(x,z,\tau) = g^{\Delta}(1;x,z,\tau)$ satisfies the generalized Hamilton–Jacobi equation. This can also be proved directly as in the following:

$$\begin{aligned} H^{\Delta}(x,z;\nabla f^{\Delta}) + \sum_{i=1}^{d} \tau_i \frac{\partial f^{\Delta}}{\partial \tau_i} &= H(x,z;\nabla f) - 1/2\sum_{k=1}^{d} \tau_k^2 + \sum_{i=1}^{d} \tau_i \frac{\partial f}{\partial \tau_i} + \sum_{k=1}^{2} \tau_k^2 \\ &= f(x,z;\tau) + 1/2\sum \tau_k^2 = f^{\Delta}(x,z;\tau). \end{aligned}$$

The Laplacian $\Delta(f^{\Delta})$ is given by

$$\Delta(f^{\Delta}) = \Delta\left(f + 1/2\sum \tau_k^2\right) = \Delta_{\text{sub}}(f).$$

We also know that the volume form $W(\tau)$ is the solution of the first-order transport equation (10.2.19):

$$\sum \tau_i \frac{\partial W}{\partial \tau_i} + \sum_j \tilde{X}_j(f^{\Delta})\tilde{X}_j(W) + \sum_k \tilde{Z}_k(f^{\Delta})\tilde{Z}_k(W) - \left(\Delta(f^{\Delta}) + \frac{d}{2}\right)\cdot W$$
$$= \sum \tau_i \frac{\partial W}{\partial \tau_i} - \left(\Delta_{\text{sub}}(f) + \frac{d}{2}\right)\cdot W = 0. \qquad (10.2.37)$$

Theorem 10.2.11. *The heat kernel K^{Δ} for the Laplacian is given by*

$$K^{\Delta}(t;(\tilde{x},\tilde{z}),(x,z)) = \frac{1}{(2\pi t)^{n/2+d}} \int_{\mathbb{R}^d} e^{-f^{\Delta}(x-\tilde{x},z-\tilde{z}-\frac{1}{2}[\tilde{x},x])} W(\tau)d\tau. \quad (10.2.38)$$

Remark 10.2.12. This integral form coincides with the one obtained in Theorem 10.2.8. In both of the expressions given by Theorem 10.2.7 (sub-Laplacian case) and Theorem 10.2.11 (Laplacian case) the integrands of $k_t(g)$ and $k_t^{\Delta}(g)$ are defined on $G \times \mathbb{R}^d$. We may identify this with the characteristic variety of the sub-Laplacian

$$Ch = \{(x,z;\xi,\theta) \mid H(x,z;\xi,\theta) = 0\}$$

through the following map:

$$G \times \mathbb{R}^d \ni (x,z,\tau) \mapsto (x,z;\Omega(\tau)x,\tau) \in T^*G.$$

The characteristic variety is a sub-bundle in T^*G and the integral formula of the heat kernel can be seen as the fiber integration of the d-forms

$$\frac{1}{(2\pi t)^{n/2+d}} e^{-\frac{f(x,z,\tau)}{t}} W(\tau)d\tau \quad \text{and} \quad \frac{1}{(2\pi t)^{n/2+d}} e^{-\frac{f^{\Delta}(x,z,\tau)}{t}} W(\tau)d\tau$$

on the characteristic variety Ch.

On the other hand, for the special two-step case given by Theorem 10.1.3, the domain of integration in the expression of the heat kernel is the dual of the center $[\mathfrak{g},\mathfrak{g}]$. It parameterizes the irreducible representation of G which appears in the description of the unitary transformation U as $L_2(G) \xrightarrow{\sim} L_2(\mathbf{X})$.

10.2.7 Examples

In this section we shall express the heat kernels for the sub-Laplacian and Laplacian on the Heisenberg group, quaternionic Heisenberg group, free two-step nilpotent

Lie group of dimension 6, and Heisenberg-type groups in general. We only need to determine the explicit formula for the matrices

$$\sqrt{-1}\Omega(\tau)\coth\sqrt{-1}\Omega(\tau) \quad \text{and} \quad \det\left(\frac{\sqrt{-1}\Omega(\tau)}{\sinh\sqrt{-1}\Omega(\tau)}\right)^{1/2}.$$

1. The three-dimensional Heisenberg group. Since the matrix $\Omega(\theta)$ for the three-dimensional Heisenberg group is

$$\begin{pmatrix} 0 & \theta \\ -\theta & 0 \end{pmatrix},$$

we have

$$\sqrt{-1}\Omega(\tau)\coth\sqrt{-1}\Omega(\tau) = \frac{\tau\coth\tau}{\sinh\tau}\begin{pmatrix} 1 & 0 \\ 0 & 1 \end{pmatrix}.$$

$$\frac{\sqrt{-1}\Omega(\tau)}{\sinh\sqrt{-1}\Omega(\tau)} = \frac{\tau}{\sinh\tau}\begin{pmatrix} 1 & 0 \\ 0 & 1 \end{pmatrix}$$

and then

$$\det\left(\frac{\sqrt{-1}\Omega(\tau)}{\sinh\sqrt{-1}\Omega(\tau)}\right)^{1/2} = \frac{\tau}{\sinh\tau}.$$

The action function $f = f(x, y, z, \tau)$ is

$$f(x, y, z, \tau) = \sqrt{-1}\tau z + \frac{\tau\cosh\tau}{2\sinh\tau}(x^2 + y^2).$$

Then the heat kernel $K(t; x, y, z, \tilde{x}, \tilde{y}, \tilde{z})$ for the sub-Laplacian $\Delta_{\text{sub}} = -\tilde{X}^2 - \tilde{Y}^2$ on the three-dimensional Heisenberg group is

$$\begin{aligned} &K(t; x, y, z, \tilde{x}, \tilde{y}, \tilde{z}) \\ &= \frac{1}{(2\pi t)^2}\int_{-\infty}^{+\infty} e^{\frac{\sqrt{-1}\tau\left(z-\tilde{z}+\frac{\tilde{y}x-\tilde{x}y}{2}\right)+\frac{\tau\cosh\tau}{2\sinh\tau}\left((x-\tilde{x})^2+(y-\tilde{y})^2\right)}{t}}\frac{\tau}{\sinh\tau}d\tau. \end{aligned} \tag{10.2.39}$$

The heat kernel $K^{\Delta}(t; x, y, z, \tilde{x}, \tilde{y}, \tilde{z})$ for the Laplacian $\Delta = -\tilde{X}^2 - \tilde{Y}^2 - \tilde{Z}^2$ is given by

$$\begin{aligned} &K^{\Delta}(t; x, y, z, \tilde{x}, \tilde{y}, \tilde{z}) \\ &= \frac{1}{(2\pi t)^2}\int_{-\infty}^{+\infty} e^{\frac{\sqrt{-1}\tau\left(z-\tilde{z}+\frac{\tilde{y}x-\tilde{x}y}{2}\right)+\frac{\tau\cosh\tau}{2\sinh\tau}\left((x-\tilde{x})^2+(y-\tilde{y})^2\right)-\frac{\tau^2}{2}}{t}}\frac{\tau}{\sinh\tau}d\tau. \end{aligned} \tag{10.2.40}$$

2. Higher-dimensional Heisenberg groups. Let H_{2n+1} be the $(2n+1)$ -dimensional Heisenberg group described in Sect. 9.2.3 with the sub-Laplacian

$$\Delta_{\text{sub}} = -\sum \tilde{X}_i^2 - \sum \tilde{Y}_i^2.$$

In this case the matrix $\Omega(\theta)_{i\,j} = a_{i\,j} \cdot \theta$ is given by

$$\Omega(\theta) = \theta \begin{pmatrix} 0 & Id_n \\ -Id_n & 0 \end{pmatrix},$$

where Id_n is the $n \times n$ identity matrix. We have the action function f expressed as

$$f(x_1, \ldots, x_n, y_1, \ldots, y_n, z, \tau) = \sqrt{-1}\tau z + \frac{\tau \cosh \tau}{2 \sinh \tau} \left(\sum x_i^2 + y_i^2 \right),$$

and the volume element $W(\tau)$ given by

$$\left(\frac{\tau}{\sinh \tau} \right)^n.$$

Hence the heat kernel $K(t; x, y, z, \tilde{x}, \tilde{y}, \tilde{z})$ is given by the formula

$$\begin{aligned} &K(t; x, y, z, \tilde{x}, \tilde{y}, \tilde{z}) \\ &\quad = \frac{1}{(2\pi t)^{n+1}} \int_{-\infty}^{+\infty} e^{\frac{\sqrt{-1}\tau\left(z - \tilde{z} + \frac{\sum \tilde{y}_i x_i - \tilde{x}_i y_i}{2}\right) + \frac{\tau \cosh \tau}{2 \sinh \tau}\left(\sum\left((x_i - \tilde{x}_i)^2 + (y_i - \tilde{y}_i)^2\right)\right)}{t}} \\ &\qquad \times \left(\frac{\tau}{\sinh \tau} \right)^n d\tau. \end{aligned} \tag{10.2.41}$$

On the other side, the heat kernel $K^{\Delta}(t; x, y, z, \tilde{x}, \tilde{y}, \tilde{z})$ for the Laplacian

$$\Delta = -\sum \tilde{X}_i^2 - \sum \tilde{Y}_i^2 - \tilde{Z}^2$$

is given by

$$\begin{aligned} &K^{\Delta}(t; x, y, z, \tilde{x}, \tilde{y}, \tilde{z}) \\ &\quad = \frac{1}{(2\pi t)^{n+1}} \int_{-\infty}^{+\infty} e^{\frac{\sqrt{-1}\tau\left(z - \tilde{z} + \frac{\sum \tilde{y}_i x_i - \tilde{x}_i y_i}{2}\right) + \frac{\tau \cosh \tau}{2 \sinh \tau}\left(\sum (x_i - \tilde{x}_i)^2 + (\sum y_i - \tilde{y}_i)^2\right) - \frac{\tau^2}{2}}{t}} \\ &\qquad \times \left(\frac{\tau}{\sinh \tau} \right)^n d\tau. \end{aligned} \tag{10.2.42}$$

3. The seven-dimensional quaternionic Heisenberg Lie group. Let $q\mathbb{H}_7$ ($\cong \mathbb{R}^7$) be the seven-dimensional quaternionic Heisenberg group described in Sect. 9.2.4. Recall the vector fields $\widetilde{X}_i$, $i = 0, 1, 2, 3$,

$$\widetilde{X}_0 = \frac{\partial}{\partial x_0} + \frac{1}{2}\sum_{i=1}^{3} x_i \frac{\partial}{\partial z_i},$$
$$\widetilde{X}_1 = \frac{\partial}{\partial x_1} + \frac{x_0}{2}\frac{\partial}{\partial z_1} + \frac{x_3}{2}\frac{\partial}{\partial z_2} - \frac{x_2}{2}\frac{\partial}{\partial z_3},$$
$$\widetilde{X}_2 = \frac{\partial}{\partial x_2} - \frac{x_3}{2}\frac{\partial}{\partial z_1} + \frac{x_0}{2}\frac{\partial}{\partial z_2} + \frac{x_1}{2}\frac{\partial}{\partial z_3},$$
$$\widetilde{X}_3 = \frac{\partial}{\partial x_3} + \frac{x_2}{2}\frac{\partial}{\partial z_1} - \frac{x_1}{2}\frac{\partial}{\partial z_2} + \frac{x_0}{2}\frac{\partial}{\partial z_3}.$$

The sub-Laplacian and Laplacian in this case are given by

$$\Delta_{\text{sub}} = -\sum_{i=0}^{3} \widetilde{X}_i^2, \tag{10.2.43}$$

$$\Delta = -\sum_{i=0}^{3} \widetilde{X}_i^2 - \sum_{j=1}^{3} \frac{\partial^2}{\partial z_j^2}. \tag{10.2.44}$$

Since the matrix $\Omega(\theta)$ has the expression

$$\Omega(\theta) = \begin{pmatrix} 0 & \theta_1 & \theta_2 & \theta_3 \\ -\theta_1 & 0 & \theta_3 & -\theta_2 \\ -\theta_2 & -\theta_3 & 0 & \theta_1 \\ -\theta_3 & \theta_2 & -\theta_1 & 0 \end{pmatrix},$$

we have

$$\sqrt{-1}\Omega(\tau)\coth\Omega(\sqrt{-1}\tau) = |\tau|\coth|\tau| Id_4.$$

Hence the action function f is given by

$$\begin{aligned} &f(x, z, \tau) \\ &\quad = f(x_0, x_1, x_2, x_3, z_1, z_2, z_3, \tau_1, \tau_2, \tau_3) = \sqrt{-1}\sum_{i=1}^{3} z_i\tau_i + \frac{1}{2}|\tau|\coth|\tau|\,|x|^2, \end{aligned}$$

where $|\tau| = \sqrt{\sum \tau_i^2}$ and $|x| = \sqrt{\sum x_i^2}$. The volume element in this case is

$$W(\tau) = \left(\frac{|\tau|}{\sinh|\tau|}\right)^2.$$

Hence the heat kernel $K(t,x,z,\tilde{x},\tilde{z})$ for the sub-Laplacian (10.2.43) is given by the formula

$$K(t,x,z,\tilde{x},\tilde{z}) = \frac{1}{(2\pi t)^5}\int_{\mathbb{R}^3} e^{-\frac{\sqrt{-1}}{t}f\left((-\tilde{x},-\tilde{z})*(x,z),\tau\right)} \times \left(\frac{|\tau|}{\sinh|\tau|}\right)^2 d\tau_1 d\tau_2 d\tau_3,$$

where $f\left((-\tilde{x},-\tilde{z})*(x,z),\tau\right)$ is given by

$$\begin{aligned} f\left((-\tilde{x},-\tilde{z})*(x,z),\tau\right) = \sqrt{-1}\Big\{ &\tau_1\left(z_1-\tilde{z}_1+\frac{\tilde{x}_1x_0-\tilde{x}_0x_1+\tilde{x}_3x_2-\tilde{x}_2x_3}{2}\right)\\ &+\tau_2\left(z_2-\tilde{z}_2+\frac{\tilde{x}_2x_0-\tilde{x}_0x_2+\tilde{x}_1x_3-\tilde{x}_3x_1}{2}\right)\\ &+\tau_3\left(z_3-\tilde{z}_3+\frac{\tilde{x}_3x_0-\tilde{x}_0x_3+\tilde{x}_2x_1-\tilde{x}_1x_2}{2}\right)\Big\}\\ &+\frac{1}{2}|\tau|\coth|\tau|\cdot|x-\tilde{x}|^2. \end{aligned}$$

The heat kernel of the Laplacian

$$\Delta = -\sum_{i=0}^{3}\widetilde{X}_i^2 - \sum_{j=1}^{3}\frac{\partial^2}{\partial z_j^2}$$

is given by

$$K^{\Delta}(t,x,z,\tilde{x},\tilde{z}) = \frac{1}{(2\pi t)^5}\int_{\mathbb{R}^3} e^{-\frac{\sqrt{-1}}{t}f^{\Delta}\left((-\tilde{x},-\tilde{z})*(x,z),\tau\right)} \times \left(\frac{|\tau|}{\sinh|\tau|}\right)^2 d\tau_1 d\tau_2 d\tau_3,$$

where the action function f^{Δ} is given by $f^{\Delta} = f(x,z,\tilde{x},\tilde{z},\tau)+1/2\left(\tau_1^2+\tau_2^2+\tau_3^2\right)$.

4. Free two-step nilpotent Lie group of dimension 6. This type of structure was introduced in Sect. 9.2.6. Let $\mathfrak{f}_g(3)$ be the free two-step nilpotent Lie algebra of dimension $3+C_3^2=6$ with an orthonormal basis $\{Z_{i\,j}\}_{1\le i<j\le 3}$ of the center and an orthonormal basis $\{X_i\}_{i=1}^3$ of the orthogonal complement of the center.

The Lie bracket relations are

$$[X_i,X_j] = -[X_j,X_i] = 2Z_{i\,j}, \text{ for } 1\le i<j\le N,$$

all others being zero.

We shall express the heat kernel of the sub-Laplacian

$$\Delta_{\text{sub}} = -\sum \tilde{X}_i^{\,2},$$

and of the Laplacian

$$\Delta = \Delta_{\text{sub}} - \sum \tilde{Z}_{i,j}^{\,2}$$

where $\tilde{X}_i$ and $\tilde{Z}_{i,j}$ are the left invariant vector fields on the free two-step nilpotent Lie group $\mathfrak{f}_G(3) \cong \mathfrak{f}_g(3)$. It this case the matrix $\Omega(\theta)$ [see also (10.2.20)] is given as the most general 3×3 antisymmetric matrix

$$\Omega(\theta) = \begin{pmatrix} 0 & \theta_1 & \theta_2 \\ -\theta_1 & 0 & \theta_3 \\ -\theta_2 & -\theta_3 & 0 \end{pmatrix},$$

where $\theta = (\theta_1, \theta_2, \theta_3) \in \mathbb{R}^3$ is arbitrary.

In the following we shall determine the volume element $W(\tau) = \det \left(\frac{\sqrt{-1}\Omega(\tau)}{\sinh \sqrt{-1}\Omega(\tau)} \right)^{1/2}$ and the action function $f(x, z, \tau) = -\sum \tau_k z_k + \frac{1}{2}\langle \sqrt{-1}\Omega(\tau) \coth \sqrt{-1}\Omega(\tau)\, x,\, x\rangle$. Since the eigenvalues of the self-adjoint matrix $\sqrt{-1}\Omega(\tau)$ are $\{0, \pm|\tau|\}$, with $|\tau| = \sqrt{\tau_1^2 + \tau_2^2 + \tau_3^2}$, we have

$$\left(\det \frac{\sqrt{-1}\Omega(\tau)}{\sinh \sqrt{-1}\Omega(\tau)} \right)^{1/2} = \frac{|\tau|}{\sinh |\tau|}.$$

Let Γ be suitably chosen such that it is enclosing the eigenvalues of the matrix $\sqrt{-1}\Omega(\tau)$. In order to calculate the expression

$$\frac{1}{2\pi\sqrt{-1}} \int_\Gamma \frac{\lambda \cosh \lambda}{\sinh \lambda} \left(\lambda - \sqrt{-1}\Omega(\tau) \right)^{-1} d\lambda,$$

it is enough to calculate the power series of the matrix

$$\left(Id + \frac{\left(\sqrt{-1}\Omega(\tau)\right)^2}{2!} + \frac{\left(\sqrt{-1}\Omega(\tau)\right)^4}{4!} + \cdots \right)$$
$$\left(Id + \frac{\left(\sqrt{-1}\Omega(\tau)\right)^2}{3!} + \frac{\left(\sqrt{-1}\Omega(\tau)\right)^4}{5!} + \cdots \right)^{-1}.$$

For that purpose we consider the unitary matrix T:

$$T = \begin{pmatrix} \frac{\tau_3}{|\tau|} & \sqrt{\frac{|\tau|^2\tau_1^2+\tau_2^2\tau_3^2}{2(\tau_1^2+\tau_2^2)|\tau|^2}} \frac{\tau_1^2+\tau_2^2}{\sqrt{-1}|\tau|\tau_1+\tau_2\tau_3} & \sqrt{\frac{|\tau|^2\tau_1^2+\tau_2^2\tau_3^2}{2(\tau_1^2+\tau_2^2)|\tau|^2}} \frac{\tau_1^2+\tau_2^2}{-\sqrt{-1}|\tau|\tau_1+\tau_2\tau_3} \\ \frac{-\tau_2}{|\tau|} & \sqrt{\frac{|\tau|^2\tau_1^2+\tau_2^2\tau_3^2}{2(\tau_1^2+\tau_2^2)|\tau|^2}} & \sqrt{\frac{|\tau|^2\tau_1^2+\tau_2^2\tau_3^2}{2(\tau_1^2+\tau_2^2)|\tau|^2}} \\ \frac{\tau_1}{|\tau|} & \sqrt{\frac{|\tau|^2\tau_1^2+\tau_2^2\tau_3^2}{2(\tau_1^2+\tau_2^2)|\tau|^2}} \frac{\sqrt{-1}|\tau|\tau_2-\tau_1\tau_3}{\sqrt{-1}|\tau|\tau_1+\tau_2\tau_3} & \sqrt{\frac{|\tau|^2\tau_1^2+\tau_2^2\tau_3^2}{2(\tau_1^2+\tau_2^2)|\tau|^2}} \frac{-\sqrt{-1}|\tau|\tau_2-\tau_1\tau_3}{-\sqrt{-1}|\tau|\tau_1+\tau_2\tau_3} \end{pmatrix},$$

which satisfies

$$T^* \sqrt{-1}\Omega(\tau) T = \begin{pmatrix} 0 & 0 & 0 \\ 0 & |\tau| & 0 \\ 0 & 0 & -|\tau| \end{pmatrix},$$

for $\tau_1 \neq 0$. Then we have

$$\begin{aligned} & TT^* \left(Id + \sum_{k=1}^{\infty} \frac{(\sqrt{-1}\Omega(\tau))^{2k}}{(2k)!} \right) \left(Id + \sum_{k=1}^{\infty} \frac{(\sqrt{-1}\Omega(\tau))^{2k}}{(2k+1)!} \right)^{-1} TT^* \\ &= T \begin{pmatrix} 1 & 0 & 0 \\ 0 & |\tau| \coth |\tau| & 0 \\ 0 & 0 & |\tau| \coth |\tau| \end{pmatrix} T^* \\ &= |\tau| \coth |\tau| Id_3 + \frac{1 - |\tau| \coth |\tau|}{|\tau|^2} \begin{pmatrix} \tau_3^2 & -\tau_2\tau_3 & \tau_1\tau_3 \\ -\tau_2\tau_3 & \tau_2^2 & -\tau_1\tau_2 \\ \tau_1\tau_3 & -\tau_1\tau_2 & \tau_1^2 \end{pmatrix}. \end{aligned}$$

We can see that the resulting expression is also valid in the case $\tau_1 = 0$. Now the action function is

$$\begin{aligned} & f(x_1, x_2, x_3, z_1, z_2, z_3, \tau_1, \tau_2, \tau_3) \\ &= -\sqrt{-1} \sum \tau_k z_k + \frac{1}{2} < T\, T^* \sqrt{-1}\Omega(\tau) \coth \sqrt{-1}\Omega(\tau)\, T\, T^*(x), x > \\ &= -\sqrt{-1} \sum \tau_k z_k + \frac{1}{2} \Big[|\tau| \coth |\tau| (x_1^2 + x_2^2 + x_3^2) \\ &\quad + \frac{1 - |\tau| \coth |\tau|}{|\tau|^2} (\tau_3 x_1 - \tau_2 x_2 + \tau_1 x_3)^2 \Big]. \end{aligned}$$

Hence the heat kernel K for the sub-Laplacian is given by the following expression:

$$\begin{aligned} & K(t, x, z, \tilde{x}, \tilde{z}) \\ &= \frac{1}{(2\pi t)^{9/2}} \int_{\mathbb{R}^3} e^{-\sqrt{-1} \frac{l\tau_1(z_1 - \tilde{z}_1 + 1/2(\tilde{x}_2 x_1 - \tilde{x}_1 x_2)) + \tau_2(z_2 - \tilde{z}_2 + 1/2(\tilde{x}_3 x_1 - \tilde{x}_1 x_3)) + \tau_3(z_3 - \tilde{z}_3 + 1/2(\tilde{x}_3 x_2 - \tilde{x}_2 x_3))}{t}} \\ &\quad \times e^{\frac{\left\langle \left(|\tau| \coth |\tau| Id_3 + \frac{1 - |\tau| \coth |\tau|}{|\tau|^2} \begin{pmatrix} \tau_3^2 & -\tau_2\tau_3 & \tau_1\tau_3 \\ -\tau_2\tau_3 & \tau_2^2 & -\tau_1\tau_2 \\ \tau_1\tau_3 & -\tau_1\tau_2 & \tau_1^2 \end{pmatrix} \right) \cdot (x - \tilde{x}), x - \tilde{x} \right\rangle}{2t}} \frac{|\tau|}{\sinh |\tau|} d\tau_1 d\tau_2 d\tau_3. \end{aligned}$$

4. Heisenberg-type group. This structure was introduced in Sect. 9.2.5. Let $V = \mathfrak{n} \oplus \mathfrak{z}^*$ be a Heisenberg-type algebra with the corresponding inner products $< \cdot, \cdot >_{\mathfrak{n}}$, $< \cdot, \cdot >_{\mathfrak{z}}$, and bilinear map $J : \mathfrak{z} \times \mathfrak{n} \to \mathfrak{n}$, with the notations $\dim \mathfrak{n} = n$ and $\dim \mathfrak{z} = d$.

The Lie bracket $[\cdot\,,\,\cdot]$ is defined by

$$\begin{aligned}&\mathfrak{n}\times\mathfrak{n}\to\mathfrak{z},\\&(X,Y)\mapsto B(X,Y)=[X,Y],\qquad B(X,Y)(Z)=<J(Z,X),Y>_{\mathfrak{n}}\,.\end{aligned}$$

We shall use the same notation for the Lie algebra V and for the Lie group V.

Let $\{X_i\}$ and $\{Z^*{}_k\}$ be orthonormal bases on $\mathfrak{n}$ and $\mathfrak{z}^*$, respectively. Denote by $\{\tilde{X}_i\}$ and $\{\tilde{Z}^*{}_k\}$ the corresponding left invariant vector fields on the Lie group V, and consider the sub-Laplacian

$$\Delta_{\text{sub}}=-\sum\tilde{X}_i^2,$$

and the Laplacian

$$\Delta=-\Delta_{\text{sub}}-\sum\tilde{Z}_k^2.$$

If the structure constants are given by $[X_i,X_j]=2\sum_{k=1}^{d}C_{i\,j}^{\,k}Z^*{}_k$, then

$$J(Z^*,\cdot)=\left(\sum_k 2C_{i\,j}^{\,k}\theta_k\right)=\Omega(\theta),\quad\text{with}\quad Z^*=\sum\theta_k Z^*{}_k\in\mathfrak{z}^*.$$

Hence we have $\Omega(\theta)^2=-|\theta|^2Id$. Now we can determine the action function f and the volume element W as in the following:

$$\begin{aligned}f(x,z,\tau)&=-\sqrt{-1}\sum\tau_k z_k+\frac{1}{2}\langle\sqrt{-1}\Omega(\tau)\coth\sqrt{-1}\Omega(\tau)x,\,x\rangle\\&=-\sqrt{-1}\sum\tau_k z_k+\frac{1}{2}|\tau|\coth|\tau|\cdot\|x\|^2,\end{aligned}$$

$$W(\tau)=\sqrt{\frac{\sqrt{-1}\Omega(\tau)}{\sinh\sqrt{-1}\Omega(\tau)}}=\left(\frac{|\tau|}{\sinh|\tau|}\right)^{n/2}.$$

With $x=\sum x_iX_i$, $z=\sum z_iZ^*{}_i$, the heat kernel $K(t,x,z,\tilde{x},\tilde{z})$ for the sub-Laplacian Δ_{sub} is given by

$$\frac{1}{(2\pi t)^{n/2+d}}\int_{\mathbb{R}^d}e^{-\sqrt{-1}\frac{\sum\tau_i\left(z_i-\tilde{z}_i-B(\tilde{x},x)\right)}{t}}e^{-\frac{|\tau|\cdot\coth|\tau|\cdot\left(|x-\tilde{x}|\right)^2}{2t}}\left(\frac{|\tau|}{\sinh|\tau|}\right)^{n/2}dz.$$

And the heat kernel for the Laplacian Δ is expressed as

$$\frac{1}{(2\pi t)^{n/2+d}}\int_{\mathbb{R}^d}e^{-\sqrt{-1}\frac{\sum\tau_i\left(z_i-\tilde{z}_i-B(\tilde{x},x)\right)}{t}}e^{-\frac{|\tau|^2}{2t}-\frac{|\tau|\cdot\coth|\tau|\cdot\left(|x-\tilde{x}|\right)^2}{2t}}\left(\frac{|\tau|}{\sinh|\tau|}\right)^{n/2}dz.$$

10.3 Grushin-Type Operators and the Heat Kernel

In this section we shall introduce the *Grushin-type operator* associated with a sub-Riemannian structure and a submersion. This operator can be seen as a generalization of the Grushin operator

$$\frac{\partial^2}{\partial x^2} + x^2 \frac{\partial^2}{\partial y^2}.$$

The heat kernel for such operators will be constructed in terms of the fiber integration of the heat kernel of a sub-Laplacian.

10.3.1 Grushin-Type Operators

Let M be a manifold. We start by recalling the definition of a sub-Riemannian structure *in a strong sense* on M, that is, a sub-bundle $\mathcal{H}$ in the tangent bundle $T(M)$ such that $\mathcal{H}$ is trivial as a vector bundle and all linear combinations of vector fields taking values in $\mathcal{H}$ and linear sums of finite numbers of their brackets span the entire tangent space $T(M)$ at each point (such a sub-bundle is called nonholonomic). Denote the space of vector fields on M taking values in $\mathcal{H}$ by $\Gamma(M, \mathcal{H})$. Usually this sub-bundle is equipped with a metric, and we call a manifold with such a sub-bundle a sub-Riemannian manifold in the strong sense.

Let M be a sub-Riemannian manifold in the strong sense and let $\varphi : M \to N$ be a (surjective) submersion. To define an operator on the manifold N in relation with the sub-Laplacian on M, we assume that not only there are vector fields $\{\tilde{X}_i\}_{i=1}^{\dim \mathcal{H}}$ that trivialize the nonholonomic sub-bundle $\mathcal{H}$, but also each X_i can be descended to the manifold N by the map φ. This means:

(Sub-1) The vector fields $\{\tilde{X}_i\}$ are linearly independent at each point on M and span the non-holonomic sub-bundle $\mathcal{H}$. So they satisfy Hörmander's condition (bracket generating property; see [35] and [65]). Here we also consider a metric on $\mathcal{H}$ in such a way that the vector fields $\{\tilde{X}_i\}$ are orthonormal at each point on M.

(Sub-2) We also assume that there is a volume form $d_M V$ on M with respect to which each vector field $\tilde{X}_i$ $(i = 1, \dots, \dim \mathcal{H})$ is skew-symmetric.

(Gru) If $\varphi(x) = \varphi(x')$, then $d\varphi_x(\tilde{X}_i) = d\varphi_{x'}(\tilde{X}_i)$ for $i = 1, \dots, \dim \mathcal{H}$.

Let Δ_{sub} be the sub-Laplacian on M; i.e.,

$$\Delta_{\text{sub}} = -\sum_{k=1}^{\dim \mathcal{H}} \tilde{X}_k^2.$$

By the assumptions **(Sub-1)** and **(Sub-2)**, Δ_{sub} is formally symmetric and hypoelliptic.

Definition 10.3.1. An operator $\mathcal{G}$ on N of the type

$$\mathcal{G} = -\sum_{k=1}^{\dim \mathcal{H}} d\varphi(\tilde{X}_k)^2$$

is called a Grushin-type operator.

Note that the commutators $[\tilde{X}_j, \tilde{X}_k]$ can also be descended to N by the map φ, so that we have the following results.

Proposition 10.3.2. *The operator $\mathcal{G}$ is hypoelliptic.*

Proposition 10.3.3. *If each fiber of the map φ is compact, then $\mathcal{G}$ is formally symmetric with respect to the volume form $\varphi_*(d_M V)$ obtained by taking the push forward of the volume form $d_M V$.*

Proof. Let f and g be in $C_0^\infty(N)$. We have

$$\begin{aligned}
\int_N \mathcal{G}(f)(y)\cdot g(y)\cdot(\varphi)_*(d_M V)(y) &= \int_M \varphi^*(\mathcal{G}(f))(x)\cdot\varphi^*(g)(x)\, d_M V(x)\\
&= \int_M \Delta_{\text{sub}}(\varphi^*(f))(x)\cdot\varphi^*(g)(x)\, d_M V(x)\\
&= \int_M \varphi^*(f)(x)\cdot\Delta_{\text{sub}}(\varphi^*(g))(x)\, d_M V(x)\\
&= \int_N f(y)\cdot\mathcal{G}(g)(y)\cdot(\varphi)_*(d_M V)(y).
\end{aligned}$$

■

Remark 10.3.4. The sub-Laplacian depends on the specific vector fields $\{\widetilde{X}_k\}$ and is not uniquely defined as it is in the case of the Laplacian on Riemannian manifolds. In concrete cases these vector fields will be defined in a geometric way and will have a certain meaning for each case.

If the fibers of the map $\varphi : M \to N$ are not compact, we will have a volume form $d_N V$ on N such that $\varphi^*(d_N V) \wedge \theta = d_M V$, with a differential form θ on M of degree $\dim M - \dim N$. In our case the Grushin-type operators will be symmetric with respect to such a volume form. Similar considerations will apply to the Heisenberg group, the Engel group and the sphere.

In the following we shall assume that the dimension $\dim \mathcal{H} = \dim N = n$, and let $\mathcal{S}$ be the submanifold on which the vector fields $\{d\varphi(\tilde{X}_k)\}$ are not linearly independent. This means that $\mathcal{H}_x \bigcap \text{Ker}(d\varphi_x) \neq \{0\}$ for $x \in \varphi^{-1}(\mathcal{S})$. So on the subset $N \backslash \mathcal{S}$ the vector fields $\{d\varphi(\tilde{X}_i)\}$ are linearly independent and span the tangent spaces there. We consider a Riemannian metric on the subset $N \backslash \mathcal{S}$ in such a way that

$$\{d\varphi(\tilde{X}_k)\} \text{ are orthonormal at each point of } N\backslash\mathcal{S}.$$

We call the manifold N with this Riemannian metric and the Grushin-type operator descended from a sub-Laplacian a *singular Riemannian manifold* (see [3]).

10.3.2 Heisenberg Group Case

We shall show that the original Grushin operator is obtained in the way explained in the last section. Consider the three-dimensional Heisenberg group H_3, which is identified with $\mathbb{R}^3$ as a manifold, together with the following noncommutative group law:

$$(x, y, z) * (\tilde{x}, \tilde{y}, \tilde{z}) = \big(x + \tilde{x}, y + \tilde{y}, z + \tilde{z} + 1/2(x\tilde{y} - \tilde{y}x)\big) \in \mathbb{R}^3.$$

The left invariant vector fields $\tilde{X}$ and $\tilde{Y}$ are given by

$$\tilde{X} = \frac{\partial}{\partial x} - \frac{y}{2}\frac{\partial}{\partial z},$$
$$\tilde{Y} = \frac{\partial}{\partial y} + \frac{x}{2}\frac{\partial}{\partial z}.$$

Then from the relation $[\tilde{X}, \tilde{Y}] = \tilde{Z}$, where $\tilde{Z} = \frac{\partial}{\partial z}$, the sub-bundle spanned by $\tilde{X}$ and $\tilde{Y}$ is nonholonomic and defines a sub-Riemannian structure on H_3 together with the metric

$$< \tilde{X}, \tilde{Y} > = 0, \qquad < \tilde{X}, \tilde{X} > = < \tilde{Y}, \tilde{Y} > = 1.$$

The sum $-(\tilde{X}^2 + \tilde{Y}^2)$ is the sub-Laplacian Δ_{sub} and is symmetric with respect to the volume form $dx \wedge dy \wedge dz$, which coincides with the volume form introduced in Sect. 9.1.3. This form is the Haar measure of the group.

We take a subgroup $N_Y = \{(0, y, 0) | y \in \mathbb{R}\}$ and consider the left quotient space $\pi_Y : H_3 \to N_Y \backslash H_3$, which is realized as

$$\pi_Y : H_3 \cong \mathbb{R}^3 \to N_Y \backslash H_3 \cong \mathbb{R}^2,$$
$$\pi_Y(x, y, z) = (u, v) = \left(x, z + \frac{xy}{2}\right).$$

The left invariance of the vector fields $\tilde{X}$ and $\tilde{Y}$ satisfies condition **(Gru)** and is descended by the projection map π_Y to the vector fields $d\pi_Y(\tilde{X}) = \frac{\partial}{\partial u}$ and $d\pi_Y(\tilde{Y}) = u\frac{\partial}{\partial v}$, respectively. The resulting operator

$$\mathcal{G} = -\left(\frac{\partial^2}{\partial u^2} + u^2 \frac{\partial^2}{\partial v^2}\right)$$

is the original Grushin operator. It is clear that this operator is symmetric with respect to the volume form $du \wedge dv$ and $\pi^*(du \wedge dv) \wedge dy = -dx \wedge dy \wedge dz$, where dy is a left N_Y-invariant 1-form (see Remark 10.3.4).

As is easy to see, the vector fields $d\pi_Y(\tilde{X})$ and $d\pi_Y(\tilde{Y})$ are linearly dependent along the line $u = 0$. We can consider a singular metric on the plane $\mathbb{R}^2$ in such a way that the two tangent vectors $d\pi_Y(\tilde{X})_{(u,v)}$ and $d\pi_Y(\tilde{Y})_{(u,v)}$ are orthonormal at each point except on the line $u = 0$. The plane endowed with this (singular) metric will be called the *Grushin plane*.

If we consider the subgroup $N_Z = \{(0,0,z)|z \in \mathbb{R}\}$, which is the center of the Heisenberg group, then the vector fields $\tilde{X}$ and $\tilde{Y}$ can be descended to the quotient space H_3/N_Z and the resulting vector fields on

$$\pi_Z : \mathbb{R}^3 \cong H_3 \to H_3/N_Z \cong \mathbb{R}^2, \qquad (x,y,z) \mapsto \pi_Z(x,y,z) = (u,v) = (x,y),$$

are

$$d\pi_Z(\tilde{X}) = \frac{\partial}{\partial u} \qquad \text{and} \qquad d\pi_Z(\tilde{Y}) = \frac{\partial}{\partial v}.$$

So in this case the Grushin-type operator is just the Euclidean Laplacian and it does not degenerate at any point.

More generally, we take a subgroup $N_{(a,b,c)}$ of the form

$$N_{(a,b,c)} = \{(at, bt, ct) \mid t \in \mathbb{R}\},$$

where $a^2 + b^2 \neq 0$. We shall investigate only the case $a \neq 0$. The other case can be treated in a similar way.

Let $\rho_0 : H_3 \to N_{(a,b,c)} \backslash H_3 \cong \mathbb{R}^2$. Then the map ρ_0 is expressed as

$$H_3 \cong \mathbb{R}^3 \ni (x,y,z) \mapsto (u,v),\ u = ay - bx,\ v = z - \frac{x}{a}\left(c + \frac{ay - bx}{2}\right).$$

We have a trivialization of the principal bundle $\rho_0 : H_3 \to N_{(a,b,c)} \backslash H_3$ given by

$$\mathbb{R} \times \mathbb{R}^2 \ni (\theta; u, v) \to (x,y,z) \in H_3,$$

where

$$x = \theta a,\ y = \frac{u}{a} + b\theta,\ z = v + \theta(c + u/2).$$

We also have the coordinate expression of the cotangent bundle $T^*(H_3)$ as

$$\mathbb{R} \times \mathbb{R}^3 \times \mathbb{R} \times \mathbb{R}^3 \cong T^*(H_3) \overset{\varphi_0}{\cong} \mathbb{R} \times \mathbb{R}^3 \times \mathbb{R} \times \mathbb{R}^3,$$

$$(\theta, u, v; \gamma, \alpha, \beta) \longleftrightarrow (x, y, z; \xi, \eta, \tau), \tag{10.3.45}$$

$$\theta = \frac{x}{a},\ u = ay - bx,\ v = z - \frac{x}{a}\left(c + \frac{ay - bx}{2}\right), \tag{10.3.46}$$

$$\gamma = a\xi + b\eta + \tau\big(c + (ay - bx)/2\big)\ \alpha = \frac{\eta}{a} + \frac{x\tau}{2a},\ \beta = \tau. \tag{10.3.47}$$

The vector fields $\tilde{X}$ and $\tilde{Y}$ are descended to the quotient space $N_{(a,b,c)} \backslash H_3$, and the resulting vector fields are given by

$$d\rho_0(\tilde{X}) = -b\frac{\partial}{\partial u} - \frac{c + u}{a}\frac{\partial}{\partial v}, \qquad d\rho_0(\tilde{Y}) = a\frac{\partial}{\partial u}.$$

It is easy to see that these vector fields are antisymmetric with respect to the volume form $du \wedge dv$. With a left $N_{(a,b,c)}$-invariant 1-form $\frac{dx}{a}$ on H_3, we have (see also Remark 10.3.4)

$$\rho_0^*(du \wedge dv) \wedge \frac{dx}{a} = dx \wedge dy \wedge dz.$$

The Grushin-type operator in this case is

$$\mathcal{G}_H = -\left(b\frac{\partial}{\partial u} + \frac{u+c}{a}\frac{\partial}{\partial v}\right)^2 - a^2\frac{\partial^2}{\partial u^2}, \tag{10.3.48}$$

and it does degenerate along the line $\{(u,v)|u = -c\}$.

Next we shall make a remark regarding the classification of one-dimensional subgroups of H_3. The group $Aut(H_3)$ of the automorphisms of the Heisenberg group H_3 is given by

$$Aut(H_3) = \left\{ S = \begin{pmatrix} g_{11} & g_{12} & 0 \\ g_{21} & g_{22} & 0 \\ a & b & \det g \end{pmatrix} \right\},$$

where $g = \begin{pmatrix} g_{11} & g_{12} \\ g_{21} & g_{22} \end{pmatrix} \in GL(2,\mathbb{R})$ and $(a,b) \in \mathbb{R}^2$ are arbitrary vectors. For any subgroup $N_{(a,b,c)}$, with $(a,b) \neq 0$, if we suitably choose an element $S \in Aut(H_3)$, then $S(N_{(a,b,c)})$ is mapped to the subgroup $N_{(0,1,0)}$.

10.3.3 Heat Kernel of the Grushin Operator

In this section we shall construct the heat kernel for Grushin operators coming from the sub-Laplacian Δ_{sub} on the three-dimensional Heisenberg group H_3 by using the method of fiber integration.

We shall use the same notations as in the last section. First, we shall construct the heat kernel of the original Grushin operator

$$\mathcal{G} = -\frac{\partial^2}{\partial u^2} - u^2\frac{\partial^2}{\partial v^2}.$$

Let $K = K(t,x,y,z,\tilde{x},\tilde{y},\tilde{z}) \in C^\infty(\mathbb{R}_+ \times H_3 \times H_3)$ be the heat kernel of the sub-Laplacian $\Delta_{\text{sub}} = -\tilde{X}^2 - \tilde{Y}^2$ on the three-dimensional Heisenberg group H_3:

$$K = \frac{1}{(2\pi t)^2}\int_{-\infty}^{+\infty} e^{-\sqrt{-1}\frac{\tau}{t}\left(z-\tilde{z}+\frac{\tilde{y}x-\tilde{x}y}{2}\right)} e^{-\frac{\tau\coth\tau}{2t}\left((x-\tilde{x})^2+(y-\tilde{y})^2\right)} \frac{\tau}{\sinh\tau}d\tau.$$

Let $K_{\mathcal{G}} du \wedge dv$ be a 2-form on $\mathbb{R}_+ \times H_3 \times (N_Y \backslash H_3)$ given as the fiber integration of the 3-form $K dx \wedge dy \wedge dz$ on $\mathbb{R}_+ \times H_3 \times H_3$ by the map

$$\pi_Y : \mathbb{R}_+ \times H_3 \times H_3 \longrightarrow \mathbb{R}_+ \times H_3 \times (N_Y \backslash H_3) \cong \mathbb{R}_+ \times \mathbb{R}^3 \times \mathbb{R}^2,$$
$$(t, x, y, z, \tilde{x}, \tilde{y}, \tilde{z}) \longmapsto (t, x, y, z, u, v), u = \tilde{x}, \; v = \tilde{z} + \frac{\tilde{x}\tilde{y}}{2};$$

that is

$$K_{\mathcal{G}}(t, x, y, z, u, v) du \wedge dv = (\pi_Y)_* \big(K(t, x, y, z, \ldots) d\tilde{x} \wedge d\tilde{y} \wedge d\tilde{z}\big)(t, x, y, z, u, v), \tag{10.3.49}$$

where the value at the point $(t, x, y, z, u, v) \in \mathbb{R}_+ \times H_3 \times (N_Y \backslash H_3)$ is given by the integration along the fiber $\pi_Y^{-1}(u, v)$ for each fixed $(t, x, y, z) \in \mathbb{R}_+ \times H_3$.

To do this integration, we shall describe the projection map π_Y as follows. Let a decomposition $\mathbb{R} \times \mathbb{R}^2$ of the group H_3 be

$$\mathcal{D} : \mathbb{R} \times \mathbb{R}^2 \cong \mathbb{R}^3 \cong H_3,$$
$$(u, v, s) \longmapsto (x, y, z) = \left(u, s, v - \frac{xs}{2}\right).$$

Then we have

$$\pi_Y \circ \mathcal{D}(u, v, s) = (u, v),$$

and

$$dx \wedge dy \wedge dz = du \wedge ds \wedge dv.$$

Hence

$$\begin{aligned}
&(\pi_Y)_* \big(K(t, x, y, z, \ldots) d\tilde{x} \wedge d\tilde{y} \wedge d\tilde{z}\big)(t, x, y, z, u, v) \\
&= \left[\int_{-\infty}^{+\infty} K(t, x, y, z, u, s, v - us/2) ds\right] du \wedge dv \\
&= \frac{1}{(2\pi t)^2} \left[\int_{-\infty}^{+\infty} \int_{-\infty}^{+\infty} e^{-\sqrt{-1}\frac{\tau}{t}(z - v + us/2 + \frac{sx - uy}{2})} \right. \\
&\qquad \left. \times\, e^{-\frac{\tau \coth \tau}{2t}\left((x-u)^2 + (y-s)^2\right)} \frac{\tau}{\sinh \tau} d\tau ds\right] du \wedge dv \\
&= \frac{1}{(2\pi t)^2} \left[\int_{-\infty}^{+\infty} \left(\int_{-\infty}^{+\infty} e^{\sqrt{-1}\frac{\tau}{2t} s(x+u)} e^{-\frac{\tau \coth \tau}{2t} s^2} ds\right) \right. \\
&\qquad \left. \times e^{-\sqrt{-1}\frac{\tau}{t}\left(z + \frac{xy}{2} - v\right)} e^{-\frac{\tau \coth \tau}{2t}(x-u)^2} \frac{\tau}{\sinh \tau} d\tau\right] du \wedge dv \\
&= \frac{1}{(2\pi t)^{3/2}} \left[\int_{-\infty}^{+\infty} \frac{1}{\sqrt{\tau \coth \tau}} e^{-\frac{\tau (u+x)^2}{8t \coth \tau}} e^{-\sqrt{-1}\frac{\tau}{t}\left(z + \frac{xy}{2} - v\right)} \right. \\
&\qquad \left. \times e^{-\frac{\tau \coth \tau}{2t}(x-u)^2} \frac{\tau}{\sinh \tau} d\tau\right] du \wedge dv
\end{aligned}$$

$$= \frac{1}{(2\pi t)^{3/2}}\left[\int_{-\infty}^{+\infty} e^{-\sqrt{-1}\frac{\tau}{t}\left(z+\frac{xy}{2}-v\right)} e^{-\frac{\tau(u+x)^2}{8t\coth\tau}}\right.$$

$$\left.\times e^{-\frac{\tau\coth\tau}{2t}\left(x-u\right)^2}\sqrt{\frac{\tau}{\cosh\tau\sinh\tau}}d\tau\right]du\wedge dv$$

$$= K_{\mathcal{G}}(t,x,y,z,u,v)du\wedge dv.$$

Since the heat kernel K of the sub-Laplacian Δ_{sub} on H_3 is invariant under the left action of the group H_3 in the sense that

$$K(t, g*(x,y,z), g*(\tilde{x},\tilde{y},\tilde{z})) = K(t,\ x,y,z,\tilde{x},\ \tilde{y},\tilde{z}),\ \forall g\in H_3,$$

and because of the commutativity of the left actions with the projection map π_Y, the function $K_{\mathcal{G}}$ obtained previously can be seen as a smooth function on $\mathbb{R}_+\times(N_Y\backslash H_3)\times(N_Y\backslash H_3)$. Also, the commutativity

$$\Delta_{\text{sub}}\circ(\pi_Y)^* = (\pi_Y)^*\circ\mathcal{G}$$

says that the function $K_{\mathcal{G}}$ is the heat kernel of the Grushin operator. This can be written as

Theorem 10.3.5. *The function $K_{\mathcal{G}}$ satisfies*

$$\left(\mathcal{G}+\frac{\partial}{\partial t}\right)K_{\mathcal{G}} = 0,$$

$$\lim_{t\to 0}\int K_{\mathcal{G}}(t,\tilde{u},\tilde{v},u,v)f(u,v)dudv = f(\tilde{u},\tilde{v}),\ \ f\in C_0^\infty(\mathbb{R}^2),$$

or, more generally, $\forall f\in L_2(\mathbb{R}^2, dudv)$.

Next we shall consider the subgroup $N_Z = \{(0,0,z)\mid z\in\mathbb{R}\}$ of H_3 and the quotient space $\pi_Z : H_3\to H_3/N_Z\cong\mathbb{R}^2$. In this case the resulting Grushin operator

$$-\pi_Z(\tilde{X})^2 - \pi_Z(\tilde{Y})^2 = -\frac{\partial^2}{\partial u^2}-\frac{\partial^2}{\partial v^2}$$

is just the Euclidean Laplacian and the heat kernel is well known in this case. This can also be deduced from the formula of the heat kernel on H_3 by integrating along the fiber, as in the previous case. Then it suffices to calculate the integral

$$\int_{-\infty}^{+\infty} K(t,x,y,z,\tilde{x},\tilde{y},\tilde{z})d\tilde{z} \tag{10.3.50}$$

by applying the inversion formula of the Fourier transform

$$\begin{aligned}
&\int_{-\infty}^{+\infty} K\big(t,x,y,z,\tilde{x},\tilde{y},\tilde{z}\big)d\tilde{z} \\
&= \frac{1}{(2\pi t)^2}\int_{-\infty}^{+\infty}\int_{-\infty}^{+\infty} e^{-\sqrt{-1}\frac{\tau}{t}(z-\tilde{z}+\frac{\tilde{y}x-\tilde{x}y}{2})}e^{-\frac{\tau\coth\tau}{2t}\left((x-\tilde{x})^2+(y-\tilde{y})^2\right)}\frac{\tau}{\sinh\tau}d\tau d\tilde{z} \\
&= \frac{1}{(2\pi t)^2}\int_{\mathbb{R}}\left(\int_{\mathbb{R}} e^{\sqrt{-1}\frac{\tau}{t}\tilde{z}}\left\{e^{-\sqrt{-1}\frac{\tau}{t}(z+\frac{\tilde{y}x-\tilde{x}y}{2})}e^{-\frac{\tau\coth\tau}{2t}\left((x-\tilde{x})^2+(y-\tilde{y})^2\right)}\frac{\tau}{\sinh\tau}\right\}d\tau\right)d\tilde{z} \\
&= \frac{1}{(2\pi t)^2}\cdot\sqrt{2\pi}\int_{\mathbb{R}}\mathcal{F}^{-1}\left(e^{-\sqrt{-1}\frac{\bullet}{t}(z+\frac{\tilde{y}x-\tilde{x}y}{2})}e^{-\frac{\bullet\coth\bullet}{2t}\left((x-\tilde{x})^2+(y-\tilde{y})^2\right)}\frac{\bullet}{\sinh\bullet}\right)\left(\frac{\tilde{z}}{t}\right)d\tilde{z} \\
&= \frac{1}{2\pi t}e^{-\frac{(x-\tilde{x})^2+(y-\tilde{y})^2}{2t}},
\end{aligned}$$

where $\mathcal{F}$ is the Fourier transform

$$\mathcal{F}(f)(\theta) = \frac{1}{\sqrt{2\pi}}\int_{\mathbb{R}} e^{-\sqrt{-1}x\theta}f(x)dx.$$

Remark 10.3.6. In the case of (10.3.49), the integrand of the heat kernel K

$$\begin{aligned}
&e^{-\sqrt{-1}\frac{\tau}{t}(z-v+us/2+\frac{sx-uy}{2})}e^{-\frac{\tau\coth\tau}{2t}\left((x-u)^2+(y-s)^2\right)}\frac{\tau}{\sinh\tau} \\
&\quad = e^{\sqrt{-1}\frac{\tau}{2t}s(x+u)}e^{-\frac{\tau\coth\tau}{2t}(y-s)^2}e^{-\sqrt{-1}\frac{\tau}{t}\left(z-\frac{uy}{2}-v\right)}e^{-\frac{\tau\coth\tau}{2t}\left(x-u\right)^2}\frac{\tau}{\sinh\tau}
\end{aligned}$$

can be integrated with respect to the variable s, keeping the other variables fixed. But in the case of (10.3.50), the order of integration of variables τ and s cannot be changed.

We shall next consider the subgroup $N_{(a,b,c)}$. We shall treat only the case $a \neq 0$, since the case $b \neq 0$ can be treated in a similar way. The heat kernel

$$K^{\mathcal{G}_H}(t,u,v,\tilde{u},\tilde{v}) \in C^\infty\big(\mathbb{R}_+\times(N_{(a,b,c)}\backslash H_3)\times(N_{(a,b,c)}\backslash H_3)\big)$$

of the Grushin-type operator $\mathcal{G}_H$ provided by (10.3.48) is given by the integral

$$\begin{aligned}
&\frac{1}{(2\pi t)^2}\int_{\mathbb{R}}\int_{\mathbb{R}} e^{-\sqrt{-1}\frac{\tau}{t}\left(z-v-\theta(c+u/2)+\frac{(u/a+b\theta)x-a\theta y}{2}\right)} \\
&\quad\times e^{-\frac{\tau\coth\tau}{2t}\left((x-a\theta)^2+(y-u/a-b\theta)^2\right)}\frac{\tau}{\sinh\tau}d\tau d\theta.
\end{aligned}$$

If we put $x = a\tilde{u}$, $y = (\tilde{u}/a + b\tilde{\theta})$ and $z = \tilde{v} + \tilde{\theta}(c + \tilde{u}/2)$, then this can be expressed in the form given in the following result.

Theorem 10.3.7. *The heat kernel of the operator* $\mathcal{G}_H$ *is*

$$K^{\mathcal{G}_H}(t,\tilde{u},\tilde{v},u,v) = \frac{1}{(2\pi t)^{3/2}}\int_{\mathbb{R}} e^{-\sqrt{-1}\frac{\tau}{t}\left(\tilde{v}-v-\frac{b(\tilde{u}-u)}{a(a^2+b^2)}\right)} e^{-\frac{\tau(c+(\tilde{u}-u)^2)}{2t(a^2+b^2)\coth\tau}-\frac{(\tilde{u}-u)^2\tau\coth\tau}{2t(a^2+b^2)}} \left(\frac{\tau}{(a^2+b^2)\sinh\tau\cosh\tau}\right)^{1/2} d\tau.$$

Chapter 11
Heat Kernel for the Sub-Laplacian on the Sphere S^3

11.1 Sub-Riemannian Structure and Sub-Laplacian on the Sphere S^3

This section deals with the study of the heat kernel of a sub-Laplacian on the three-dimensional sphere, and a Grushin-type operator on S^2, called the *spherical Grushin operator*. This is a hypoelliptic operator on S^2 and is defined by the method explained in the previous chapter. The method of investigation is the explicit determination of eigenvalues and eigenfunctions of the Laplacian and sub-Laplacian in terms of harmonic polynomials (see [7–9, 97]).

11.1.1 S^3 in the Quaternion Number Field

Let $\mathbb{H}$ be the quaternion number field over $\mathbb{R}$, with the basis $\{\mathbf{1}, \mathbf{i}, \mathbf{j}, \mathbf{k}\}$ specified in Sect. 9.2.4. We consider the three-dimensional sphere as the set of unit elements [$\cong Sp(1)$] in $\mathbb{H}$. Let $h = x_0\mathbf{1} + x_1\mathbf{i} + x_2\mathbf{j} + x_3\mathbf{k} \in S^3 \subset \mathbb{H}$ and consider the curves $\{h \cdot \exp\, t\mathbf{i}\}_{t\in\mathbb{R}}$, $\{h \cdot \exp\, t\mathbf{j}\}_{t\in\mathbb{R}}$ and $\{h \cdot \exp\, t\mathbf{k}\}_{t\in\mathbb{R}}$ in S^3. Then these define left invariant vector fields $X_\mathbf{i}$, $X_\mathbf{j}$ and $X_\mathbf{k}$ on S^3 (in fact, defined on the whole $\mathbb{H}$ and tangent to S^3) and are expressed as follows:

$$
\begin{aligned}
X_\mathbf{i} &= -x_1\frac{\partial}{\partial x_0} + x_0\frac{\partial}{\partial x_1} + x_3\frac{\partial}{\partial x_2} - x_2\frac{\partial}{\partial x_3},\\
X_\mathbf{j} &= -x_2\frac{\partial}{\partial x_0} - x_3\frac{\partial}{\partial x_1} + x_0\frac{\partial}{\partial x_2} + x_1\frac{\partial}{\partial x_3},\\
X_\mathbf{k} &= -x_3\frac{\partial}{\partial x_0} + x_2\frac{\partial}{\partial x_1} - x_1\frac{\partial}{\partial x_2} + x_0\frac{\partial}{\partial x_3}.
\end{aligned}
$$

It is easy to check that the previous vector fields are orthonormal at each point $x \in S^3$ with respect to the Euclidean metric. Considered first-order differential

O. Calin et al., *Heat Kernels for Elliptic and Sub-elliptic Operators*, Applied and Numerical Harmonic Analysis, DOI 10.1007/978-0-8176-4995-1_11,

operators, the vector fields $X_{\mathbf{i}}$, $X_{\mathbf{j}}$ and $X_{\mathbf{k}}$ are skew-symmetric with respect to the following Riemannian volume form $dS(x)$ on S^3:

$$\begin{aligned} dS(x) &= x_0 dx_1 \wedge dx_2 \wedge x_3 - x_1 dx_0 \wedge dx_2 \wedge dx_3 \\ &\quad + x_2 dx_0 \wedge dx_1 \wedge dx_3 - x_3 dx_0 \wedge dx_1 \wedge dx_2. \end{aligned} \tag{11.1.1}$$

The Lie algebra of the three-dimensional sphere S^3, considered a Lie group $Sp(1)$, is generated by the basis $\mathbf{i}, \mathbf{j}, \mathbf{k}$. The Lie brackets are defined to be the commutators of the elements. In particular, we have $[\mathbf{k}, \mathbf{i}] = \mathbf{ki} - \mathbf{ik} = 2\mathbf{j}$ and then

$$[X_{\mathbf{i}}, X_{\mathbf{k}}] = -2X_{\mathbf{j}}.$$

Therefore, the sub-bundle $\mathcal{H}$ in $T(S^3)$ generated by the two vector fields $\{X_{\mathbf{i}}, X_{\mathbf{k}}\}$ defines a *minimal* sub-Riemannian structure in the strong sense. Consequently, the operator Δ_{sub} on S^3 defined by

$$\Delta_{\text{sub}} = -X_{\mathbf{i}}^2 - X_{\mathbf{k}}^2 \tag{11.1.2}$$

is the sub-Laplacian. Also, the operator

$$\Delta = -X_{\mathbf{i}}^2 - X_{\mathbf{j}}^2 - X_{\mathbf{k}}^2$$

is the Laplacian on S^3 with respect to the standard Riemannian metric.

Remark 11.1.1. The volume form constructed in Sect. 9.1.3 coincides with the Riemannian volume form (11.1.1) up to a multiplicative constant.

Remark 11.1.2. The vector field $\mathcal{N}$ normal to S^3 is given by

$$\mathcal{N} := x_0 \frac{\partial}{\partial x_0} + x_1 \frac{\partial}{\partial x_1} + x_2 \frac{\partial}{\partial x_2} + x_3 \frac{\partial}{\partial x_3}.$$

We identify the set of quaternions $\mathbb{H} \cong \mathbb{R}^4$ with $\mathbb{C}^2$ by the following correspondence:

$$\begin{aligned} \mathbb{H} \ni x &= x_0\mathbf{1} + x_1\mathbf{i} + x_2\mathbf{j} + x_3\mathbf{k} = (x_0\mathbf{1} + x_2\mathbf{j}) + \mathbf{i}(x_1\mathbf{1} + x_3\mathbf{j}) \\ &\leftrightarrow (x_0 + x_2\sqrt{-1}, x_1 + x_3\sqrt{-1}) =: \big(w_0(x), w_1(x)\big) = (w_0, w_1) \in \mathbb{C}^2, \end{aligned} \tag{11.1.3}$$

and we consider the space $\mathbb{H}$ as a (right-)complex vector space with the complex multiplication

$$\begin{aligned} \mathbb{H} \times \mathbb{C} \ni (x, \lambda) &= (x_0\mathbf{1} + x_1\mathbf{i} + x_2\mathbf{j} + x_3\mathbf{k},\ a + \sqrt{-1}b) = (w_0, w_1, \lambda) \\ &\mapsto (w_0\lambda, w_1\lambda) = \big(x_0\mathbf{1} + x_2\mathbf{j} + \mathbf{i}(x_1\mathbf{1} + x_3\mathbf{j})\big) \cdot (a\mathbf{1} + b\mathbf{j}) \in \mathbb{H}. \end{aligned}$$

Under this identification, the one-parameter transformation group generated by the vector field $X_{\mathbf{j}}$ is the scalar multiplication by $\lambda = a\mathbf{1} + b\mathbf{j} \in U(1)$ from the right. The orbit space of this action on S^3 is the complex projective space $P^1\mathbb{C}$, and we have the following principal bundle, called the *Hopf bundle*:

$$\pi_R : S^3 \to P^1\mathbb{C}.$$

Let $\chi_\ell : U(1) \to U(1)$ be the character $\chi_\ell(\lambda) := \lambda^\ell$. We denote the complex line bundle on $P^1\mathbb{C}$ associated with the character χ_ℓ by $\mathbf{L}^\ell$. Let $\mathcal{P}_N[\,x_0, x_1, x_2, x_3\,] = \mathcal{P}_N[\,x\,]$ be the space of homogeneous polynomials of degree N in the variables x_0, x_1, x_2 and x_3. We denote by $\mathcal{H}_N = \mathcal{H}_N[\,x\,]$ the subspace of harmonic polynomials in $\mathcal{P}_N[\,x\,]$:

$$\mathcal{H}_N[\,x\,] = \left\{ p \in \mathcal{P}_N[\,x\,] \,\middle|\, \sum_{i=0}^{3} \frac{\partial^2 p(x)}{\partial x_i^2} = 0 \right\}. \tag{11.1.4}$$

Proposition 11.1.3. *The following decomposition of $\mathcal{P}_N[\,x\,]$ holds:*

$$\mathcal{P}_N[\,x\,] = \mathcal{H}_N[\,x\,] + \left(\sum_{i=0}^{3} x_i^2\right) \cdot \mathcal{P}_{N-2}[\,x\,].$$

The space $\mathcal{H}_N[\,x\,]$ restricted to the sphere S^3, denoted by $\mathbf{H}_N$, is the eigenspace of the Laplacian $\Delta = -(X_{\mathbf{i}}^2 + X_{\mathbf{j}}^2 + X_{\mathbf{k}}^2)$ with respect to the eigenvalue $\lambda_N = N(N+2)$. Moreover, its dimension is given by

$$\dim \mathbf{H}_N = (N+1)^2 = \dim \mathcal{H}_N[\,x\,].$$

We shall express the elements of $\mathcal{P}_N[\,x\,]$ in terms of the complex and anticomplex variables $w_0, w_1, \overline{w_0}$ and $\overline{w_1}$ given by (11.1.3). We shall denote by $\mathcal{P}_{(n,m)}$ the subspace of $\mathcal{P}_N[\,x\,] = \mathcal{P}_N[\,w_0, w_1, \overline{w_0}, \overline{w_1}\,]$ (with $n + m = N$) defined by

$$\begin{aligned}\mathcal{P}_{(n,m)} = \Big\{ & p(\,w_0, w_1, \overline{w_0}, \overline{w_1}\,) \in \mathcal{P}_N[\,w_0, w_1, \overline{w_0}, \overline{w_1}\,] \,\Big| \\ & p\big(w_0 \cdot e^{\sqrt{-1}t}, w_1 \cdot e^{\sqrt{-1}t}, \overline{w_0 \cdot e^{\sqrt{-1}t}}, \overline{w_1 \cdot e^{\sqrt{-1}t}}\big) \\ & = e^{\sqrt{-1}(n-m)t} p\big(w_0, w_1, \overline{w_0}, \overline{w_1}\big),\ \forall\, t \in \mathbb{R}\Big\}.\end{aligned}$$

Let $N = n + m$ and $\mathcal{H}_{(n,m)} := \mathcal{P}_{(n,m)} \bigcap \mathcal{H}_N$. Then we have

$$\mathcal{H}_N[\,x\,] = \mathcal{H}_N[\,w_0, w_1, \overline{w_0}, \overline{w_1}\,] = \sum_{\substack{n+m=N \\ n\ge 0, m\ge 0}} \oplus\, \mathcal{H}_{(n,m)},$$

and the following decomposition holds:

$$\mathcal{P}_{(n,m)} = \mathcal{H}_{(n,m)} + \left(w_0\overline{w_0} + w_1\overline{w_1}\right) \cdot \mathcal{P}_{(n-1,m-1)}.$$

For any $p \in \mathcal{H}_{(n,m)}$, the following relation holds:

$$\Delta p = -\left(X_{\mathbf{i}}^2 + X_{\mathbf{k}}^2\right)p + (n-m)^2 p = \Delta_{\text{sub}}\, p + (n-m)^2 p. \qquad (11.1.5)$$

The restriction of $\mathcal{H}_{(n,m)}$ to S^3 is denoted by $\mathbf{H}_{(n,m)}$. Relation (11.1.5) implies the following result.

Proposition 11.1.4 ([7, 9]). *The space $\mathbf{H}_{(n,m)}$ is the eigenspace of the sub-Laplacian Δ_{sub} corresponding to the eigenvalue*

$$N(N+2) - \ell^2 = 4m^2 + 4m(1+|\ell|) + 2|\ell|,$$

with multiplicity

$$\dim \mathcal{H}_{(n,m)} = |\ell| + 2m + 1,$$

where $\ell = n - m$, $N = n + m$, $n \geq 0, m \geq 0$.

Let $\mathcal{F}^\ell$ be the subspace of $C^\infty(S^3)$ defined by

$$\mathcal{F}^\ell = \Big\{ f \in C^\infty(S^3) \mid f(x \cdot (a + b\mathbf{j})) = (a + b\sqrt{-1})^{-\ell} f(x),\ a^2 + b^2 = 1,\ a, b \in \mathbb{R} \Big\}.$$

Then $\mathcal{F}^\ell$ can be identified with the space of smooth sections $\Gamma(\mathbf{L}^\ell)$ of the complex line bundle $\mathbf{L}^\ell$. Due to this identification and since $c_{\text{sub}}(\mathcal{F}^\ell) \subset \mathcal{F}^\ell$, the sub-Laplacian Δ_{sub} can be seen as a second-order differential operator acting on $\Gamma(\mathbf{L}^\ell)$:

$$D_\ell : \Gamma(\mathbf{L}^\ell) \to \Gamma(\mathbf{L}^\ell).$$

The operator D_ℓ is elliptic and is called the *horizontal Laplacian*; see [7]. In fact, the principal symbol $\sigma(D_\ell)(x, \xi) : \mathbf{L}^\ell \to \mathbf{L}^\ell$, where $(x, \xi) \in T^*(P^1\mathbb{C})\backslash\{0\}$, is given by

$$\sigma(D_\ell)(x,\xi) = \left([\, X_{\mathbf{i}}(\pi_R^*(f))(\tilde{x})\,]^2 + [\, X_{\mathbf{k}}(\pi_R^*(f))(\tilde{x})\,]^2 \right) \pi^*(s)(\tilde{x}),$$

with $s \in \Gamma(\mathbf{L}^\ell)$ and $f \in C^\infty(P^1\mathbb{C})$, such that $df(x) = \xi \in T_x^*(P^1\mathbb{C})$, $f(x) = 0$, and $\pi_R(\tilde{x}) = x$. In other words, the principal symbol is given as the multiplication by the nonzero number

$$[\, X_{\mathbf{i}}(\pi_R^*(f))(\tilde{x})\,]^2 + [\, X_{\mathbf{k}}(\pi_R^*(f))(\tilde{x})\,]^2.$$

Remark 11.1.5. For each $\ell \in \mathbb{Z}$, the minimal eigenvalue of the horizontal Laplacian D_ℓ is $2|\ell|$ with multiplicity $|\ell| + 1$.

Proposition 11.1.6. *For fixed $\ell \in \mathbb{Z}$, the sum*

$$\sum_{\substack{m-n=\ell \\ n\geq 0, m\geq 0}} \oplus \mathcal{H}_{(n,m)} = \Gamma(\mathbf{L}^\ell) \cong \mathcal{F}^\ell$$

is the eigenspace decomposition of $\Gamma(\mathbf{L}^\ell)$ with respect to the operator D_ℓ.

11.1.2 Heat Kernel of the Sub-Laplacian on S^3

We start this section by recalling certain operators discussed in [97] (see also [8]) that are used to express the heat kernels of the horizontal Laplacian D_ℓ and the sub-Laplacian Δ_{sub} and also the Laplacian on S^3.

Define the map

$$\tau_S : T(S^3)\backslash\{0\} = \Big\{(x,y) \in \mathbb{R}^4 \times \mathbb{R}^4 \mid |x| = 1,\ < x, y >= 0,\ y \neq 0\Big\} \to \mathbb{C}^4\backslash\{0\}$$

by

$$T(\,S^3\,)\backslash\{\,0\,\} \ni (x,y) \mapsto \tau_S(x,y) = z = |\,y\,|x + \sqrt{-1}y \in \mathbb{C}^4\backslash\{\,0\,\}.$$

The image $X_S = \tau_S\big(\,T(S^3)\backslash\{\,0\,\}\,\big)$ is a *quadric hypersurface*

$$X_S = \Big\{ z = (z_0, z_1, z_2, z_3) \in \mathbb{C}^4\backslash\{0\} \,\Big|\, z^2 = \sum_{i=0}^{4} z_i{}^2 = 0 \Big\}.$$

We identify the tangent bundle $T(S^3)$ and the cotangent bundle $T^*(S^3)$ through the standard Riemannian metric. Then under the isomorphism τ_S, the symplectic form ω_S on $T^*(S^3)\backslash\{0\}$ is expressed as a Kähler form

$$\omega_S = \tau_S^* \left(\sqrt{2}\sqrt{-1}\ \bar{\partial}\,\partial \sqrt{\sum_{i=0}^{3} |z_i|^2} \right).$$

Let $P_N[z] = P_N[\,z_0, z_1, z_2, z_3\,]$ denote the homogeneous polynomials of degree N in the complex variables (z_0, z_1, z_2, z_3). We fix an inner product $< \cdot, \cdot >_{X_S}$ on the space $\sum_{N\geq 0} \oplus P_N[\,z\,]$ of all polynomials restricted to the quadric hypersurface X_S by

$$< f, g >_{X_S} = \int_{X_S} f(z)\overline{g(z)} \exp\big(-\hbar|z|\big) \cdot |z|^{\mathrm{q}}\, \Omega_S,$$

where $\hbar > 0$, $\mathbf{q} > -3$ are fixed constants and

$$\Omega_S = -\frac{1}{3!}\,\omega_S \wedge \omega_S \wedge \omega_S$$

is the *Liouville volume form* on $T^*(S^3)\backslash\{0\}$ [we identify $T^*(S^3)\backslash\{0\}$ with X_S via τ_S]. With respect to this inner product, the spaces $P_k[z]$ and $P_l[z]$ (restricted to X_S) are orthogonal for $k \neq l$. On the space $\mathbf{H}_N$ (the space of harmonic polynomials of degree N restricted to the sphere S^3), we consider the map

$$\begin{aligned} \mathcal{A}_N : \mathbf{H}_N &\longrightarrow P_N[z], \\ \varphi &\mapsto \int_{S^3} \varphi(x) < x, z >^N \, dS(x), \end{aligned} \tag{11.1.6}$$

where $z \in X_S$ and $< x, z > = \sum_{i=0}^{3} x_i z_i$ is a bilinear form on $\mathbb{R}^4 \times \mathbb{C}^4$.

Since $\sum \frac{\partial^2}{\partial x_i^2} < x, z >^N = < x, z >^{N-2} \sum z_i^2$ for each fixed $z \in X_S$, the polynomial $\varphi_z(x) = < x, z >^N$ of the variable x is harmonic. Hence, if $\psi \in \mathbf{H}_k$ $(k \neq N)$, then

$$\int_{S^3} \psi(x) < x, z >^N \, dS(x) = 0.$$

Next, we define a map $\mathcal{B}_N$ by

$$\begin{aligned} \mathcal{B}_N : P_N[z] &\longrightarrow \mathbf{H}_N, \\ \psi &\mapsto \int_{X_S} \psi(z) < x, \bar{z} >^N \exp\big(-\hbar|z|\big) \cdot |z|^{\mathbf{q}}\, \Omega_S. \end{aligned} \tag{11.1.7}$$

Since the integration in (11.1.7) is taken over X_S, it can easily be checked that $\mathcal{B}_N(\psi) \in \mathbf{H}_N$.

Proposition 11.1.7 ([8,97]). *With the identity operator Id on $\mathbf{H}_N$, the composition $\mathcal{B}_N \circ \mathcal{A}_N$ has the form*

$$\mathcal{B}_N \circ \mathcal{A}_N = \mathbf{a}_N Id,$$

where the real constant $\mathbf{a}_N$ is given by

$$\mathbf{a}_N = \mathbf{V} \cdot \frac{\Gamma(2N + \mathbf{q} + 3)}{y} 2^{N+1} \cdot \hbar^{2N} (N+1)^3 \quad \textit{with}$$

$$\mathbf{V} := \frac{4\pi \mathrm{Vol}(S^3) \cdot \mathrm{Vol}(S^3_{(1/\sqrt{2})})}{\hbar^{3+\mathbf{q}}}. \tag{11.1.8}$$

Here we denote by $\mathrm{Vol}(S^3)$ [respectively, $\mathrm{Vol}(S^3_{(1/\sqrt{2})})$] the volume of the unit sphere S^3 [resp. the sphere $S^3_{(1/\sqrt{2})}$ of radius $1/\sqrt{2}$].

From now on we shall work in the coordinates $(w_0, w_1) = (w_0(x), w_1(x))$ described in (11.1.3) rather than using $(x_0, x_1, x_2, x_3) \in \mathbb{R}^4$. We also define new coordinates (u_0, u_1, v_0, v_1) in $\mathbb{C}^4$ [instead of (z_0, z_1, z_2, z_3)] by the coordinate transformation

$$u_i = \frac{z_i - \sqrt{-1}z_{i+2}}{2} \quad \text{and} \quad v_i = \frac{z_i + \sqrt{-1}z_{i+2}}{2}, \qquad i = 0, 1.$$

Then the bilinear map $\mathbb{R}^4 \times \mathbb{C}^4 \to \mathbb{C}\ (x, z) \mapsto \sum x_i z_i =< x, z >$ is expressed as

$$< x, z > = w_0 u_0 + \overline{w_0} v_0 + w_1 u_1 + \overline{w_1} v_1,$$

and $\sum_{i=0}^{3} {z_i}^2 = 0$ holds if and only if $u_0 v_0 + u_1 v_1 = 0$. The Laplacian Δ on $\mathbb{R}^4$ in coordinates (w_0, w_1) is expressed in the form

$$-4\Big(\frac{\partial^2}{\partial w_0 \partial \overline{w_0}} + \frac{\partial^2}{\partial w_1 \partial \overline{w_1}} \Big).$$

Therefore, with the notations (11.1.4), we have

Lemma 11.1.8.

$$\mathcal{H}_N[\,x\,] = \left\{ f \in \mathcal{P}_N[\,w_0, w_1, \overline{w_0}, \overline{w_1}\,] \;\middle|\; \sum_{i=0}^{1} \frac{\partial^2 f}{\partial w_i \partial \overline{w_i}} = 0 \right\}.$$

With $(r, s) \in \mathbb{N}_0 \times \mathbb{N}_0$, let $A_{r,s} = A_{r,s}(w_0, w_1, \overline{w_0}, \overline{w_1}; u_0, u_1, v_0, v_1)$ be the function defined by

$$A_{r,s}(w_0, w_1, \overline{w_0}, \overline{w_1}; u_0, u_1, v_0, v_1) = \Big(w_0 u_0 + w_1 u_1\Big)^r \Big(\overline{w_0} v_0 + \overline{w_1} v_1\Big)^s.$$

Considered a function of the variables w_0, w_1 and $\overline{w}_0, \overline{w}_1$, for any fixed (u_0, u_1, v_0, v_1) with $u_0 v_0 + u_1 v_1 = 0$, it can be checked that $A_{r,s}$ is a harmonic polynomial of degree $r + s$.

Let $\mathcal{A}_{r,s}$ be the operator

$$\begin{aligned} \mathcal{A}_{r,s} : \sum \oplus \mathcal{H}_N[\,x\,] &\to \sum \oplus \mathcal{P}_N[\,u, v\,] \\ f &\mapsto \int_{S^3} f(w_0, w_1, \overline{w_0}, \overline{w_1}) \Big(w_0 u_0 + w_1 u_1\Big)^r \\ &\quad \times \Big(\overline{w_0} v_0 + \overline{w_1} v_1\Big)^s \, dS(x). \end{aligned}$$

Proposition 11.1.9. *Using the notations of Sect. 11.1.1, we have*

(i) $\mathcal{A}_{r,s}(\,\mathcal{H}_N\,) = 0$ *if* $N \neq r + s$

(ii) $\mathcal{A}_{r,s}(\,\mathcal{H}_{(k,l)}\,) = 0$ *for* $r + s = k + l$, *but* $r - s \neq k - l$

The image $\mathcal{A}_{r,s}(\mathcal{H}_{(r,s)})$ is spanned by the monomials of the form $u_0^{\alpha_0} \cdot u_1^{\alpha_1} \cdot v_0^{\beta_0} \cdot v_1^{\beta_1}$ with $\sum \alpha_i = r$ and $\sum \beta_i = s$.

Put $P_{r,s}[u,v] := \mathcal{A}_{r,s}(\mathcal{H}_{(r,s)})$. Then it follows that $P_{r,s}[u,v]$ and $P_{k,l}[u,v]$ are orthogonal when $r+s \neq k+l$ or $r-s \neq k-l$ in case $r+s = k+l$. We conclude these properties with the following result.

Proposition 11.1.10. *The map $\mathcal{A}_N$ in (11.1.6) can be decomposed in the form*

$$\sum_{r=0}^{N} C_N^r \mathcal{A}_{N-r,r} = \mathcal{A}_N, \quad \text{where} \quad C_N^r = \frac{N!}{(N-r)!r!}.$$

Moreover, we have

Proposition 11.1.11. *The inverse operator $\mathcal{A}_{r,s}^{-1} : P_{r,s}[u,v] \to \mathcal{H}_{(r,s)}$ is given by a constant $C_{r,s}$ times the integral operator*

$$\mathcal{B}_{r,s} : P_{r,s}[u,v] \to \mathcal{H}_{(r,s)}.$$

If we put $|(u,v)| := \left(|u_0|^2 + |u_1|^2 + |v_0|^2 + |v_1|^2 \right)^{1/2}$, then $\mathcal{B}_{r,s}$ is defined by

$$\mathcal{B}_{r,s} : \psi \mapsto 2^{\frac{q+3}{2}} \int_{X_S} \psi(u,v) \left(\sum_{i=0}^{1} \overline{w_i} \bar{u}_i \right)^r \cdot \left(\sum_{i=0}^{1} w_i \bar{v}_i \right)^s |(u,v)|^q e^{-\sqrt{2\hbar}\,|(u,v)|}\, \Omega_S(u,v).$$

Again, we have a decomposition $\sum_{r=0}^{N} C_N^r \mathcal{B}_{N-r,r} = \mathcal{B}_N$ and

$$\mathcal{B}_N \circ \mathcal{A}_N = \sum_{r=0}^{N} {C_N^r}^2 \mathcal{B}_{N-r,r} \circ \mathcal{A}_{N-r,r} = \mathbf{a}_N Id.$$

Moreover, on each subspace $\mathcal{H}_{(N-r,r)}$ we have $\mathcal{B}_{N-r,r} \circ \mathcal{A}_{N-r,r} = \frac{\mathbf{a}_N}{(C_N^r)^2}$.

We shall express the heat kernel $k^\ell(t;x,y) = k^\ell(t; w(x), w(y))$ of the operator Δ_{sub} acting on the space $\mathcal{F}^\ell$ in terms of the variables

$$w(x) = \left(w_0(x), w_1(x) \right), \; w_i(x) = x_i + \sqrt{-1}x_{i+2}, \quad \text{where } \; i = 0, 1.$$

Since the vector fields $X_\mathbf{i}$ and $X_\mathbf{k}$ are left invariant, the kernel function k^ℓ satisfies the invariance

$$k^\ell(t; g\cdot x, g\cdot y) = k^\ell(t; x, y) \quad \text{with} \quad g,\ x,\ y \in S^3. \tag{11.1.9}$$

The function space $\Gamma(\mathbf{L}^\ell)$ is left invariant as well. In the following proposition we give an expression of the kernel $k^\ell(t;x,y)$ (with $x,\ y \in S^3$) of the operator e^{-tD_ℓ} acting on the complex line bundle $\mathbf{L}^\ell$.

Proposition 11.1.12. *Let $\ell \geq 0$ and $x, y \in S^3$. Then the heat kernel $k^\ell(t; x, y)$ can be expressed as*

$$
\begin{aligned}
k^\ell(t; x, y)
&= 2^{\frac{\mathfrak{q}+3}{2}} \sum_{m=0}^{\infty} \frac{\left(C_{2m+\ell}^m\right)^2}{\mathbf{a}_{2m+\ell}} e^{-t\left(4m^2+4m(1+\ell)+2\ell\right)} \\
&\quad \times \int_{X_S} \left(\sum \overline{w_i(x)}\,\overline{u}_i\right)^{m+\ell} \cdot \left(\sum w_i(x)\,\overline{v_i}\right)^m \cdot \left(\sum w_i(y)\,u_i\right)^{m+\ell} \cdot \left(\sum \overline{w_i(y)}\,v_i\right)^m \\
&\quad \times |(u, v)|^{\mathfrak{q}} \cdot e^{-\sqrt{2}\hbar|(u,v)|}\, \Omega_S(u, v) \\
&= 2^{\frac{\mathfrak{q}+3}{2}} \sum_{m=0}^{\infty} \frac{\left(C_{2m+\ell}^m\right)^2}{\mathbf{a}_{2m+\ell}} e^{-t\left(4m^2+4m(1+\ell)+2\ell\right)} \\
&\quad \times \int_{X_S} \left(\sum \overline{w_i(y^{-1}x)}\,\overline{u}_i\right)^{m+\ell} \cdot \left(\sum w_i(y^{-1}x)\,\overline{v_i}\right)^m \cdot u_0^{m+\ell} \cdot v_0^m \\
&\quad \times |(u, v)|^{\mathfrak{q}} \cdot e^{-\sqrt{2}\hbar|(u,v)|}\, \Omega_S(u, v).
\end{aligned}
$$

Note that

$$
\begin{aligned}
w_0(y^{-1}x) &= \overline{w_0(y)}w_0(x) + \overline{w_1(y)}w_1(x), \\
w_1(y^{-1}x) &= w_0(y)w_1(x) - w_1(y)w_0(x).
\end{aligned}
$$

The kernel function k^ℓ satisfies the identities

(i) $k^\ell\left(t; x \cdot e^{\sqrt{-1}\theta}, y\right) = e^{-\sqrt{-1}\theta\ell} \cdot k^\ell\left(t; x, y\right)$,

(ii) $k^\ell\left(t; x, y \cdot e^{\sqrt{-1}\theta}\right) = e^{\sqrt{-1}\theta\ell} \cdot k^\ell\left(t; x, y\right)$.

In the case $\ell < 0$ one has

$$
\begin{aligned}
k^\ell(t;\ x, y)
&= 2^{\frac{\mathfrak{q}+3}{2}} \sum_{m=0}^{\infty} \frac{\left(C_{2m+|\ell|}^m\right)^2}{\mathbf{a}_{2m+|\ell|}} e^{-t\left(4m^2+4m(1+|\ell|)+2|\ell|\right)} \\
&\quad \times \int_{X_S} \left(\sum \overline{w_i(x)}\,\overline{u_i}\right)^{m} \cdot \left(\sum w_i(x)\,\overline{v_i}\right)^{m+|\ell|} \cdot \left(\sum w_i(y)\,u_i\right)^{m} \cdot \left(\sum \overline{w_i(y)}\,v_i\right)^{m+|\ell|} \\
&\quad \times |(u, v)|^{\mathfrak{q}} \cdot e^{-\sqrt{2}\hbar|(u,v)|}\, \Omega_S(u, v) \\
&= 2^{\frac{\mathfrak{q}+3}{2}} \sum_{m=0}^{\infty} \frac{\left(C_{2m+|\ell|}^m\right)^2}{\mathbf{a}_{2m+|\ell|}} e^{-t\left(4m^2+4m(1+|\ell|)+2|\ell|\right)} \\
&\quad \times \int_{X_S} \left(\sum \overline{w_i(y^{-1}x)}\,\overline{u_i}\right)^{m} \cdot \left(\sum w_i(y^{-1}x)\,\overline{v_i}\right)^{m+|\ell|} \cdot u_0^m \cdot v_0^{m+|\ell|} \\
&\quad \times |(u, v)|^{\mathfrak{q}} \cdot e^{-\sqrt{2}\hbar|(u,v)|}\, \Omega_S(u, v).
\end{aligned}
$$

For $\ell < 0$, the kernel k^ℓ satisfies equivalence properties similar to (i) and (ii) above.

The heat kernel $K_{\Delta_{\text{sub}}}(t;x,y)$ of the sub-Laplacian can be obtained as the sum over the kernels $k^\ell(t;x,y)$, $\ell \in \mathbb{Z}$.

Theorem 11.1.13.

$$\begin{aligned}
&K_{\Delta_{\text{sub}}}(t;x,y)\\
&= 2^{\frac{\mathfrak{q}+3}{2}} \sum_{\ell=0}^{\infty}\sum_{m=0}^{\infty} \frac{\left(C^m_{2m+\ell}\right)^2}{\mathbf{a}_{2m+\ell}} e^{-t\left(4m^2+4m(1+\ell)+2\ell\right)}\\
&\times \int_{X_S} \Big(\sum \overline{w_i(x)}\overline{u}_i\Big)^{m+\ell} \cdot \Big(\sum w_i(x)\overline{v}_i\Big)^m \cdot \Big(\sum w_i(y)u_i\Big)^{m+\ell} \cdot \Big(\sum \overline{w_i(y)}v_i\Big)^m\\
&\times |(u,v)|^{\mathfrak{q}} \cdot e^{-\sqrt{2}\hbar|(u,v)|}\, \Omega_S(u,v)\\
&+ 2^{\frac{\mathfrak{q}+3}{2}} \sum_{\ell=1}^{\infty}\sum_{m=0}^{\infty} \frac{\left(C^m_{2m+\ell}\right)^2}{\mathbf{a}_{2m+\ell}} e^{-t\left(4m^2+4m(1+\ell)+2\ell\right)}\\
&\times \int_{X_S} \Big(\sum \overline{w_i(x)}\overline{u}_i\Big)^{m} \cdot \Big(\sum w_i(x)\overline{v_i}\Big)^{m+\ell} \cdot \Big(\sum w_i(y)u_i\Big)^{m} \cdot \Big(\sum \overline{w_i(y)}v_i\Big)^{m+\ell}\\
&\times |(u,v)|^{\mathfrak{q}} \cdot e^{-\sqrt{2}\hbar|(u,v)|}\, \Omega_S(u,v).
\end{aligned}$$

11.1.3 Spherical Grushin Operator and the Heat Kernel

In this section we consider the *left action* of the group $U(1) \cong \{e^{t\mathbf{j}}\}_{t\in\mathbb{R}}$ on S^3, and define a Grushin-type operator on $P^1(\mathbb{C})$; i.e., we consider the *Hopf bundle* $\pi_L : S^3 \to U(1)\backslash S^3 \cong P^1\mathbb{C}$ [= the orbit space by the left action of $U(1)$]. Then, by the associative law of the quaternion number field, the operators $X_{\mathbf{i}}$ and $X_{\mathbf{k}}$ commute with the left action and one obtains vector fields $d\pi_L(X_{\mathbf{i}})$ and $d\pi_L(X_{\mathbf{k}})$ on $P^1\mathbb{C}$. Considered first-order operators, these vector fields are skew-symmetric with respect to the volume form $(\pi_L)_*(dS(x))$ [the forward push of the volume form $dS(x)$ by the projection map π_L; see Sect. 9.1.3]. This volume form also coincides with the standard volume form on S^2 except for a multiplicative constant. Since the vector fields $X_{\mathbf{i}}$ and $X_{\mathbf{k}}$ satisfy conditions **(Sub-1)**, **(Sub-2)** and **(Gru)**, we have

Proposition 11.1.14. *The operator*

$$\mathcal{G}_S := -d\pi_L(X_{\mathbf{i}})^2 - d\pi_L(X_{\mathbf{k}})^2$$

is hypoelliptic on $P^1\mathbb{C}$.

We shall call the operator $\mathcal{G}_S$ the *spherical Grushin operator*, since it is constructed in a similar way as the Grushin operator $\mathcal{G}_H := -\left(\frac{\partial^2}{\partial x^2} + x^2\frac{\partial^2}{\partial y^2}\right)$.

Remark 11.1.15. All elements $h \in \mathbb{H}$ with $h^2 = -\mathbf{1}$ are conjugate to each other. Therefore, taking another generator g_0 ($g_0{}^2 = -\mathbf{1}$) of the group $U(1)$, the resulting operator on $P^1\mathbb{C}$ is conjugate to $\mathcal{G}_S$ as well.

We determine the subset in $P^1\mathbb{C}$ on which the operator $\mathcal{G}_S$ degenerates, i.e., the subset on which the vector fields $d\pi_L(X_\mathbf{i})$ and $d\pi_L(X_\mathbf{k})$ are linearly dependent. This can be done by determining all points $x = (x_0, x_1, x_2, x_3) \in S^3$ for which a solution $(a, b) \in \mathbb{R}^2$, $a^2 + b^2 = 1$, of

$$\mathbf{j}\left(x_0 + x_1\mathbf{i} + x_2\mathbf{j} + x_3\mathbf{k}\right) = \left(x_0 + x_1\mathbf{i} + x_2\mathbf{j} + x_3\mathbf{k}\right)\left(a\mathbf{i} + b\mathbf{k}\right)$$

exists. A necessary condition for solving this equation turns out to be

$$x_0^2 + x_2^2 = x_1^2 + x_3^2 = \frac{1}{2}$$

and the solution (a, b) is given by

$$\begin{aligned} a &= 2(x_1x_2 + x_0x_3), \\ b &= 2(x_2x_3 - x_0x_1). \end{aligned}$$

If we put $\sqrt{2}x_0 = \cos\theta$, $\sqrt{2}x_2 = \sin\theta$, $\sqrt{2}x_1 = \cos\eta$ and $\sqrt{2}x_3 = \sin\eta$, then the numbers a and b can be expressed in the following form:

$$\begin{aligned} a &= \sin(\theta + \eta), \\ b &= -\cos(\theta + \eta). \end{aligned}$$

Let $\mathfrak{p}(\mathbf{i}, \mathbf{k}) = \{x \in S^3 \mid x_0^2 + x_2^2 = 1/2 = x_1^2 + x_3^2\}$. It can be checked that $\mathfrak{p}(\mathbf{i}, \mathbf{k})$ is invariant under the action of the group $U(1)$ to the left (and also to the right).

Proposition 11.1.16. *The operator $\mathcal{G}_S$ degenerates on the set $\pi_L(\mathfrak{p}(\mathbf{i}, \mathbf{k})) \cong S^1$; i.e., the vector fields $d\pi_L(X_\mathbf{i})$ and $d\pi_L(X_\mathbf{k})$ are linearly dependent (but do not vanish simultaneously).*

In order to make a comparison with the *Heisenberg–Grushin operator* $\mathcal{G}_H$, we shall present the spherical Grushin operator $\mathcal{G}_S$ explicitly in two ways.

First, we consider the map π_L given by

$$\begin{aligned} &\pi_L : \mathbb{H} \cong \mathbb{R}^4 \to \mathbb{R}^3, \\ &(x_0, x_1, x_2, x_3) \mapsto (u_1, u_2, u_3), \\ &\qquad u_1 = x_0x_1 - x_2x_3,\ u_2 = x_0x_3 + x_1x_2,\ u_3 = x_0^2 + x_2^2 - 1/2, \end{aligned}$$

which does realize $P^1\mathbb{C}$ as a sphere $S^2_{(1/2)}$ of radius $1/2$ in $\mathbb{R}^3$. This map is characterized by the equation

$$u_1^2 + u_2^2 + u_3^2 - 1/4 = \left(x_0^2 + x_2^2\right)\left(x_0^2 + x_1^2 + x_2^2 + x_3^2 - 1\right). \tag{11.1.10}$$

Using (11.1.10) it can be easily checked that $\pi_L(\mathfrak{p}(\mathbf{i},\mathbf{k}))$ coincides with the equator. Moreover, S^3 is divided into two connected components having the two-dimensional torus $\mathfrak{p}(\mathbf{i},\mathbf{k})$ as the common boundary. As stated in the beginning of this section, we can descend the right actions of $\{e^{t\mathbf{i}}\}$ and $\{e^{t\mathbf{k}}\}$ onto $\mathbb{R}^3$ through the map π_L. The resulting actions are the rotations with respect to the u_2-axis and u_1-axis, respectively. The vector fields $d\pi_L(X_\mathbf{i})$ and $d\pi_L(X_\mathbf{k})$ are of the form

$$d\pi_L(X_\mathbf{i}) = -2u_1\frac{\partial}{\partial u_3} + 2u_3\frac{\partial}{\partial u_1}, \tag{11.1.11}$$

$$d\pi_L(X_\mathbf{k}) = -2u_2\frac{\partial}{\partial u_3} + 2u_3\frac{\partial}{\partial u_2}. \tag{11.1.12}$$

It is obvious that $d\pi_L(X_\mathbf{i})$ and $d\pi_L(X_\mathbf{k})$ are tangent to the sphere and are linearly dependent only along the equator $\{\ (u_1,u_2,0) \mid u_1^2+u_2^2 = 1/4\ \}$. The *spherical Grushin operator* $\mathcal{G}_S$ is the restriction of

$$\begin{aligned}
-\frac{1}{4}\mathcal{G}_S &= \frac{1}{4}\{\, d\pi_L(X_\mathbf{i})^2 + d\pi_L(X_\mathbf{k})^2 \,\} \\
&= (u_1^2+u_2^2)\frac{\partial^2}{\partial u_3^2} + u_3^2\left(\frac{\partial^2}{\partial u_1^2}+\frac{\partial^2}{\partial u_2^2}\right) - 2u_3\frac{\partial}{\partial u_3} - u_1\frac{\partial}{\partial u_1} \\
&\quad - u_2\frac{\partial}{\partial u_2} - 2u_1u_3\frac{\partial^2}{\partial u_1\,\partial u_3} - 2u_2u_3\frac{\partial^2}{\partial u_2\,\partial u_3}
\end{aligned}$$

to the sphere $u_1^2+u_2^2+u_3^2 = 1/4$.

Next, we shall provide an expression in local coordinates, with the exception of one point. Consider the following local trivialization:

$$\Phi_L : U(1)\times\mathbb{C}\to U_0, \tag{11.1.13}$$

$$\begin{aligned}
&\phi_L : (\lambda, z) = (\lambda, x+\sqrt{-1}y) \mapsto \\
&Re\left(\frac{\lambda}{\sqrt{1+|z|^2}}\right)\mathbf{1} + Re\left(\frac{\lambda z}{\sqrt{1+|z|^2}}\right)\mathbf{i} \\
&\quad + Im\left(\frac{\lambda}{\sqrt{1+|z|^2}}\right)\mathbf{j} - Im\left(\frac{\lambda z}{\sqrt{1+|z|^2}}\right)\mathbf{k} \in S^3 \subset \mathbb{H}.
\end{aligned} \tag{11.1.14}$$

Then the variable $z = x+\sqrt{-1}y$ is a local coordinate on the subset $\pi_L(U_0)$, and with this coordinate the vector fields $d\pi_L(X_\mathbf{i})$ and $d\pi_L(X_\mathbf{k})$ are expressed as

$$d\pi_L(X_\mathbf{i}) = \left(1+x^2-y^2\right)\frac{\partial}{\partial x} + 2xy\frac{\partial}{\partial y},$$

$$d\pi_L(X_\mathbf{k}) = -2xy\frac{\partial}{\partial x} + \left(x^2-y^2-1\right)\frac{\partial}{\partial y}.$$

Only on the unit circle $x^2 + y^2 = 1$ are these vectors linearly dependent, and the spherical Grushin operator has the form

$$
\begin{aligned}
-\mathcal{G}_S &= \Big(1 + 2(x^2 - y^2) + (x^2 + y^2)^2\Big)\frac{\partial^2}{\partial x^2} \\
&\quad + \Big(1 - 2(x^2 - y^2) + (x^2 + y^2)^2\Big)\frac{\partial^2}{\partial y^2} + 8xy\frac{\partial^2}{\partial x \partial y} + 4x\frac{\partial}{\partial x} + 4y\frac{\partial}{\partial y}.
\end{aligned}
$$

Remark 11.1.17. Recently, in [3] a structure on a manifold which is called *almost-Riemannian* was defined. Roughly speaking, this structure is given through a set of Lie bracket generating vector fields without assuming the linear independence at every point (the number of the vector fields will be less than the dimension of the manifold). Moreover, a nondegeneracy condition for the sub-manifold on which the vector fields are linearly dependent is assumed. The S^2-case carrying the vector fields $\{ d\pi_L(X_{\mathbf{i}}),\ d\pi_L(X_{\mathbf{k}}) \}$ is mentioned as an example.

Now using the heat kernel expression for the sub-Laplacian on S^3 given in the last section, we give an integral expression of the heat kernel for the spherical Grushin operator.

Since $K_{\Delta_{\mathrm{sub}}}(t; x, y)$ satisfies the following invariance:

$$
K_{\Delta_{\mathrm{sub}}}\big(t; g\cdot x, g\cdot y\big), \qquad \forall g,\ x,\ y \in S^3,
$$

according to the construction of the spherical Grushin operator $\mathcal{G}_S$, the heat kernel

$$
K_{\mathcal{G}_S}(t, \pi_L(x), \pi_L(y)) \in C^\infty\big(\mathbb{R}_+ \times P^1\mathbb{C} \times P^1\mathbb{C}\big)
$$

of $\mathcal{G}_S$ can be expressed in the following integral form.

Theorem 11.1.18.

$$
K_{\mathcal{G}_S}\big(t, \pi_L(x), \pi_L(y)\big) = \frac{1}{\sqrt{-1}}\int_{U(1)} K_{\Delta_{\mathrm{sub}}}\big(t; \lambda\cdot x, y\big)\,\frac{d\lambda}{\lambda}.
$$

We also have

Theorem 11.1.19. *The following trace formula holds:*

$$
\begin{aligned}
&\operatorname{tr}\left(e^{-t\mathcal{G}_S}\right) \\
&= \frac{1}{\sqrt{-1}}\int_{S^3}\int_{U(1)} K_{\Delta_{\mathrm{sub}}}\big(t; x^{-1}\cdot\lambda\cdot x, \mathbf{1}\big)\,\frac{d\lambda}{\lambda}\,dS(x) \\
&= \frac{2^{\frac{q+3}{2}}}{\sqrt{-1}}\sum_{\ell=0}^{\infty}\sum_{m=0}^{\infty}\frac{\left(C^m_{2m+\ell}\right)^2}{\mathbf{a}_{2m+\ell}}e^{-t\left(4m^2+4m(1+\ell)+2\ell\right)} \\
&\quad\times\int_{U(1)}\int_{S^3}\int_{X_S}\Big[\big(\overline{\lambda}|w_0(x)|^2 + \lambda|w_1(x)|^2\big)\overline{u_0} + \big(\overline{w_0(x)}(\lambda - \overline{\lambda})\overline{w_1(x)}\big)\overline{u_1}\Big]^{m+\ell}
\end{aligned}
$$

$$\times\Big[\big(\lambda|w_0(x)|^2+\overline{\lambda}|w_1(x)|^2\big)\overline{v_0}+\big(w_0(x)(\overline{\lambda}-\lambda)w_1(x)\big)\overline{v_1}\Big]^m\cdot u_0^{m+\ell}\cdot v_0^m$$

$$\times|(u,v)|^{\mathfrak{q}}\cdot e^{-\sqrt{2}\hbar|(u,v)|}\,\Omega_S(u,v)\,dS(x)\,\frac{d\lambda}{\lambda}$$

$$+\frac{2^{\frac{\mathfrak{q}+3}{2}}}{\sqrt{-1}}\sum_{\ell=1}^{\infty}\sum_{m=0}^{\infty}\frac{\left(C_{2m+\ell}^m\right)^2}{\mathfrak{a}_{2m+\ell}}e^{-t\left(4m^2+4m(1+\ell)+2\ell\right)}$$

$$\times\int_{U(1)}\int_{S^3}\int_{X_S}\Big[\big(\overline{\lambda}|w_0(x)|^2+\lambda|w_1(x)|^2\big)\overline{u_0}+\big(\overline{w_0(x)}(\lambda-\overline{\lambda})\overline{w_1(x)}\big)\overline{u_1}\Big]^m$$

$$\times\Big[\big(\lambda|w_0(x)|^2+\overline{\lambda}|w_1(x)|^2\big)\overline{v_0}+\big(w_0(x)(\overline{\lambda}-\lambda)w_1(x)\big)\overline{v_1}\Big]^{m+\ell}\cdot u_0^m\cdot v_0^{m+\ell}$$

$$\times|(u,v)|^{\mathfrak{q}}\cdot e^{-\sqrt{2}\hbar|(u,v)|}\,\Omega_S(u,v)\,dS(x)\,\frac{d\lambda}{\lambda},$$

where we have used the identities

$$\begin{aligned}w_0(x^{-1}\cdot\lambda\cdot x)&=\lambda|w_0(x)|^2+\overline{\lambda}|w_1(x)|^2,\\ w_1(x^{-1}\cdot\lambda\cdot x)&=w_0(x)(\overline{\lambda}-\lambda)w_1(x).\end{aligned}$$

Based on the previous information, the zeta-regularized determinant of the sub-Laplacian and other related operators is computed; see [10].

Similar computations can be found in [10] for the case of a two-step, codimension-1 sub-Riemannian structure on the seven-dimensional sphere S^7. The problem of finding the heat kernel in the case of the two-step, codimension-3 sub-Riemannian structure on S^7 still remains an open problem. In the case of the sphere S^7, expressing the zeta-regularized determinant is not easy. However, we know that the spectral zeta function is holomorphic at the origin.

Part III
Laguerre Calculus and the Fourier Method

Chapter 12
Finding Heat Kernels Using the Laguerre Calculus

12.1 Introduction

In this chapter, we are going to use a harmonic analysis method to construct the heat kernels and fundamental solutions of the *sub-Laplacian* on the *Heisenberg group*. This method relies on *Laguerre calculus*. We shall start with a beautiful idea of Mikhlin from his 1936 study of convolution operators on $\mathbb{R}^2$. Let $\mathbf{K}$ be a *principal value (P.V.) convolution operator* on $\mathbb{R}^2$:

$$\mathbf{K}(f)(x) = \lim_{\varepsilon \to 0} \int_{|y|>\varepsilon} K(y) f(x-y) dy,$$

where $f \in C_0^\infty(\mathbb{R}^2)$ and $K \in C^\infty(\mathbb{R}^2 \setminus \{(0,0)\})$ is homogeneous of degree -2 with vanishing mean value; i.e.,

$$\int_{|y|=1} K(y) dy = 0.$$

Thus we can write

$$K(x) = \frac{f(\theta)}{r^2}, \qquad x = x_1 + i x_2 = r e^{i\theta},$$

where

$$f(\theta) = \sum_{m \in \mathbb{Z}, m \neq 0} f_m e^{im\theta}.$$

Suppose that g is another smooth function on $[0, 2\pi]$ with

$$g(\theta) = \sum_{m \in \mathbb{Z}, m \neq 0} g_m e^{im\theta}.$$

Then g induces a principal value convolution operator $\mathbf{G}$ on $\mathbb{R}^2$ with kernel $\frac{g(\theta)}{r^2}$. In [91], we found the following identity.

O. Calin et al., *Heat Kernels for Elliptic and Sub-elliptic Operators*, Applied and Numerical Harmonic Analysis, DOI 10.1007/978-0-8176-4995-1_12,

Proposition 12.1.1.

$$\frac{|m|i^{-|m|}}{2\pi}\frac{e^{im\theta}}{r^2} *_{\mathbb{R}^2} \frac{|k|i^{-|k|}}{2\pi}\frac{e^{ik\theta}}{r^2} = \frac{|m+k i^{-|m+k|}}{2\pi}\frac{e^{i(m+k)\theta}}{r^2}.$$

Here $*_{\mathbb{R}^2}$ stands for the standard convolution on $\mathbb{R}^2$.

Definition 12.1.2. Let

$$f(\theta) = \sum_{m\in\mathbb{Z}, m\neq 0} f_m e^{im\theta},$$

inducing the principal value convolution operator $\mathbf{K}$, with kernel $\frac{f(\theta)}{r^2}$ on $C_0^\infty(\mathbb{R}^2)$. Then the symbol $\sigma(\mathbf{K})$ of $\mathbf{K}$ is defined by

$$\sigma(\mathbf{K}) = \sum_{m\in\mathbb{Z}, m\neq 0}\left(\frac{|m|i^{-|m|}}{2\pi}\right)^{-1} f_m e^{im\theta}.$$

With this definition we may rewrite Proposition 12.1.1 as follows.

Theorem 12.1.3. *Let $\mathbf{K}$ and $\mathbf{G}$ be two principal value convolution operators on $C_0^\infty(\mathbb{R}^2)$. Then*

$$\sigma\big(\mathbf{K} *_{\mathbb{R}^2} \mathbf{G}\big) = \sigma(\mathbf{K})\cdot\sigma(\mathbf{G}).$$

Now we shall generalize the result of Theorem 12.1.3 to the Heisenberg group. The *nonisotropic Heisenberg group* H_n is the Lie group with the underlying manifold

$$\mathbb{C}^n\times\mathbb{R} = \{[\mathbf{z}, t] : \mathbf{z}\in\mathbb{C}^n,\ t\in\mathbb{R}\}$$

and the multiplication law

$$[\mathbf{z}, t]\cdot[\mathbf{w}, s] = \left[\mathbf{z}+\mathbf{w}, t+s+2\mathrm{Im}\sum_{j=1}^n a_j z_j \bar{w}_j\right], \tag{12.1.1}$$

where $a_1, a_2, \ldots, a_n$ are positive numbers.

It is easy to check that the multiplication (12.1.1) does indeed make $\mathbb{C}^n\times\mathbb{R}$ into a group whose identity is the origin $e = [\mathbf{0}, 0]$, and where the inverse is given by $[\mathbf{z}, t]^{-1} = [-\mathbf{z}, -t]$.

The Lie algebra of H_n is a vector space which, together with a Lie bracket operation defined on it, represents the infinitesimal action of H_n. Let $\mathfrak{h}_n$ denote the vector space of left invariant vector fields on H_n. Note that this linear space is closed with respect to the bracket operation

$$[\mathbf{V}_1,\ \mathbf{V}_2] = \mathbf{V}_1\mathbf{V}_2 - \mathbf{V}_2\mathbf{V}_1.$$

The space $\mathfrak{h}_n$, equipped with this bracket, is referred to as the *Lie algebra* of H_n.

The Lie algebra structure of $\mathfrak{h}_n$ is most readily understood by describing it in terms of the following basis:

$$\mathbf{X}_j = \frac{\partial}{\partial x_j} + 2a_j y_j \frac{\partial}{\partial t}, \quad \mathbf{Y}_j = \frac{\partial}{\partial y_j} - 2a_j x_j \frac{\partial}{\partial t} \quad \text{and} \quad \mathbf{T} = \frac{\partial}{\partial t}, \tag{12.1.2}$$

where $j = 1, 2, \ldots, n$, $\mathbf{z} = (z_1, z_2, \ldots, z_n) \in \mathbb{C}^n$ with $z_j = x_j + iy_j$; $t \in \mathbb{R}$.
Note that we have the commutation relations

$$[\mathbf{Y}_j, \mathbf{X}_k] = 4a_j \delta_{jk} \mathbf{T} \quad \text{for} \quad j, k = 1, 2, \ldots, n. \tag{12.1.3}$$

Next, we define the complex vector fields

$$\bar{\mathbf{Z}}_j = \frac{1}{2}(\mathbf{X}_j + i\mathbf{Y}_j) = \frac{\partial}{\partial \bar{z}_j} - ia_j z_j \frac{\partial}{\partial t} \quad \text{and} \quad \mathbf{Z}_j = \frac{1}{2}(\mathbf{X}_j - i\mathbf{Y}_j) = \frac{\partial}{\partial z_j} + ia_j \bar{z}_j \frac{\partial}{\partial t} \tag{12.1.4}$$

for $j = 1, 2, \ldots, n$. Here, as usual,

$$\frac{\partial}{\partial z_j} = \frac{1}{2}\left(\frac{\partial}{\partial x_j} - i\frac{\partial}{\partial y_j}\right) \quad \text{and} \quad \frac{\partial}{\partial \bar{z}_j} = \frac{1}{2}\left(\frac{\partial}{\partial x_j} + i\frac{\partial}{\partial y_j}\right).$$

The commutation relations (12.1.3) then become

$$[\bar{\mathbf{Z}}_j, \mathbf{Z}_k] = 2ia_j \delta_{jk} \mathbf{T}$$

with all other commutators among the $\mathbf{Z}_j$, $\bar{\mathbf{Z}}_k$ and $\mathbf{T}$ vanishing.

12.2 Laguerre Calculus

Laguerre calculus is the symbolic tensor calculus on the Heisenberg group H_n. It was first introduced on H_1 by Greiner [57] and extended to H_n and $\mathrm{H}_n \times \mathbb{R}^d$ by Beals, Gaveau, Greiner and Vauthier [17]. The Laguerre functions have been used in the study of the twisted convolution, or equivalently, the Heisenberg convolution for several decades (see Geller [51] and Peetre [96]). The Laguerre functions also played an important role in the Fock–Bargmann and Schrödinger representations of the Heisenberg group (see Folland [43] for details). But it was in Greiner [57] that for the first time Laguerre functions were connected with left invariant convolution operators on H_1, and were used to invert some basic differential operators on H_1, namely, the Lewy operator and the Heisenberg sub-Laplacian.

In this chapter we shall use the Laguerre calculus to find the heat kernel and fundamental solution of the *Kohn Laplacian* and *Paneitz operator*. Of course, this method has some limitations since it depends heavily on the group structure and orthogonality of the Laguerre functions. However, this method allows us to construct heat kernels for higher-order partial differential operators, like powers of the sub-Laplacian and Paneitz operator on the group. In order to introduce the Laguerre calculus, we first recall the definitions of the twisted and P.V. convolutions on the Heisenberg group.

Consider the functions $f,\ g \in C_0^\infty(\mathrm{H}_n)$. The Heisenberg convolution is given by

$$f * g(\mathbf{x}) = \int_{\mathrm{H}_n} f(\mathbf{y}) g(\mathbf{y}^{-1}\mathbf{x}) dV(\mathbf{y}); \tag{12.2.5}$$

here $dV(\mathbf{y})$ is the Haar measure on H_n and is exactly the Euclidean measure on $\mathbb{R}^{2n+1}$.

12.2.1 Twisted Convolution

We focus our attention on the phase space $\mathbb{R}^n \times \mathbb{R}^n$, which we identify with $\mathbb{C}^n$ via $\zeta \in \mathbb{C}^n, \zeta = u + iv \leftrightarrow (\mathbf{u}, \mathbf{v}) \in \mathbb{R}^n \times \mathbb{R}^n$.

Let

$$A = \begin{bmatrix} a_1 & & & \\ & a_2 & & \\ & & \ddots & \\ & & & a_n \end{bmatrix}$$

be a positive definite diagonal $n \times n$ real matrix. Consider the symplectic form $< \cdot, \cdot >$ given by the Heisenberg group multiplication law (12.1.1) and defined by

$$< \mathbf{z}, \mathbf{w} > = 2\mathrm{Im}(A\mathbf{z} \cdot \bar{\mathbf{w}}) = 2\mathrm{Im}\left(\sum_{j=1}^{n} a_j z_j \bar{w}_j \right),$$

where $\mathbf{z}, \mathbf{w} \in \mathbb{C}^n$. With τ a fixed real constant, we can define the *twisted convolution* of two functions F and G by

$$(F *_\tau G)(\mathbf{z}) = \int_{\mathbb{C}^n} F(\mathbf{z} - \mathbf{w}) G(\mathbf{w}) e^{-i\tau<\mathbf{z},\mathbf{w}>} d\mathbf{w}; \tag{12.2.6}$$

here $d\mathbf{w}$ is the Euclidean measure on $\mathbb{C}^n$. Notice that, in view of the antisymmetry of $< \cdot, \cdot >$, we have $< \mathbf{z} - \mathbf{w}, \mathbf{w} > = - < \mathbf{w}, \mathbf{z} >$; thus

$$G *_\tau F = F *_{-\tau} G,$$

so the twisted product is not commutative.

The twisted convolution arises when we analyze the convolution of functions on the Heisenberg group in terms of the Fourier transform in the t-variable. To see this, let $f(\mathbf{z}, t)$ be a test function on H_n. Define

$$\widetilde{f}_\tau(\mathbf{z}) = \int_{\mathbb{R}} f(\mathbf{z}, t) e^{-i\tau t} dt. \tag{12.2.7}$$

Similarly define $\widetilde{g}_\tau$ when g is another test function on H_n. Suppose $f * g$ is the convolution of f and g on H_n. Then

$$\widetilde{(f * g)}_\tau = \widetilde{f}_\tau *_\tau \widetilde{g}_\tau. \tag{12.2.8}$$

12.2.2 P.V. Convolution Operators

In order to show the regularity of the solution operator $S(f) = f * \Psi_\alpha$, we need to introduce principal value convolution operators on H_n. These operators are the analog of Calderón–Zygmund principal value convolution operators on $\mathbb{R}^n$. As we know, the underlying manifold of H_n is $\mathbb{R}^{2n+1}$; but the role of the additive structure in $\mathbb{R}^{2n+1}$ is supplied by the Heisenberg group multiplication law (12.2.4). Moreover, the group law forces us to use nonisotropic dilations on H_n; i.e.,

$$\mathbf{x} \mapsto \delta \circ \mathbf{x} = \delta \circ [\mathbf{z}, t] = [\delta \mathbf{z}, \delta^2 t]$$

for all $\delta > 0$. These dilations are automorphisms of the group H_n:

$$\delta \circ (\mathbf{x} \cdot \mathbf{y}) = (\delta \circ \mathbf{x}) \cdot (\delta \circ \mathbf{y});$$

but the standard isotropic dilations of $\mathbb{R}^{2n+1}$ are not automorphisms of H_n.

A function f defined on H_n is said to be H-homogeneous of degree m on H_n if

$$f(\delta \circ [\mathbf{z}, t]) = f(\delta \mathbf{z}, \delta^2 t) = \delta^m f(\mathbf{z}, t)$$

for all $\delta > 0$. For example, the fundamental solution Ψ_m of $\mathcal{L}_\alpha^m$ is H-homogeneous of degree $-2n + 2m - 2$. Next we introduce the norm function ρ given by

$$\rho(\mathbf{x}) = \left(\|\mathbf{z}\|^4 + t^2\right)^{1/4}, \quad \text{where} \quad \|\mathbf{z}\|^2 = \sum_{j=1}^{n} a_j |z_j|^2. \tag{12.2.9}$$

Obviously, we have $\rho(\mathbf{x}^{-1}) = \rho(-\mathbf{x}) = \rho(\mathbf{x})$ and $\rho(\delta \circ \mathbf{x}) = \delta \rho(\mathbf{x})$. In addition, the function ρ satisfies the triangle inequality:

$$\rho(\mathbf{x} \cdot \mathbf{y}) \leq C_1 \{\rho(\mathbf{x}) + \rho(\mathbf{y})\}$$

for some universal constant C_1. The distance function, $d(\mathbf{x}, \mathbf{y})$ of points $\mathbf{x}, \mathbf{y} \in \mathrm{H}_n$, is defined to be

$$d(\mathbf{x}, \mathbf{y}) = \rho(\mathbf{y}^{-1} \cdot \mathbf{x}).$$

It is clear that $d(\mathbf{x}, \mathbf{y})$ satisfies the symmetric property $d(\mathbf{x}, \mathbf{y}) = d(\mathbf{y}, \mathbf{x})$.

Suppose that $K \in C^{\infty}(\mathrm{H}_n \setminus \{0\})$ is H-homogeneous of degree γ. Then K is locally integrable near the origin if $\gamma > -2n - 2$. See Folland and Stein [45] for the proof of this statement.

Definition 12.2.1. Let $K \in C^{\infty}(\mathrm{H}_n \setminus \{0\})$ be H-homogeneous of degree $-2n - 2$. Then K is said to have mean value zero property if

$$\int_{\rho(\mathbf{x})=1} K(\mathbf{x}) d\sigma(\mathbf{x}) = 0, \tag{12.2.10}$$

where $d\sigma(\mathbf{x})$ is the induced measure on the Heisenberg unit sphere $\rho(\mathbf{x}) = 1$.

Using Theorem 3 and Corollary 5.24 of Chap. XII in Stein [105], we can get the basic estimate concerning P.V convolution operators on H_n:

Theorem 12.2.2. *Let $K \in C^{\infty}(H_n \setminus \{0\})$, H-homogeneous of degree $-2n - 2$ with mean value zero property. Then K induces a principal value convolution operator, given by*

$$\mathbf{K}(f)(\mathbf{x}) = (f * K)(\mathbf{x}) = \lim_{\varepsilon \to 0} \int_{d(\mathbf{x},\mathbf{y})>\varepsilon} f(\mathbf{y}) K(\mathbf{y}^{-1} \cdot \mathbf{x}) dV(\mathbf{y}), \tag{12.2.11}$$

for $f \in C_0^{\infty}(H_n)$. Moreover, the operator $\mathbf{K}$ given by (12.2.11) can be extended to a bounded operator from the L^p-Sobolev space $L^p_k(H_n)$ into itself, for $1 < p < \infty$ and $k \in \mathbb{Z}^+$.

12.2.3 Laguerre Functions

Consider the generalized *Laguerre polynomials* $L_k^{(\alpha)}(x)$ defined by their usual generating function formula

$$\sum_{k=1}^{\infty} L_k^{(\alpha)}(x) w^k = \frac{1}{(1-w)^{\alpha+1}} \exp\left\{-\frac{xw}{1-w}\right\}, \text{ for } \alpha = 0, 1, 2, \ldots, \quad x \geq 0,$$

$$\text{and } |w| < 1. \tag{12.2.12}$$

Definition 12.2.3. Let $z = |z|e^{i\theta}$ and $k,\ p = 0, 1, 2, \ldots$. Then we define

$$\widetilde{\mathcal{W}}_k^{(p)}(z,\tau) = \frac{2|\tau|}{\pi}\left[\frac{\Gamma(k+1)}{\Gamma(k+p+1)}\right]^{1/2}(2|\tau||z|^2)^{p/2}e^{ip\theta}e^{-|\tau||z|^2}L_k^{(p)}(2|\tau||z|^2), \tag{12.2.13}$$

$$\widetilde{\mathcal{W}}_k^{(-p)}(z,\tau) = \frac{2|\tau|}{\pi}(-1)^p\left[\frac{\Gamma(k+1)}{\Gamma(k+p+1)}\right]^{1/2} \times (2|\tau||z|^2)^{p/2}e^{-ip\theta}e^{-|\tau||z|^2}L_k^{(p)}(2|\tau||z|^2). \tag{12.2.14}$$

12.2.4 Laguerre Calculus on H_1

The most important property of $\widetilde{\mathcal{W}}_k^{(p)}(z,\tau)$ is the following theorem of Greiner [57]:

Theorem 12.2.4. *Let $p,\ k,\ q,\ m = 1, 2, \ldots$. Then*

$$\begin{aligned}\widetilde{\mathcal{W}}^{(p-k)}_{(p\wedge k)-1} *_{|\tau|} \widetilde{\mathcal{W}}^{(q-m)}_{(q\wedge m)-1} &= \delta_k^{(q)}\cdot\widetilde{\mathcal{W}}^{(p-m)}_{(p\wedge m)-1},\\ \widetilde{\mathcal{W}}^{(p-k)}_{(p\wedge k)-1} *_{-|\tau|} \widetilde{\mathcal{W}}^{(q-m)}_{(q\wedge m)-1} &= \delta_m^{(p)}\cdot\widetilde{\mathcal{W}}^{(q-k)}_{(q\wedge k)-1},\end{aligned} \tag{12.2.15}$$

where $a\wedge b = \min(a,b)$ and $\delta_k^{(q)}$ denotes the Kronecker delta function; i.e., $\delta_k^{(q)} = 1$ if $k = q$ and vanishes otherwise.

Thus the twisted convolution of two functions of $\widetilde{\mathcal{W}}_k^{(p)}$ is another function of the same type. This surprising result justifies the use of Laguerre function expansion on the Heisenberg group in analogy with Mikhlin's use of the spherical harmonics on $\mathbb{R}^n$.

Let $\mathcal{W}_k^{(p)}(\mathbf{z},t),\ \pm\ p,\ k = 0, 1, 2, \ldots$, be the inverse Fourier transform of $\widetilde{\mathcal{W}}_k^{(p)}(\mathbf{z},\tau)$ with respect to τ i.e.,

$$\mathcal{W}_k^{(p)}(\mathbf{z},t) = \frac{1}{2\pi}\int_{-\infty}^{\infty} e^{it\tau}\widetilde{\mathcal{W}}_k^{(p)}(\mathbf{z},\tau)d\tau.$$

These are the kernels of the generalized Cauchy–Szegö operators on H_1. In particular,

$$\mathcal{W}_0^{(0)}(\mathbf{z},t) = \mathcal{W}_{+,0}^{(0)}(\mathbf{z},t)+\mathcal{W}_{-,0}^{(0)}(\mathbf{z},t) = S_+ + S_-, \quad \text{where } S_\pm = \frac{1}{\pi^2}\cdot\frac{1}{(|z|^2\mp it)^2} \tag{12.2.16}$$

denotes the Cauchy–Szegö kernels in H_1.

The following result implies that the generalized Cauchy–Szegö kernels indeed induce principal value convolution operators.

Theorem 12.2.5. *The generalized Cauchy–Szegö kernels $\mathcal{W}_k^{(p)}(\mathbf{z},t)$ are in C^∞ $(H_1 \setminus \{0\})$ and have zero mean value property.*

Formula (12.2.15) is reminiscent of matrix multiplication. To show this similarity, Greiner [57] introduced the Laguerre matrix:

Definition 12.2.6. We define the positive Laguerre matrix

$$\mathcal{M}_+\left(\widetilde{\mathcal{W}}^{(p-k)}_{(p\wedge k)-1}\right)$$

of $\widetilde{\mathcal{W}}^{(p-k)}_{(p\wedge k)-1}$ to be the infinite matrix which has one at the intersection of the pth row and kth column and zeros everywhere else. The negative Laguerre matrix $\mathcal{M}_-\left(\widetilde{\mathcal{W}}^{(p-k)}_{(p\wedge k)-1}\right)$ can be defined as the transpose of the positive Laguerre matrix.

Following this definition, Theorem 12.2.4 takes the following form:

$$\mathcal{M}_+\left(\widetilde{\mathcal{W}}^{(p-k)}_{(p\wedge k)-1} *_{|\tau|} \widetilde{\mathcal{W}}^{(q-m)}_{(q\wedge m)-1}\right) = \mathcal{M}_+\left(\widetilde{\mathcal{W}}^{(p-k)}_{(p\wedge k)-1}\right)\cdot \mathcal{M}_+\left(\widetilde{\mathcal{W}}^{(q-m)}_{(q\wedge m)-1}\right). \tag{12.2.17}$$

The Laguerre matrix for any left invariant convolution operator can be defined as in [57] and [17]. In particular, we can define the Laguerre matrix for any left invariant differential operator on H_1, since we can write it in the form of convolution operators. We omit the details here, and the interested reader can consult [17] and Berenstein–Chang–Tie [19]. We only list the following results:

Theorem 12.2.7. *If* **F** *and* **G** *are two P.V. convolution operators on* $\mathbf{H}_1$, *and* $\mathcal{M}(\tilde{\mathbf{F}})$ *and* $\mathcal{M}(\tilde{\mathbf{G}})$ *denote the Laguerre matrices of* **F** *and* **G**, *respectively, then*

$$\mathcal{M}(\widetilde{\mathbf{F}} *_\tau \widetilde{\mathbf{G}}) = \mathcal{M}(\widetilde{\mathbf{F}})\cdot\mathcal{M}(\widetilde{\mathbf{G}}) = \mathcal{M}_+(\widetilde{\mathbf{F}})\cdot\mathcal{M}_+(\widetilde{\mathbf{G}}) \oplus \mathcal{M}_-(\widetilde{\mathbf{F}})\cdot\mathcal{M}_-(\widetilde{\mathbf{G}}). \tag{12.2.18}$$

A simple consequence of this theorem is

Corollary 12.2.8. *Let* **I** *denote the identity operator on* $C_0^\infty(H_1)$. *Then* **I** *is induced by the identity Laguerre matrix* $\mathcal{W}_\pm(\widetilde{\mathbf{I}}) = (\delta_k^{(p)})$.

12.2.5 Laguerre Calculus on H_n

We define the n-dimensional version of the exponential *Laguerre functions* on H^n by the n-fold product:

$$\widetilde{\mathcal{W}}^{(\mathbf{p})}_{\mathbf{k}}(\mathbf{z},\tau) = \prod_{j=1}^{n} a_j \widetilde{\mathcal{W}}^{(p_j)}_{k_j}(\sqrt{a_j}z_j,\tau), \tag{12.2.19}$$

where $\widetilde{\mathcal{W}}^{(p_j)}_{k_j}(\sqrt{a_j}z_j,\tau)$ is given by (12.2.13).

Composing two functions of the type (12.2.19) via twisted convolution yields

$$\widetilde{\mathcal{W}}_{\mathbf{k}}^{(\mathbf{p})} *_\tau \widetilde{\mathcal{W}}_{\mathbf{m}}^{(\mathbf{q})} = \left[\prod_{j=1}^{n} a_j \widetilde{\mathcal{W}}_{k_j}^{(p_j)}(\sqrt{a_j} z_j, \tau)\right] *_\tau \left[\prod_{j=1}^{n} a_j \widetilde{\mathcal{W}}_{m_j}^{(q_j)}(\sqrt{a_j} z_j, \tau)\right]$$

$$= \prod_{j=1}^{n} a_j^2 \int_{\mathbb{R}^2} e^{2ia_j\tau \mathrm{Im}(z_j \bar{w}_j)} \widetilde{\mathcal{W}}_{k_j}^{(p_j)}(\sqrt{a_j}(z_j - w_j), \tau) \widetilde{\mathcal{W}}_{m_j}^{(q_j)}$$
$$\times (\sqrt{a_j} w_j, \tau) dw_j$$
$$= \prod_{j=1}^{n} a_j \widetilde{\mathcal{W}}_{k_j}^{(p_j)} *_\tau \widetilde{\mathcal{W}}_{m_j}^{(q_j)}(\sqrt{a_j} z_j, \tau).$$

Consequently, Theorem 12.2.4 implies that

Theorem 12.2.9. *Let k_j, p_j, m_j and $q_j = 1, 2, 3, \ldots$ for $j = 1, 2, \ldots, n$. Then*

$$\widetilde{\mathcal{W}}_{(\mathbf{k}\wedge\mathbf{p})-1}^{(\mathbf{p}-\mathbf{k})} *_{|\tau|} \widetilde{\mathcal{W}}_{(\mathbf{m}\wedge\mathbf{q})-1}^{(\mathbf{q}-\mathbf{m})} = \delta_{\mathbf{k}}^{(\mathbf{q})} \cdot \widetilde{\mathcal{W}}_{(\mathbf{p}\wedge\mathbf{m})-1}^{(\mathbf{p}-\mathbf{m})},$$
$$\widetilde{\mathcal{W}}_{(\mathbf{k}\wedge\mathbf{p})-1}^{(\mathbf{p}-\mathbf{k})} *_{-|\tau|} \widetilde{\mathcal{W}}_{(\mathbf{m}\wedge\mathbf{q})-1}^{(\mathbf{q}-\mathbf{m})} = \delta_{\mathbf{m}}^{(\mathbf{p})} \cdot \widetilde{\mathcal{W}}_{(\mathbf{q}\wedge\mathbf{k})-1}^{(\mathbf{q}-\mathbf{k})}, \tag{12.2.20}$$

where $\mathbf{k} = (k_1, k_2, \ldots, k_n)$, $\mathbf{p} = (p_1, p_2, \ldots, p_n)$, $\mathbf{m} = (m_1, m_2, \ldots, m_n)$ and $\mathbf{q} = (q_1, q_2, \ldots, q_n)$. Here

$$(\mathbf{k} \wedge \mathbf{p}) - \mathbf{1} = (\min(p_1, k_1) - 1, \ldots, \min(p_n, k_n) - 1)$$

and $\delta_{\mathbf{k}}^{(\mathbf{p})} = \prod_{j=1}^{n} \delta_{k_j}^{(p_j)}$ is the n-fold Kronecker delta function.

Instead of the Laguerre matrix, Beals et al. [17] introduced the definition of a Laguerre tensor for the convolution operators on H_n. We omit the details here.

Then we have the n-fold version of Theorem 12.2.7:

Theorem 12.2.10. *(The Laguerre calculus on H_n) Let $\mathbf{F}$ and $\mathbf{G}$ induce the convolution operators on H_n. Let $\mathcal{M}(\widetilde{\mathbf{F}})$ and $\mathcal{M}(\widetilde{\mathbf{G}})$ denote the Laguerre tensors of $\mathbf{F}$ and $\mathbf{G}$, respectively. Then $\mathcal{M}(\widetilde{\mathbf{F}} *_\tau \widetilde{\mathbf{G}}) = \mathcal{M}(\widetilde{\mathbf{F}}) \cdot \mathcal{M}(\widetilde{\mathbf{G}})$.*

Corollary 12.2.11. *The identity operator $\mathbf{I}$ on $C_0^\infty(H_n)$ is induced by the identity Laguerre tensor:*

$$\mathcal{M}_\pm(\widetilde{\mathbf{I}}) = (\delta_{k_1}^{(p_1)} \cdots \delta_{k_n}^{(p_n)}).$$

12.2.6 Left Invariant Differential Operators

A *left invariant differential operator* $\mathcal{P}$ on H_n is a polynomial $\mathcal{P}(\mathbf{X}, \mathbf{Y}, \mathbf{T})$ with constant coefficients, or in complex coordinates, a polynomial in vector fields $\mathbf{T}, \mathbf{Z}_j$, and $\bar{\mathbf{Z}}_j$. We can have the following representation for $\mathcal{P}$ as a convolution operator on H_n:

$$\mathcal{P} = \mathcal{P}\mathbf{I} = \sum_{|\mathbf{k}|=0}^{\infty} \mathcal{P}\mathcal{W}_{k_1,\ldots,k_n}^{(0,\ldots,0)} *, \quad \text{where} \quad \mathbf{I} = \sum_{|\mathbf{k}|=0}^{\infty} \mathcal{W}_{k_1,\ldots,k_n}^{(0,\ldots,0)} * \tag{12.2.21}$$

is the identity operator on $C_0^\infty(\mathrm{H}_n)$. In particular, $\mathbf{T}$, $\mathbf{Z}_j$, and $\bar{\mathbf{Z}}_j$, $j = 1, 2, \ldots, n$, can be represented as convolution operators and written in the Laguerre tensor forms. This is contained in the next proposition:

Proposition 12.2.12. *(1) $\mathcal{M}(\widetilde{\mathbf{T}})$ is the $i\tau$ multiple of the identity Laguerre tensor:*

$$\mathcal{M}_\pm(\widetilde{\mathbf{T}}) = i\tau(\delta_{k_1}^{(p_1)} \ldots \delta_{k_n}^{(p_n)}).$$

(2) $\mathbf{Z}_j$, $j = 1, 2, \ldots, n$, has the following Laguerre tensor representation:

$$\mathcal{M}(\widetilde{\mathbf{Z}}_j) = \mathcal{M}_+(\widetilde{\mathbf{Z}}_j) \oplus \mathcal{M}_-(\widetilde{\mathbf{Z}}_j), \quad \textit{where}$$

$$\mathcal{M}_-(\widetilde{\mathbf{Z}}_j)_{k_1,\ldots,k_n}^{(p_1,\ldots,p_n)} = \sqrt{2a_j p_j |\tau|}\delta_{k_1}^{(p_1)} \ldots \delta_{k_j}^{(p_j+1)} \ldots \delta_{k_n}^{(p_n)}$$

$$\textit{and } \mathcal{M}_+(\widetilde{\mathbf{Z}}_j) = \mathcal{M}_-(\widetilde{\mathbf{Z}}_j)^t.$$

(3) $\bar{\mathbf{Z}}_j$, $j = 1, 2, \ldots, n$, has the following Laguerre tensor representation: $\mathcal{M}(\widetilde{\bar{\mathbf{Z}}}_j) = -\mathcal{M}(\widetilde{\mathbf{Z}}_j)^t$.

Theorem 12.2.13. *Let $\mathcal{P} = \mathcal{P}(\mathbf{Z}, \bar{\mathbf{Z}}, \mathbf{T}) = \mathcal{P}(\mathbf{Z}_1, \ldots, \mathbf{Z}_n, \bar{\mathbf{Z}}_1, \ldots, \bar{\mathbf{Z}}_n, \mathbf{T})$ denote a left invariant differential operator on H_n i.e., $\mathcal{P}$ is a polynomial in the vector fields $\mathbf{T}$, $\mathbf{Z}_j$, and $\bar{\mathbf{Z}}_j$, $j = 1, 2, \ldots, n$. Then*

$$\mathcal{M}(\widetilde{\mathcal{P}}) = \mathcal{P}(\mathcal{M}(\widetilde{\mathbf{Z}}), \mathcal{M}(\widetilde{\bar{\mathbf{Z}}}), i\tau), \tag{12.2.22}$$

where we set $\mathcal{M}(\widetilde{\mathbf{Z}}) = (\mathcal{M}(\widetilde{\mathbf{Z}}_1), \ldots, \mathcal{M}(\widetilde{\mathbf{Z}}_n))$ and $\mathcal{M}(\widetilde{\bar{\mathbf{Z}}}) = (\mathcal{M}(\widetilde{\bar{\mathbf{Z}}}_1), \ldots, \mathcal{M}(\widetilde{\bar{\mathbf{Z}}}_n))$.

Example 12.2.14. Assume that $n = 1$ and $a_1 = 1$. Then we have

$$\mathcal{M}_+(\widetilde{\mathbf{Z}}_1) = \sqrt{2|\tau|} \begin{bmatrix} 0 & \sqrt{1} & 0 & 0 & \cdots \\ 0 & 0 & \sqrt{2} & 0 & \cdots \\ 0 & 0 & 0 & \sqrt{3} & \cdots \\ \vdots & \vdots & \vdots & \vdots & \ddots \end{bmatrix},$$

and

$$\mathcal{M}_-(\widetilde{\mathbf{Z}}_1) = [\mathcal{M}_+(\widetilde{\mathbf{Z}}_1)]^t.$$

Now we may set

$$\mathcal{M}_+(K) = \frac{1}{\sqrt{2|\tau|}} \begin{bmatrix} 0 & 0 & 0 & 0 & \cdots \\ \frac{1}{\sqrt{1}} & 0 & 0 & 0 & \cdots \\ 0 & \frac{1}{\sqrt{2}} & 0 & 0 & \cdots \\ 0 & 0 & \frac{1}{\sqrt{3}} & 0 & \cdots \\ \vdots & \vdots & \vdots & \vdots & \ddots \end{bmatrix},$$

and

$$\mathcal{M}_-(K) = \left[\mathcal{M}_+(K)\right]^t.$$

Thus

$$\widetilde{K}_\pm(z,\tau) = \frac{1}{\sqrt{2|\tau|}} \sum_{k=0}^\infty \frac{1}{\sqrt{k+1}} \widetilde{\mathcal{W}}^{(1)}_{\pm,k}(z,\tau).$$

Using the definition of $\widetilde{W}^{(1)}_{\pm,k}(z,\tau)$, we sum the series

$$\widetilde{K}(z,\tau) = \frac{2|\tau| z e^{-|\tau||z|^2}}{\pi} \int_0^1 \sum_{k=0}^\infty r^k L_k^{(1)}(2|\tau||z|^2)\,dr.$$

Since

$$\sum_{k=0}^\infty r^k L_k^{(1)}(x) = \frac{e^x}{(1-r)^2} e^{-\frac{x}{1-r}},$$

therefore,

$$\widetilde{K}(z,\tau) = \frac{1}{\pi} \cdot \frac{e^{-|\tau||z|^2}}{\bar{z}},$$

and

$$K(z,t) = \frac{1}{2\pi^2 \bar{z}} \int_{-\infty}^{+\infty} e^{it\tau - |\tau||z|^2}\,d\tau = \frac{z}{\pi^2(|z|^4 + t^2)}.$$

Hence, we recover the Greiner, Kohn and Stein theorem [60]:

$$\mathbf{Z}_1 K = I - \mathcal{W}^{(0)}_{-,0} = I - S_-,$$

$$K\mathbf{Z}_1 = I - \mathcal{W}^{(0)}_{+,0} = I - S_+.$$

Here $S_\pm$ are the Cauchy–Szegö kernels which were defined by (12.2.16).

12.3 The Heisenberg Sub-Laplacian

The *Heisenberg sub-Laplacian* is the differential operator

$$\mathcal{L}_\alpha = -\frac{1}{2}\sum_{j=1}^{n}(\mathbf{Z}_j\bar{\mathbf{Z}}_j + \bar{\mathbf{Z}}_j\mathbf{Z}_j) + i\alpha\mathbf{T} \tag{12.3.23}$$

with $\mathbf{Z}_j$ and $\bar{\mathbf{Z}}_j$ given by (12.1.4). In the case of $a_j = 1$ for all j, the operator $\mathcal{L}_\alpha$ was first introduced by Folland and Stein [45] in the study of $\bar{\partial}_b$ complex on a nondegenerate CR manifold. They found the fundamental solution of $\mathcal{L}_\alpha$. Beals and Greiner [18] solved the case when the a_j are different.

Next we consider two operators related to $\mathcal{L}_\alpha$. The first one is the powers of $\mathcal{L}_\alpha$. In the special case of $a_j = 1/4$ for all $j = 1, 2, \ldots, n$ and $\alpha = 0$, $\mathcal{L}_0^m$, $1 \le m \le n$, has been studied by Dolley, Benson and Ratcliff [40]. If all the a_j are equal, the unitary group $\mathcal{U}(n)$ acts on H_n via

$$u[\mathbf{z}, t] = [u(\mathbf{z}), t] \quad \text{for} \quad u \in \mathcal{U}(n), \quad [\mathbf{z}, t] \in \mathrm{H}_n. \tag{12.3.24}$$

$\mathcal{L}_\alpha$ is invariant under the $\mathcal{U}(n)$-action. The key idea in [40] is to exploit this invariance by expanding the fundamental solution in terms of $\mathcal{U}(n)$-spherical functions $\phi_{\lambda,m}$ which are given by

$$\phi_{\lambda,m}(\mathbf{z}, t) = e^{i\lambda t}e^{-|\lambda||\mathbf{z}|^2/4}L_{m-1}^{(n-1)}(|\lambda||\mathbf{z}|^2/2)$$

with $\lambda \in \mathbb{R} \setminus \{0\}$, $m \in \mathbb{Z}^+$. Here $L_m^{(n-1)}$ denotes the generalized Laguerre polynomial.

However, the invariance (12.3.24) does not hold for the nonisotropic case. Instead, we shall apply the Laguerre calculus to solve the general case. Since the Laguerre tensor of $\mathcal{L}_\alpha^m$ is simply diagonal, its inverse is also simple and diagonal.

As usual, taking the partial Fourier transform with respect to the t-variable, one has

$$\widetilde{\mathcal{L}}_\alpha = \widetilde{\mathcal{L}}_\alpha\widetilde{\mathbf{I}} = \sum_{|\mathbf{k}|=0}^{\infty}\left[-\frac{1}{2}\sum_{j=1}^{n}(\widetilde{\mathbf{Z}}_j\widetilde{\bar{\mathbf{Z}}}_j + \widetilde{\bar{\mathbf{Z}}}_j\widetilde{\mathbf{Z}}_j) - \alpha\tau\right]\prod_{j=1}^{n} a_j\widetilde{\mathcal{W}}_{k_j}^{(0)}(\sqrt{a_j}z_j, \tau) *_\tau. \tag{12.3.25}$$

A simple calculation yields

$$-\frac{1}{2}(\widetilde{\mathbf{Z}}_j\widetilde{\bar{\mathbf{Z}}}_j + \widetilde{\bar{\mathbf{Z}}}_j\widetilde{\mathbf{Z}}_j)\widetilde{\mathcal{W}}_k^{(0)}(\sqrt{a_j}z_j, \tau) = (2k+1)|\tau|a_j\widetilde{\mathcal{W}}_k^{(0)}(\sqrt{a_j}z_j, \tau). \tag{12.3.26}$$

Thus (12.3.25) and (12.3.26) imply

$$\widetilde{\mathcal{L}}_\alpha = \sum_{|\mathbf{k}|=0}^{\infty}\left(\sum_{j=1}^{n}(2k_j+1)|\tau|a_j - \alpha\tau\right)\prod_{j=1}^{n} a_j\widetilde{\mathcal{W}}_{k_j}^{(0)}(\sqrt{a_j}z_j, \tau) *_\tau. \tag{12.3.27}$$

Consequently, the Laguerre tensor of the convolution operator induced by $\mathcal{L}_\alpha$ is

$$\mathcal{M}(\widetilde{\mathcal{L}_\alpha}) = |\tau| \left(\left[\sum_{j=1}^{n} (2k_j + 1)a_j - \alpha \mathrm{sgn}(\tau) \right] \delta_{k_1}^{(p_1)} \cdots \delta_{k_n}^{(p_n)} \right), \tag{12.3.28}$$

which is invertible as long as α does not belong to the exceptional set Λ_α, where

$$\Lambda_\alpha = \sum_{j=1}^{n} (2k_j + 1)a_j, \quad \mathbf{k} = (k_1, k_2, \ldots, k_n) \in (\mathbb{Z}_+)^n.$$

According to Theorem 12.2.10, the inverse Laguerre tensor of (12.3.28) is

$$\mathcal{M}(\widetilde{\mathcal{L}_\alpha^{-1}}) = |\tau|^{-1} \left(\left[\sum_{j=1}^{n} (2k_j + 1)a_j - \alpha \mathrm{sgn}(\tau) \right]^{-1} \delta_{k_1}^{(p_1)} \cdots \delta_{k_n}^{(p_n)} \right). \tag{12.3.29}$$

If we write it in the Laguerre series expansion:

$$\widetilde{\mathcal{L}_\alpha^{-1}}(\mathbf{z}, \tau) = |\tau|^{-1} \sum_{|\mathbf{k}|=0}^{\infty} \left(\sum_{j=1}^{n} (2k_j + 1)a_j - \alpha \mathrm{sgn}(\tau) \right)^{-1} \prod_{j=1}^{n} a_j \widetilde{\mathcal{W}}_{k_j}^{(0)}(\sqrt{a_j} z_j, \tau), \tag{12.3.30}$$

then we can sum this series in the sense of Abel, take the inverse Fourier transform with respect to τ, and find the fundamental solution of $\mathcal{L}_\alpha$. In fact, the above calculation makes sense for any polynomial of $\mathcal{L}_\alpha$. We will carry out this extension, and find the fundamental solution of $\mathcal{L}_\alpha^m$ and the Paneitz operator $\mathcal{L}_\alpha \bar{\mathcal{L}}_\alpha$, in the rest of this chapter.

Remark 12.3.1. Being concerned about any possible confusions regarding the formulas (12.3.25) and (12.3.27), we shall clarify the notation $*_\tau$ one more time. The operators $\mathcal{L}_\alpha$ and P_α are left invariant. So when they act on a function f, we may write $\mathcal{L}_\alpha(f)$ and $P_\alpha(f)$. Taking the partial Fourier transform with respect to the t-variable (whose dual variable is τ), one has $\widetilde{\mathcal{L}_\alpha} *_\tau \tilde{f}$, and $\widetilde{P_\alpha} *_\tau \tilde{f}$. Here $*_\tau$ means "twisted convolution," concept defined in this chapter. So, removing the function variable $\tilde{f}$, the operators become $\widetilde{\mathcal{L}_\alpha} *_\tau$ and $\widetilde{P_\alpha} *_\tau$. We may decompose these operators into Laguerre tensor expansions, which are given by (12.3.25) and (12.3.27).

Before going further, we shall say a few words about the exceptional set Λ_α of the operator $\mathcal{L}_\alpha$. Basically, we look at the exceptional set from a different point of view. From the calculations of this section we know that

$$\alpha \neq \pm \sum_{j=1}^{n} (2k_j + 1)a_j : \quad \mathbf{k} = (k_1, k_2, \ldots, k_n) \in \mathbb{Z}_+^n.$$

Let us consider the "real part" of the operator $\mathcal{L}_\alpha$:

$$\mathcal{L} = 4\mathcal{L}_0 = -\sum_{j=1}^{2n} X_j^2$$

of the Kohn Laplacian and the operator $T = \frac{\partial}{\partial t}$. Since $-\mathcal{L}$ and iT are essentially self-adjoint strongly commuting operators, there is a well-defined joint spectrum. We define the joint spectrum of the pair $(\mathcal{L}, iT)$ as the complement of the set of $(\lambda, \mu) \in \mathbb{C}^2$ for which there exists L^p-bounded operators A and B with

$$A(\lambda\mathbf{I} - \mathcal{L}) + B(\mu\mathbf{I} - iT) = \mathbf{I}.$$

This implies that the spectrum should be the set of $(\lambda, \mu) \in \mathbb{C}^2$ for which neither $\lambda\mathbf{I} - \mathcal{L}$ nor $\mu\mathbf{I} - iT$ is invertible. From the Laguerre calculus, we know that a convolution operator is invertible if and only if its Laguerre tensor is invertible. So we reduce the invertibility of an operator to that of its Laguerre tensor. In fact, we know that the Laguerre tensor of the operator $\lambda\mathbf{I} - \mathcal{L}$ is

$$\mathcal{M}(\lambda\widetilde{\mathbf{I}} - \widetilde{\mathcal{L}}) = \left(\left[\lambda - |\tau| \sum_{j=1}^{n} a_j(2k_1 + 1)\right] \delta_{k_1}^{(p_1)} \cdots \delta_{k_n}^{(p_n)} \right)$$

and the Laguerre tensor of the operator $\mu\mathbf{I} - iT$ is

$$\mathcal{M}(\mu\widetilde{\mathbf{I}} - i\widetilde{T}) = \left([\mu + \tau] \delta_{k_1}^{(p_1)} \cdots \delta_{k_n}^{(p_n)} \right).$$

Therefore,

$$(\lambda\mathbf{I} - \mathcal{L}) \quad \text{is invertible} \quad \Leftrightarrow \quad \lambda - |\tau| \sum_{j=1}^{n} a_j(2k_1 + 1) \neq 0$$

for all $\mathbf{k} \in (\mathbb{Z}_+)^n$, $\tau \in \mathbb{R}$ and

$$(\mu\mathbf{I} - iT) \quad \text{is invertible} \quad \Leftrightarrow \quad \mu + \tau \neq 0.$$

Hence the spectrum of $\mathcal{L}$ alone is the set of nonnegative numbers $\{\lambda \in \mathbb{R} : \lambda \geq 0\}$, the spectrum of iT is the set of real numbers $\mathbb{R}$, and the joint spectrum of $(\mathcal{L}, iT)$ is the union of

$$\left\{ (\lambda, \mu) \in \mathbb{C}^2 : \lambda = |\mu| \sum_{j=1}^{n} (2k_j + 1)a_j \ \text{ and } \ \mu \in \mathbb{R} \right\}$$

over the set $\mathbf{k} \in (\mathbb{Z}_+)^n$. We can also write the joint spectrum in the form

$$\sigma(\mathcal{L}, iT) = \bigcup_{\mathbf{k} \in (\mathbb{Z}_+)^n} \left\{ (\lambda, \mu) \in \mathbb{C}^2 : \lambda \geq 0, \ \mu = \frac{\pm\lambda}{\sum_{j=1}^{n} (2k_j + 1)a_j} \right\}. \tag{12.3.31}$$

The set (12.3.31) (see Fig. 12.1) is called the *Heisenberg brush* (see [104]).

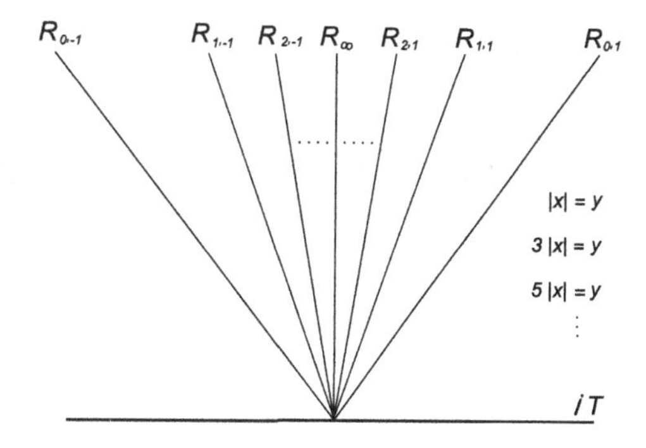

Fig. 12.1 The Heisenberg brush in the isotropic case $a_1 = \cdots = a_n = a$

12.4 Powers of the Sub-Laplacian

In this section we consider the powers of the Heisenberg sub-Laplacian $\mathcal{L}_\alpha^m$. For $1 \le m \le n$, this work has been done by Dolley, Benson and Ratcliff [40] in the case when all the a_j are equal. They used a different method.

First we will find the Laguerre tensor of the operator $\mathcal{L}_\alpha^m$ from the Laguerre tensor of $\mathcal{L}_\alpha$. Similarly to (12.3.25), we can take the Fourier transform with respect to t, and write $\widetilde{\mathcal{L}}_\alpha^m$ as a twisted convolution form:

$$\widetilde{\mathcal{L}}_\alpha^m = \widetilde{\mathcal{L}}_\alpha^m \widetilde{\mathbf{I}} = \sum_{|\mathbf{k}|=0}^{\infty} \left[-\frac{1}{2} \sum_{j=1}^{n} (\widetilde{\mathbf{Z}}_j \widetilde{\bar{\mathbf{Z}}}_j + \widetilde{\bar{\mathbf{Z}}}_j \widetilde{\mathbf{Z}}_j) - \alpha\tau \right]^m \prod_{j=1}^{n} a_j \widetilde{\mathcal{W}}_{k_j}^{(0)}(\sqrt{a_j} z_j, \tau) *_\tau . \tag{12.4.32}$$

Then (12.3.26) yields

$$\widetilde{\mathcal{L}}_\alpha^m = \sum_{|\mathbf{k}|=0}^{\infty} \left(\sum_{j=1}^{n} (2k_j + 1)|\tau| a_j - \alpha\tau \right)^m \prod_{j=1}^{n} a_j \widetilde{\mathcal{W}}_{k_j}^{(0)}(\sqrt{a_j} z_j, \tau) *_\tau . \tag{12.4.33}$$

Consequently, the Laguerre tensor of the convolution operator induced by $\mathcal{L}_\alpha^m$ is

$$\mathcal{M}(\widetilde{\mathcal{L}}_\alpha^m) = |\tau|^m \left(\left[\sum_{j=1}^{n} (2k_j + 1) a_j - \alpha \operatorname{sgn}(\tau) \right]^m \delta_{k_1}^{(p_1)} \cdots \delta_{k_n}^{(p_n)} \right), \tag{12.4.34}$$

which is invertible as long as α does not belong to the singular set Λ_α, where

$$\Lambda_\alpha = \left\{ \pm \sum_{j=1}^{n} (2k_j + 1) a_j : \quad \mathbf{k} = (k_1, k_2, \ldots, k_n) \in \mathbb{Z}_+^n \right\}.$$

According to Theorem 12.2.10, the inverse Laguerre tensor of (12.6.54) is

$$\mathcal{M}(\widetilde{\mathcal{L}}_\alpha^{-m}) = |\tau|^{-m}\left(\left[\sum_{j=1}^{n}(2k_j+1)a_j - \alpha\operatorname{sgn}(\tau)\right]^{-m}\delta_{k_1}^{(p_1)}\cdots\delta_{k_n}^{(p_n)}\right), \tag{12.4.35}$$

and we write its kernel $\widetilde{\Psi}_m(\mathbf{z},\tau)$ in the following Laguerre series expansion:

$$\widetilde{\Psi}_m(\mathbf{z},\tau) = |\tau|^{-m}\sum_{|\mathbf{k}|=0}^{\infty}\left(\sum_{j=1}^{n}(2k_j+1)a_j - \alpha\operatorname{sgn}(\tau)\right)^{-m}\prod_{j=1}^{n}a_j\widetilde{\mathcal{W}}_{k_j}^{(0)}(\sqrt{a_j}z_j,\tau). \tag{12.4.36}$$

To find the fundamental solution of $\mathcal{L}_\alpha^m$, we can sum this series and take the inverse Fourier transform with respect to τ. First we introduce the following integral representation of A^{-m}:

$$\frac{1}{A^m} = \frac{1}{\Gamma(m)}\int_0^\infty s^{m-1}e^{-As}\,ds \quad \text{for} \quad \operatorname{Re}(A) > 0. \tag{12.4.37}$$

Then, we can write (12.6.56) in the following form:

$$\widetilde{\Psi}_m(\mathbf{z},\tau) = \frac{|\tau|^{-m}}{\Gamma(m)}\sum_{|\mathbf{k}|=0}^{\infty}\int_0^\infty s^{m-1}e^{-(\sum_{j=1}^{n}(2k_j+1)a_j-\alpha\operatorname{sgn}(\tau))s}\,ds$$

$$\times\prod_{j=1}^{n}a_j\widetilde{\mathcal{W}}_{k_j}^{(0)}(\sqrt{a_j}z_j,\tau). \tag{12.4.38}$$

Next, we interchange the summation and the integration, and use the definition of $\widetilde{\mathcal{W}}_{k_j}^{(0)}$:

$$\begin{aligned}
&\widetilde{\Psi}_m(\mathbf{z},\tau)\\
&= \frac{|\tau|^{-m}}{\Gamma(m)}\int_0^\infty s^{m-1}\sum_{|\mathbf{k}|=0}^{\infty}e^{-(\sum_{j=1}^{n}(2k_j+1)a_j-\alpha\operatorname{sgn}(\tau))s}\,ds\prod_{j=1}^{n}a_j\widetilde{\mathcal{W}}_{k_j}^{(0)}(\sqrt{a_j}z_j,\tau)\\
&= \frac{|\tau|^{n-m}}{\pi^n\Gamma(m)}\int_0^\infty s^{m-1}\sum_{|\mathbf{k}|=0}^{\infty}e^{-(\sum_{j=1}^{n}(2k_j+1)a_j-\alpha\operatorname{sgn}(\tau))s}\,ds\\
&\quad\times\prod_{j=1}^{n}2a_je^{-a_j|\tau||z_j|^2}L_{k_j}^{(0)}(2a_j|\tau||z_j|^2)
\end{aligned}$$

$$= \frac{|\tau|^{n-m}}{\pi^n \Gamma(m)} \int_0^\infty s^{m-1} e^{\alpha \mathrm{sgn}(\tau) s} \prod_{j=1}^n 2a_j e^{-a_j s - a_j |\tau||z_j|^2}$$

$$\times \sum_{k_j=0}^\infty (e^{-2a_j s})^{k_j} L_{k_j}^{(0)}(2a_j|\tau||z_j|^2) ds,$$

Applying the generating formula for the Laguerre polynomials

$$\sum_{k=0}^\infty L_k^{(p)}(x) z^k = \frac{1}{(1-z)^{p+1}} \exp\left\{-\frac{xz}{1-z}\right\}$$

to the last formula for $\widetilde{\Psi}_m(\mathbf{z}, \tau)$, we obtain

$$\widetilde{\Psi}_m(\mathbf{z}, \tau) = \frac{|\tau|^{n-m}}{\pi^n \Gamma(m)} \int_0^\infty s^{m-1} e^{\alpha \mathrm{sgn}(\tau) s} \prod_{j=1}^n \frac{2a_j e^{-a_j s}}{1 - e^{-2a_j s}}$$

$$\times \exp\left\{-a_j |\tau||z_j|^2 \left[1 + \frac{2e^{-2a_j s}}{1 - e^{-2a_j s}}\right]\right\} ds$$

$$= \frac{|\tau|^{n-m}}{\pi^n \Gamma(m)} \int_0^\infty s^{m-1} e^{\alpha \mathrm{sgn}(\tau) s} \left[\prod_{j=1}^n \frac{a_j}{\sinh(a_j s)}\right]$$

$$\times \exp\left\{-|\tau| \sum_{j=1}^n a_j |z_j|^2 \coth(a_j s)\right\} ds.$$

Letting $\gamma(s; \mathbf{z}) = \sum_{j=1}^n a_j |z_j|^2 \coth(a_j s)$ and taking the inverse Fourier transform with respect to τ, we obtain for any $0 \le m \le n$

$$\Psi_m(\mathbf{z}, t) = \frac{\Gamma(n-m+1)}{2\pi^{n+1}\Gamma(m)} \left[\int_0^\infty s^{m-1} e^{\alpha s} \left(\prod_{j=1}^n \frac{a_j}{\sinh(a_j s)}\right) \frac{ds}{[\gamma(s; \mathbf{z}) - it]^{n-m+1}}\right.$$

$$\left. + \int_0^\infty s^{m-1} e^{-\alpha s} \left(\prod_{j=1}^n \frac{a_j}{\sinh(a_j s)}\right) \frac{ds}{[\gamma(s; \mathbf{z}) + it]^{n-m+1}}\right] \tag{12.4.39}$$

$$= \frac{\Gamma(n-m)}{2\pi^{n+1}\Gamma(m)} \int_{-\infty}^\infty \left(\prod_{j=1}^n \frac{a_j}{\sinh(a_j s)}\right) \frac{e^{\alpha s} s^{m-1} ds}{[\gamma(s; \mathbf{z}) - it]^{n-m+1}}. \tag{12.4.40}$$

In the above calculation, we applied the following integral formula:

$$\int_{-\infty}^{\infty} |\tau|^{n-m} e^{\alpha \operatorname{sgn}(\tau)s + it\tau - |\tau|\gamma(s;\mathbf{z})} d\tau = \frac{\Gamma(n-m+1)e^{\alpha s}}{[\gamma(s;\mathbf{z}) - it]^{n-m+1}} + \frac{\Gamma(n-m+1)e^{-\alpha s}}{[\gamma(s;\mathbf{z}) + it]^{n-m+1}} \tag{12.4.41}$$

to get (12.4.39), and then we substituted s with $-s$ in the second integral of (12.4.39) to obtain (12.4.40) using $\gamma(-s;\mathbf{z}) = -\gamma(s;\mathbf{z})$.

Following [16] and [28], we introduce the complex distance and volume element on the Heisenberg group:

$$g(s;\mathbf{z},t) = \gamma(2s;\mathbf{z}) - it \quad \text{and} \quad v(s) = \prod_{j=1}^{n} \frac{2a_j}{\sinh(2a_j s)}. \tag{12.4.42}$$

We can write (12.4.40) in the closed form

$$\Psi_m(\mathbf{z},t) = \frac{2^m(n-m)!}{(2\pi)^{n+1}\Gamma(m)} \int_{-\infty}^{\infty} e^{2\alpha s} s^{m-1} \frac{v(s)ds}{[g(s;\mathbf{z},t)]^{n-m+1}}. \tag{12.4.43}$$

When $|\mathbf{z}| = 0$, $t \neq 0$, then $g(s;\mathbf{z},t) = -it$. The integrand of (12.4.43) is not integrable at $s = 0$. To regularize the integration, we must deform its path of integration from $(-\infty, \infty)$ to

$$(-\infty + i\varepsilon \operatorname{sgn} t, \infty + i\varepsilon \operatorname{sgn} t), \quad \text{where} \quad 0 < \varepsilon < \min_{1 \le j \le n} \frac{\pi}{2a_j}.$$

We refer to [18] for the exact definition of this path. Finally, we have the formula

$$\Psi_m(\mathbf{z},t) = \frac{2^m(n-m)!}{(2\pi)^{n+1}\Gamma(m)} \int_{-\infty+i\varepsilon \operatorname{sgn} t}^{\infty+i\varepsilon \operatorname{sgn} t} e^{2\alpha s} s^{m-1} \frac{v(s)ds}{[g(s;\mathbf{z},t)]^{n-m+1}}.$$

Remark 12.4.1. As we noted in the introduction, if all the a_j are equal, the unitary group $\mathcal{U}(n)$ acts on H_n via

$$u(\mathbf{z},t) = (u\mathbf{z},t) \quad \text{for} \quad u \in \mathcal{U}(n), \quad (\mathbf{z},t) \in \mathrm{H}_n.$$

The operator $\mathcal{L}_\alpha^m$ is invariant under the $\mathcal{U}(n)$-action. The key idea in [40] is to exploit this invariance.

12.5 Heat Kernel for the Operator $\mathcal{L}_\alpha^m$

In this section we shall compute the kernel of the fundamental solution for the powers of the *Kohn Laplacian* via the heat kernel $h_s(\mathbf{z},t) = \exp\{-s\mathcal{L}_\alpha\}\delta_0$. In the isotropic case, the heat kernel was independently studied by Gaveau [48] via the probability method and Hulanicki [68] using the Fourier transform on H_n and the basis of Laguerre functions. Later, Beals and Greiner [18] solved the general case by a different method. We will see that $h_s(\mathbf{z}, t)$ can be obtained easily using the Laguerre calculus. Our approach is more closely related to the method of Hulanicki.

We first take the Fourier transform with respect to the t-variable and write the heat kernel $\widetilde{h}_s(\mathbf{z}, t)$ as a twisted convolution operator:

$$\widetilde{h}_s(\mathbf{z},\tau) = \exp\{-s\widetilde{\mathcal{L}}_\alpha\}\widetilde{\mathbf{I}} = \sum_{|\mathbf{k}|=0}^{\infty} \exp\{-s\widetilde{\mathcal{L}}_\alpha\}\left[\prod_{j=1}^{n} a_j \widetilde{\mathcal{W}}_{k_j}^{(0)}(\sqrt{a_j}z_j,\tau)\right]$$

$$= \sum_{|\mathbf{k}|=0}^{\infty} e^{-s\sum_{j=1}^{n} a_j|\tau|(2k_j+1)+s\alpha\tau} \prod_{j=1}^{n} a_j \widetilde{\mathcal{W}}_{k_j}^{(0)}(\sqrt{a_j}z_j,\tau).$$

Next, a similar computation as in the last section leads to

$$\widetilde{h}_s(\mathbf{z},\tau) = \frac{e^{\alpha\tau s}}{\pi^n}\left[\prod_{j=1}^{n} \frac{a_j|\tau|}{\sinh(a_j|\tau|s)}\right] \exp\left\{-|\tau|\sum_{j=1}^{n} a_j|z_j|^2\coth(a_j|\tau|s)\right\}. \tag{12.5.44}$$

Since

$$\prod_{j=1}^{n} \frac{a_j|\tau|}{\sinh(a_j|\tau|s)} = \prod_{j=1}^{n} \frac{a_j\tau}{\sinh(a_j\tau s)} \quad \text{and} \quad |\tau|\coth(a_j|\tau|s) = \tau\coth(a_j\tau s),$$

we can simplify (12.5.44) by removing the absolute sign for τ and have

$$\widetilde{h}_s(\mathbf{z},\tau) = \frac{e^{\alpha\tau s}}{\pi^n}\left[\prod_{j=1}^{n} \frac{a_j\tau}{\sinh(a_j\tau s)}\right] e^{-\tau\gamma(\tau s;\mathbf{z})}, \tag{12.5.45}$$

where, as before, $\gamma(\tau s;\mathbf{z}) = \sum_{j=1}^{n} a_j|z_j|^2\coth(a_j\tau s)$.

Now we take the inverse Fourier transform with respect to the τ-variable and obtain the heat kernel in the integral form

$$h_s(\mathbf{z},t) = \frac{1}{2\pi^{n+1}}\int_{-\infty}^{+\infty}\left[\prod_{j=1}^{n} \frac{a_j\tau}{\sinh(a_j\tau s)}\right] e^{\alpha\tau s+it\tau-\tau\gamma(\tau s;\mathbf{z})}d\tau$$

$$= \frac{1}{2(\pi s)^{n+1}}\int_{-\infty}^{+\infty}\left[\prod_{j=1}^{n} \frac{a_j\tau}{\sinh(a_j\tau)}\right] e^{\alpha\tau+i\frac{\tau}{s}t-\frac{\tau}{s}\gamma(\mathbf{z},\tau)}d\tau.$$

We substitute τ by 2τ and rewrite the heat kernel in terms of the complex distance g and volume element ν of the Heisenberg group:

$$h_s(\mathbf{z},t) = \frac{1}{(\pi s)^{n+1}} \int_{-\infty}^{+\infty} \nu(\tau)\tau^n e^{2\alpha\tau - \frac{2\tau}{s} g(\tau;\mathbf{z},t)} d\tau. \tag{12.5.46}$$

Now we may combine the heat kernel $h_s(\mathbf{z},t)$ and the formula

$$\lambda^{-m} = \frac{1}{\Gamma(m)} \int_0^\infty s^{m-1} e^{-s\lambda} ds, \quad \mathrm{Re}(\lambda) > 0 \quad \text{and} \quad \mathrm{Re}(m) > 0$$

to give another derivation of the fundamental solution of $\mathcal{L}_\alpha^m$, $0 \le m \le n$. We can first write the inverse formally as

$$\begin{aligned}\mathcal{L}_\alpha^{-m}(\mathbf{z},t) &= \frac{1}{\Gamma(m)} \int_0^\infty s^{m-1} e^{-s\mathcal{L}_\alpha} ds \\ &= \frac{1}{\pi^{n+1}\Gamma(m)} \int_0^\infty s^{m-n-2} \int_{-\infty}^{+\infty} \nu(\tau)\tau^n e^{2\alpha\tau - \frac{2\tau}{s} g(\tau;\mathbf{z},t)} d\tau ds.\end{aligned}$$

By Fubini's theorem, if $\mathbf{z} \neq \mathbf{0}$, we may interchange the order of integration. Moreover, if $\mathrm{Re}(m) > 0$, one has

$$\int_0^\infty s^{m-n} e^{-\frac{\alpha}{s}} \frac{ds}{s^2} = \int_0^\infty s^{n-m} e^{-\alpha s} ds = \frac{\Gamma(n-m+1)}{\alpha^{n-m+1}} = \frac{(n-m)!}{\alpha^{n-m+1}},$$

so that we may rewrite $\mathcal{L}_\alpha^{-m}(\mathbf{z},t)$ as

$$\Psi_m(\mathbf{z},t) = \frac{2^m(n-m)!}{(2\pi)^{n+1}\Gamma(m)} \int_{-\infty}^\infty \frac{\nu(\tau)\tau^{m-1}e^{2\alpha\tau}}{[g(\tau;\mathbf{z},t)]^{n-m+1}} d\tau. \tag{12.5.47}$$

This is exactly (12.4.43) obtained in the last section. When $m > n$, we cannot apply Fubini's theorem since the integral in the variable s diverges as $s \to \infty$.

Similarly, we may consider the kernel induced by the operator $\Psi_\beta(\mathbf{z},t) = \mathcal{L}_\alpha^{i\beta}\delta_0$. We may combine the heat kernel in (12.5.46) with the following formula:

$$w^{i\beta} = \frac{1}{\Gamma(-i\beta)} \int_0^\infty s^{-i\beta-1} e^{-sw} ds, \qquad \mathrm{Re}(w) > 0 \quad \text{and} \quad \mathrm{Im}(\beta) > 0.$$

This leads to

$$\mathcal{L}_\alpha^{i\beta}(\mathbf{z},t) = \frac{\Gamma(n+i\beta+1)}{2^{n+i\beta+1}\Gamma(-i\beta)} \int_{-\infty}^\infty \frac{\nu(s)s^{-i\beta-1}e^{2\alpha\tau}}{[g(s;\mathbf{z},t)]^{n+i\beta+1}} ds. \tag{12.5.48}$$

By analytic continuation, formula (12.5.48) remains true in the case of real β. This operator was studied by Müller and Stein [93] in the case of $a_j = 1$ for all $j = 1, \ldots, n$ and $\alpha = 0$. They have shown that the weak-type (1,1)-norm of the operator $\mathcal{L}_0^{i\beta}$ is bounded from below by $C|\beta|^{(2n+1)/2}$ as $|\beta| \to \infty$.

We shall mention another application of formulas (12.5.47) and (12.4.43). Replacing m in (12.5.47) by $\frac{1}{2}$, we obtain the kernel for $\mathcal{L}_0^{-1/2}$. Then kernels for the Riesz transforms $\mathcal{R}_j = \mathbf{Z}_j\mathcal{L}_0^{-\frac{1}{2}}$ and $\mathcal{R}_{n+j} = \bar{\mathbf{Z}}_j\mathcal{L}_0^{-\frac{1}{2}}$, $j = 1, \ldots, n$, which were first defined by Christ and Geller [36], can be obtained by composing $\mathbf{Z}_j$ or $\bar{\mathbf{Z}}_j$ to $\mathcal{L}_0^{-12}$. Here are the formulas:

$$\mathbf{Z}_j\mathcal{L}_0^{-\frac{1}{2}} = \frac{\Gamma(n+\frac{1}{2})a_j\bar{z}_j}{2^{n+\frac{1}{2}}\pi^{n+\frac{3}{2}}}\int_{-\infty}^{\infty}\frac{v(s)s^{n+1}[\coth(2a_js)+1]}{[sg(s;\mathbf{z},t)]^{n+3/2}}ds \tag{12.5.49}$$

and

$$\bar{\mathbf{Z}}_j\mathcal{L}_0^{-\frac{1}{2}} = \frac{\Gamma(n+\frac{1}{2})a_jz_j}{2^{n+\frac{1}{2}}\pi^{n+\frac{3}{2}}}\int_{-\infty}^{\infty}\frac{v(s)s^{n+1}[\coth(2a_js)-1]}{[sg(s;\mathbf{z},t)]^{n+3/2}}ds. \tag{12.5.50}$$

It is easy to see that these kernels are H-homogeneous of degree $-2n-2$ and satisfy the mean value zero property because the factor $\bar{z}_j$ or z_j appears (see the next section for details).

12.6 Fundamental Solution of the Paneitz Operator

In this section, we consider the Paneitz operator $\mathcal{P}_\alpha$ on the Heisenberg group, which is defined as

$$\mathcal{P}_\alpha = \mathcal{L}_\alpha\bar{\mathcal{L}}_\alpha = \frac{1}{4}\left[\sum_{j=1}^{n}(\mathbf{Z}_j\bar{\mathbf{Z}}_j + \bar{\mathbf{Z}}_j\mathbf{Z}_j)\right]^2 + \alpha^2\mathbf{T}^2. \tag{12.6.51}$$

We shall first recall some background about the Paneitz operator. Consider the coframe $\{\theta, \omega_j, \bar{\omega}_j\}_{j=1}^n$ dual to $\{\mathbf{T}, \mathbf{Z}_j, \bar{\mathbf{Z}}_j\}_{j=1}^n$. Let $\varphi \in C_0^\infty(\mathrm{H}_n)$ be a smooth function with compact support. It is easy to see that

$$\mathcal{L}_n\varphi = -\frac{1}{2}\sum_{j=1}^{n}(\mathbf{Z}_j\bar{\mathbf{Z}}_j + \bar{\mathbf{Z}}_j\mathbf{Z}_j) + (in\mathbf{T})(\varphi) = -\sum_{j=1}^{n}\bar{\mathbf{Z}}_j\mathbf{Z}_j(\varphi).$$

It follows that

$$
\begin{aligned}
\mathcal{P}_n\varphi &= \mathcal{L}_n\bar{\mathcal{L}}_n\varphi \\
&= \frac{1}{4}\Bigg[\sum_{j,k=1}^{n}(\mathbf{Z}_j\bar{\mathbf{Z}}_j+\bar{\mathbf{Z}}_j\mathbf{Z}_j)+2in\mathbf{T}\Bigg]\Big[(\mathbf{Z}_k\bar{\mathbf{Z}}_k+\bar{\mathbf{Z}}_k\mathbf{Z}_k)-2in\mathbf{T}\Big]\varphi \\
&= \left\{\frac{1}{4}\Bigg[\sum_{j=1}^{n}(\mathbf{Z}_j\bar{\mathbf{Z}}_j+\bar{\mathbf{Z}}_j\mathbf{Z}_j)\Bigg]^2+n^2\mathbf{T}^2\right\}\varphi \\
&= 4\sum_{k=1}^{n}\bar{\mathbf{Z}}_k(P_k\varphi).
\end{aligned}
$$

Here (see [87])

$$
P_k\varphi=\sum_{j=1}^{n}\mathbf{Z}_k(\mathbf{Z}_j\bar{\mathbf{Z}}_j\varphi),\quad k=1,\dots,n,
$$

and

$$
P\varphi=\sum_{k=1}^{n}(P_k\varphi)\omega_k
$$

is an operator that characterizes the *CR-pluriharmonic functions*. Note that a smooth real-valued function u on $\mathbf{H}_n$ is said to be a CR-pluriharmonic function if for any point $p\in\mathbf{H}_n$, there are an open neighborhood U of p in $\mathbf{H}_n$ and a smooth real-valued function v on U such that $\bar{\partial}_b(u+iv)=0$. Here $\bar{\partial}_b u=\sum_{j=1}^{n}(\bar{\mathbf{Z}}_j u)\bar{\omega}_j$.

Moreover, one has

$$
-\int_{\mathbf{H}_n}\langle P\varphi+\bar{P}\varphi, d_b\varphi\rangle\,d\mu=\int_{\mathbf{H}_n}\mathcal{P}_n\varphi\cdot\varphi\,d\mu,\qquad \forall\varphi\in C_0^\infty(\mathbf{H}_n).
$$

Here

$$
d_b=\pi\circ d : C^\infty(\mathbf{H}_n)\to\operatorname{span}\{\omega_j,\bar{\omega}_j\}_{j=1}^{n}
$$

and π is the orthogonal projection onto the subspace $\operatorname{span}\{\omega_j,\bar{\omega}_j\}_{j=1}^{n}$ in the cotangent space. Moreover, a smooth real-valued function $\varphi\in L^2(\mathbf{H}_n)$ satisfies $\mathcal{P}_n\varphi=0$ on $\mathbf{H}_n$ if and only if $P\varphi=0$ on $\mathbf{H}_n$.

Note that via the CR-Paneitz operator $\mathcal{P}_n$, we are able to get a sub-gradient estimate and establish Liouville-type theorems of the CR-heat equation on $\mathbf{H}_n$ (see [32] and [34]). For complex geometric aspects about these operators, the reader is referred to Graham and Lee [56], Hirachi [64] and Lee [87].

12.6.1 Laguerre Tensor of the Paneitz Operator

First we will find the Laguerre tensor of the operator $\mathcal{P}_\alpha$ from the Laguerre tensor of $\mathcal{L}_\alpha$. Similarly to (12.3.25), we can take the Fourier transform with respect to t, and write $\widetilde{\mathcal{P}}_\alpha$ as a twisted convolution form:

$$\begin{aligned}\widetilde{\mathcal{P}}_\alpha &= \widetilde{\mathcal{L}}_\alpha \widetilde{\overline{\mathcal{L}}}_\alpha \widetilde{\mathbf{I}} \\ &= \sum_{|\mathbf{k}|=0}^{\infty} \left[-\frac{1}{2} \sum_{j=1}^{n} (\widetilde{\mathbf{Z}}_j \widetilde{\overline{\mathbf{Z}}}_j + \widetilde{\overline{\mathbf{Z}}}_j \widetilde{\mathbf{Z}}_j) - \alpha\tau \right] \left[-\frac{1}{2} \sum_{j=1}^{n} (\widetilde{\mathbf{Z}}_j \widetilde{\overline{\mathbf{Z}}}_j + \widetilde{\overline{\mathbf{Z}}}_j \widetilde{\mathbf{Z}}_j) + \alpha\tau \right] \\ &\quad \times \prod_{j=1}^{n} a_j \widetilde{\mathcal{W}}_{k_j}^{(0)}(\sqrt{a_j} z_j, \tau) *_\tau. \end{aligned} \tag{12.6.52}$$

Then (12.3.26) yields

$$\widetilde{\mathcal{P}}_\alpha = \sum_{|\mathbf{k}|=0}^{\infty} \left\{ \left[\sum_{j=1}^{n} (2k_j + 1)|\tau| a_j \right]^2 - \alpha^2 \tau^2 \right\} \prod_{j=1}^{n} a_j \widetilde{\mathcal{W}}_{k_j}^{(0)}(\sqrt{a_j} z_j, \tau) *_\tau. \tag{12.6.53}$$

Consequently, the Laguerre tensor of the convolution operator induced by $\mathcal{P}_\alpha$ is

$$\mathcal{M}(\widetilde{\mathcal{P}}_\alpha) = \tau^2 \left(\left[\left(\sum_{j=1}^{n} (2k_j + 1) a_j \right)^2 - \alpha^2 \right] \delta_{k_1}^{(p_1)} \cdots \delta_{k_n}^{(p_n)} \right), \tag{12.6.54}$$

which is invertible as long as α does not belong to the exceptional set Λ_α, where

$$\Lambda_\alpha = \left\{ \pm \sum_{j=1}^{n} (2k_j + 1) a_j; \quad \mathbf{k} = (k_1, k_2, \ldots, k_n) \in \mathbb{Z}_+^n \right\}.$$

According to Theorem 12.2.10, the inverse Laguerre tensor of (12.6.54) is

$$\mathcal{M}(\widetilde{\mathcal{P}}_\alpha^{-1}) = \tau^{-2} \left(\left[\left(\sum_{j=1}^{n} (2k_j + 1) a_j \right)^2 - \alpha^2 \right]^{-1} \delta_{k_1}^{(p_1)} \cdots \delta_{k_n}^{(p_n)} \right), \tag{12.6.55}$$

and we write its kernel $\widetilde{\Psi}_\alpha(\mathbf{z}, \tau)$ in the Laguerre series expansion:

$$\widetilde{\Psi}_\alpha(\mathbf{z}, \tau) = \tau^{-2} \sum_{|\mathbf{k}|=0}^{\infty} \left[\left(\sum_{j=1}^{n} (2k_j + 1) a_j \right)^2 - \alpha^2 \right]^{-1} \prod_{j=1}^{n} a_j \widetilde{\mathcal{W}}_{k_j}^{(0)}(\sqrt{a_j} z_j, \tau). \tag{12.6.56}$$

In order to find the fundamental solution of $\mathcal{P}_\alpha$, we may sum this series and take the inverse partial Fourier transform with respect to the τ-variable. Consider the integral representation of A^{-1} given by (12.6.57):

$$\frac{1}{A} = \frac{1}{\Gamma(m)} \int_0^\infty e^{-As} ds \quad \text{for} \quad \mathrm{Re}(A) > 0. \tag{12.6.57}$$

Let $A(\mathbf{k}) = \sum_{j=1}^n (2k_j + 1)a_j$. Note that

$$\frac{1}{A^2(\mathbf{k}) - \alpha^2} = \frac{1}{2\alpha}\left(\frac{1}{A(\mathbf{k}) - \alpha} - \frac{1}{A(\mathbf{k}) + \alpha}\right).$$

Assume that $|\alpha| < a_1 + a_2 + \cdots + a_n$. Then we apply (12.6.57) and write (12.6.56) in the following form:

$$\widetilde{\Psi}_\alpha(\mathbf{z}, \tau) = \frac{1}{2\alpha\tau^2} \sum_{|\mathbf{k}|=0}^\infty \int_0^\infty \left[e^{-(\sum_{j=1}^n (2k_j+1)a_j - \alpha)s} - e^{-(\sum_{j=1}^n (2k_j+1)a_j + \alpha)s}\right] ds$$
$$\times \prod_{j=1}^n a_j \widetilde{\mathcal{W}}_{k_j}^{(0)}(\sqrt{a_j} z_j, \tau). \tag{12.6.58}$$

Next, we interchange the summation and integration, and use the definition of $\widetilde{\mathcal{W}}_{k_j}^{(0)}$:

$$\widetilde{\Psi}_\alpha(\mathbf{z}, \tau) = \frac{1}{2\alpha|\tau|^2} \int_0^\infty \sum_{|\mathbf{k}|=0}^\infty \left[e^{-(\sum_{j=1}^n (2k_j+1)a_j - \alpha)s} - e^{-(\sum_{j=1}^n (2k_j+1)a_j + \alpha)s}\right]$$
$$\times \prod_{j=1}^n a_j \widetilde{\mathcal{W}}_{k_j}^{(0)}(\sqrt{a_j} z_j, \tau)$$
$$= \frac{|\tau|^{n-2}}{2\alpha\pi^n} \int_0^\infty \sum_{|\mathbf{k}|=0}^\infty \left[e^{-(\sum_{j=1}^n (2k_j+1)a_j - \alpha)s} - e^{-(\sum_{j=1}^n (2k_j+1)a_j + \alpha)s}\right] ds,$$
$$\times \prod_{j=1}^n 2a_j e^{-a_j|\tau||z_j|^2} L_{k_j}^{(0)}(2a_j|\tau||z_j|^2)$$
$$= \frac{|\tau|^{n-2}}{2\alpha\pi^n} \int_0^\infty [\Phi_\alpha(\mathbf{z}, \tau; s) - \Phi_{-\alpha}(\mathbf{z}, \tau; s)]\, ds,$$

where

$$\Phi_\alpha(\mathbf{z}, \tau; s) = e^{\alpha s} \prod_{j=1}^n 2a_j e^{-a_j s - a_j|\tau||z_j|^2} \sum_{k_j=0}^\infty (e^{-2a_j s})^{k_j} L_{k_j}^{(0)}(2a_j|\tau||z_j|^2).$$

Applying the generating formula for the Laguerre polynomials

$$\sum_{k=0}^{\infty} L_k^{(p)}(x) z^k = \frac{1}{(1-z)^{p+1}} \exp\left\{-\frac{xz}{1-z}\right\}$$

to $\Phi(\mathbf{z}, \tau; s)$, we obtain

$$\begin{aligned}\Phi_\alpha(\mathbf{z}, \tau; s) &= e^{\alpha s} \prod_{j=1}^{n} \frac{2a_j e^{-a_j s}}{1-e^{-2a_j s}} \exp\left\{-a_j|\tau||z_j|^2\left[1+\frac{2e^{-2a_j s}}{1-e^{-2a_j s}}\right]\right\} \\ &= e^{\alpha s}\left[\prod_{j=1}^{n} \frac{a_j}{\sinh(a_j s)}\right] \exp\left\{-|\tau| \sum_{j=1}^{n} a_j |z_j|^2 \coth(a_j s)\right\}.\end{aligned}$$

This yields

$$\begin{aligned}\widetilde{\Psi}_\alpha(\mathbf{z}, \tau) = \frac{|\tau|^{n-2}}{\alpha\pi^n} \int_0^\infty & \sinh(\alpha s)\left[\prod_{j=1}^{n} \frac{a_j}{\sinh(a_j s)}\right] \\ & \times \exp\left\{-|\tau| \sum_{j=1}^{n} a_j |z_j|^2 \coth(a_j s)\right\} ds. \qquad (12.6.59)\end{aligned}$$

12.6.2 *Fundamental Solution: The Case n ≥ 2*

To simplify the notation, we introduce the function

$$\gamma(\mathbf{z}, s) = \sum_{j=1}^{n} a_j |z_j|^2 \coth(a_j s).$$

We next take the inverse Fourier transform with respect to τ and find

$$\Psi_\alpha(\mathbf{z}, t) = \frac{1}{2\alpha\pi^{n+1}} \int_{-\infty}^{\infty} |\tau|^{n-2} \int_0^\infty \sinh(\alpha s)\left[\prod_{j=1}^{n} \frac{a_j}{\sinh(a_j s)}\right] e^{it\tau - |\tau|\gamma(\mathbf{z},s)}\, ds\, d\tau.$$

Changing the order of integration, we obtain

$$\begin{aligned}\Psi_\alpha(\mathbf{z}, t) = \frac{\Gamma(n-1)}{2\alpha\pi^{n+1}} \int_0^\infty & \sinh(\alpha s)\left[\prod_{j=1}^{n} \frac{a_j}{\sinh(a_j s)}\right] \\ & \times \left\{\frac{1}{[\gamma(\mathbf{z}, s) - it]^{n-1}} + \frac{1}{[\gamma(\mathbf{z}, s) + it]^{n-1}}\right\} ds.\end{aligned}$$

For the second part of the integral, we substitute s with $-s$, and note that $\sinh(-\alpha s) = -\sinh(\alpha s)$ and $\gamma(\mathbf{z}, -s) = -\gamma(\mathbf{z}, s)$ since sinh and coth are odd functions. Hence

$$\begin{aligned}\int_0^{\infty}\left[\prod_{j=1}^{n}\frac{a_j}{\sinh(a_j s)}\right]\cdot\frac{\sinh(\alpha s)}{[\gamma(\mathbf{z},t)+it]^{n-1}}ds\\ &= \int_0^{-\infty}\left[\prod_{j=1}^{n}\frac{a_j}{\sinh(-a_j s)}\right]\cdot\frac{\sinh(-\alpha s)}{[-\gamma(\mathbf{z},t)+it]^{n-1}}(-1)ds\\ &= \int_{-\infty}^{0}\frac{(-1)}{(-1)^{n+n-1}}\left[\prod_{j=1}^{n}\frac{a_j}{\sinh(a_j s)}\right]\cdot\frac{\sinh(\alpha s)}{[\gamma(\mathbf{z},t)-it]^{n-1}}ds\\ &= \int_{-\infty}^{0}\left[\prod_{j=1}^{n}\frac{a_j}{\sinh(a_j s)}\right]\cdot\frac{\sinh(\alpha s)}{[\gamma(\mathbf{z},t)-it]^{n-1}}ds.\end{aligned}$$

We can write $\Psi_\alpha(\mathbf{z}, t)$ in the compact form

$$\Psi_\alpha(\mathbf{z},t) = \frac{\Gamma(n-1)}{2\alpha\pi^{n+1}}\int_{-\infty}^{\infty}\left[\prod_{j=1}^{n}\frac{a_j}{\sinh(a_j s)}\right]\frac{\sinh(\alpha s)}{[\gamma(\mathbf{z},s)-it]^{n-1}}ds. \tag{12.6.60}$$

As we mentioned before, we may rewrite the fundamental solution in terms of the modified complex action $g(s; \mathbf{z}, t)$ and volume element $v(s)$ which were defined in (12.4.42):

$$g(s;\mathbf{z},t) = \gamma(2s;\mathbf{z}) - it = \sum_{j=1}^{n} a_j|z_j|^2\coth(2a_j s) - it \quad \text{and}$$

$$v(s) = \prod_{j=1}^{n}\frac{2a_j}{\sinh(2a_j s)}.$$

Substituting s with $2s$, one obtains

$$\begin{aligned}\Psi_\alpha(\mathbf{z},t) &= \frac{\Gamma(n-1)}{2^n\alpha\pi^{n+1}}\int_{-\infty}^{\infty}\left[\prod_{j=1}^{n}\frac{2a_j}{\sinh(2a_j s)}\right]\frac{\sinh(2\alpha s)}{[g(s;\mathbf{z},t)]^{n-1}}ds\\ &= \frac{\Gamma(n-1)}{2^n\alpha\pi^{n+1}}\int_{-\infty}^{\infty}\frac{v(s)\sinh(2\alpha s)}{[g(s;\mathbf{z},t)]^{n-1}}ds.\end{aligned}$$

It seems impossible to find the exact formula for the above integral in general. We will consider the special case of $a_j = a$ for all j. In this case,

$$\Psi_\alpha(\mathbf{z},t) = \frac{\Gamma(n-1)}{2\alpha\pi^{n+1}}\int_{-\infty}^{\infty}\left[\frac{a}{\sinh(as)}\right]^n\frac{\sinh(\alpha s)}{[a|\mathbf{z}|^2\coth(as)-it]^{n-1}}ds.$$

Differentiating with respect to t, we obtain

$$\frac{\partial \Psi_\alpha}{\partial t}(\mathbf{z},t) = \frac{\Gamma(n) i a^n}{2\alpha\pi^{n+1}} \int_{-\infty}^{\infty} \frac{\sinh(\alpha s)}{[a|\mathbf{z}|^2 \cosh(as) - it \sinh(as)]^n} ds.$$

Denote

$$\rho = (a^2|\mathbf{z}|^4 + t^2)^{\frac{1}{4}} \quad \text{and} \quad e^{-i\phi} = \rho^{-2}(a|\mathbf{z}|^2 - it)$$

with $\phi \in \left(-\frac{\pi}{2}, \frac{\pi}{2}\right)$. Using the identity

$$\cosh(s + i\phi) = \cosh(s)\cos\phi + i\sinh(s)\sin\phi,$$

we can write

$$\frac{\partial \Psi_\alpha}{\partial t}(\mathbf{z},t) = \frac{\Gamma(n) i a^n}{2\alpha\pi^{n+1}\rho^{2n}} \int_{-\infty}^{\infty} \frac{\sinh(\alpha s)}{[\cosh(as - i\phi)]^n} ds.$$

Next, we apply the integral formula:

$$\int_{-\infty}^{\infty} \frac{e^{\alpha s}}{\cosh^\nu(as+b)} dx = \frac{2^{\nu-1}}{a\Gamma(\nu)} e^{\alpha\frac{b}{a}} \Gamma\left(\frac{1}{2}\nu - \frac{\alpha}{2a}\right)\Gamma\left(\frac{1}{2}\nu + \frac{\alpha}{2a}\right).$$

Hence,

$$\begin{aligned}
\int_{-\infty}^{\infty} \frac{\sinh(\alpha s)}{[\cosh(as - i\phi)]^n} ds &= \frac{1}{2}\int_{-\infty}^{\infty}\left[\frac{e^{\alpha s}}{[\cosh(as - i\phi)]^n} - \frac{e^{-\alpha s}}{[\cosh(as - i\phi)]^n}\right] ds \\
&= \frac{2^{n-1}\Gamma\left(\frac{n}{2} - \frac{\alpha}{2a}\right)\Gamma\left(\frac{n}{2} + \frac{\alpha}{2a}\right)}{2a\Gamma(n)}\left(e^{-i\frac{\alpha}{a}\phi} - e^{i\frac{\alpha}{a}\phi}\right) \\
&= \frac{2^{n-1}\Gamma\left(\frac{n}{2} - \frac{\alpha}{2a}\right)\Gamma\left(\frac{n}{2} + \frac{\alpha}{2a}\right)}{a\Gamma(n)}\left[-i\sin\left(\frac{\alpha}{a}\phi\right)\right].
\end{aligned}$$

This yields

$$\begin{aligned}
\frac{\partial \Psi_\alpha}{\partial t}(\mathbf{z},t) &= \frac{\Gamma(n) a^n}{2\alpha\pi^{n+1}\rho^{2n}} \cdot \frac{2^{n-1}\Gamma\left(\frac{n}{2} - \frac{\alpha}{2a}\right)\Gamma\left(\frac{n}{2} + \frac{\alpha}{2a}\right)}{a\Gamma(n)} \sin\left(\frac{\alpha}{a}\phi\right) \\
&= \frac{2^{n-2}a^{n-1}\Gamma\left(\frac{n}{2} - \frac{\alpha}{2a}\right)\Gamma\left(\frac{n}{2} + \frac{\alpha}{2a}\right)}{\alpha\pi^{n+1}\rho^{2n}} \sin\left(\frac{\alpha}{a}\phi\right).
\end{aligned}$$

Recall that $\rho^4 = (a^2|\mathbf{z}|^4 + t^2)^{1/4}$ and $\rho^2 e^{i\phi} = a|\mathbf{z}|^2 + it$. We can write $\Psi_\alpha(\mathbf{z},t)$ explicitly in terms of $\mathbf{z}$ and t by integrating with respect to t, requiring

$\lim_{t\to-\infty}\Psi_\alpha(\mathbf{z},t)=0$. In the case $a=\alpha$, we can find a more explicit formula for the fundamental solution, noting that for $a=\alpha$,

$$\begin{aligned}\Psi_\alpha(\mathbf{z},t)&=\frac{\Gamma(n-1)}{2\pi^{n+1}}\int_{-\infty}^{\infty}\left[\frac{a}{\sinh(as)}\right]^{n-1}\frac{1}{[a|\mathbf{z}|^2\coth(as)-it]^{n-1}}ds\\&=\frac{\Gamma(n-1)a^{n-1}}{2\pi^{n+1}}\int_{-\infty}^{\infty}\frac{1}{[a|\mathbf{z}|^2\cosh(as)-it\sinh(as)]^{n-1}}ds\\&=\frac{\Gamma(n-1)a^{n-1}}{2\pi^{n+1}\rho^{2n-2}}\int_{-\infty}^{\infty}\frac{1}{[\cosh(as-i\phi)]^{n-1}}ds\\&=\frac{\Gamma(n-1)a^{n-1}}{2\pi^{n+1}\rho^{2n-2}}\cdot\frac{2^{n-2}\left[\Gamma(\frac{n-1}{2})\right]^2}{a\Gamma(n-1)}\\&=\frac{a^{n-2}2^{n-3}\left[\Gamma(\frac{n-1}{2})\right]^2}{\pi^{n+1}\left(a^2|\mathbf{z}|^4+t^2\right)^{\frac{n-1}{2}}}.\end{aligned}$$

12.6.3 Fundamental Solution: The Case n = 1

In this case

$$\widetilde{\Psi}_\alpha(\mathbf{z},\tau)=\frac{1}{\alpha\pi|\tau|}\int_0^{\infty}\sinh(\alpha s)\left[\frac{a}{\sinh(as)}\right]\exp\left\{-|\tau|a|z|^2\coth(as)\right\}ds. \tag{12.6.61}$$

We note that if we take the inverse Fourier transform directly, the integral will diverge, so the fundamental solution is not a regular function. We will provide a formula for the fundamental solution in the form of a generalized function. Recall the Fourier transform with respect to t and its inverse:

$$\widetilde{f}(\mathbf{z},\tau)=\int_{\mathbb{R}}e^{-it\tau}f(\mathbf{z},\tau)\quad\text{and}\quad f(\mathbf{z},t)=\frac{1}{2\pi}\int_{\mathbb{R}}e^{it\tau}\widetilde{f}(\mathbf{z},\tau)d\tau.$$

We need to compute the inverse Fourier transform of

$$\widetilde{f}(\mathbf{z},\tau)=\frac{1}{|\tau|}e^{-\gamma(\mathbf{z},s)|\tau|}.$$

The following computation will yield the formula for the inverse Fourier transform. We apply the following formula for the inverse Fourier transform (see (33c), p. 153, of Kanwal [76]):

$$\mathcal{F}^{-1}\left(\frac{1}{|\tau|}\right)=-\frac{\varepsilon}{\pi}-\frac{\ln|t|}{\pi}\quad\text{and}\quad\mathcal{F}^{-1}(e^{-\gamma|\tau|})=\frac{\gamma}{\pi(\gamma^2+t^2)},$$

where ε is the Euler's constant:

$$\varepsilon = \int_0^1 \frac{1-\cos y}{y}dy - \int_1^\infty \frac{\cos y}{y}dy.$$

Then, the inverse Fourier transform of $\widetilde{f}(\mathbf{z},\tau)$ is the convolution

$$\Big[-\frac{\varepsilon}{\pi} - \frac{\ln|t|}{\pi}\Big] * \Big[\frac{\gamma}{\pi(\gamma^2+t^2)}\Big] = -\frac{\gamma}{\pi^2}\int_{-\infty}^{\infty}\frac{\varepsilon+\ln|u|}{\gamma^2+(t-u)^2}du.$$

We start by computing the integral:

$$\int_{-\infty}^{\infty}\frac{\varepsilon}{\gamma^2+(t-u)^2}du = \int_{-\infty}^{\infty}\frac{\varepsilon}{\gamma^2+u^2}du = \frac{\varepsilon\pi}{\gamma}.$$

We have

$$\int_{-\infty}^{\infty}\frac{\ln|u|}{\gamma^2+(t-u)^2}du = \int_0^\infty \frac{\ln u}{(u-t)^2+\gamma^2}du + \int_0^\infty \frac{\ln u}{(u+t)^2+\gamma^2}du.$$

Applying the residue theorem (see formula (7.2.11) of Antimirov, Kolyshkin and Vaillancourt [4]), yields

$$\begin{aligned}\int_0^\infty \frac{\ln u}{(u-t)^2+\gamma^2}du &= -\frac{1}{2}\mathrm{Re}\left\{\Bigg(\underbrace{\mathrm{Res}}_{z=t+i\gamma} + \underbrace{\mathrm{Res}}_{z=t-i\gamma}\Bigg)\Big[\frac{(\ln z)^2}{(z-t)^2+\gamma^2}\Big]\right\}\\ &= -\frac{1}{2}\mathrm{Re}\left\{\frac{\ln^2(t+i\gamma)}{2i\gamma} + \frac{\ln^2(t-i\gamma)}{-2i\gamma}\right\}\\ &= \frac{1}{4\gamma}\mathrm{Re}\{i[\ln^2(t+i\gamma) - \ln^2(t-i\gamma)]\}.\end{aligned}$$

Since $0 \le \arg(z) < 2\pi$, then for $t > 0$,

$$\ln(t+i\gamma) = \ln\sqrt{t^2+\gamma^2} + i\theta \quad \text{and} \quad \ln(t-i\gamma) = \ln\sqrt{t^2+\gamma^2} + i(2\pi-\theta),$$

where $\theta = \arctan(\gamma/t) \in [0,\pi/2]$. Hence

$$\int_0^\infty \frac{\ln u}{(u-t)^2+\gamma^2}du = \frac{\pi-\theta}{2\gamma}\ln(t^2+\gamma^2).$$

Similarly, for $t < 0$,

$$\ln(t+i\gamma) = \ln\sqrt{t^2+\gamma^2} + i(\pi-\theta) \quad \text{and} \quad \ln(t-i\gamma) = \ln\sqrt{t^2+\gamma^2} + i(\pi+\theta),$$

where $\theta = \arctan(\gamma/|t|) \in [0, \pi/2]$. Then

$$\int_0^\infty \frac{\ln u}{(u-t)^2+\gamma^2} du = \frac{\theta}{2\gamma} \ln(t^2+\gamma^2).$$

Similarly, we have

$$\int_0^\infty \frac{\ln u}{(u+t)^2+\gamma^2} du = \begin{cases} \frac{\pi-\theta}{2\gamma} \ln(t^2+\gamma^2), & t<0, \\ \frac{\theta}{4\gamma} \ln(t^2+\gamma^2), & t>0. \end{cases}$$

Hence

$$\int_0^\infty \frac{\ln u}{(u-t)^2+\gamma^2} du + \int_0^\infty \frac{\ln u}{(u+t)^2+\gamma^2} du = \frac{2\pi-\theta}{4\gamma} \ln(t^2+\gamma^2),$$

where $\theta = \arctan(\gamma/|t|)$. Summarizing the calculations, we have

$$\left[-\frac{\varepsilon}{\pi} - \frac{\ln|t|}{\pi}\right] * \left[\frac{\gamma}{\pi(\gamma^2+t^2)}\right] = -\frac{1}{\pi^2}\left[\varepsilon\pi + \left(\frac{\pi}{2} - \frac{\theta}{4}\right)\ln(t^2+\gamma^2)\right].$$

Hence the fundamental solution is

$$\begin{aligned} \Psi_\alpha(\mathbf{z},t) &= -\frac{a}{\alpha\pi^3}\int_0^\infty \frac{\sinh(\alpha s)}{\sinh(as)}\gamma(\mathbf{z},s)\int_{-\infty}^\infty \frac{\varepsilon+\ln|u|}{\gamma^2(\mathbf{z},s)+(t-u)^2}duds \\ &= -\frac{a^2|z|^2}{\alpha\pi^3}\int_0^\infty\int_{-\infty}^\infty \frac{\sinh(\alpha s)\cosh(as)}{\sinh^2(as)} \frac{\varepsilon+\ln|u|}{a^2|z|^4\coth^2(as)+(t-u)^2}duds \\ &= -\frac{a^2|z|^2}{\alpha\pi^3}\int_0^\infty\int_{-\infty}^\infty \frac{\sinh(\alpha s)\cosh(as)(\varepsilon+\ln|u|)}{a^2|z|^4\cosh^2(as)+(t-u)^2\sinh^2(as)}duds \\ &= -\frac{a}{\alpha\pi^3}\int_0^\infty \frac{\sinh(\alpha s)}{\sinh(as)}\left[\varepsilon\pi + \left(\frac{\pi}{2}-\frac{\theta}{4}\right)\ln[a^2\coth^2(as)|z|^4+t^2]\right]ds. \end{aligned}$$

12.7 Heat Kernel of the Paneitz Operator

In this section we will compute the heat kernel $h_s(\mathbf{z},t) = \exp\{-s\mathcal{P}_\alpha\}\delta_0$. We first take the Fourier transform with respect to the t-variable and write the heat kernel $\widetilde{h}_s(\mathbf{z},t)$ as a twisted convolution operator:

$$\begin{aligned} \widetilde{h}_s(\mathbf{z},\tau) = \exp\{-s\widetilde{\mathcal{P}}_\alpha\}\widetilde{\mathbf{I}} &= \sum_{|\mathbf{k}|=0}^\infty \exp\{-s\widetilde{\mathcal{P}}_\alpha\}\left[\prod_{j=1}^n a_j\widetilde{\mathcal{W}}_{k_j}^{(0)}(\sqrt{a_j}z_j,\tau)\right] \\ &= \sum_{|\mathbf{k}|=0}^\infty e^{-s\tau^2[\sum_{j=1}^n a_j(2k_j+1)]^2+s\alpha^2\tau^2}\prod_{j=1}^n a_j\widetilde{\mathcal{W}}_{k_j}^{(0)}(\sqrt{a_j}z_j,\tau). \end{aligned}$$

Then we apply the Fourier integral formula,

$$e^{-s\xi^2} = \frac{1}{\sqrt{4\pi s}} \int_{-\infty}^{\infty} e^{-\frac{x^2}{4s} - ix\xi} dx,$$

for $\xi = \sum_{j=1}^n (2k_j + 1)a_j$ and $s\tau^2$ for s:

$$e^{-s\tau^2[\sum_{j=1}^n (2k_j+1)a_j]^2} = \frac{1}{\sqrt{4\pi s\tau^2}} \int_{-\infty}^{\infty} e^{-\frac{x^2}{4s\tau^2} - ix\sum_{j=1}^n (2k_j+1)a_j} dx.$$

We can write

$$\begin{aligned}
&\widetilde{h}_s(\mathbf{z}, \tau) \\
&= \sum_{|\mathbf{k}|=0}^{\infty} e^{s\alpha^2\tau^2} \frac{1}{\sqrt{4\pi s\tau^2}} \left[\int_{-\infty}^{\infty} e^{-\frac{x^2}{4s\tau^2} - ix\sum_{j=1}^n (2k_j+1)a_j} dx \right] \\
&\quad \times \prod_{j=1}^n a_j \widetilde{\mathcal{W}}_{k_j}^{(0)}(\sqrt{a_j} z_j, \tau) \\
&= \frac{e^{s\alpha^2\tau^2}}{\sqrt{4\pi s\tau^2}} \int_{-\infty}^{\infty} e^{-\frac{x^2}{4s\tau^2}} \left[\sum_{|\mathbf{k}|=0}^{\infty} \prod_{j=1}^n a_j e^{-i(2k_j+1)a_j x} \widetilde{\mathcal{W}}_{k_j}^{(0)}(\sqrt{a_j} z_j, \tau) \right] dx \\
&= \frac{e^{s\alpha^2\tau^2}}{\sqrt{4\pi s\tau^2}} \int_{-\infty}^{\infty} e^{-\frac{x^2}{4s\tau^2}} \left[\prod_{j=1}^n a_j \frac{|\tau|}{\pi} e^{-a_j|\tau||z_j|^2 - ia_j x} \sum_{k_j=0}^{\infty} e^{-2ik_j a_j x} \right. \\
&\quad \left. \times L_{k_j}^{(0)}(2a_j|\tau||z_j|^2) \right] dx.
\end{aligned}$$

We cannot sum up the series of Laguerre polynomials since $|e^{-2ia_j x}| = 1$. We can insert a convergence factor $e^{-\epsilon k_j}$ and let

$$g_{s,\epsilon}(\mathbf{z}, \tau, x) = \left[\prod_{j=1}^n a_j \frac{|\tau|}{\pi} e^{-a_j|\tau||z_j|^2 - ia_j x} \sum_{k_j=0}^{\infty} e^{-(\epsilon+2ia_j x)k_j} L_{k_j}^{(0)}(2a_j|\tau||z_j|^2) \right].$$

Then, applying the generating formula, we obtain

$$\begin{aligned}
g_{s,\epsilon}(\mathbf{z}, \tau, x) &= \prod_{j=1}^n a_j \frac{|\tau|}{\pi} e^{-a_j|\tau||z_j|^2 - ia_j x} \frac{1}{1 - e^{-(\epsilon+2ia_j x)}} \\
&\quad \times \exp\left\{ -\frac{e^{-(\epsilon+2ia_j x)}}{1 - e^{-(\epsilon+2ia_j x)}} 2a_j|\tau||z_j|^2 \right\}
\end{aligned}$$

$$= \frac{|\tau|^n}{\pi^n} \prod_{j=1}^{n} \frac{a_j e^{-ia_j x}}{1 - e^{-(\epsilon + 2ia_j x)}} \exp\left\{ -\frac{1 + e^{-(\epsilon + 2ia_j x)}}{1 - e^{-(\epsilon + 2ia_j x)}} a_j |\tau| |z_j|^2 \right\}$$

$$= \frac{|\tau|^n e^{n\epsilon/2}}{2^n \pi^n} \left[\prod_{j=1}^{n} \frac{a_j}{\sinh(\frac{\epsilon}{2} + ia_j x)} \right]$$

$$\times \exp\left\{ -|\tau| \sum_{j=1}^{n} a_j |z_j|^2 \coth\left(\frac{\epsilon}{2} + ia_j x\right) \right\}.$$

Taking the limit $\epsilon \to 0^+$, we have

$$\lim_{\epsilon \to 0^+} g_{s,\epsilon}(\mathbf{z}, \tau, x) = \frac{|\tau|^n}{2^n \pi^n} \left[\prod_{j=1}^{n} \frac{a_j}{\sinh(ia_j x)} \right] \exp\left\{ -|\tau| \sum_{j=1}^{n} a_j |z_j|^2 \coth(ia_j x) \right\}.$$

Since $\sinh(ix) = i \sin x$ and $\coth(ix) = -i \cot x$, we have

$$\widetilde{h}_s(\mathbf{z}, \tau) = \frac{e^{s\alpha^2\tau^2}}{\sqrt{4\pi s}} \frac{|\tau|^{n-1}}{2^n \pi^n} \int_{-\infty}^{\infty} e^{-\frac{x^2}{4s\tau^2}} \left[\prod_{j=1}^{n} \frac{a_j}{i \sin(a_j x)} \right]$$

$$\times \exp\left\{ i|\tau| \sum_{j=1}^{n} a_j |z_j|^2 \cot(a_j x) \right\} dx.$$

Next, we take the inverse Fourier transform with respect to τ and obtain the heat kernel associated with the Paneitz operator:

$$h_s(\mathbf{z}, t) = \frac{(-i)^n}{(2\pi)^{n+1}\sqrt{\pi s}} \int_0^{\infty} \tau^{n-1} e^{\alpha^2 s\tau^2} \cos(t\tau) \int_{-\infty}^{\infty} \left[\prod_{j=1}^{n} \frac{a_j}{\sin(a_j x)} \right]$$

$$\times \exp\left\{ -\frac{x^2}{4s\tau^2} + i\tau \sum_{j=1}^{n} a_j |z_j|^2 \cot(a_j x) \right\} dx d\tau.$$

It seems impossible to find either of the integrals explicitly. In the special case of $a_j = a$ for all $j = 1, 2, \ldots, n$, we obtain

$$h_s(\mathbf{z}, t) = \frac{(-i)^n}{(2\pi)^{n+1}\sqrt{\pi s}} \int_0^{\infty} \tau^{n-1} e^{\alpha^2 s\tau^2} \cos(t\tau) \int_{-\infty}^{\infty} \left[\frac{a}{\sin(ax)} \right]^n$$

$$\times \exp\left\{ -\frac{x^2}{4s\tau^2} + i\tau a |\mathbf{z}|^2 \cot(ax) \right\} dx d\tau.$$

The integration with respect to τ is the following:

$$\int_0^\infty \tau^{n-1} e^{\alpha^2 s\tau^2 - \frac{x^2}{4s\tau^2} + i\tau a|\mathbf{z}|^2 \cot(ax)} \cos(t\tau) d\tau.$$

The above integral can eventually be written in terms of some special functions.

12.8 Projection and Relative Fundamental Solution

In [61], Greiner and Stein have shown that

$$-\frac{1}{2}(\widetilde{\mathbf{Z}}_j\bar{\widetilde{\mathbf{Z}}}_j + \bar{\widetilde{\mathbf{Z}}}_j\widetilde{\mathbf{Z}}_j)\widetilde{\mathcal{W}}_k^{(\pm p)}(\sqrt{a_j}z_j, \tau) = \big[(2k+1) + p(1 \pm \operatorname{sgn}\tau)\big]|\tau|a_j\widetilde{\mathcal{W}}_k^{(\pm p)} \times(\sqrt{a_j}z_j, \tau) \qquad (12.8.62)$$

for $k, p = 0, 1, 2, \ldots$. This implies

$$\begin{aligned}
-\frac{1}{2}\sum_{j=1}^n(\widetilde{\mathbf{Z}}_j\bar{\widetilde{\mathbf{Z}}}_j + \bar{\widetilde{\mathbf{Z}}}_j\widetilde{\mathbf{Z}}_j)\prod_{j=1}^n \widetilde{\mathcal{W}}_{k_j}^{(\pm p_j)}(\sqrt{a_j}z_j, \tau) \\
= \left\{\sum_{j=1}^n \big[(2k_j+1) + p_j(1 \pm \operatorname{sgn}\tau)\big]|\tau|a_j\right\}\prod_{j=1}^n \widetilde{\mathcal{W}}_{k_j}^{(\pm p_j)}(\sqrt{a_j}z_j, \tau),
\end{aligned}$$

and the Paneitz operator $\mathcal{P}_\alpha = \mathcal{L}_\alpha\bar{\mathcal{L}}_\alpha$ satisfies

$$\begin{aligned}
&\widetilde{\mathcal{P}}_\alpha\left(\prod_{j=1}^n \widetilde{\mathcal{W}}_{k_j}^{(\pm p_j)}(\sqrt{a_j}z_j, \tau)\right) \\
&= \tau^2\left\{\left(\sum_{j=1}^n \big[(2k_j+1)+p_j(1 \pm \operatorname{sgn}\tau)\big]a_j\right)^2 - \alpha^2\right\}\prod_{j=1}^n \widetilde{\mathcal{W}}_{k_j}^{(\pm p_j)}(\sqrt{a_j}z_j, \tau).
\end{aligned}$$

Hence the spectrum of the operator $\widetilde{\mathcal{P}}_\alpha$ is the set

$$\begin{aligned}
\sigma(\widetilde{\mathcal{P}}_\alpha) = \Bigg\{\lambda_{\mathbf{k},\mathbf{p}} = \tau^2\left[\left(\sum_{j=1}^n \big[(2k_j+1) + (|p_j| + p_j \cdot \operatorname{sgn}\tau)\big]a_j\right)^2 - \alpha^2\right], \\
\mathbf{k} \in \mathbb{Z}_+^n \quad \text{and} \quad \mathbf{p} \in \mathbb{Z}^n\Bigg\}.
\end{aligned}$$

The operator $\mathcal{P}_\alpha$ is not invertible if $\alpha \in \Lambda$ with

$$\Lambda = \left\{ \pm \sum_{j=1}^{n} \big[(2k_j+1) + (|p_j| + p_j \cdot \operatorname{sgn}\tau)\big] a_j,\ \mathbf{k} \in \mathbb{Z}_+^n \quad \text{and} \quad \mathbf{p} \in \mathbb{Z}^n \right\}.$$

This implies that if $\alpha_{\mathbf{k},\mathbf{p}} = \pm \sum_{j=1}^{n} \big[(2k_j+1) + (|p_j| + p_j \cdot \operatorname{sgn}\tau)\big] a_j$ for some $\mathbf{k} \in \mathbb{Z}_+^n$ and $\mathbf{p} \in \mathbb{Z}^n$, then

$$\widetilde{\mathcal{P}}_{\alpha_{\mathbf{k},\mathbf{p}}} \left(\prod_{j=1}^{n} \widetilde{\mathcal{W}}_{k_j}^{(p_j)}(\sqrt{a_j} z_j, \tau) \right) = 0.$$

Let $\widetilde{\mathcal{P}}_\alpha^{(+)} = \widetilde{\mathcal{P}}_\alpha$ for $\tau > 0$ and $\widetilde{\mathcal{P}}_\alpha^{(-)} = \widetilde{\mathcal{P}}_\alpha$ for $\tau < 0$. Then for $\tau > 0$,

$$\begin{aligned} \alpha_{\mathbf{k},\mathbf{p}} &= \pm \sum_{j=1}^{n} \big[(2k_j+1) + (|p_j| + p_j)\big] a_j \\ &= \begin{cases} \pm \sum_{j=1}^{n} \big[2(k_j+p_j)+1\big] a_j, & p_j \geq 0, \\ \pm \sum_{j=1}^{n} (2k_j+1) a_j, & p_j \leq 0. \end{cases} \end{aligned}$$

And for $\tau < 0$,

$$\begin{aligned} \alpha_{\mathbf{k},\mathbf{p}} &= \pm \sum_{j=1}^{n} \big[(2k_j+1) + (|p_j| - p_j)\big] a_j \\ &= \begin{cases} \pm \sum_{j=1}^{n} (2k_j+1) a_j, & p_j \geq 0, \\ \pm \sum_{j=1}^{n} [2(k_j-p_j)+1] a_j, & p_j \leq 0. \end{cases} \end{aligned}$$

We summarize this result in the following proposition.

Proposition 12.8.1. *(i) If $\tau > 0$, then $\widetilde{\mathcal{P}}_\alpha u = 0$ has the following linearly independent set of L^2-solutions:*

$$\left\{ \prod_{j=1}^{n} \widetilde{\mathcal{W}}_{k_j}^{(p_j)}(\sqrt{a_j} z_j, \tau),\ \mathbf{k} \in \mathbb{Z}_+^n \text{ and } \mathbf{p} \in \mathbb{Z}^n \text{ satisfy} \right.$$

$$\left. \pm \sum_{j=1}^{n} \big[(2k_j+1) + (|p_j| + p_j)\big] a_j = \alpha \right\}.$$

(ii) *If* $\tau < 0$, *then* $\widetilde{\mathcal{P}}_\alpha u = 0$ *has the following linearly independent set of* L^2-*solutions:*

$$\left\{ \prod_{j=1}^{n} \widetilde{W}_{k_j}^{(p_j)}(\sqrt{a_j} z_j, \tau),\ \mathbf{k} \in \mathbb{Z}_+^n \text{ and } \mathbf{p} \in \mathbb{Z}^n \text{ satisfy} \right.$$

$$\left. \pm \sum_{j=1}^{n} \big[(2k_j + 1) + (|p_j| - p_j)\big] a_j = \alpha \right\}.$$

Given α, the set that indexes L^2-solutions of $\widetilde{P}_\alpha u = 0$ is quite complicated and is given by

$$\Sigma_\alpha = \left\{ \pm \sum_{j=1}^{n} \big[(2k_j + 1) + (|p_j| + p_j \operatorname{sgn}(\tau))\big] a_j = \alpha,\ \mathbf{k} \in \mathbb{Z}_+^n \text{ and } \mathbf{p} \in \mathbb{Z}^n \right\}.$$

In the case when $a_j = 1$ for all j, the set Σ_α is empty if n is odd and α is not an odd integer or n is even and α is not an even integer. In general, this set cannot be classified by some simple rules.

We now consider the case of $n = 1$ and $a_1 = 1$. Then Σ_α is empty if α is not an odd integer. So we let $\alpha = 2m + 1$. Then for $\tau > 0$,

$$2m + 1 = \pm[(2k + 1) + (|p| + p)] = \begin{cases} \pm[2(k + p) + 1], & p \geq 0, \\ \pm(2k + 1), & p < 0. \end{cases}$$

Similarly, for $\tau < 0$,

$$2m + 1 = \pm[(2k + 1) + (|p| - p)] = \begin{cases} \pm[2(k - p) + 1], & p < 0, \\ \pm(2k + 1), & p \geq 0. \end{cases}$$

Let $m \in \mathbb{N}$. Then we have

(i) If $\tau > 0$ and $\alpha = \pm(2m + 1)$, then the set of linearly independent L^2-solutions of $\widetilde{\mathcal{P}}^{(+)}_{\pm(2m+1)} u = 0$ is

$$\left\{ \widetilde{W}_m^{(-p)}(z, \tau),\ p = 0, 1, 2, \ldots \right\} \cup \left\{ \widetilde{W}_k^{(m-k)}(z, \tau),\ k = 0, 1, 2, \ldots, m \right\}.$$

(ii) If $\tau < 0$ and $\alpha = \pm(2m + 1)$, then the set of linearly independent L^2-solutions of $\widetilde{\mathcal{P}}^{(-)}_{\pm(2m+1)} u = 0$ is

$$\left\{ \widetilde{W}_m^{(p)}(z, \tau),\ p = 0, 1, 2, \ldots \right\} \cup \left\{ \widetilde{W}_k^{(k-m)}(z, \tau),\ k = 0, 1, 2, \ldots, m \right\}.$$

The Laguerre matrix of $\widetilde{P}_\alpha$ given by $\mathcal{M}(\widetilde{\mathcal{P}}_\alpha) = \tau^2\left(\left[(2j-1)^2-\alpha^2\right]\delta_j^k\right)$ is diagonal. If $\alpha = \pm(2m+1)$, we let

$$\widetilde{M}_{2m+1} = \frac{1}{\tau^2}\sum_{k\neq m}\frac{1}{(2j+1)^2-(2m+1)^2}\widetilde{\mathcal{W}}_k^{(0)}(z,\tau).$$

Then Laguerre calculus will yield

$$\widetilde{P}_{2m+1}\widetilde{M}_{2m+1}+\widetilde{S}_m = I,$$

where $\widetilde{S}_m = \widetilde{\mathcal{W}}_m(z,\tau)$. We now sum up $\widetilde{M}_{2m+1}$:

$$\begin{aligned}
\widetilde{M}_{2m+1} &= \frac{1}{\tau^2}\sum_{k\neq m}\frac{1}{(2k+1)^2-(2m+1)^2}\widetilde{\mathcal{W}}_k^{(0)}(z,\tau)\\
&= \frac{1}{\tau^2}\cdot\frac{2|\tau|}{\pi}e^{-|\tau||z|^2}\sum_{k\neq m}\frac{L_k^{(0)}(2|\tau||z|^2)}{4(k-m)(k+m+1)}\\
&= \frac{e^{-|\tau||z|^2}}{2|\tau|\pi}\cdot\frac{1}{2m+1}\sum_{k\neq m}\left(\frac{1}{k-m}-\frac{1}{k+m+1}\right)L_k^{(0)}(2|\tau||z|^2).
\end{aligned}$$

We first consider the case of $m = 0$. We need to sum up the series

$$F(\omega) = \sum_{k=1}^{\infty}\left(\frac{1}{k}-\frac{1}{k+1}\right)L_k^{(0)}(2|\tau||z|^2).$$

Let $\omega = 2|\tau||z|^2$. Then

$$\begin{aligned}
F(\omega) &= \sum_{k=1}^{\infty}\left(\frac{1}{k}-\frac{1}{k+1}\right)L_k^{(0)}(\omega) = \sum_{k=1}^{\infty}\frac{L_k^{(0)}(\omega)}{k}-\frac{L_k^{(0)}(\omega)}{k+1}\\
&= \sum_{k=0}^{\infty}\frac{L_{k+1}^{(0)}(\omega)}{k+1}-\sum_{k=1}^{\infty}\frac{L_k^{(0)}(\omega)}{k+1}\\
&= L_1^{(0)}(\omega)+\sum_{k=1}^{\infty}\frac{1}{k+1}[L_k^{(0)}(\omega)-L_{k+1}^{(0)}(\omega)].
\end{aligned}$$

We apply the formula

$$\omega L_k^{(p+1)}(\omega) = (k+p+1)L_k^{(p)}(\omega)-(k+1)L_{k+1}^{(p)}(\omega)$$

with $p = 0$ to obtain

$$F(\omega) = L_1^{(0)}(\omega) - \omega \sum_{k=1}^{\infty} \frac{L_k^{(0)}(\omega)}{(k+1)^2}.$$

Applying the formula

$$\frac{1}{(k+1)^2} = \int_o^{\infty} e^{-(k+1)s} sds$$

and the generating formula of the Laguerre polynomials yields

$$\begin{aligned} F(\omega) &= L_1^{(0)}(\omega) - \omega \sum_{k=1}^{\infty} L_k^{(0)} \int_0^{\infty} e^{-s(k+1)} sds \\ &= L_1^{(0)}(\omega) - \omega \int_0^{\infty} se^{-s} \left[\sum_{k=1}^{\infty} L_k^{(0)}(\omega) e^{-sk} \right] ds \\ &= L_1^{(0)}(\omega) - \omega \int_0^{\infty} se^{-s} \left[\frac{1}{(1-e^{-s})^2} \exp\left\{ -\frac{e^{-s}}{1-e^{-s}} \omega \right\} - L_0^{(1)}(\omega) \right] ds \\ &= L_1^{(0)}(\omega) + \omega L_0^{(1)}(\omega) \int_0^{\infty} se^{-s} ds - \omega \int_0^{\infty} \frac{se^{-s}}{(1-e^{-s})^2} \exp\left\{ -\frac{e^{-s}}{1-e^{-s}} \omega \right\} ds. \end{aligned}$$

Note that

$$\int_0^{\infty} se^{-s} ds = 1 \quad \text{and} \quad L_1^{(0)}(\omega) + \omega L_0^{(1)}(\omega) = 1.$$

We obtain

$$F(\omega) = 1 - \omega \int_0^{\infty} \frac{se^{-s}}{(1-e^{-s})^2} \exp\left\{ -\frac{e^{-s}}{1-e^{-s}} \omega \right\} ds.$$

This yields

$$\begin{aligned} \widetilde{M}_1(z,\tau) &= \frac{e^{-\omega/2}}{2|\tau|\pi} F(\omega) = \frac{e^{-\omega/2}}{2|\tau|\pi} - \frac{\omega}{2|\tau|\pi} \int_0^{\infty} \frac{se^{-s}}{(1-e^{-s})^2} \exp\left\{ -\frac{1+e^{-s}}{1-e^{-s}} \frac{\omega}{2} \right\} ds \\ &= \frac{e^{-|\tau||z|^2}}{2|\tau|\pi} - \frac{|z|^2}{\pi} \int_0^{\infty} \frac{s}{(\sinh s)^2} e^{-|\tau||z|^2 \coth s} ds. \end{aligned}$$

Taking the inverse Fourier transform with respect to τ yields

$$\begin{aligned} M_1(z,t) &= \frac{1}{2\pi} \int_{-\infty}^{\infty} e^{it\tau} \widetilde{M}_1(z,\tau) d\tau = \frac{1}{(2\pi)^2} \int_{-\infty}^{\infty} |\tau|^{-1} e^{it\tau - |z|^2|\tau|} d\tau \\ &\quad - \frac{|z|^2}{2\pi^2} \int_0^{\infty} \frac{s}{(\sinh s)^2} \int_{-\infty}^{\infty} e^{it\tau - |\tau||z|^2 \coth s} d\tau ds. \end{aligned}$$

The second part can be integrated with respect to τ:

$$\int_{-\infty}^{\infty} e^{it\tau-|\tau||z|^2\coth s}d\tau = \frac{1}{|z|^2\coth s - it} + \frac{1}{|z|^2\coth s + it} = \frac{2|z|^2\coth s}{|z|^4\coth^2 s + t^2}.$$

Hence, we have

$$M_1(z,t) = \frac{1}{(2\pi)^2}\int_{-\infty}^{\infty}|\tau|^{-1}e^{it\tau-|z|^2|\tau|}d\tau - \frac{|z|^4}{\pi^2}\int_0^{\infty}\frac{s\coth s}{|z|^4\cosh^2 s + t^2\sinh^2 s}ds.$$

We can compute the first integral using the convolution. It is similar to the integral we computed when we derived the fundamental solution.

$$\begin{aligned}\frac{1}{(2\pi)^2}\int_{-\infty}^{\infty}|\tau|^{-1}e^{it\tau-|z|^2|\tau|}d\tau &= \frac{1}{2\pi}\left[-\frac{\varepsilon}{\pi} - \frac{\ln|t|}{\pi}\right] * \left[\frac{|z|^2}{\pi(|z|^4+t^2)}\right]\\ &= -\frac{1}{2\pi^3}\left[\varepsilon\pi + \left(\frac{\pi}{2}-\frac{\theta}{4}\right)\ln(t^2+|z|^4)\right],\end{aligned}$$

where $\theta = \arctan(|z|^2/|t|)$. Summarizing the computation, we have the relative fundamental solution:

$$M_1(z,t) = -\frac{1}{2\pi^3}\left[\varepsilon\pi + \left(\frac{\pi}{2} - \frac{\arctan(|z|^2/|t|)}{4}\right)\ln(t^2+|z|^4)\right] \qquad (12.8.63)$$

$$- \frac{|z|^4}{\pi^2}\int_0^{\infty}\frac{s\coth s}{|z|^4\cosh^2 s + t^2\sinh^2 s}ds. \qquad (12.8.64)$$

The integral converges for $z \neq 0$.

12.9 The Kernel $\Psi_m(z,t)$ for m > n

In the case of $m > n$, we do not have a simple formula like (12.4.41). We shall calculate the following integral in the sense of distributions:

$$\begin{aligned}&\int_{-\infty}^{\infty}|\tau|^{n-m}e^{it\tau+\alpha\,\mathrm{sgn}(\tau)s-|\tau|\gamma(s;\mathbf{z})}d\tau\\ &= e^{\alpha s}\int_0^{\infty}\tau^{n-m}e^{it\tau-\tau\gamma(s;\mathbf{z})}d\tau + e^{-\alpha s}\int_0^{\infty}\tau^{n-m}e^{-it\tau-\tau\gamma(s;\mathbf{z})}d\tau. \quad (12.9.65)\end{aligned}$$

We apply the following formula (see p. 177, equation (14) of Gelfand and Shilov [50]):

$$\int_0^\infty \tau^{-n} e^{i\sigma\tau} d\tau = a_0^{(n)} \sigma^{n-1} - a_{-1}^{(n)} \sigma^{n-1} \log(\sigma + i0), \quad \text{with} \tag{12.9.66}$$

$$a_{-1}^{(n)} = \frac{i^{n-1}}{(n-1)!} \quad \text{and} \quad a_0^{(n)} = \frac{i^{n-1}}{(n-1)!}\left[1 + \frac{1}{2} + \cdots + \frac{1}{n-1} + \Gamma'(1) + i\frac{\pi}{2}\right]. \tag{12.9.67}$$

This yields

$$\int_0^\infty \tau^{n-m} e^{it\tau - \tau\gamma(s;\mathbf{z})} d\tau = (i\gamma(s;\mathbf{z}) + t)^{m-n-1} \times [a_0^{(m-n)} - a_{-1}^{(m-n)} \log(i\gamma(s;\mathbf{z}) + t)], \tag{12.9.68}$$

$$\int_0^\infty \tau^{n-m} e^{-it\tau - \tau\gamma(s;\mathbf{z})} d\tau = (i\gamma(s;\mathbf{z}) - t)^{m-n-1} \times [a_0^{(n)} - a_{-1}^{(m-n)} \log(i\gamma(s;\mathbf{z}) - t)]. \tag{12.9.69}$$

Taking the inverse Fourier transform of (12.4.43) and applying (12.9.65), (12.9.68) and (12.9.69), we obtain

$$\begin{aligned}\Psi_m(\mathbf{z}, t) &= \frac{1}{2\pi^{n+1}\Gamma(m)} \int_0^\infty s^{m-1} e^{\alpha s} \prod_{j=1}^n \frac{a_j}{\sinh(a_j s)} \\ &\quad \times (i\gamma(s;\mathbf{z}) + t)^{m-n-1} \left[a_0^{(m-n)} - a_{-1}^{(m-n)} \log(i\gamma(s;\mathbf{z}) + t)\right] ds \\ &\quad + \frac{1}{2\pi^{n+1}\Gamma(m)} \int_0^\infty s^{m-1} e^{-\alpha s} \prod_{j=1}^n \frac{a_j}{\sinh(a_j s)} \\ &\quad \times (i\gamma(s;\mathbf{z}) - t)^{m-n-1} \left[a_0^{(m-n)} - a_{-1}^{(m-n)} \log(i\gamma(s;\mathbf{z}) - t)\right] ds,\end{aligned}$$

where $a_0^{(m-n)}$ and $a_{-1}^{(m-n)}$ are given by (12.9.67). We can also rewrite the above equation in terms of the complex distance and volume element (12.4.42), but here we shall omit the details of algebraic manipulation:

$$\begin{aligned}\Psi_m(\mathbf{z}, t) &= -\frac{2^{m-1} i^{m-n} a_{-1}^{(m-n)}}{(2\pi)^n \Gamma(m)} \int_{-\infty}^0 s^{m-1} e^{-2\alpha s} v(s) [g(s;\mathbf{z}, t)]^{m-n-1} \\ &\quad \times \log(i \cdot g(s;\mathbf{z}, t)) ds + \frac{2^m i^{m-n-1}}{(2\pi)^{n+1}\Gamma(m)} \int_{-\infty}^\infty s^{m-1} e^{2\alpha s} v(s) \\ &\quad \times [g(s;\mathbf{z}, t)]^{m-n-1} \left[a_0^{(m-n)} - a_{-1}^{(m-n)} \log(i \cdot g(s;\mathbf{z}, t))\right] ds.\end{aligned} \tag{12.9.70}$$

12.10 The Isotropic Heisenberg Group

The special case of $\alpha = 0$ and $a_j = 1/4$ for all $j = 1, 2, \ldots, n$ has been studied by Doley, Benson and Ratcliff [40], though they defined the $\mathbf{Z}_j$ and $\bar{\mathbf{Z}}_j$ differently. We shall summarize their result here and show that it coincides with our result in the special case. Let

$$\mathbf{Z}'_j = 2\frac{\partial}{\partial z_j} - \frac{1}{2} i \bar{z}_j \frac{\partial}{\partial t} \quad \text{and} \quad \bar{\mathbf{Z}}'_j = 2\frac{\partial}{\partial \bar{z}_j} + \frac{1}{2} i z_j \frac{\partial}{\partial t}, \quad j = 1, 2, \ldots, n.$$

Note that $\mathbf{Z}'_j = 2\mathbf{Z}_j$ and $\bar{\mathbf{Z}}'_j = 2\bar{\mathbf{Z}}_j$ with $a_j = 1/4$. They defined the *Heisenberg sub-Laplacian* to be

$$\Delta_{\mathrm{H}_n} = \frac{1}{2} \sum_{j=1}^{n} (\bar{\mathbf{Z}}'_j \mathbf{Z}'_j + \mathbf{Z}'_j \bar{\mathbf{Z}}'_j). \tag{12.10.71}$$

It is easy to see that $\Delta_{\mathrm{H}_n} = -4\mathcal{L}_0$. Let

$$\frac{|\mathbf{z}|^2}{4} - it = re^{i\theta}, \quad \text{where } r = \left(\frac{|\mathbf{z}|^4}{16} + t^2\right)^{1/2} \text{ and } -\frac{\pi}{2} < \theta < \frac{\pi}{2}.$$

Then they introduced the functions

$$G_s(\theta) = e^{i(n-m+1)\theta} \int_0^s \frac{1}{s_m} \cdots \int_0^{s_3} \frac{1}{s_2} \int_0^{s_2} \frac{s_1^{n-1}}{(1 - s_1^2)^{m-1}(s_1^2 + e^{2i\theta})^{n-m+1}} k \times ds_1 \cdots ds_m \tag{12.10.72}$$

and

$$\psi_m(\theta) = \lim_{s \to 1^-} \mathrm{Re}(G_s(\theta)).$$

Finally, they showed that

$$\Psi_m(\mathbf{z}, t) = \frac{(-1)^m (n-m)!}{2^n \pi^{n+1} r^{n-m+1}} \psi_m(\theta) \tag{12.10.73}$$

is a tempered fundamental solution for $\Delta^m_{\mathrm{H}_n}$ with singular support $\{(\mathbf{0}, 0)\}$.

Remark 12.10.1. In [40], the Haar measure of H_n is defined to be $(2\pi)^{-n-1}$ times the Euclidean measure on $\mathbb{R}^{2n+1}$. Formula (12.10.73) is different from the one given in Theorem A of [40] since we normalize the Haar measure to be the Euclidean measure. The integration in the definition of $G_s(\theta)$ should be $ds_1 \cdots ds_m$, not $ds_1 \cdots ds_n$. Note that our m is equal to their p.

We will show that if we set $a_j = a = \frac{1}{4}$ and $\alpha = 0$ in (12.4.40), we can obtain the result of [40]. The relation $\Delta_{\mathrm{H}_n}^m = -4\mathcal{L}_0$ implies that we must prove

$$\mathcal{L}_0^{-m}(\mathbf{z},t) = (-4)^m \Psi_m(\mathbf{z},t). \tag{12.10.74}$$

We first rewrite formula (12.10.73). The crucial observation is that

$$\lim_{s\to 1^-} G_s(\theta) = \frac{(-1)^{m-1}e^{i(n-m+1)\theta}}{(m-1)!}\int_0^1 \left(\frac{\log s_1}{1-s_1^2}\right)^{m-1} \times \frac{s_1^{n-1}}{(s_1^2+e^{2i\theta})^{n-m+1}}ds_1. \tag{12.10.75}$$

Formula (12.10.75) can be proved via integration by parts. Equations (12.10.73) and (12.10.75) yield

$$\Psi_m(\mathbf{z},t) = -\frac{(n-m)!}{2^n\pi^{n+1}\Gamma(m)r^{n-m+1}} \times \mathrm{Re}\left\{e^{i(n-m+1)\theta} \times \int_0^1 \frac{(\log u)^{m-1}u^{n-1}}{(1-u^2)^{m-1}(u^2+e^{2i\theta})^{n-m+1}}du\right\}. \tag{12.10.76}$$

In the case of $a_j = a$ for all $j \geq 1$, then (12.4.40) becomes

$$\begin{aligned}\mathcal{L}_0^{-m}(\mathbf{z},t) &= \frac{(n-m)!}{2\pi^{n+1}\Gamma(m)}\int_{-\infty}^{\infty}\left(\frac{a}{\sinh(as)}\right)^n \frac{s^{m-1}ds}{[a|\mathbf{z}|^2\coth(as)-it]^{n-m+1}} \\ &= \frac{a^{n-m+1}(n-m)!}{2\pi^{n+1}\Gamma(m)}\int_{-\infty}^{\infty}\left(\frac{as}{\sinh(as)}\right)^{m-1} \\ &\quad \times [a|\mathbf{z}|^2\cosh(as) - it\sinh(as)]^{m-n-1}ds.\end{aligned}$$

To evaluate $\mathcal{L}_0^{-m}(\mathbf{z},t)$ when $|\mathbf{z}| \neq 0$, we set

$$re^{-i\theta} = a|\mathbf{z}|^2 - it \quad \text{with } r = \left(a^2|\mathbf{z}|^4+t^2\right)^{1/2} \text{ and } |\theta| < \frac{\pi}{2}.$$

Then the identity $\cosh(as-i\theta) = \cosh(as)\cos\theta - i\sinh(as)\sin\theta$ yields

$$a|\mathbf{z}|^2[\cosh(as) - it\sinh(as)]^{m-n-1} = r^{m-n-1}[\cosh(as-i\theta)]^{m-n-1}.$$

This implies

$$\mathcal{L}_0^{-m}(\mathbf{z},t) = \frac{a^{n-m+1}(n-m)!}{2\pi^{n+1}\Gamma(m)r^{n-m+1}}\int_{-\infty}^{\infty}\left(\frac{as}{\sinh(as)}\right)^{m-1}[\cosh(as-i\theta)]^{m-n-1}ds. \tag{12.10.77}$$

If we let

$$F(\theta) = \int_0^\infty \left(\frac{as}{\sinh(as)}\right)^{m-1} [\cosh(as + i\theta)]^{m-n-1} ds,$$

then $\mathcal{L}_0^{-m}$ can be written as

$$\mathcal{L}_0^{-m}(\mathbf{z}, t) = \frac{a^{n-m+1}(n-m)!}{\pi^{n+1}\Gamma(m) r^{n-m+1}} \times \mathrm{Re}(F(\theta)). \tag{12.10.78}$$

We now use the definitions of sinh and cosh to rewrite $F(\theta)$ as

$$F(\theta) = 2^n e^{i(n-m+1)\theta} \int_0^\infty \left(\frac{as}{1 - e^{-2as}}\right)^{m-1} A \frac{e^{-nas}}{[e^{-2as} + e^{2i\theta}]^{n-m+1}} ds. \tag{12.10.79}$$

Introducing the new variable $u = e^{-as}$, we reduce (12.10.79) to the form

$$F(\theta) = \frac{2^n(-1)^{m-1} e^{i(n-m+1)\theta}}{a} \int_0^1 \frac{(\log u)^{m-1} u^{n-1}}{(1-u^2)^{m-1}(u^2 + e^{2i\theta})^{n-m+1}} du. \tag{12.10.80}$$

Combining (12.10.78) and (12.10.80), we obtain

$$\mathcal{L}_0^{-m}(\mathbf{z}, t) = \frac{2^n(-1)^{m-1} a^{n-m}(n-m)!}{\pi^{n+1}\Gamma(m) r^{n-m+1}} \times \mathrm{Re}\left\{ e^{i(n-m+1)\theta} \times \int_0^1 \frac{(\log u)^{m-1} u^{n-1}}{(1-u^2)^{m-1}(u^2 + e^{2i\theta})^{n-m+1}} du \right\}. \tag{12.10.81}$$

By comparison with (12.10.76) and (12.10.81), the relation (12.10.74) holds by setting $a = 1/4$ in (12.10.81).

12.11 Conclusions

Laguerre calculus is a powerful tool for harmonic analysis on the Heisenberg group. Many sub-elliptic partial differential operators can be inverted by Laguerre calculus. In this chapter we used the Laguerre calculus to find explicit kernels of the fundamental solution for the Paneitz operator and its heat equation. The Paneitz operator, which plays an important role in CR-geometry, can be written as

$$\mathcal{P}_\alpha = \mathcal{L}_\alpha \bar{\mathcal{L}}_\alpha = \frac{1}{4}\left[\sum_{j=1}^n (\mathbf{Z}_j \bar{\mathbf{Z}}_j + \bar{\mathbf{Z}}_j \mathbf{Z}_j)\right]^2 + \alpha^2 \mathbf{T}^2.$$

Here $\{\mathbf{Z}_j\}_{j=1}^n$ is an orthonormal basis for the sub-bundle $T^{(1,0)}$ of the complex tangent bundle $T_{\mathbb{C}}(\mathrm{H}_n)$ and $\mathbf{T}$ is the "missing direction." The operator $\mathcal{L}_\alpha$ is the sub-Laplacian on the Heisenberg group, which is sub-elliptic if α does not belong to an exceptional set Λ_α. We also constructed projection operators and a relative fundamental solution for the operator $\mathcal{L}_\alpha$ for $\alpha \in \Lambda_\alpha$.

Chapter 13
Constructing Heat Kernels for Degenerate Elliptic Operators

13.1 Introduction

In this chapter we describe a method that was first studied by Beals [12] and Aarão [1, 2] to construct heat kernels for a large class of operators that may or may not be group invariant, including certain degenerate elliptic and kinetic operators. Once again, we shall start with the Heisenberg group. Consider first the n-dimensional Hermite operator

$$\Delta - \lambda^2 |x|^2 = \sum_{k=1}^{n} \frac{\partial^2}{\partial x_k^2} - \lambda^2 \sum_{k=1}^{n} x_k^2. \tag{13.1.1}$$

Inspired by the Gaussian, we assume that the heat kernel for (13.1.1) has the form

$$P_t(x, y) = \phi(t) e^{-Q_t(x,y)}, \qquad t > 0, \tag{13.1.2}$$

where $Q_t(x, y)$ is a quadratic form in $2n$ variables. Taking symmetries into account, we expect

$$Q_t(x, y) = \frac{1}{2}\alpha(t)\langle x, x\rangle + \beta(t)\langle x, y\rangle + \frac{1}{2}\alpha(t)\langle y, y\rangle,$$

where $\langle \cdot, \cdot \rangle$ denotes the inner product. We apply the operator

$$-\frac{\partial}{\partial t} + \Delta - \lambda^2 |x|^2$$

to (13.1.2), multiply by $e^{Q_t(x,y)}$, and set the coefficients of $\langle x, x\rangle$, $\langle x, y\rangle$, $\langle y, y\rangle$ and also the constant term equal to zero. The conditions on the coefficients are

$$-\frac{1}{2}\dot{\alpha} = \alpha^2 - \lambda^2, \qquad -\dot{\beta} = 2\alpha\beta, \qquad -\frac{1}{2}\dot{\alpha} = \beta^2 \tag{13.1.3}$$

and

$$\phi^{-1}\dot{\phi} = -n\alpha. \tag{13.1.4}$$

O. Calin et al., *Heat Kernels for Elliptic and Sub-elliptic Operators*, Applied and Numerical Harmonic Analysis, DOI 10.1007/978-0-8176-4995-1_13,

Since P must act like an approximation to the identity as $t \to 0$, we need α to blow up, and $\beta \sim -\alpha$, as $t \to 0$.

The equation $\dot{\alpha} = 2\alpha^2 - 2\lambda^2$ is a Riccati equation which can be solved by means of the usual linearization $\alpha = u^{-1} + \lambda$. Then the solution that blows up at $\lambda = 0$ is

$$\alpha(t) = \frac{\lambda \cosh(2\lambda t)}{\sinh(2\lambda t)}.$$

Then $-\dot{\beta} = 2\alpha\beta$, with boundary condition $\beta \sim \alpha$ as $t \to 0$, and $\phi^{-1}\dot{\phi} = -n\alpha$, normalized so that P is an approximation to the identity, has the solutions

$$\beta(t) = -\frac{\lambda}{\sinh(2\lambda t)} \qquad \text{and} \qquad \phi(t) = -\frac{\lambda^{\frac{n}{2}}}{(2\pi)^{\frac{n}{2}} \sinh^{\frac{n}{2}}(2\lambda t)}.$$

The remaining equation $-\frac{1}{2}\dot{\alpha} = \beta^2$ is a consequence.

The sub-Laplacian associated with the $(2n+1)$-dimensional Heisenberg group has the form

$$\mathcal{L} = \sum_{k=1}^{2n} \frac{\partial^2}{\partial x_k^2} + a^2 \left(\sum_{k=1}^{n} x_k^2 \right) \frac{\partial^2}{\partial x_0^2} + 2 \left\langle Bx, \frac{\partial}{\partial x} \right\rangle \frac{\partial}{\partial x_0}, \tag{13.1.5}$$

where B is a skew-symmetric $2n \times 2n$ matrix. We take the partial Fourier transform with respect to the x_0-variable to obtain in $\mathbb{R}^{2n}$ the operator

$$\widetilde{\mathcal{L}}_\tau = \Delta - a^2\tau^2|x|^2 + 2i\tau \left\langle Bx, \frac{\partial}{\partial x} \right\rangle,$$

where $x = (x_1, \ldots, x_{2n})$ and, as usual, $|x|^2 = \sum_{k=1}^{n} x_k^2$. Because of the group structure, we may use the same method which has been discussed above to compute the kernel at the origin $y = 0$. The skew symmetry of B reduces the associated equation to the same form as (13.1.3) and (13.1.4). We take the inverse partial Fourier transform with respect to the τ-variable and rescale the variable of integration to obtain the heat kernel for the Heisenberg sub-Laplacian:

$$P(x_0, x; t) = \frac{1}{(2\pi t)^{n+1}} \int_{\mathbb{R}} e^{\frac{\tau(2ix_0 - a\coth(2a\tau)|x|^2)}{2t}} \left(\frac{2a\tau}{\sinh(2a\tau)} \right)^n d\tau. \tag{13.1.6}$$

It can be shown that despite the use of the partial Fourier transform and its inverse, formula (13.1.6) is an integral that is absolutely convergent when $|x| \neq 0$. We also know that the absolute convergence when $|x| = 0$, $x_1 \neq 0$ can be obtained by a change in the contour integration.

This is a very powerful method which can be used to handle a class of operators that arise in fluid dynamics and statistical physics (see Aarão [1]) either directly as in the case of (13.1.1) or via a partial Fourier transform as in the case of (13.1.5).

This class of operators includes operators associated to general step-2 two nilpotent Lie groups and Grushin operators like

$$L_1 = \frac{\partial^2}{\partial x_1^2} + x_1^2 \frac{\partial^2}{\partial x_2^2}, \tag{13.1.7}$$

and Kolmogorov-type operators like

$$L_2 = \frac{\partial^2}{\partial x_1^2} - bx_1 \frac{\partial}{\partial x_2} - c\frac{\partial}{\partial x_1} + \kappa, \tag{13.1.8}$$

$$L_3 = \frac{\partial^2}{\partial x_1^2} - x_1 \frac{\partial}{\partial x_2} + x_2 \frac{\partial}{\partial x_1} + \kappa, \tag{13.1.9}$$

$$L_4 = \left(\frac{\partial^2}{\partial x_1^2} + \frac{\partial^2}{\partial x_2^2}\right) - bx_1 \frac{\partial}{\partial x_1} - bx_2 \frac{\partial}{\partial x_2} + \kappa, \tag{13.1.10}$$

$$L_5 = \left(\frac{\partial^2}{\partial x_1^2} + \frac{\partial^2}{\partial x_2^2} + \frac{\partial^2}{\partial x_3^2}\right) + bx_2 \frac{\partial}{\partial x_3} + \kappa. \tag{13.1.11}$$

We will use this method to compute heat kernels for this class of operators in the next section.

13.2 Finding Heat Kernels for Operators L_j, $j = 1, \ldots, 5$

In general, we want to find the heat kernel for the following operator:

$$\begin{aligned} L &= -\left(a_1 \frac{\partial^2}{\partial x_1^2} + a_2 \frac{\partial^2}{\partial x_2^2} + a_3 \frac{\partial^2}{\partial x_3^2}\right) + (b_1 x_1 + b_2 x_2 + b_3 x_3) \frac{\partial}{\partial x_1} \\ &\quad + (b_4 x_1 + b_5 x_2 + b_6 x_3) \frac{\partial}{\partial x_2} + (b_7 x_1 + b_8 x_2 + b_9 x_3) \frac{\partial}{\partial x_3} + \kappa \\ &= \left\langle A \frac{\partial}{\partial x}, \frac{\partial}{\partial x} \right\rangle + \left\langle Bx, \frac{\partial}{\partial x} \right\rangle + \kappa, \end{aligned} \tag{13.2.12}$$

where

$$A = \begin{bmatrix} a_1 & 0 & 0 \\ 0 & a_2 & 0 \\ 0 & 0 & a_3 \end{bmatrix}, \qquad B = \begin{bmatrix} b_1 & b_2 & b_3 \\ b_4 & b_5 & b_6 \\ b_7 & b_8 & b_9 \end{bmatrix},$$

with $a_1, a_2, a_3, b_1, \ldots, b_9$ and κ constants. Our goal is to obtain a formula for the heat kernel $P_t(x, y)$ which satisfies

$$\frac{\partial}{\partial t} P_t(x, y) = L P_t(x, y), \qquad t > 0,$$
$$\lim_{t \to 0+} P_t(x, y) = \delta(x - y), \tag{13.2.13}$$

where $x = (x_1, x_2, x_3)$ and $y = (y_1, y_2, y_3)$. For convenience, we discuss this system in matrix notation. Since the situations for $n = 2$ and $n = 3$ are similar, we give a detailed discussion just for the case $n = 2$. We first apply conjugation by $e^{\kappa t}$ to remove the constant term κ. As we mentioned in Sect. 13.1, we assume the kernel $P_t(x, y)$ has the following form:

$$P_t(x, y) = \phi(t) e^{-Q_t(x,y)} = \phi(t) e^{-\frac{1}{2}\langle \mathcal{M}\vec{v}, \vec{v}\rangle} = \phi(t) e^{-\frac{1}{2}\langle \alpha(t)x, x\rangle - \langle \beta x, y\rangle - \frac{1}{2}\langle \gamma(t) y, y\rangle}, \tag{13.2.14}$$

where

$$\mathcal{M} = \begin{bmatrix} \alpha & \beta \\ \beta^T & \gamma \end{bmatrix}$$

and β^T is the transpose of the matrix β. Here $\vec{v} = [x, y]^T = [x_1, x_2, y_1, y_2]^T$ is a vector in $\mathbb{R}^{2\times 2}$ and α, β, γ are 2×2 matrices, with $\alpha \geq 0$, $\gamma \geq 0$.

We shall calculate the derivatives of $P_t(x, y)$ first:

$$\begin{aligned} \frac{\partial P_t}{\partial t} &= \dot{\phi} e^{-\frac{1}{2}\langle \mathcal{M}\vec{v}, \vec{v}\rangle} - \frac{\phi}{2}\langle \dot{\mathcal{M}}\vec{v}, \vec{v}\rangle e^{-\frac{1}{2}\langle \mathcal{M}\vec{v}, \vec{v}\rangle} \\ &= \left(\frac{\dot{\phi}}{\phi} - \frac{\phi}{2}\langle \dot{\mathcal{M}}\vec{v}, \vec{v}\rangle \right) P_t, \end{aligned}$$

and

$$\frac{\partial P_t}{\partial x_1} = -\frac{1}{2}\left(\frac{\partial}{\partial x_1}\langle \mathcal{M}\vec{v}, \vec{v}\rangle \right) P_t.$$

Here the dot denotes the differentiation $\frac{d}{dt}$. Since x_1 is the first coordinate of $\vec{v}$, then $\frac{\partial \vec{v}}{\partial x_1} = \vec{e}_1$. Using the symmetry of $\mathcal{M}$, we have

$$\frac{\partial}{\partial x_1}\langle \mathcal{M}\vec{v}, \vec{v}\rangle = \langle \mathcal{M}\vec{e}_1, \vec{v}\rangle + \langle \mathcal{M}\vec{v}, \vec{e}_1\rangle = 2\langle \mathcal{M}\vec{v}, \vec{e}_1\rangle.$$

It follows that

$$\frac{\partial P_t}{\partial x_1} = -\langle \mathcal{M}\vec{v}, \vec{e}_1\rangle P_t.$$

Similarly, one has

$$\frac{\partial P_t}{\partial x_2} = -\langle \mathcal{M}\vec{v}, \vec{e}_2 \rangle P_t,$$

and

$$\begin{aligned}\frac{\partial^2 P_t}{\partial x_1^2} &= -\langle \mathcal{M}\vec{e}_1, \vec{e}_1 \rangle P_t + \langle \mathcal{M}\vec{v}, \vec{e}_1 \rangle^2 P_t \\ &= -\alpha_{11} + \left(\vec{v}^T \mathcal{M}\vec{e}_1 \vec{e}_1^T \mathcal{M}\vec{v}, \vec{e}_1\right) P_t = -\alpha_{11} + \langle \mathcal{M} E_1 \mathcal{M}\vec{v}, \vec{v} \rangle P_t,\end{aligned}$$

where α_{11} is the $(1, 1)$-component of the matrix $\mathcal{M}$ and

$$E_1 = \vec{e}_1 \vec{e}_1^T = \begin{bmatrix} 1 & 0 & 0 & 0 \\ 0 & 0 & 0 & 0 \\ 0 & 0 & 0 & 0 \\ 0 & 0 & 0 & 0 \end{bmatrix}.$$

Similarly, we have

$$\frac{\partial^2 P_t}{\partial x_2^2} = -\alpha_{22} + \langle \mathcal{M} E_2 \mathcal{M}\vec{v}, \vec{v} \rangle P_t,$$

where

$$E_2 = \vec{e}_2 \vec{e}_2^T = \begin{bmatrix} 0 & 0 & 0 & 0 \\ 0 & 1 & 0 & 0 \\ 0 & 0 & 0 & 0 \\ 0 & 0 & 0 & 0 \end{bmatrix}.$$

Applying $-\frac{\partial}{\partial t} + L$ to (13.2.14) gives

$$\frac{\dot{\phi}}{\phi} - \frac{1}{2}\langle \dot{\mathcal{M}}\vec{v}, \vec{v} \rangle - \langle \mathcal{M}\vec{v}, \widetilde{B}\vec{v} \rangle + \left(a_1\alpha_{11} + a_2\alpha_{22}\right) - \langle \mathcal{M}\widetilde{A}\mathcal{M}\vec{v}, \widetilde{B}\vec{v} \rangle = 0, \quad (13.2.15)$$

where the matrices $\widetilde{A}$ and $\widetilde{B}$ are given by

$$\widetilde{A} = \begin{bmatrix} a_1 & 0 & 0 & 0 \\ 0 & a_2 & 0 & 0 \\ 0 & 0 & 0 & 0 \\ 0 & 0 & 0 & 0 \end{bmatrix} = \begin{bmatrix} A & 0 \\ 0 & 0 \end{bmatrix}, \qquad \widetilde{B} = \begin{bmatrix} b_1 & b_2 & 0 & 0 \\ b_3 & b_4 & 0 & 0 \\ 0 & 0 & 0 & 0 \\ 0 & 0 & 0 & 0 \end{bmatrix} = \begin{bmatrix} B & 0 \\ 0 & 0 \end{bmatrix}.$$

Since

$$\langle \mathcal{M}\vec{v}, \widetilde{B}\vec{v} \rangle = \langle \widetilde{B}^T \mathcal{M}\vec{v}, \vec{v} \rangle = \langle \mathcal{M}\widetilde{B}\vec{v}, \vec{v} \rangle,$$

we see that if we require that

$$\dot{\mathcal{M}} + \mathcal{M}\widetilde{B} + \widetilde{B}^T \mathcal{M} + 2\mathcal{M}\widetilde{A}\mathcal{M} = 0,$$

and that

$$\frac{\dot{\phi}}{\phi} + \left(a_1\alpha_{11} + a_2\alpha_{22}\right) = 0,$$

then (13.2.15) will hold for all $\vec{v}$. More precisely, we have the following system:

$$\dot{\alpha} = -2\alpha \begin{bmatrix} a_1 & 0 \\ 0 & a_2 \end{bmatrix} \alpha - \alpha \begin{bmatrix} b_1 & b_2 \\ b_3 & b_4 \end{bmatrix} - \begin{bmatrix} b_1 & b_3 \\ b_2 & b_4 \end{bmatrix} \alpha = -2\alpha A \alpha - \left(\alpha B + B^T \alpha\right),$$

$$\dot{\beta} = -2\beta \begin{bmatrix} a_1 & 0 \\ 0 & a_2 \end{bmatrix} \alpha - \begin{bmatrix} b_1 & b_3 \\ b_2 & b_4 \end{bmatrix} \beta = -2\alpha A \beta - B^T \beta,$$

$$\dot{\gamma} = -2\beta \begin{bmatrix} a_1 & 0 \\ 0 & a_2 \end{bmatrix} \beta^T = -2\beta A \beta^T,$$

$$\phi^{-1}\dot{\phi} = -(a_1\alpha_{11} + a_2\alpha_{22}) = -\text{trace}\,(A\alpha). \tag{13.2.16}$$

We need to solve the first equation in (13.2.16). Once this is done, the solutions β and γ are obtained in a simple manner. Indeed, let χ be a matrix solving $\dot{\chi} = \begin{bmatrix} b_1 & b_2 \\ b_3 & b_4 \end{bmatrix} \chi = B\chi$ satisfying the initial condition $\chi(0) = \chi_0$. Set $\beta = \alpha\chi$ and $\gamma = \chi^T \alpha \chi + M$, with M a constant symmetric matrix. It is easy to see that as long as α satisfies the first equation of (13.2.16), then β and γ will satisfy the second and third equations of (13.2.16), respectively. So, the problem reduces to finding the matrices χ_0 and M.

Since $P_t(0,0) \to \infty$ as $t \to 0^+$, we know that $\phi(t) \to \infty$ as $t \to 0^+$. Moreover, the second equation of (13.2.13) provides us with the boundary conditions

$$\alpha \to +\infty \cdot I, \qquad \alpha \sim -\beta \sim \gamma, \qquad \text{as } t \to 0^+. \tag{13.2.17}$$

Assume that

$$\langle \mathcal{M}\vec{v}, \vec{v} \rangle = \langle \alpha(x + \chi y), x + \chi y \rangle + \langle My, y \rangle.$$

Again, the second equation of (13.2.13) tells us that

$$\int_{\mathbb{R}^2} P_t(x,y)dx = \phi(t) e^{-\frac{1}{2}\langle My, y \rangle} \int_{\mathbb{R}^2} e^{-\frac{1}{2}\langle \alpha(x+\chi y), x+\chi y \rangle} dx \to 1$$

as $t \to 0^+$. Hence we conclude that $M = 0$. Finally, we want to conclude that $\chi_0 = \chi(0) = -I$. If not, then for some x we have $x + \chi_0 x \neq 0$, and hence

$$\langle \alpha(x + \chi x), x + \chi x \rangle \to \infty, \qquad \text{as} \quad t \to 0^+.$$

This implies that $P_t(x, x)$ converges to zero, when it should be going to infinity. This contradiction leads to the conclusion that $\chi_0 = -I$. As mentioned in [1], this argument can only be fully justified after we find α. Furthermore, χ_0 is nonsingular. If it were, then for some $w \neq 0$, one has $\chi_0 w = 0$, and then

$$\chi(t)w = \exp\left(t \begin{bmatrix} b_1 & b_2 \\ b_3 & b_4 \end{bmatrix}\right) \chi_0 w = 0, \qquad \forall t > 0,$$

and this would imply that $\gamma(t)$ is singular for all $t > 0$.

Now we shall solve the Riccati matrix equation in the system (13.2.16) for α:

$$\dot{\alpha} = -2\alpha \begin{bmatrix} a_1 & 0 \\ 0 & a_2 \end{bmatrix} \alpha - \alpha \begin{bmatrix} b_1 & b_2 \\ b_3 & b_4 \end{bmatrix} - \begin{bmatrix} b_1 & b_3 \\ b_2 & b_4 \end{bmatrix} \alpha = -2\alpha A \alpha - \left(\alpha B + B^T \alpha\right). \tag{13.2.18}$$

Conjugating the (13.2.18) by the matrix χ, we have

$$\chi^T \dot{\alpha} \chi + \chi^T B^T \alpha \chi + \chi^T \alpha B \chi = -2\chi^T \alpha A \alpha \chi.$$

Because $\dot{\chi} = B\chi$, we can rewrite the above equation as

$$\frac{d}{dt}\left(\chi^T \alpha \chi\right) = -2\chi^T \alpha A \alpha \chi.$$

Since $\gamma = \chi^T \alpha \chi$, the above equation becomes

$$\dot{\gamma} = -2\gamma A' \gamma,$$

where $A' = \chi^{-1} A (\chi^T)^{-1}$. Denote $\omega = \gamma^{-1}$. Then one has $\dot{\gamma} = -\gamma \dot{\omega} \gamma$, and so

$$\dot{\omega} = 2A'.$$

If we can solve the above equation with initial condition $\omega(0) = 0$ (since $\omega = \gamma^{-1}$), we find that

$$\alpha = (\chi^T)^{-1} \omega^{-1} \chi^{-1} = \left(\chi \omega \chi^T\right)^{-1}.$$

In this case, the general self-adjoint solution of (13.2.18) is (see Beals [12])

$$\alpha(t) = e^{2tB} \left(\mathcal{W} + 2\int_0^t e^{2\tau B} A e^{2\tau B^T} d\tau\right)^{-1} e^{2tB^T}. \tag{13.2.19}$$

Since $\beta = \alpha\chi$, then one obtains

$$\beta(t) = \alpha(t)e^{-2tB} = \left(\mathcal{W} + 2\int_0^t e^{2\tau B} A e^{2\tau B^T} d\tau\right)^{-1} e^{-2tB}, \tag{13.2.20}$$

where $\mathcal{W}$ will be determined later. Moreover, it is immediate from (13.2.16) and (13.2.20) that the solution to the third equation $\dot{\gamma} = -2\beta A \beta^T$ in (13.2.16) that satisfies the boundary condition is

$$\gamma(t) = e^{-2tB}\alpha(t)e^{-2tB^T} = \left(\mathcal{W} + 2\int_0^t e^{2\tau B} A e^{2\tau B^T} d\tau\right)^{-1}. \tag{13.2.21}$$

Combining (13.2.19)–(13.2.21), we may write the quadratic form as

$$Q_t(x, y) = \left(-\frac{1}{2}\right)\left\langle \gamma(t)\left[e^{2tB}x - y\right], e^{2tB}x - y\right\rangle.$$

Any solution of (13.2.16) fulfills the first requirement in (13.2.13). The delicate part is to choose the constants of integration: the matrix $\mathcal{W}$ in (13.2.19) to (13.2.21) so that it satisfies the boundary conditions in (13.2.13). In particular, $\mathcal{W}$ must be singular.

We expect the heat kernel to provide a two-sided inverse for the heat equation for L. Therefore, P should satisfy analogous conditions in the y-variables for the adjoint operator L^* of L. These conditions provide another set of equations dual to (13.2.16). The latter equations can be shown to be compatible with (13.2.13). We illustrate these points with a few comments on examples.

13.3 Some Explicit Calculations

We are very interested in finding the heat kernel for the operator L defined in (13.2.12):

$$\begin{aligned} L = &-\left(a_1\frac{\partial^2}{\partial x_1^2} + a_2\frac{\partial^2}{\partial x_2^2} + a_3\frac{\partial^2}{\partial x_3^2}\right) + (b_1x_1 + b_2x_2 + b_3x_3)\frac{\partial}{\partial x_1} \\ &+ (b_4x_1 + b_5x_2 + b_6x_3)\frac{\partial}{\partial x_2} + (b_7x_1 + b_8x_2 + b_9x_3)\frac{\partial}{\partial x_3} + \kappa. \end{aligned}$$

Let us point out that for each of the operators (13.1.8)–(13.1.11), the major difficulty is choosing the correct constants of integration in order to achieve the initial conditions (13.2.17). Let us demonstrate this point by a detailed discussion of operators L_j, $j = 1, \dots, 5$.

13.3.1 Laplace Operator

We start with the Laplace operator $L = -\Delta$ on $\mathbb{R}^3$. In this case, $A = I_3$ is the 3×3 identity matrix, and $B = 0$ is the zero matrix. Likewise, $\kappa = 0$. Hence, from (13.2.16), one has

$$\dot{\alpha} = -2\langle \alpha, \alpha \rangle.$$

Let $\eta = \alpha^{-1}$. As we have seen before, the initial condition $\eta(0)$ of Λ must be zero. Moreover, $\alpha\eta = I$ implies that $\dot{\eta} = 2I$, which provides $\eta(t) = 2tI$. Therefore,

$$\alpha(t) = -\frac{1}{2t} I.$$

Since $\dot{\chi} = B\chi = 0$, we have $\chi(t) = \chi(0) = \chi_0$. Since $\langle \mathcal{M}\vec{v}, \vec{v} \rangle = \langle \alpha(x + \chi y), x + \chi y \rangle$, we obtain

$$P_t(x, y) = \phi(t) e^{-\frac{1}{4t}|x+\chi y|^2}.$$

However, we know that $\chi_0 = -I$ from the discussion in Sect. 13.2. This implies that $P_t(x, y) = \phi(t) e^{-\frac{1}{4t}|x-y|^2}$. Finally, let us calculate ϕ. From (13.2.16), we need to solve

$$\frac{\dot{\phi}}{\phi} = -\mathrm{trace}\,(A\alpha) = -\frac{3}{2t}.$$

Hence $\phi(t) = ct^{-\frac{3}{2}}$ for some constant c. Since

$$\int_{\mathbb{R}^3} e^{-\frac{1}{4t}|x-y|^2} dx = (4\pi t)^{\frac{3}{2}}$$

independently of y, one concludes that $c = \frac{1}{(4\pi)^{3/2}}$. Therefore,

$$P_t(x, y) = \frac{1}{(4\pi t)^{3/2}} e^{-\frac{1}{4t}|x-y|^2}.$$

In fact, a similar formula works for any n. In this case, α, β and γ are all $n \times n$ matrices. The system (13.2.16) would be the same and hence the above analysis will be the same. The only difference is that $\dot{\phi}(t) = -\frac{n}{2t}\phi$. Moreover, the operator $-\Delta$ is invariant under translations. Thus it suffices to consider $y = 0$, and the heat kernel for the n-dimensional Laplace operator becomes $P_t(x) = \frac{1}{(4\pi t)^{n/2}} e^{-\frac{1}{4t}|x|^2}$, which is the well-known Gaussian distribution.

Similarly, we may deal with hypoelliptic second-order operators associated with two-step nilpotent groups which lead to (13.1.5), whose solutions are similar to

$$\alpha(t) = \frac{\tau \cosh(2\tau t)}{\sinh(2\tau t)},$$
$$\beta(t) = -\frac{\tau}{\sinh(2\tau t)},$$
$$\phi(t) = -\frac{\tau^{\frac{n}{2}}}{[2\pi \sinh(2\tau t)]^{\frac{n}{2}}}.$$

These kernels are similar to the formula

$$P_t(x_0, x; 0, 0) = \frac{1}{(2\pi t)^{n+1}} \int_{\mathbb{R}} e^{\frac{\tau(2ix_0 - a\coth(2a\tau)|x|^2)}{2t}} \left(\frac{2a\tau}{\sinh(2a\tau)}\right)^n d\tau,$$

which was obtained by several different methods.

13.3.2 Kolmogorov Operator

Now let us turn to the operator $L_2 = \frac{\partial^2}{\partial x_1^2} - bx_1 \frac{\partial}{\partial x_2} - c\frac{\partial}{\partial x_1}$ given in (13.1.8). We first conjugate it by $\exp\{\frac{2cx - c^2 t}{4}\}$ to eliminate the term $c\frac{\partial}{\partial x}$. We may assume $b \neq 0$ and rescale in x_1 so that $b = 1$. This equation was first studied by Kolmogorov in [81]. Starting at a point x in the plane, a particle describes a motion $(X_1(t), X_2(t))$ in the following way. In the first coordinate, $X_1(t)$, it is just a Brownian motion starting at x_1. In the second coordinate, $X_2(t)$ obeys the equation $dX_2 = X_1(t)dt$ starting at x_2. In particular, if $X_1 < 0$, then X_2 decreases. In this situation, after a time $t > 0$ elapses, the probability that the particle is inside the open set $\Omega \subset \mathbb{R}^2$ is given by $\int_\Omega p_t(x, y)dx$. In [81], Kolmogorov used probability techniques to compute the fundamental solution for the operator L_2. Here we use the method which was discussed in the previous section to find its heal kernel (and hence the fundamental solution). Now, the operator L_2 reduces to (13.2.16) with

$$A = \begin{bmatrix} 1 & 0 \\ 0 & 0 \end{bmatrix}, \qquad B = \begin{bmatrix} 0 & 0 \\ 1 & 0 \end{bmatrix}.$$

We set $\eta = \alpha^{-1}$ and find

$$\dot{\eta} = 2A + \left(\eta B + B^T \eta\right).$$

More precisely, this gives us

$$\begin{bmatrix} \dot{\eta}_{11} & \dot{\eta}_{12} \\ \dot{\eta}_{21} & \dot{\eta}_{22} \end{bmatrix} = \begin{bmatrix} 2 & \eta_{11} \\ \eta_{11} & 2\eta_{12} \end{bmatrix}.$$

It follows that

$$\eta_{11} = 2t + \kappa_1, \qquad \eta_{12} = t^2 + \kappa_1 t + \kappa_2, \qquad \eta_{22} = \frac{2}{3}t^3 + \kappa_1 t^2 + 2\kappa_2 t + \kappa_3.$$

Here κ_1, κ_2 and κ_3 are three constants. As before, we set $\eta(0) = 0$. This implies that $\kappa_1 = \kappa_2 = \kappa_3 = 0$. Hence

$$\alpha(t) = \eta^{-1} = \frac{1}{t}\begin{bmatrix} 2 & -\frac{3}{t} \\ -\frac{3}{t} & \frac{6}{t^2} \end{bmatrix} = \begin{bmatrix} \frac{2}{t} & -\frac{3}{t^2} \\ -\frac{3}{t^2} & \frac{6}{t^3} \end{bmatrix}. \tag{13.3.22}$$

Since $B^2 = 0$, the exponential

$$e^{2\tau B} = I + 2\tau B = \begin{bmatrix} 1 & 0 \\ -\tau & 1 \end{bmatrix}$$

and thus

$$\gamma(t) = e^{-2\tau B^T}\alpha(t)e^{2\tau B} = \begin{bmatrix} \frac{2}{t} & \frac{3}{t^2} \\ \frac{3}{t^2} & \frac{6}{t^3} \end{bmatrix},$$

and

$$\beta(t) = -\gamma(t)e^{2\tau B} = \begin{bmatrix} \frac{1}{t} & -\frac{3}{t^2} \\ \frac{3}{t^2} & -\frac{6}{t^3} \end{bmatrix}.$$

It follows that

$$\begin{aligned}
Q_t(x, y) &= -\frac{1}{2}\langle\alpha(t)x, x\rangle - \langle\beta(t)x, y\rangle - \frac{1}{2}\langle\gamma(t)y, y\rangle \\
&= -\frac{1}{2}\left\{\left\langle \begin{bmatrix} \frac{2}{t} & -\frac{3}{t^2} \\ -\frac{3}{t^2} & \frac{6}{t^3} \end{bmatrix}\begin{bmatrix} x_1 \\ x_2 \end{bmatrix}, \begin{bmatrix} x_1 \\ x_2 \end{bmatrix}\right\rangle + 2\left\langle \begin{bmatrix} \frac{1}{t} & -\frac{3}{t^2} \\ \frac{3}{t^2} & -\frac{6}{t^3} \end{bmatrix}\begin{bmatrix} x_1 \\ x_2 \end{bmatrix}, \begin{bmatrix} y_1 \\ y_2 \end{bmatrix}\right\rangle\right. \\
&\qquad \left. + \left\langle \begin{bmatrix} \frac{2}{t} & \frac{3}{t^2} \\ \frac{3}{t^2} & \frac{6}{t^3} \end{bmatrix}\begin{bmatrix} y_1 \\ y_2 \end{bmatrix}, \begin{bmatrix} y_1 \\ y_2 \end{bmatrix}\right\rangle\right\} \\
&= -\frac{1}{2t^3}\left\{\left\langle \begin{bmatrix} 2t^2x_1 - 3tx_2 \\ -3tx_1 + 6x_2 \end{bmatrix}, \begin{bmatrix} x_1 \\ x_2 \end{bmatrix}\right\rangle + 2\left\langle \begin{bmatrix} t^2x_1 - 3tx_2 \\ 3tx_1 - 6x_2 \end{bmatrix}, \begin{bmatrix} y_1 \\ y_2 \end{bmatrix}\right\rangle\right. \\
&\qquad \left. + \left\langle \begin{bmatrix} 2t^2y_1 + 3ty_2 \\ 3ty_1 + 6y_2 \end{bmatrix}, \begin{bmatrix} y_1 \\ y_2 \end{bmatrix}\right\rangle\right\} \\
&= -\frac{1}{2t^3}\Big\{\big(2t^2x_1^2 - 5tx_1x_2 + 6x_2^2\big) + 2\big(t^2x_1y_1 - 3tx_2y_1 + 3tx_1y_2 \\
&\qquad -6x_2y_2\big) + \big(2t^2y_1^2 + 6ty_1y_2 + 6y_2\big)\Big\}
\end{aligned}$$

$$= -\frac{1}{2t^3}\left\{2t^2(x_1^2+x_1y_1+y_1^2)+6t(x_1+y_1)(y_2-x_2)+6(x_2-y_2)^2\right\}$$
$$= -\frac{1}{2t^3}\left\{\frac{t^2}{2}(x_1-y_1)^2+6\big[(x_2-y_2)-\frac{1}{2}(x_1+y_1)t\big]^2\right\}.$$

We also know that

$$A\alpha(t) = \begin{bmatrix} 1 & 0 \\ 0 & 0 \end{bmatrix}\begin{bmatrix} \frac{2}{t} & -\frac{3}{t^2} \\ -\frac{3}{t^2} & \frac{6}{t^3} \end{bmatrix} = \begin{bmatrix} \frac{2}{t} & 0 \\ 0 & 0 \end{bmatrix}.$$

Hence

$$\operatorname{trace}\big(A\alpha(t)\big) = \frac{2}{t}.$$

It follows that $\phi^{-1}\dot{\phi} = -\operatorname{trace}\big(A\alpha(t)\big) = -\frac{2}{t}$, which implies that $\phi(t) = ct^{-2}$. Now we need to calculate the constant c. However, a straightforward computation shows that

$$\int_{\mathbb{R}^2} e^{-Q_t(x,y)}\,dx = \frac{\pi t^2}{\sqrt{3}}.$$

Hence we have proved the following theorem.

Theorem 13.3.1. *The heat kernel of the Kolmogorov operator* $-\frac{\partial^2}{\partial x_1^2} - x_1\frac{\partial}{\partial x_2}$ *is*

$$P_t(x_1,x_2,y_1,y_2) = \frac{\sqrt{3}}{\pi t^2}e^{-\frac{(x_1-y_1)^2}{4t}-\frac{3}{t^3}\left[(x_2-y_2)-\frac{1}{2}(x_1+y_1)t\right]^2}.$$

13.3.3 The Operator L_3

Now we are going to study the operator

$$L_3 = -\frac{\partial^2}{\partial x_1^2} + x_1\frac{\partial}{\partial x_2} - x_2\frac{\partial}{\partial x_1} + \kappa.$$

Here we have

$$A = \begin{bmatrix} 1 & 0 \\ 0 & 0 \end{bmatrix}, \qquad B = \begin{bmatrix} -1 & 0 \\ 0 & 1 \end{bmatrix}.$$

Once again, set $\eta = \alpha^{-1}$ and find

$$\begin{bmatrix} \dot{\eta}_{11} & \dot{\eta}_{12} \\ \dot{\eta}_{21} & \dot{\eta}_{22} \end{bmatrix} = \begin{bmatrix} 2-2\eta_{12} & \eta_{11}-\eta_{22} \\ \eta_{11}-\eta_{22} & 2\eta_{12} \end{bmatrix}.$$

It follows that

$$\ddot{\eta}_{12} + 4\eta_{12} = 2,$$

and we have $\eta_{12} = \frac{1}{2} + \kappa_1 \cos(2t) + \kappa_2 \sin(2t)$. Hence

$$\eta_{11} = \kappa_3 + t - \kappa_1 \cos(2t) + \kappa_2 \sin(2t),$$
$$\eta_{22} = \kappa_4 + t + \kappa_1 \sin(2t) - \kappa_2 \cos(2t).$$

Here κ_j, $j = 1, \ldots, 4$, are constants. Since $\dot{\eta}_{12} = \eta_{11} - \eta_{22}$, we conclude that $\kappa_3 = \kappa_4$. The initial condition $\eta(0) = 0$ tells us that $\eta_{11}(0) = \kappa_3 - \kappa_1 = 0$ and $\eta_{22}(0) = \kappa_4 - \kappa_2 = 0$. Hence, $\kappa_1 = \kappa_2 = \kappa_3 = \kappa_4 = c$, which provides

$$\alpha(t) = \frac{1}{2(t^2 - \sin^2 t)} \begin{bmatrix} 2t - \sin(2t) & -1 + \cos(2t) \\ -1 + \cos(2t) & 2t + \sin(2t) \end{bmatrix}.$$

Next, we need to find χ. Since the matrix B is diagonal, we know that χ is also diagonal and can be calculated as

$$\chi(t) = \begin{bmatrix} e^{-t} & 0 \\ 0 & e^t \end{bmatrix}.$$

Finally, the equation $\dot{\phi} + \alpha_{11}\phi = 0$ can be written as

$$\dot{\phi} + \frac{1}{2} \frac{2t - \sin(2t)}{t^2 - \sin^2 t} \phi = 0,$$

which gives

$$\phi(t) = \frac{c}{\sqrt{t^2 - \sin^2 t}}.$$

A straightforward computation shows that

$$\int_{\mathbb{R}^2} e^{-Q_t(x,y)} dx = 2\pi \sqrt{t^2 - \sin^2 t}$$

and so we conclude that $c = \frac{1}{2\pi}$. We have proved the following theorem.

Theorem 13.3.2. *The heat kernel of the operator*

$$L_3 = -\frac{\partial^2}{\partial x_1^2} + x_1 \frac{\partial}{\partial x_2} - x_2 \frac{\partial}{\partial x_1}$$

is

$$P_t(x,y) = \frac{1}{2\pi\sqrt{t^2-\sin^2 t}} \exp\left\{-\frac{1}{2(t^2-\sin^2 t)}\Big[(2t-\sin(2t))^2\left(x_1+e^{-t}y_1\right)^2 - 2(1-\cos(2t))(x_1+e^{-t}y_1)(x_2+e^t y_2) + (2t+\sin(2t))^2 \times (x_2+e^t y_2)^2\Big]\right\}.$$

13.3.4 The Operator L_4

The operator L_4 is defined by

$$L_4 = -\left(\frac{\partial^2}{\partial x_1^2} + \frac{\partial^2}{\partial x_2^2}\right) + bx_1\frac{\partial}{\partial x_1} - bx_2\frac{\partial}{\partial x_1} + \kappa.$$

Here we have

$$A = \begin{bmatrix} 1 & 0 \\ 0 & 1 \end{bmatrix}, \qquad B = \begin{bmatrix} b & 0 \\ 0 & -b \end{bmatrix}.$$

Obviously, both matrices A and B are diagonal. Set $\eta = \alpha^{-1}$ and get

$$\begin{bmatrix} \dot\eta_{11} & \dot\eta_{12} \\ \dot\eta_{21} & \dot\eta_{22} \end{bmatrix} = \begin{bmatrix} 2+2b\eta_{11} & 0 \\ 0 & 2-2b\eta_{22} \end{bmatrix}.$$

It follows that

$$\eta_{11} = \frac{1}{b}\left(e^{2bt}-1\right), \qquad \eta_{12} = 0, \qquad \eta_{22} = \frac{1}{b}\left(1-e^{-2bt}\right).$$

Hence

$$\alpha(t) = b\begin{bmatrix} \frac{1}{e^{2bt}-1} & 0 \\ 0 & \frac{1}{1-e^{-2bt}} \end{bmatrix}.$$

Using the same method as before, we know that

$$\phi(t) = \frac{b}{4\pi\sinh(bt)}.$$

The matrix χ is diagonal because B is diagonal. In fact, one finds that

$$\chi = \begin{bmatrix} -e^{bt} & 0 \\ 0 & -e^{-2bt} \end{bmatrix}.$$

As a consequence, we have the following theorem.

Theorem 13.3.3. *The heat kernel of the operator*

$$L_4 = -\left(\frac{\partial^2}{\partial x_1^2} + \frac{\partial^2}{\partial x_2^2}\right) + bx_1\frac{\partial}{\partial x_1} - bx_2\frac{\partial}{\partial x_1} + \kappa$$

is

$$P_t(x,y) = \frac{b}{2\pi(e^{bt} - e^{-bt})} \exp\left\{-\frac{b}{2}\left[\frac{(x_1 - e^{bt}y_1)^2}{e^{2bt} - 1} + \frac{(x_2 - e^{-bt}y_2)^2}{1 - e^{-2bt}}\right]\right\}.$$

This recovers a result of Lingevitch and Bernoff in [88].

13.3.5 The Operator L_5

The operator L_5 is defined as follows:

$$L_5 = -\left(\frac{\partial^2}{\partial x_1^2} + \frac{\partial^2}{\partial x_2^2} + \frac{\partial^2}{\partial x_3^2}\right) - 2bx_2\frac{\partial}{\partial x_3}.$$

Here we have

$$A = \begin{bmatrix} 1 & 0 & 0 \\ 0 & 1 & 0 \\ 0 & 0 & 1 \end{bmatrix}, \qquad B = \begin{bmatrix} 0 & 0 & 0 \\ 0 & 0 & 0 \\ 0 & -2b & 0 \end{bmatrix}.$$

Again set $\eta = \alpha^{-1}$. With the initial condition $\eta(0) = 0$, we find that

$$\begin{bmatrix} \dot\eta_{11} & \dot\eta_{12} & \dot\eta_{13} \\ \dot\eta_{21} & \dot\eta_{22} & \dot\eta_{23} \\ \dot\eta_{31} & \dot\eta_{32} & \dot\eta_{33} \end{bmatrix} = \begin{bmatrix} 2 & 0 & -2b\eta_{12} \\ 0 & 2 & -2b\eta_{22} \\ -2b\eta_{12} & -2b\eta_{22} & 2 - 4b\eta_{23} \end{bmatrix}.$$

It follows that

$$\eta_{11} = \eta_{22} = 2t, \qquad \eta_{12} = \eta_{21} = \eta_{13} = \eta_{32} = 0, \qquad \eta_{23} = -2bt^2,$$
$$\eta_{33} = 2t + 8\frac{b^2t^3}{3}.$$

Thus

$$\alpha(t) = \frac{1}{2t\left(1 + \frac{b^2t^2}{3}\right)} \begin{bmatrix} 1 + \frac{b^2t^2}{3} & 0 & 0 \\ 0 & 1 + 4\frac{b^2t^2}{3} & bt \\ 0 & bt & 1 \end{bmatrix}.$$

Furthermore,

$$\chi = \begin{bmatrix} -1 & 0 & 0 \\ 0 & -1 & 0 \\ 0 & 2bt & -1 \end{bmatrix}.$$

One can also show that

$$\phi(t) = \frac{1}{8\sqrt{\pi^3 t^3}\sqrt{1 + \frac{b^2t^2}{3}}}.$$

Hence, we proved the following theorem.

Theorem 13.3.4. *The heat kernel of the operator*

$$L_5 = -\left(\frac{\partial^2}{\partial x_1^2} + \frac{\partial^2}{\partial x_2^2} + \frac{\partial^2}{\partial x_3^2}\right) - 2bx_2\frac{\partial}{\partial x_3}$$

is

$$\begin{aligned} P_t(x, y) = \frac{1}{8\sqrt{\pi^3 t^3}\sqrt{1 + \frac{b^2t^2}{3}}} \exp \Bigg\{ & - \frac{1}{2t\left(1 + \frac{b^2t^2}{3}\right)} \Bigg[\left(1 + \frac{b^2t^2}{3}\right)(x_1 - y_1)^2 \\ & + \left(1 + 4\frac{b^2t^2}{3}\right)(x_2 - y_2)^2 + (x_3 - y_3)^2 \\ & + 2bt(x_2 - y_2)(x_3 - y_3) \\ & + 2b^2t^2 y_2(x_2 - y_2) + 2bty_2(x_3 - y_3) \Bigg] \Bigg\}. \end{aligned}$$

Chapter 14
Heat Kernel for the Kohn Laplacian on the Heisenberg Group

14.1 The Kohn Laplacian on the Heisenberg Group

We shall deal next with the nonsymmetric form of the Heisenberg group. The Heisenberg group considered in this section will be the set $\mathbb{H}_n = \mathbb{R}^n \times \mathbb{R}^n \times \mathbb{R}$ with the following group law:

$$(x, y, t) * (x', y', t') = (x + x', y + y', t + t' + x \cdot y'),$$

where $(x, y, t), (x', y', t') \in \mathbb{R}^n \times \mathbb{R}^n \times \mathbb{R}$ and

$$x \cdot y' = \sum_{k=1}^{n} x_k y'_k.$$

The leftinvariant vector fields for this group structure are

$$X_j = \frac{\partial}{\partial x_j} + y_j \frac{\partial}{\partial t}, \qquad Y_j = \frac{\partial}{\partial y_j}, \qquad 1 \leq j \leq n,$$

and

$$T = \frac{\partial}{\partial t}. \tag{14.1.1}$$

The group $\mathbb{H}_n$ can also be identified with the following hypersurface in $\mathbb{C}^{n+1}$:

$$\mathcal{H}_n = \left\{ (z_1, \ldots, z_n, z_{n+1}) \in \mathbb{C}^{n+1} : \operatorname{Im}(z_{n+1}) = \frac{1}{2} \sum_{j=1}^{n} [\operatorname{Im}(z_j)]^2 \right\}.$$

Here we identify $\left(z_1, \ldots, z_n, t + \frac{i}{2} \sum_{j=1}^{n} [\operatorname{Im}(z_j)]^2\right) \in \mathcal{H}_n$ with

$$(z_1, \ldots, z_n, t) = (x_1, \ldots, x_n, y_1, \ldots, y_n, t),$$

O. Calin et al., *Heat Kernels for Elliptic and Sub-elliptic Operators*,
Applied and Numerical Harmonic Analysis, DOI 10.1007/978-0-8176-4995-1_14,

where $z_j = x_j + iy_j \in \mathbb{C}$. With this identification, the leftinvariant vector fields of type $(1,0)$ and $(0,1)$ are, respectively,

$$\begin{aligned} Z_j &= \frac{1}{2}(X_j - iY_j) = \frac{\partial}{\partial z_j} - i\frac{z_j - \bar{z}_j}{4}\frac{\partial}{\partial t}, \\ \bar{Z}_j &= \frac{1}{2}(X_j + iY_j) = \frac{\partial}{\partial \bar{z}_j} - i\frac{z_j - \bar{z}_j}{4}\frac{\partial}{\partial t} \end{aligned} \tag{14.1.2}$$

for $1 \leq j \leq n$. Hence, one has

$$[\bar{Z}_j, Z_k] = \frac{i}{2}\delta_{jk}T, \qquad j,k = 1,\ldots,n.$$

We make a further assumption that $\{Z_j,\ldots,Z_n,\bar{Z}_1,\ldots,\bar{Z}_n\}$ form an orthonormal basis for the $T^{(1,0)} \oplus T^{(0,1)}$ on $\mathbb{H}_n$.

Let ϑ_q be the set of all increasing q-tuples $J = (j_1,\ldots,j_q)$, $1 \leq j_k \leq n$. The Kohn Laplacian

$$\Box_b = \bar{\partial}_b\bar{\partial}_b^* + \bar{\partial}_b^*\bar{\partial}_b$$

acting on $(0,q)$-forms on $\mathcal{H}_n \approx \mathbb{H}_n$ can be calculated in terms of the vector fields (14.1.1) and (14.1.2). Suppose

$$g = \sum_{J\in\vartheta_q} g_J\bar{\omega}^J = \sum_{J\in\vartheta_q} g_{j_1,\ldots,j_q}\bar{\omega}_{j_1} \wedge \cdots \wedge \bar{\omega}_{j_q}$$

is a $(0,q)$-form. Then it can be shown that (see Chap. 12 in Stein's book [103])

$$\Box_b g = \sum_{J\in\vartheta_q} \mathcal{L}_{n-2q}g_J\bar{\omega}^J, \tag{14.1.3}$$

where

$$\mathcal{L}_\alpha = -\frac{1}{4}\left(\sum_{j=1}^{n}(X_j^2 + Y_j^2) + i\alpha T\right). \tag{14.1.4}$$

The heat operator is defined on $(0,q)$-forms g on $\mathbb{H}_n$ with coefficient functions that depend on the time variable $u \in (0,\infty)$ and $(x,y,t) \in \mathbb{H}_n$. This is given by

$$\frac{\partial g}{\partial u} + \Box_b g = 0.$$

As we can see from (14.1.3), the Kohn Laplacian $\Box_b$ acts diagonally; we can look at the heat operator acting on each component of g; i.e., we are looking for the heat kernel K_u that satisfies

$$\begin{cases} \frac{\partial K_u}{\partial u} = -\mathcal{L}_\alpha K_u \qquad u > 0, \quad (x, y, t) \in \mathbb{H}_n, \\ \lim_{u \to 0^+} K_u(x, y, t) = \delta_0(x, y, t). \end{cases} \tag{14.1.5}$$

Here $\delta_0(x, y, t)$ is the Dirac delta function at the origin.

14.2 Full Fourier Transform on the Group

We will use the full Fourier transform in the variables (x, y, t). Let (ξ, ζ, τ) be the dual variables of (x, y, t) and define

$$\widehat{f}(\xi, \zeta, \tau) = \int_{\mathbb{H}_n} f(x, y, t) e^{-i(x\cdot\xi + y\cdot\zeta + t\tau)} dx\, dy\, dt.$$

Then one may prove the following theorem.

Theorem 14.2.1. *For any $\alpha \in \mathbb{C}$, the Fourier transform of the heat kernel satisfying* (14.1.5) *is given by*

$$\widehat{K}_u(\xi, \zeta, \tau) = \frac{e^{-\alpha \frac{u\tau}{4}}}{\cosh^{\frac{n}{2}}\left(\frac{u\tau}{2}\right)} e^{-\frac{A}{2}(|\xi|^2 + |\zeta|^2) + iB\xi\cdot\zeta}, \tag{14.2.6}$$

where

$$A = \frac{\sinh\left(\frac{u\tau}{2}\right)}{\tau \cosh\left(\frac{u\tau}{2}\right)}, \qquad B = \frac{2\sinh^2\left(\frac{u\tau}{4}\right)}{\tau \cosh\left(\frac{u\tau}{2}\right)}. \tag{14.2.7}$$

Proof. Recall that

$$\begin{aligned} \widehat{X_j^2 f}(\xi, \zeta, \tau) &= \left(-\xi^2 - 2i\xi_j\tau\frac{\partial}{\partial\zeta_j} + \tau^2\frac{\partial^2}{\partial\zeta_j^2}\right)\widehat{f}, \\ \widehat{Y_j^2 f}(\xi, \zeta, \tau) &= -\zeta_j^2\widehat{f}, \\ \widehat{Tf}(\xi, \zeta, \tau) &= i\tau\widehat{f}. \end{aligned} \tag{14.2.8}$$

Since the heat kernel on the n-dimensional space is the product of n copies of the one-dimensional heat kernel, we may first reduce the problem to the one-dimensional case. Denote by $\widehat{K_u^{\alpha,1}}$ the one-dimensional version of (14.2.6); i.e., for $1 \le j \le n$,

$$\widehat{K_u^{\alpha,j}}(\xi_j, \zeta_j, \tau) = \frac{e^{-\alpha\frac{u\tau}{4}}}{\sqrt{\cosh(\frac{u\tau}{2})}} e^{-\frac{A}{2}(\xi_j^2+\zeta_j^2)+iB\xi_j\zeta_j},$$

where A and B are given by (14.2.7). Then from (14.2.6), one has

$$\widehat{K}_u(\xi, \zeta, \tau) = \prod_{j=1}^{n} \widehat{K_u^{\frac{\alpha}{n},j}}(\xi_j, \zeta_j, \tau), \tag{14.2.9}$$

where

$$\xi = (\xi_1, \dots, \xi_n), \qquad \zeta = (\zeta_1, \dots, \zeta_n) \in \mathbb{R}^n.$$

Then the problem reduces to showing that $\widehat{K_u^{\alpha,j}}$ satisfies the one-dimensional transformed heat equation; i.e., for $1 \le j \le n$,

$$\left(\frac{\partial}{\partial u} - \frac{1}{4}\left(\widehat{X_j^2} + \widehat{Y_j^2} + i\alpha\widehat{T}\right)\right) \widehat{K_u^{\alpha,j}}(\xi_j, \zeta_j, \tau) = 0, \tag{14.2.10}$$

with the initial condition $\widehat{K_0^{\alpha,j}} = 1$. Then by using formula (14.2.9), we can conclude the result of this theorem. ∎

From now on, we are working on the one-dimensional Heisenberg group $\mathbb{H}_1$. We may drop the index j from the kernel. Therefore, x, y, t and ξ, ζ, τ are all real variables. Define

$$\mathcal{K}_\alpha(\xi, \zeta, \tau; u) = \widehat{K_u^\alpha}(\xi, \zeta, \tau)e^{-i\frac{\xi\zeta}{\tau}}. \tag{14.2.11}$$

Then, it is easy to verify the equations

$$\mathcal{K}_\alpha(\xi, \zeta, \tau; 0) = e^{-i\frac{\xi\zeta}{\tau}} \tag{14.2.12}$$

and

$$\frac{\partial \mathcal{K}_\alpha}{\partial u} = \frac{1}{4}\left(\tau^2\frac{\partial^2}{\partial \zeta^2} - \zeta^2 - \alpha\tau\right)\mathcal{K}_\alpha. \tag{14.2.13}$$

Equation (14.2.12) follows from the fact that the Fourier transform of the Dirac delta function is the constant 1. Equation (14.2.13) follows from (14.2.10) and (14.2.8). As in the paper of Boggess and Raich [20], we call (14.2.13) the *transformed heat equation*. It is easy to see that the differential operator (14.2.13) is a Hermite operator. Therefore, we may use the Hermite function to solve the problem.

14.3 Solving the Transformed Heat Equation Using Hermite Functions

For $k = 0, 1, 2, \ldots$ and $x \in \mathbb{R}$, denote

$$h_k(x) = \frac{(-1)^k}{\sqrt{2^k k! \sqrt{\pi}}} e^{\frac{x^2}{2}} \frac{d^k}{dx^k}(e^{-x^2}).$$

The functions $h_k(x)$ satisfy the following relations:

$$\begin{aligned}
L(h_k)(x) &= \left(-\frac{d}{dx} + x\right) h_k(x) \\
&= \left(-\frac{d}{dx} + x\right) \left[\frac{(-1)^k}{\sqrt{2^k k! \sqrt{\pi}}} e^{\frac{x^2}{2}} \frac{d^k}{dx^k}(e^{-x^2})\right] \\
&= \frac{(-1)^k}{\sqrt{2^k k! \sqrt{\pi}}} \left[-x e^{\frac{x^2}{2}} \frac{d^{k+1}}{dx^{k+1}}\left(e^{-x^2}\right) - e^{\frac{x^2}{2}} \frac{d^k}{dx^k}\left(e^{-x^2}\right)\right. \\
&\qquad \left. + x e^{\frac{x^2}{2}} \frac{d^k}{dx^k}\left(e^{-x^2}\right)\right] \\
&= \sqrt{2(k+1)} \frac{(-1)^{k+1}}{\sqrt{2^{k+1}(k+1)! \sqrt{\pi}}} e^{\frac{x^2}{2}} \frac{d^{k+1}}{dx^{k+1}}\left(e^{-x^2}\right) \\
&= \sqrt{2(k+1)} h_{k+1}(x)
\end{aligned} \qquad (14.3.14)$$

and

$$L^*(h_k)(x) = \left(\frac{d}{dx} + x\right) h_k(x) = \sqrt{2k} h_{k-1}(x). \qquad (14.3.15)$$

The operators $L = -\frac{d}{dx} + x$ and $L^* = \frac{d}{dx} + x$ are called the creation and annihilation operators in quantum mechanics. An easy calculation shows that

$$\mathbf{H} = \frac{1}{2}(LL^* + L^*L) = -\frac{d^2}{dx^2} + x^2,$$

the Hermite operator. Hence, by (14.3.14) and (14.3.15), one has (see (1.1.28) in Thangavelu [107])

$$\mathbf{H}(h_k)(x) = -h_k''(x) + x^2 h_k(x) = (2k+1) h_k(x). \qquad (14.3.16)$$

For $\alpha \in \mathbb{R}$, let

$$H_k^\tau(x) = \frac{1}{|\tau|^{\frac{1}{4}}} h_k\left(\frac{x}{\sqrt{|\tau|}}\right).$$

It is known that $\{h_k(x)\}$ and hence $\{H_k^\tau(x)\}$ form an orthonormal system in $L^2(\mathbb{R})$ (see Chap. 1 in Thangavelu [107]). We first assume that $\tau > 0$. Replacing x by $\frac{\zeta}{\sqrt{\tau}}$ in (14.3.16) yields

$$\left(\tau^2 \frac{\partial^2}{\partial \zeta^2} - \zeta^2 - \alpha\tau\right) H_k^\tau = -(2k+1+\alpha)\tau H_k^\tau(\zeta). \tag{14.3.17}$$

In other words, H_k^τ is an eigenfunction of the differential operator on the right-hand side of (14.2.13) with eigenvalue $-(2k+1+\alpha)\tau$.

Since $\{H_k^\tau(x)\}$ is an orthonormal basis for $L^2(\mathbb{R})$, $\mathcal{K}_\alpha(\xi,\zeta,\tau;u)$ can be expressed as

$$\mathcal{K}_\alpha(\xi,\zeta,\tau;u) = \sum_{k=0}^{\infty} a_k(\xi,\tau) e^{-\frac{1}{4}(2k+1+\alpha)\tau u} H_k^\tau(\zeta), \tag{14.3.18}$$

where $a_k(\xi,\tau)$ are the Fourier–Hermite coefficients which will be determined later. Differentiating (14.3.18) with respect to the time variable u and using (14.3.17), one has

$$\begin{aligned}
\frac{\partial \mathcal{K}_\alpha}{\partial u}(\xi,\zeta,\tau;u) &= \sum_{k=0}^{\infty} a_k(\xi,\tau) e^{-\frac{1}{4}(2k+1+\alpha)\tau u}\left(-\frac{1}{4}(2k+1+\alpha)\right)\tau \cdot H_k^\tau(\zeta) \\
&= \frac{1}{4}\sum_{k=0}^{\infty} a_k(\xi,\tau) e^{-\frac{1}{4}(2k+1+\alpha)\tau u}\left(\tau^2\frac{\partial^2}{\partial\zeta^2} - \zeta^2 - \alpha\tau\right) H_k^\tau(\zeta) \\
&= \frac{1}{4}\left(\tau^2\frac{\partial^2}{\partial\zeta^2} - \zeta^2 - \alpha\tau\right)\left[\sum_{k=0}^{\infty} a_k(\xi,\tau) e^{-\frac{1}{4}(2k+1+\alpha)\tau u} H_k^\tau(\zeta)\right] \\
&= \frac{1}{4}\left(\tau^2\frac{\partial^2}{\partial\zeta^2} - \zeta^2 - \alpha\tau\right)\mathcal{K}_\alpha(\xi,\zeta,\tau;u).
\end{aligned}$$

Hence, $\mathcal{K}_\alpha$ satisfies the transformed heat equation (14.2.13). If the kernel $\mathcal{K}_\alpha$ satisfies the initial condition (14.2.12), one must have

$$e^{-i\frac{\xi\zeta}{\tau}} = \mathcal{K}_\alpha(\xi,\zeta,\tau;0) = \sum_{k=0}^{\infty} a_k(\xi,\tau) H_k^\tau(\zeta).$$

Using the fact that the $\{H_k^\tau(x)\}$ is an orthonormal system, we have

$$a_k(\xi,\tau) = \int_{\mathbb{R}} e^{-i\frac{\xi\zeta}{\tau}} H_k^\tau(\zeta) d\zeta = \tau^{\frac{1}{4}} \int_{\mathbb{R}} e^{-i\frac{\xi}{\sqrt{\tau}}\zeta} h_k(\zeta) d\zeta.$$

The integral on the right-hand side is just the Fourier transform of the Hermite function $h_k(\zeta)$ at the point $\frac{\xi}{\sqrt{\tau}}$. At this point, we need to prove the following auxiliary lemma, which can be found in Thangavelu [107]. For self-containing reasons, we shall repeat the proof here.

Lemma 14.3.1. *The Hermite functions are eigenfunctions of the Fourier transform:*

$$\widehat{h}_k(\xi) = (-i)^k \sqrt{2\pi} h_k(\xi).$$

Proof. From (14.3.14), we know that

$$L(h_k)(x) = \left(-\frac{d}{dx} + x\right) h_k(x) = \sqrt{2k+2} h_k(x).$$

It is known that

$$\widehat{(xf)}(\xi) = i\sqrt{2\pi}\frac{d}{d\xi}\widehat{f}(\xi), \qquad \widehat{\left(-\frac{d}{dx}\right)}(\xi) = -i\sqrt{2\pi}\xi\widehat{f}(\xi).$$

Now taking the Fourier transform on both sides of (14.3.14), we have

$$-i\left(-\frac{d}{d\xi} + \xi\right)\widehat{h}_k(\xi) = \sqrt{4\pi(k+1)}\widehat{h}_k(\xi).$$

If we assume that the lemma is true for h_k, then it follows that

$$\begin{aligned} \sqrt{4\pi(k+1)}\widehat{h}_{k+1}(\xi) &= (-i)^{k+1}\sqrt{2\pi}\left(-\frac{d}{d\xi} + \xi\right) h_k(\xi) \\ &= (-i)^{k+1}\sqrt{4\pi(k+1)} h_{k+1}(\xi). \end{aligned}$$

Therefore, it is enough to show that $\widehat{h}_0(\xi) = \sqrt{2\pi} h_0(\xi)$. But $h_0(x) = e^{-\frac{1}{2}x^2}/\sqrt[4]{\pi}$, and hence by a straightforward calculations, we conclude that $\widehat{h}_0 = \sqrt{2\pi} h_0$. ∎

Therefore, by Lemma 14.3.1, we know that

$$a_k(\xi,\tau) = (-i)^k \sqrt{2\pi}\tau^{\frac{1}{4}} h_k\left(\frac{\xi}{\sqrt{\tau}}\right).$$

Substituting this value of a_k into the expression for $\mathcal{K}_\alpha$, we have

$$\mathcal{K}_\alpha(\xi,\zeta,\tau;u) = \sqrt{2\pi} e^{-\frac{1}{4}(1+\alpha)\tau u} \sum_{k=0}^{\infty} (-i)^k h_k\left(\frac{\xi}{\sqrt{\tau}}\right) h_k\left(\frac{\zeta}{\sqrt{\tau}}\right) e^{-\frac{1}{2}k\tau u}.$$

Now by (14.2.11), we have

$$\begin{aligned}\widehat{K_u^\alpha}(\xi,\zeta,\tau) &= \mathcal{K}_\alpha(\xi,\zeta,\tau;u)e^{i\frac{\xi\zeta}{\tau}}\\ &= \sqrt{2\pi}e^{-\frac{1}{4}(1+\alpha)\tau u}\sum_{k=0}^{\infty}(-i)^k h_k\left(\frac{\xi}{\sqrt{\tau}}\right)h_k\left(\frac{\zeta}{\sqrt{\tau}}\right)e^{-\frac{1}{2}k\tau u}e^{i\frac{\xi\zeta}{\tau}}.\end{aligned}$$

Now let $\Theta = e^{-\frac{1}{2}\tau u}$. Since $|i\Theta| < 1$, we obtain

$$\begin{aligned}\widehat{K_u^\alpha}(\xi,\zeta,\tau) &= \sqrt{2\pi}\Theta^{\frac{1}{2}(1+\alpha)}\left[\sum_{k=0}^{\infty}(-i\Theta)^k h_k\left(\frac{\xi}{\sqrt{\tau}}\right)h_k\left(\frac{\zeta}{\sqrt{\tau}}\right)\right]\cdot e^{i\frac{\xi}{\sqrt{\tau}}\frac{\zeta}{\sqrt{\tau}}}\\ &= \frac{\sqrt{2}\Theta^{\frac{1}{2}(1+\alpha)}}{\sqrt{1+\Theta^2}}e^{-\frac{1}{2}\frac{1-\Theta^2}{1+\Theta^2}(\frac{\xi^2+\zeta^2}{\tau})}e^{i\frac{\xi\zeta}{\tau}(\frac{-2\Theta}{1+\Theta^2}+1)}.\end{aligned}$$

Now

$$\sqrt{1+\Theta^2} = \sqrt{1+e^{-\tau u}} = e^{-\frac{\tau u}{4}}\sqrt{e^{\frac{\tau u}{2}}+e^{-\frac{\tau u}{2}}} = \sqrt{2}e^{-\frac{\tau u}{4}}\sqrt{\cosh\left(\frac{u\tau}{2}\right)}.$$

Then we have

$$\frac{\sqrt{2}\Theta^{\frac{1}{2}(1+\alpha)}}{\sqrt{1+\Theta^2}} = \frac{e^{-\alpha\frac{u\tau}{4}}}{\sqrt{\cosh\left(\frac{u\tau}{2}\right)}}.$$

Next,

$$\frac{1-\Theta^2}{1+\Theta^2} = \frac{1-e^{-\tau u}}{1+e^{-\tau u}} = \frac{\sinh\left(\frac{u\tau}{2}\right)}{\cosh\left(\frac{u\tau}{2}\right)}$$

and

$$\frac{-2\Theta}{1+\Theta^2}+1 = \frac{(1-e^{-\frac{\tau u}{2}})^2}{1+e^{-\tau u}} = \frac{e^{-\frac{\tau u}{2}}(e^{\frac{\tau u}{4}}-e^{-\frac{\tau u}{4}})^2}{e^{-\frac{\tau u}{2}}(e^{\frac{\tau u}{2}}+e^{-\frac{\tau u}{2}})} = \frac{2\sinh^2\left(\frac{u\tau}{4}\right)}{\cosh\left(\frac{u\tau}{2}\right)}.$$

It follows that

$$\widehat{K_u^\alpha}(\xi,\zeta,\tau) = \frac{e^{-\alpha\frac{u\tau}{4}}}{\sqrt{\cosh\left(\frac{u\tau}{2}\right)}}e^{-\frac{A}{2}(|\xi|^2+|\zeta|^2)+iB\xi\cdot\zeta},$$

where

$$A = \frac{\sinh\left(\frac{u\tau}{2}\right)}{\tau\cosh\left(\frac{u\tau}{2}\right)}, \qquad B = \frac{2\sinh^2\left(\frac{u\tau}{4}\right)}{\tau\cosh\left(\frac{u\tau}{2}\right)}.$$

Note that $\widehat{K_0}(\xi,\zeta,\tau) = 1$. The proof of the theorem for $\tau > 0$ is therefore complete.

If $\tau = 0$, the kernel in (14.2.6) becomes

$$\widehat{K_u^\alpha}(\xi, \zeta, 0) = e^{-\frac{u}{4}(\xi^2+\zeta^2)},$$

which is easily shown to satisfy

$$\left(\frac{\partial}{\partial u} - \frac{1}{4}(\widehat{X^2} + \widehat{Y^2} + i\alpha\widehat{T})\right)\widehat{K_u^\alpha}(\xi, \zeta, 0) = 0.$$

If $\tau < 0$, then τ is replaced by $|\tau|$ on the right side of (14.3.17), which will slightly change the previous calculation. However, the formula for the solution given in Theorem 14.2.1 also remains true in this case.

If $\alpha = 0$, then the operator $\mathcal{L}_\alpha$ becomes the sum of the square of vector fields:

$$\mathcal{L} = -\frac{1}{4}\sum_{j=1}^{n}(X_j^2 + Y_j^2). \tag{14.3.19}$$

In this case, by Theorem 14.2.1, we know that the Fourier transform of the heat kernel of the operator $\mathcal{L}$ is

$$\widehat{K}_u(\xi, \zeta, \tau) = \frac{1}{\cosh^{\frac{1}{2}}\left(\frac{u\tau}{2}\right)} e^{-\frac{A}{2}(|\xi|^2+|\zeta|^2)+iB\xi\cdot\zeta}, \tag{14.3.20}$$

where A and B are given by (14.2.7):

$$A = \frac{\sinh\left(\frac{u\tau}{2}\right)}{\tau\cosh\left(\frac{u\tau}{2}\right)}, \qquad B = \frac{2\sinh^2\left(\frac{u\tau}{4}\right)}{\tau\cosh\left(\frac{u\tau}{2}\right)}.$$

It follows that the heat kernel for the sub-Laplacian is

$$\begin{aligned} K_u(\xi, \zeta, \tau) &= \frac{1}{(2\pi)^3}\int_{\mathbb{R}^3}\widehat{K}_u(\xi, \zeta, \tau)e^{i(x\xi+y\zeta+t\tau)}d\xi\, d\zeta\, d\tau \\ &= \frac{1}{2\pi}\int_{\mathbb{R}}\frac{1}{\cosh^{\frac{1}{2}}\left(\frac{u\tau}{2}\right)\sqrt{A^2+B^2}} e^{-\frac{A}{2(A^2+B^2)}(x^2+y^2)-i\frac{B}{2(A^2+B^2)}xy}\, d\tau. \end{aligned}$$

By (14.2.7), it is an easy exercise to show that the coefficients of the real and imaginary parts of the exponent of the above integrand are

$$\frac{B}{2(A^2+B^2)} = \frac{\tau}{2} \quad \text{and} \quad \frac{A}{B} = \frac{\cosh\left(\frac{u\tau}{4}\right)}{\sinh\left(\frac{u\tau}{4}\right)} = \coth\left(\frac{u\tau}{4}\right).$$

Consequently, the heat kernel should be

$$
\begin{aligned}
K_u(\xi, \zeta, \tau) &= \frac{1}{8\pi^2} \int_{\mathbb{R}} \frac{\tau}{\sinh(\frac{u\tau}{4})} e^{-\frac{1}{4}\coth(\frac{u\tau}{4})(x^2+y^2) - i\frac{\tau}{2}xy + i\tau t}\, d\tau \\
&= \frac{4}{(2\pi u)^2} \int_{\mathbb{R}} e^{-\frac{2}{u}\left\{\tau\coth(2\tau)(x^2+y^2) + 2i\tau xy - 4i\tau t\right\}} \frac{2\tau}{\sinh(2\tau)}\, d\tau.
\end{aligned}
$$

14.4 Conclusions

Even if the method described in this chapter is not universally applicable, however, it is widely applicable. Not every operator of the form (13.2.12) has a well-behaved heat kernel. So far, we have not imposed any positivity conditions. For operators of the form (13.2.13) one has an additional problem of tracking behavior with respect to the Fourier transform variable and translating it into the behavior of the heat kernel itself. It may or may not be possible to express the integration with respect to the time variable t for obtaining the fundamental solution in closed form, and reading off the properties of the fundamental solution may pose new difficulties.

Part IV
Pseudo-Differential Operators

Chapter 15
The Pseudo-Differential Operator Technique

The pseudo-differential operator theory emerged from the theory of singular integrals and Fourier analysis, having Kohn and Nirenberg as initiators. The theory was later extended and developed by Hörmander and has become an important tool in the theory of modern PDEs.

In this chapter we shall study the construction of the fundamental solution for heat operators using the symbolic calculus of pseudo-differential operators. After we provide the definition of pseudo-differential operators, we shall deal with the symbol of the product of pseudo-differential operators and provide an estimate for their multi-product. In this chapter we use pseudo-differential operators with both the usual symbols and Weyl symbols.

The main part is dedicated to the construction of the fundamental solution as a pseudo-differential operator with parameter t, for both nondegenerate and degenerate parabolic operators. In the case of the quadratic polynomial symbol of (x, ξ), the exact symbol of the fundamental solution is obtained. These results are then applied to Grushin, sub-Laplacian, and Kolmogorov operators.

It is worth noting that the fundamental solution $E(t)$ obtained in this chapter is a smooth operator for any positive t and $\int_0^c E(t)\, dt$ is a parametrix for any positive c.

This method has proved useful in proving the index theorem in its local version; see Gilkey [53] and Iwasaki [70, 71].

15.1 Basic Results of Pseudo-Differential Operators

We shall start with the definition of pseudo-differential operators following Hörmander [66] and Kumano-go [82], and then we shall provide the estimation of the symbol of multi-product. These computations will play an important role in the construction of the fundamental solution.

O. Calin et al., *Heat Kernels for Elliptic and Sub-elliptic Operators*, Applied and Numerical Harmonic Analysis, DOI 10.1007/978-0-8176-4995-1_15,

15.1.1 Definition of Pseudo-Differential Operators

In the following we shall define the set $S^m_{\rho,\delta}$ of symbols of order m and type (ρ, δ), and organize it as a Fréchet space with respect to some semi-norms; see Hörmander [66] and Kumano-go [82]. We recall the well-known notations

$$|\alpha| = \alpha_1 + \cdots + \alpha_n,$$
$$\partial^\alpha_x f(x) = \frac{\partial^{\alpha_1}}{\partial x_1^{\alpha_1}} \cdots \frac{\partial^{\alpha_n}}{\partial x_1^{\alpha_n}} f(x_1, \ldots, x_n),$$

for any multi-index $\alpha = (\alpha_1, \ldots, \alpha_n) \in \mathbb{R}^n$ and $x = (x_1, \ldots, x_n) \in \mathbb{R}^n$.

Definition 15.1.1. Let m, ρ, δ be real numbers such that $0 \leq \delta \leq \rho \leq 1$ and $\delta < 1$. Then we denote by $S^m_{\rho,\delta}$ the set of all C^∞-functions $p(x, \xi)$ defined on $\mathbb{R}^n \times \mathbb{R}^n$ satisfying the following estimates for any multi-indices $\alpha = (\alpha_1, \alpha_2, \ldots, \alpha_n)$ and $\beta = (\beta_1, \beta_2, \ldots, \beta_n)$;

$$|\partial^\alpha_\xi \partial^\beta_x p(x, \xi)| \leq C_{\alpha,\beta} < \xi >^{m - \rho|\alpha| + \delta|\beta|};$$

with a positive constant $C_{\alpha,\beta}$, where $< \xi >= (1 + |\xi|^2)^{1/2}$.

For simplicity we use the notation

$$p^{(\alpha)}_{(\beta)}(x, \xi) = \partial^\alpha_\xi \partial^\beta_x p(x, \xi).$$

For any symbol $p(x, \xi) \in S^m_{\rho,\delta}$, we define the semi-norms $|p|^{(m)}_\ell$, $\ell = 0, 1, 2, \ldots$, by

$$|p|^{(m)}_\ell = \max_{|\alpha| + |\beta| \leq \ell} \sup_{(x,\xi) \in \mathbb{R}^n \times \mathbb{R}^n} \left\{ |p^{(\alpha)}_{(\beta)}(x, \xi)| < \xi >^{-m + \rho|\alpha| - \delta|\beta|} \right\}. \tag{15.1.1}$$

Then $S^m_{\rho,\delta}$ becomes a Fréchet space with respect to the semi-norms (15.1.1). The following notation will also be used:

$$S^{-\infty} = \bigcap_{-\infty < m < \infty} S^m_{1,0}.$$

In the following definition the prefix *Os*- will denote an *oscillatory integral*. More precisely,

$$Os - \int_{\mathbb{R}^n \times \mathbb{R}^n} e^{-iy \cdot \xi} f(y, \eta) \, dy \, d\eta = \lim_{\epsilon \to 0} \int_{\mathbb{R}^n \times \mathbb{R}^n} e^{-iy \cdot \xi} f(y, \eta) \chi_\varepsilon(y, \eta) \, dy d\eta,$$

where $\chi_\varepsilon(y, \eta) = \chi(\varepsilon y, \varepsilon \eta)$, with $\chi(y, \eta)$ a rapidly decreasing function such that $\chi(0, 0) = 1$. It is left as an exercise for the reader to show that the previous definition is independent on the cut function χ.

A *pseudo-differential operator* P of symbol $\sigma(P) = p(x,\xi) \in S^m_{\rho,\delta}$ is an oscillatory integral defined for all $u \in \mathcal{S}(\mathbb{R}^n) = \{f \in C^\infty(\mathbb{R}^n)\,;\, \sup_{x\in\mathbb{R}^n} |x^\alpha \partial_x^\beta f(x)| < \infty\}$, the set of rapidly decreasing functions, as

$$\begin{aligned} Pu(x) &= p(x,D)u(x) \\ &= Os - (2\pi)^{-n} \int_{\mathbb{R}^n\times\mathbb{R}^n} e^{-iy\cdot\xi} p(x,\xi)u(x+y)dy\,d\xi \\ &= Os - (2\pi)^{-n} \int_{\mathbb{R}^n\times\mathbb{R}^n} e^{i(x-y)\cdot\xi} p(x,\xi)u(y)dy\,d\xi. \end{aligned}$$

15.1.2 Calculus with Pseudo-Differential Operators

This section will deal with some elements of pseudo-differential operator calculus, such as the product formula, the multi-product formula, and the estimates for symbols of the product of pseudo-differential operators.

The product of two pseudo-differential operators $P = p(x,D)$ and $Q = q(x,D)$ is also a pseudo-differential operator, whose symbol is denoted by $p \circ q$ and is given by

$$(p\circ q)(x,D) = p(x,D)q(x,D).$$

In fact, $p \circ q$ is given as an oscillatory integral of the form

$$(p\circ q)(x,\xi) = Os - (2\pi)^{-n} \int_{\mathbb{R}^n\times\mathbb{R}^n} e^{-iy\cdot\eta} p(x,\xi+\eta)q(x+y,\xi)dy\,d\eta.$$

The following result can be found in Kumano-go [82].

Theorem 15.1.2. *Let $p \in S^{m_1}_{\rho,\delta}$ and $q \in S^{m_2}_{\rho,\delta}$ be two symbols. Then for any integer N, we have the expansion*

$$p\circ q = \sum_{j=0}^{N-1} s_j(p,q) + r_N(p,q),$$

where

$$s_j(p,q) = \sum_{|\alpha|=j} \frac{(-i)^{|\alpha|}}{\alpha!} p^{(\alpha)}(x,\xi)q_{(\alpha)}(x,\xi) \in S^{m-(\rho-\delta)j}_{\rho,\delta},$$

$r_N(p,q) \in S^{m-(\rho-\delta)N}_{\rho,\delta}$, and there exist ℓ_0 and C such that the following estimate holds for any ℓ:

$$|r_N|_\ell^{m-(\rho-\delta)N} \le C \sum_{|\alpha|=N} |p^{(\alpha)}|_{\ell+\ell_0}^{(m_1-\rho|\alpha|)} |q_{(\alpha)}|_{\ell+\ell_0}^{(m_2+\delta|\alpha|)}.$$

The following lemma is the key result for proving both the previous theorem and Theorem 15.1.4. We shall provide its proof at the end of this section by following Iwasaki [69].

Lemma 15.1.3. *Let $f(x^1,\xi^1,\dots,x^\nu,\xi^\nu)$ be a C^∞-function on $\mathbb{R}^{2n\nu}$ such that*

$$\left|\prod_{j=1}^{\nu}\partial_{\xi_i}^{\alpha_j}\prod_{j=1}^{\nu}\partial_{x_i}^{\beta_j}f(x^1,\xi^1,\dots,x^\nu,\xi^\nu)\right|$$
$$\le M(\alpha^1,\alpha^2,\dots,\alpha^\nu,\beta^1,\beta^2,\dots,\beta^\nu)\prod_{j=1}^{\nu}<\xi^j>^{m_j-\rho|\alpha^j|+\delta|\beta^j|}$$

for any sequence of multi-indices $\alpha^1,\alpha^2,\dots,\alpha^\nu,\beta^1,\beta^2,\dots,\beta^\nu$ and constants

$$M(\alpha^1,\alpha^2,\dots,\alpha^\nu,\beta^1,\beta^2,\dots,\beta^\nu).$$

Set for $0\le\theta_j\le 1$, with $j=1,\dots,\nu-1$,

$$I_\theta=Os-\int_{\mathbb{R}^n\times\mathbb{R}^n}\cdots\int_{\mathbb{R}^n\times\mathbb{R}^n}\exp\left(-i\sum_{j=1}^{\nu-1}y^j\eta^j\right)\times f\left(x,\xi+\theta_1\eta^1,x+y^1,\xi\right.$$
$$\left.+\theta_2\eta^2,x+y^1+y^2,\dots,\xi+\theta_{\nu-1}\eta^{\nu-1},x+\sum_{j=1}^{\nu-1}y^j,\xi\right)dV,$$

where

$$dV=dy^1d\eta^1dy^2d\eta^2\cdots dy^{\nu-1}d\eta^{\nu-1}.$$

Then we can find ℓ_0 and a constant C_0 that depend on $\sum_{j=1}^{\nu}|m_j|$ but independent of ν and θ, such that the following estimate holds:

$$|I_\theta|\le(C_0)^\nu M(\ell_0)<\xi>^{m_0},$$

where

$$m_0=\sum_{j=1}^{\nu}m_j,\quad M(\ell_0)=\max_{|\alpha^j|+|\beta^j|\le\ell_0}M(\alpha^1,\alpha^2,\dots,\alpha^\nu,\beta^1,\beta^2,\dots,\beta^\nu).$$

Proof of Theorem 15.1.2: By the Taylor expansion we can write

$$\begin{aligned}(p\circ q)(x,\xi)&=Os-(2\pi)^{-n}\int_{\mathbb{R}^n\times\mathbb{R}^n}e^{-iy\cdot\eta}p(x,\xi+\eta)q(x+y,\xi)dy\,d\eta\\&=Os-(2\pi)^{-n}\int_{\mathbb{R}^n\times\mathbb{R}^n}e^{-iy\cdot\eta}\sum_{|\alpha|<N}\frac{1}{\alpha!}p^{(\alpha)}(x,\xi)\eta^\alpha\end{aligned}$$

$$\times q(x+y,\xi)dy\,d\eta + Os - (2\pi)^{-n}\int_{\mathbb{R}^n\times\mathbb{R}^n} e^{-iy\cdot\eta}\sum_{|\gamma|=N}\frac{1}{\gamma!}\eta^{\gamma}$$

$$\times\int_0^1 N(1-\theta)^{N-1}p^{(\gamma)}(x,\xi+\theta\eta)d\theta q(x+y,\xi)\,dy\,d\eta$$

$$= \sum_{|\alpha|<N}\frac{(-i)^{|\alpha|}}{\alpha!}p^{(\alpha)}(x,\xi)q_{(\alpha)}(x,\xi) + r_N(x,\xi),$$

where

$$r_N(x,\xi) = (2\pi)^{-n}(-i)^N N\sum_{|\gamma|=N}\int_0^1\frac{(1-\theta)^{N-1}}{\gamma!}$$

$$\times Os - \int_{\mathbb{R}^n\times\mathbb{R}^n} e^{-iy\cdot\eta}p^{(\gamma)}(x,\xi+\theta\eta)q_{(\gamma)}(x+y,\xi)\,dy\,d\eta\,d\theta.$$

Applying Lemma 15.1.3 for $r_N(x,\xi)$ with $\nu = 2$, $\theta_1 = \theta$, $\theta_2 = 0$, and $f(x^1,\xi^1,x^2,\xi^2) = p^{(\gamma)}(x^1,\xi^1)q_{(\gamma)}(x^2,\xi^2)$, we get the assertion. ■

Next, we shall deal with a basic theorem of the symbol of the multi-product of pseudo-differential operators see [69].

Theorem 15.1.4. *If $p_j \in S^{m(j)}_{\rho,\delta}$ for $j = 1,\ldots,\nu$, then the product of pseudo-differential operators $p_1(x.D)\cdots p_\nu(x,D)$ is also a pseudo-differential operator whose symbol*

$$p(x,\xi) = \sigma(p_1(x.D)\cdots p_\nu(x,D)) \in S^m_{\rho,\delta},$$

with $m = \Sigma_{j=1}^{\nu} m(j)$, satisfies the following estimate for any ℓ:

$$|p|_\ell^{(m)} \le C^\nu\prod_{j=1}^{\nu}|p_j|_{\ell+\ell_0}^{m(j)},$$

where the constants C and ℓ_0 are independent of ν.

Proof. We can write

$$p_1(x,D)\cdots p_\nu(x,D)u(x)$$

$$= Os - (2\pi)^{-n\nu}\int_{\mathbb{R}^n\times\mathbb{R}^n}\cdots\int_{\mathbb{R}^n\times\mathbb{R}^n}\exp\left(-i\sum_{j=1}^{\nu}y^j\xi^j\right)$$

$$\times p_1(x,\xi^1)p_2(x+y^1,\xi^2)\cdots p_\nu\left(x+\sum_{j=1}^{\nu-1}y^j,\xi^\nu\right)$$

$$\times u\left(x+\sum_{j=1}^{\nu}y^j\right)dy^1d\xi^1dy^2d\xi^2\cdots dy^\nu d\xi^\nu.$$

So the symbol $p(x,\xi)$ is given by

$$p(x,\xi) = Os - (2\pi)^{-n(\nu-1)} \int_{\mathbb{R}^n\times\mathbb{R}^n} \cdots \int_{\mathbb{R}^n\times\mathbb{R}^n} \exp\left(-i\sum_{j=1}^{\nu-1} y^j\eta^j\right) \times \prod_{j=1}^{\nu} p_j\left(x + \sum_{k=1}^{j-1} y^k, \xi + \eta^j\right) dV,$$

where

$$y^0 = 0, \quad \eta^\nu = 0, \quad dV = dy^1\, d\eta^1\, dy^2\, d\eta^2 \cdots dy^{\nu-1}\, d\eta^{\nu-1}.$$

Now we can apply Lemma 15.1.3 to this formula, setting $f(x^1,\xi^1,\ldots,x^\nu,\xi^\nu) = \prod_{j=1}^{\nu} p_j(x^j,\xi^j)$ and $\theta_j = 1$. Then we get

$$|p|_0^{(m)} \le C^\nu \prod_{j=1}^{\nu} |p_j|_{\ell_0}^{m(j)}.$$

The proof for any ℓ can be done by using a similar argument. ∎

Remark 15.1.5. Even if it might look tempting to prove the previous estimation using the induction over ν, this method was proven not feasible for this case.

15.1.3 Proof of Lemma 15.1.3

We shall prepare the ground with a few propositions. In order to avoid repeating the same argument for each variable, for simplicity we may assume $\rho_j = \rho$ and $\delta_j = \delta$ for any j.

For the proof of Lemma 15.1.3, we use integration by parts with respect to the variables η^j, $j = 1,\ldots,\nu-1$. In this case we need the following propositions regarding $F(\eta; y)$.

Set for $y, \eta \in \mathbb{R}^n$

$$F(\eta; y) = (1 + <\eta>^{2\bar{\delta}n_0} |y|^{2n_0})^{-1},$$

where $\bar{\delta} = \max(\delta, 0)$ and $n_0 = [n/2] + 1$. Then using the fact

$$\left|\left(\frac{\partial}{\partial\eta_1}\right)^{\alpha_1}\left(\frac{\partial}{\partial\eta_2}\right)^{\alpha_2}\cdots\left(\frac{\partial}{\partial\eta_n}\right)^{\alpha_n} <\eta>\right| \le c_\alpha <\eta>^{1-|\alpha|}, \tag{15.1.2}$$

we obtain the following result:

Proposition 15.1.6. *$F(\eta; y)$ satisfies the inequality with constants $C_{\alpha,\beta}$*

$$\left|\left(\frac{\partial}{\partial\eta}\right)^{\alpha}\left(\frac{\partial}{\partial y}\right)^{\beta} F(\eta; y)\right| \leq C_{\alpha,\beta} F(\eta; y) < \eta >^{-|\alpha|+\bar{\delta}|\beta|}$$

for all α and β.

Proposition 15.1.7. *The following inequality holds for some positive constant C:*

$$\int_{\mathbb{R}^n} F(\xi + \eta^1; z^1 - z^0) F(\xi + \eta^2; z^2 - z^1)\, dz^l$$
$$\leq CF(\xi + \eta^2; z^2 - z^0)\left\{< \xi + \eta^1 >^{-n\bar{\delta}} + \frac{F(\xi + \eta^1; z^2 - z^0)}{F(\xi + \eta^2; z^2 - z^0)} < \xi + \eta^2 >^{-n\bar{\delta}}\right\}.$$

Proof. For each fixed pair $(z^0, z^2) \in \mathbb{R}^n \times \mathbb{R}^n$ we divide $\mathbb{R}^n$ into two parts:

$$\Omega_1 = \{z^1 \in R^n; |z^1 - z^2| \geq |z^0 - z^2|/2\} \text{ and } \Omega_2 = R^n \backslash \Omega_1.$$

For $z^1 \in \Omega_1$, we have

$$F(\xi + \eta^2; z^2 - z^1) \leq 2^{2n_0} F(\xi + \eta^2; z^2 - z^0) \qquad \text{in } \Omega_1.$$

For $z^1 \in \Omega_2$, we get

$$F(\xi + \eta^1; z^1 - z^0) \leq 2^{2n_0} F(\xi + \eta^1; z^2 - z^0) \qquad \text{in } \Omega_2.$$

Then we have

$$\int_{\mathbb{R}^n} F(\xi + \eta^1; z^1 - z^0) F(\xi + \eta^2; z^2 - z^1) dz^1$$
$$\leq 2^{2n_0} F(\xi + \eta^2; z^2 - z^0) \int_{\Omega_1} F(\xi + \eta^1; z^1 - z^0) dz^1$$
$$+ 2^{2n_0} F(\xi + \eta^1; z^2 - z^0) \int_{\Omega_2} F(\xi + \eta^2; z^2 - z^0) dz^2. \quad (15.1.3)$$

Since $2n_0 > n$, it is clear that

$$\int_{R^n} F(\eta; y) dy = c_1 < \eta >^{-n\bar{\delta}}. \quad (15.1.4)$$

Then by (15.1.3) and (15.1.4), we get

$$\int_{\mathbb{R}^n} F(\xi + \eta^1; z^1 - z^0) F(\xi + \eta^2; z^2 - z^1) dz^1 \leq C\{F(\xi + \eta^2; z^2 - z^0)$$
$$\times < \xi + \eta^1 >^{-n\bar{\delta}} + F(\xi + \eta^1; z^2 - z^0) < \xi + \eta^2 >^{-n\bar{\delta}}\}$$

with $C = 2^{n_0} c_1$. ■

By (15.1.2), if $|\eta| \le c_0 < \xi >$, there exists a constant $c_0 > 0$ such that

$$| < \xi + \eta > - < \xi > | \le \frac{1}{2} < \xi >. \tag{15.1.5}$$

For the proof of Lemma 15.1.3, we also use integration by parts with respect to the variables y^j, $j = 1, \dots, \nu - 1$. The following two results deal with the integrand.

Proposition 15.1.8. *Let $\xi \in \mathbb{R}^n$ be fixed and consider $K \ge 0$. Set*

$$I(K)(\eta) = |\eta|^{-2K} < \xi + \eta >^m \{< \xi + \eta > + < \xi >\}^{2K\bar{\delta}} \times \left\{ < \xi + \eta >^{-n\bar{\delta}} + \frac{F(\xi + \eta; y)}{F(\xi; y)} < \xi >^{-n\bar{\delta}} \right\}$$

and let

$$\begin{aligned} I_1 &= \{\eta \in \mathbb{R}^n; |\eta| \le c_0 < \xi >^{\bar{\delta}}\}, \\ I_2 &= \{\eta \in \mathbb{R}^n; c_0 < \xi >^{\bar{\delta}} \le |\eta| \le c_0 < \xi >\}, \\ I_3 &= \{\eta \in \mathbb{R}^n; |\eta| \ge c_0 < \xi >\}. \end{aligned} \tag{15.1.6}$$

Then we have the constants $b(K_k)$, which depend on K_k, such that

$$\int_{I_k} I(K_k)(\eta) d\eta \le b(K_k) < \xi >^m, \qquad k = 1, 2, 3, \tag{15.1.7}$$

with

$$K_1 = 0, \quad K_2 \ge n/2, \quad K_3 \ge (|m| + 1 + 2\bar{\delta} n_0 + n)/2(1 - \bar{\delta}). \tag{15.1.8}$$

Proof. If η belongs to I_1 or I_2, by (15.1.5), we have for some constant $\tilde{b}(K)$

$$I(K)(\eta) \le \tilde{b}(K) |\eta|^{-2K} < \xi >^{(2K-n)\bar{\delta}+m}, \qquad K \ge 0.$$

Hence (15.1.7) is proved for $k = 1$ and 2 because we have

$$\begin{aligned} \int_{I_1} I(K_1)(\eta) d\eta &\le \tilde{b}(K_1) < \xi >^{-n\bar{\delta}+m} \int_{I_1} d\eta, \\ \int_{I_2} I(K_2)(\eta) d\eta &\le \tilde{b}(K_2) < \xi >^{(2K_2-n)\bar{\delta}+m} \int_{I_2} |\eta|^{-2K_2} d\eta. \end{aligned}$$

If η belongs to I_3, there is a constant $b^*(K)$ such that

$$I(K) \le b^*(K) |\eta|^{-2K+(\bar{m}+2\bar{\delta}K+2\bar{\delta}n_0)}, \qquad \bar{m} = \max(m, 0), \tag{15.1.9}$$

since the following system of inequalities holds:

$$\begin{cases} <\xi+\eta> \leq \left(1+\frac{1}{c_0}\right)|\eta|, \qquad <\xi> \leq \frac{1}{c_0}|\eta|, \\ \frac{F(\xi+\eta;y)}{F(\xi;y)} \leq (\frac{|\eta|}{c_0})^{2\bar{\delta}n_0+1}. \end{cases}$$

Using (15.1.9), we get

$$\int_{I_3} I(K_3)(\eta)d\eta \leq b^*(K_3)\int_{I_3} |\eta|^{-2(1-\delta)K_3+\bar{m}+1+2\bar{\delta}n_0}d\eta,$$

and hence inequality (15.1.7) holds for k = 3, with K_3 chosen as above. ■

Proposition 15.1.9. *For* $k=1,2,3$, *let*

$$J_k = |\eta|^{-2K_k}\{<\xi+\eta>+<\xi>\}^{2\bar{\delta}K_k} <\xi+\eta>^m \\ \times F(\xi+\eta;z^1-z^0)F(\xi;z^2-z^1).$$

Then we have

$$\int_{I_k}\int_{\mathbb{R}^n} J_k dz^1 d\eta \leq B(K_k) <\xi>^m\ F(\xi;z^2-z^0)$$

with $B(K)=Cb(K)$ *if* I_k *and* K_k *are defined as in (15.1.6) and (15.1.8).*

Proof. By means of Proposition 15.1.7, for $\eta^1=\eta,\eta^2=0$, we get

$$\begin{aligned}\int_{\mathbb{R}^n} J_k dz^1 &\leq C|\eta|^{-2K_k}\{<\xi+\eta>+<\xi>\}^{2\bar{\delta}K_k} \\ &\times\left\{<\xi+\eta>^{-\bar{\delta}n}+\frac{F(\xi+\eta;z^2-z^0)}{F(\xi;z^2-z^0)}<\xi>^{-\bar{\delta}n}\right\} \\ &\times <\xi+\eta>^m\ F(\xi;z^2-z^0) \\ &= CI(K_k)(\eta)F(\xi;z^2-z^0), \qquad k=1,2,3.\end{aligned}$$

Now, by Proposition 15.1.8, we get the assertion. ■

Now are able to provide the proof for Lemma 15.1.3.

Proof of Lemma 15.1.3. Set $n_0=[n/2]+1$, $M=\sum_{j=1}^{\nu}|m_j|$, $K=[(M+2\bar{\delta}n_0+n+1)/2(1-\bar{\delta})]+1$, and consider the functions $K_j=K_j(\eta^j;\eta^{j+1})$ with $j=1,\ldots,\nu-1$, given by $K_j=0$ on $I_{j,1}$, $K_j=n_0$ on $I_{j,2}$ and $K_j=K$ on $I_{j,3}$, where

$$\begin{aligned} I_{j,1} &= \{\eta^j\in\mathbb{R}^n; |\eta^j-\eta^{j+1}|\leq c_0<\xi+\theta\eta^{j+1}>^{\bar{\delta}}\}, \\ I_{j,2} &= \{\eta^j\in\mathbb{R}^n; c_0<\xi+\theta\eta^{j+1}>^{\bar{\delta}}\leq|\eta^j-\eta^{j+1}|\leq c_0<\xi+\theta\eta^{j+1}>\}, \\ I_{j,3} &= \{\eta^j\in\mathbb{R}^n; |\eta^j-\eta^{j+1}|\geq c_0<\xi+\theta\eta^{j+1}>\}, \qquad \eta^\nu=0. \end{aligned}$$

Set $B = \max\{B(0), B(n_0), B(K)\}$. Integrating by parts, we obtain

$$I_\theta = Os - \int_{\mathbb{R}^n\times\mathbb{R}^n} \cdots \int_{\mathbb{R}^n\times\mathbb{R}^n} \exp\left\{-i\sum_{j=1}^{\nu-1} y^j \cdot \eta^j\right\}$$

$$\times \prod_{j=1}^{\nu-1}\{1 + (-\Delta_{\eta^j})^{n_0} < \xi + \theta\eta^j >^{2\bar{\delta}n_0}\}$$

$$\times\{1+ < \xi + \theta\eta^j >^{2\bar{\delta}n_0} |y^j|^{2n_0}\}^{-1} f\left(x, \xi + \theta_1\eta^1, x + y^1,\right.$$

$$\left.\xi + \theta_2\eta^2, x + y^1 + y^2, \ldots, \xi + \theta_{\nu-1}\eta^{\nu-1}, x + \sum_{j=1}^{\nu-1} y^j, \xi\right) dV,$$

where $y^0 = 0$. Then by the change of variables

$$x + \sum_{k=1}^{j} y^k = z^j, \qquad j = 1, \ldots, \nu - 1,$$

we get

$$I_\theta = \int_{\mathbb{R}^n\times\mathbb{R}^n} \cdots \int_{\mathbb{R}^n\times\mathbb{R}^n} \exp\left\{-i\sum_{j=1}^{\nu-1} z^j \cdot (\eta^j - \eta^{j+1})\right\}$$

$$\times \prod_{j=1}^{\nu-1} |\eta^j - \eta^{j+1}|^{-2K_j} (-\Delta_{z^j})^{K_j} \mathbf{r} dV,$$

where

$$\mathbf{r} = \prod_{j=1}^{\nu-1}\{1 + (-\Delta_{\eta^j})^{n_0} < \xi + \theta\eta^j >^{2\bar{\delta}n_0}\}$$

$$\times \prod_{j=1}^{\nu-1} F(\xi + \theta\eta^j; z^j - z^{j-1}) f(x, \xi + \theta_1\eta^1,$$

$$z^1, \xi + \theta_2\eta^2, z^2, \ldots\ldots, \xi + \theta_{\nu-1}\eta^{\nu-1}, z^{\nu-1}, \xi),$$

$z^0 = x$ and $\eta^\nu = 0$. Then from Proposition 15.1.6 and (15.1.2), we have with a constant C_1 that

$$\left|\prod_{j=1}^{\nu-1}(-\Delta_{z^j})^{K_j}\mathbf{r}\right| \le (C_1)^\nu M(2(K+n_0)) \tag{15.1.10}$$

$$\times \prod_{j=1}^{\nu-1}\{<\xi+\theta\eta^j> + <\xi+\theta\eta^{j+1}>\}^{2\bar{\delta}K_j}$$

$$\times F(\xi+\theta\eta^j; z^j - z^{j-1}) \prod_{j=1}^{\nu} <\xi+\theta\eta^j>^{m_j}, \tag{15.1.11}$$

where $z^0 = z^\nu = x, \eta^\nu = 0$. From (15.1.10) and Proposition 15.1.9, we get for $k = 1, 2, 3$

$$\int_{I_{1,k}}\int_{\mathbb{R}^n} |\eta^1 - \eta^2|^{-2K_1}\left|\prod_{j=1}^{\nu-1}(-\Delta_{z^j})^{K_j}\mathbf{r}\right| dz^1 d\eta^1$$

$$\le (C_1)^\nu BM(2(K+n_0)) \prod_{j=2}^{\nu-1}\{<\xi+\theta\eta^j> + <\xi+\theta\eta^{j+1}>\}^{2\bar{\delta}K_j}$$

$$\times \prod_{j=3}^{\nu-1} F(\xi+\theta\eta^j; z^j - z^{j-1}) <\xi+\theta\eta^j>^{m_j}$$

$$\times F(\xi+\theta\eta^2; z^2 - z^0) <\xi+\theta\eta^2>^{\tilde{m}_2} <\xi>^{m_\nu},$$

where $\tilde{m_2} = m_1 + m_2$.

If we repeat the same argument, then we obtain

$$\int_{\mathbb{R}^n\times\mathbb{R}^n}\cdots\int_{\mathbb{R}^n\times\mathbb{R}^n}\prod_{j=1}^{\nu-1} |\eta^j - \eta^{j+1}|^{-2K_j}|(-\Delta_{z^j})^{K_j}\mathbf{r}| dz^1 dz^2\cdots dz^{\nu-2} d\eta^1 d\eta^2 \cdots d\eta^{\nu-2}$$

$$\le (C_1)^\nu B^{\nu-1} M(2(K+n_0))|\eta^\nu - \eta^{\nu+1}|^{-2K_{\nu-1}}$$

$$\times\{<\xi+\theta\eta^{\nu-1}> + <\xi+\theta\eta^\nu>\}^{2\bar{\delta}K_{\nu-1}} F(\xi+\theta\eta^{\nu-1}; z^{\nu-1} - z^0)$$

$$\times <\xi+\theta\eta^{\nu-1}>^{\tilde{m}_{\nu-1}} <\xi>^{m_\nu} \quad (z^0 = z^\nu = x, \eta^\nu = 0),$$

where $\tilde{m}_{\nu-1} = \sum_{j=1}^{\nu-1} m_j$. Using Proposition 15.1.8, we get

$$\int_{\mathbb{R}^n\times\mathbb{R}^n}\cdots\int_{\mathbb{R}^n\times\mathbb{R}^n}\prod_{j=1}^{\nu-1} |\eta^j - \eta^{j+1}|^{-2K_j}| - \Delta_{z^j}^{K_j}\mathbf{r}| dV$$

$$\le (C_1)^\nu B^\nu M(2(K+n_0)) <\xi>^{m_0}.$$

Taking $l_0 = 2(K+n_0)$ and $C_0 = C_1 B$ leads to the desired result.

15.2 Fundamental Solution by Symbolic Calculus: The Nondegenerate Case

This section deals with the construction of the fundamental solution for heat equations by the method of symbolic calculus in the nondegenerate case. We start reviewing the construction of the fundamental solution to the following Cauchy problem on $(0,T)\times\mathbb{R}^n$ following Tsutsumi [108] and Iwasaki [69]:

$$LU = \left(\frac{d}{dt} + P\right)U(t) = 0 \quad \text{on } (0,T)\times\mathbb{R}^n,$$
$$U(0) = I \quad \text{on } \mathbb{R}^n,$$

Here T is a finite positive number and P is a strongly elliptic differential operator of order m defined on $\mathbb{R}^n$ with symbol $p(x,\xi)$. Let $p(x,\xi) = \sum_{j=1}^{m} p_j(x,\xi)$, where $p_j(x,\xi)$ are homogeneous functions of order j with respect to ξ.

The fundamental solution $U(t)$ is constructed as a pseudo-differential operator with parameter t. For the construction of the fundamental solution, we need the estimate of the multi-product of pseudo-differential operators.

Theorem 15.2.1. *If there exist positive constants C and R such that*

$$p_m(x,\xi) \geq C <\xi>^m \quad \text{for } |\xi| > R,$$

the fundamental solution $U(t)$ is constructed as a pseudo-differential operator of a symbol $u(t)$ belonging to $S^0_{1,0}$ with parameter t and $u(t)\in S^{-\infty}$ for $t>0$. Moreover, $u(t)$ has the following expansion for any N:

$$u(t) - \sum_{j=0}^{N-1} u_j(t) \in S^{-N}_{1,0},$$

$$u_0(t) = e^{-pt}, \quad u_j(t) = f_j(t)u_0(t) \in S^{-j}_{1,0},$$

where $f_j(t)$ are polynomials with respect to ξ and t.

Proof. Let $f_j(t;x,\xi)$ for any $j\geq 1$ be the solutions of the following ordinary differential equations with parameters (x,ξ):

$$\begin{cases} (\frac{d}{dt} + p(x,\xi))(f_j u_0) = -q_j(t;x,\xi), \\ f_j|_{t=0} = 0, \end{cases}$$

where

$$q_j(t;x,\xi) = \sum_{\substack{k+\ell=j \\ 0\leq k\leq j-1}} s_\ell(p, f_k u_0).$$

For example, we have

$$\begin{cases} f_1 = -tp_1 + \frac{t^2}{2} s_1(p_2, p_2), \\ f_2 = -tp_0 + \frac{t^2}{2} \{(p_1)^2 + s_1(p_1, p_2) + s_1(p_2, p_1) + s_2(p_2, p_2)\} \\ \quad + \frac{t^3}{6} \left\{ \sum_{j,k=1}^{n} \left(\frac{\partial}{\partial x_j} p_2 \right) \left(\frac{\partial}{\partial x_k} p_2 \right) \left(\frac{\partial^2}{\partial \xi_j \partial \xi_k} p_2 \right) - s_1(p_2, s_1(p_2, p_2)) \right. \\ \quad \left. - 3 p_1 s_1(p_2, p_2) \right\} + \frac{t^4}{8} \{s_1(p_2, p_2)\}^2. \end{cases}$$

We note that

$$u_j(t) = f_j(t; x, \xi) u_0(t; x, \xi) = -\int_0^t u_0(t-s) q_j(s) ds, \qquad j \geq 1.$$

Set

$$\begin{cases} a_{j,\alpha,\beta}(t; x, \xi) = u_{j(\beta)}^{(\alpha)}(t; x, \xi) e^{tp}, & j \geq 0, \\ b_{j,\alpha,\beta}(t; x, \xi) = q_{j(\beta)}^{(\alpha)}(t; x, \xi) e^{tp}, & j \geq 1. \end{cases}$$

Then we have the following estimation for $a_{j,\alpha,\beta}(t; x, \xi)$ and $b_{j,\alpha,\beta}(t; x, \xi)$ by induction with respect to j and making use of the previous representation of $q_j(t; x, \xi)$ and $u_j(t; x, \xi)$.

Estimation I:

$$\begin{cases} |a_{j,\alpha,\beta}| \leq C_{j,\alpha,\beta} < \xi >^{-j-|\alpha|} \omega_{j,\alpha,\beta}, & j \geq 0, \\ |b_{j,\alpha,\beta}| \leq C_{j,\alpha,\beta} < \xi >^{m-j-|\alpha|} \omega'_{j,\alpha,\beta} & j \geq 1, \end{cases}$$

where $\omega_{j,\alpha,\beta}$ and $\omega'_{j,\alpha,\beta}$ are defined by

$$\begin{aligned} \omega_{0,0,0} &= 1, \quad \omega = < \xi >^m t, \\ \omega_{0,\alpha,\beta} &= \max\{\omega, \omega^{|\alpha|+|\beta|}\}, & |\alpha| + |\beta| \neq 0, \\ \omega_{j,\alpha,\beta} &= \max\{\omega^2, \omega^{2+|\alpha|+|\beta|}\}, & j \geq 1, \\ \omega'_{j,\alpha,\beta} &= \max\{\omega^2, \omega^{2j-1+|\alpha|+|\beta|}\} & j \geq 1. \end{aligned}$$

By *Estimation I*, we have $u_j \in S_{1,0}^{-j}$, since we have

$$|e^{-tp}| \leq C_0 e^{-C<\xi>^m t}$$

with a positive constant C_0.

For any N, we can write

$$p \circ u_j(t) = pu_j(t) + \sum_{\ell=1}^{N-j-1} s_\ell \left(p, u_j(t) \right) + r_{N,j}(t),$$

with $r_{N,j}(t) \in S^{m-N}_{1,0}$. So we have

$$\begin{aligned}
&\left(\frac{d}{dt} + p\right) \circ \left(\sum_{j=0}^{N-1} u_j(t)\right) \\
&= \left(\frac{d}{dt} + p\right)\left(\sum_{j=0}^{N-1} u_j(t)\right) + \sum_{j=0}^{N-1}\sum_{\ell=1}^{N-j-1} s_\ell\left(p, u_j(t)\right) + \sum_{j=0}^{N-1} r_{N,j}(t) \\
&= \sum_{j=0}^{N-1}\left(\frac{d}{dt} + p\right) u_j(t) + \sum_{j=1}^{N-1} q_j(t) + \sum_{j=0}^{N-1} r_{N,j}(t) \\
&= \sum_{j=0}^{N-1} r_{N,j}(t).
\end{aligned}$$

Set

$$g_N(t; x, \xi) = \sum_{j=0}^{N-1} u_j(t; x, \xi), r_N(t; x, \xi) = \sum_{j=0}^{N-1} r_{N,j}(t; x, \xi)$$

for any positive integer N. Then $g_N(t; x, D) = G_N(t)$ satisfies following equation with $r_N(t; x, D) = R_N(t)$:

$$\begin{cases} \left[\frac{d}{dt} + p(x, D)\right] G_N(t) = R_N(t), \\ G_N(0) \qquad\qquad\qquad = I. \end{cases}$$

Now we will construct $u(t; x, \xi)$ of the form

$$u(t; x, D) = \int_0^t G_N(t-s)\varphi(s; x, D) ds.$$

Then $\varphi(t; x, D) = \Phi(t)$ must satisfy Volterra's integral equation

$$R_N(t) + \Phi(t) + \int_0^t R_N(t-s)\Phi(s) ds = 0.$$

Set $\Phi_1(t) = -R_N(t)$ and for $j \geq 2$,

$$\begin{aligned}
\Phi_j(t) &= \int_0^t \Phi_1(t-s)\Phi_{j-1}(s) ds \\
&= \int_0^t \int_0^{s_1} \cdots \int_0^{s_{j-2}} \Phi_1(t-s_1)\Phi_1(s_1-s_2)\cdots\Phi_1(s_{j-1}) ds_{j-1} \cdots ds_2\, ds_1.
\end{aligned}$$

Then we have

$$\sum_{j=1}^{l} \Phi_j(t) = \Phi_1(t) + \sum_{j=2}^{l} \Phi_j(t)$$

$$= -R_N(t) - \int_0^t R_N(t-s) \sum_{j=1}^{l-1} \Phi_j(s)\, ds.$$

For $\sigma(\Phi_j(t)) = \varphi_j(t;x,\xi)$, we have the following estimate that helps in solving the previous integral equation.

Estimation II: If $m - N \leq 0$, then we have some constants $B_{\alpha,\beta}$ independent of j such that

$$\left|\varphi_{j(\beta)}^{(\alpha)}(t;x,\xi)\right| \leq |\varphi_1|_{l_0+|\alpha|+|\beta|}^{(m-N)} \left(B_{\alpha,\beta}\right)^{j-1} \frac{t^{j-1}}{(j-1)!} < \xi >^{m-N-\rho|\alpha|}.$$

Proof of Estimation II: Note that $\varphi_1(t;x,\xi) = -r_N(t;x,\xi)$, We can apply Theorem 15.1.4 for the multi-product $\Phi_1(s_{j-1} - s_j)$. For any l, there exists l_0 such that

$$|\varphi_j(t)|_l^{(m-N)} \leq C^j |\varphi_1|_{l_0}^{(m-N)} \left(|\varphi_1|_{l_0}^{(0)}\right)^{j-1} \int_0^t \int_0^{s_1} \cdots \int_0^{s_{j-2}} ds_{j-1} \cdots ds_2 ds_1$$

$$\leq C |\varphi_1|_{l_0}^{(m-N)} \left(C|\varphi_1|_{l_0}^{(0)}\right)^{j-1} \frac{t^{j-1}}{(j-1)!}.$$

Proof of Theorem 15.2.1: By *Estimation II*, we can define $\varphi(t) \in S_{1,0}^{m-N}$ by

$$\sum_{j=0}^{\infty} \varphi_j(t) = \varphi(t).$$

Then $\varphi(t;x,D) = \Phi(t)$ satisfies

$$R_N(t) + \Phi(t) + \int_0^t R_N(t-s)\Phi(s)ds = 0,$$

and

$$\left|\varphi_{(\beta)}^{(\alpha)}(t)\right| \leq C|\varphi_1|_{l_0}^{(m-N)} < \xi >^{m-N-|\alpha|} \exp[B_{\alpha,\beta} t].$$

For any N, choose $\tilde{N} = N + m$. Then for $r_{\tilde{N}}(t) \in S_{1,0}^{m-\tilde{N}} = S_{1,0}^{-N}$, we can write

$$u(t) = \sum_{j=0}^{\tilde{N}-1} u_j(t) + r_{\tilde{N}}(t)$$

$$= \sum_{j=0}^{N-1} u_j(t) + \sum_{j=N}^{\tilde{N}-1} u_j(t) + r_{\tilde{N}}(t)$$

$$= g_N(t) + r_N(t),$$

where $r_N(t) = \sum_{j=N}^{\tilde{N}-1} u_j(t) + r_{\tilde{N}}(t)$. It is clear that $r_N(t)$ belongs to $S_{1,0}^{-N}$, and hence we have arrived at the desired result. ■

Remark 15.2.2. The kernel of the operator $U(t) = u(t; x, D)$ is given by the integral

$$K(t,x,y) = (2\pi)^{-n} \int_{\mathbb{R}^n} e^{i(x-y)\cdot\xi} u(t;x,\xi) d\xi.$$

For instance, in the case of the Laplacian $P = \sum_{j=1}^{n} \frac{\partial^2}{\partial x_j{}^2}$, the symbol is $p = |\xi|^2$, and $u(t) = e^{-|\xi|^2 t}$.

15.3 Basic Results for Pseudo-Differential Operators of Weyl Symbols

We shall start with the definition of pseudo-differential operators of Weyl symbols, which will play an important role in the sequel.

Definition 15.3.1. Let $0 \le \delta \le \rho \le 1, \delta < 1$. A pseudo-differential operator P on $\mathbb{R}^n$ of Weyl symbol $p(x,\xi) \in S^m_{\rho,\delta}$ is defined by the formula

$$\begin{aligned} Pu(x) &= p^w(x,D)u(x) \\ &= Os - (2\pi)^{-n} \int_{\mathbb{R}^n \times \mathbb{R}^n} e^{-iy\cdot\xi} p\left(x + \frac{y}{2}, \xi\right) u(x+y) dy\, d\xi. \\ &= Os - (2\pi)^{-n} \int_{\mathbb{R}^n \times \mathbb{R}^n} e^{i(x-y)\cdot\xi} p\left(\frac{x+y}{2}, \xi\right) u(y) dy\, d\xi. \end{aligned}$$

In the rest of this chapter we shall use pseudo-differential operators of Weyl symbols. The reader can find details in the references Hörmander [67]. and Iwasaki and Iwasaki [72].

15.3.1 Calculus with Pseudo-Differential Operators

In this section we shall deal with the product formula, multi-product formula and estimates for symbols. The product of two pseudo-differential operators of Weyl symbols $P = p^w(x,D)$ and $Q = q^w(x,D)$ is also a pseudo-differential operator of Weyl symbols given by $\sigma^w(p^w(x,D)q^w(x,D)) = p \circ_w q$; that is,

$$p^w(x,D)q^w(x,D) = (p \circ_w q)^w(x,D).$$

Expressed as an oscillatory integral, $p \circ_w q$ is given by

$$(p \circ_w q)(x,\xi) = Os - (2\pi)^{-2n} \int_{\mathbb{R}^n \times \mathbb{R}^n} \int_{\mathbb{R}^n \times \mathbb{R}^n} e^{-i(y^1 \cdot \eta^1 + y^2 \cdot \eta^2)}$$
$$\times\, p\left(x - \frac{y^2}{2}, \xi + \eta^1\right) q\left(x + \frac{y^1}{2}, \xi + \eta^2\right) dy^1 d\eta^1 dy^2 d\eta^2.$$

In fact, $p \circ_w q$ can be given in the form provided by the following theorem.

Theorem 15.3.2. *Let $p \in S^{m_1}_{\rho,\delta}$ and $q \in S^{m_2}_{\rho,\delta}$. Then for any integer N, we have the expansion*

$$p \circ_w q = \sum_{j=0}^{N-1} \left(\frac{1}{2i}\right)^j \sigma_j(p,q) + r^w_N(p,q),$$

where

$$\sigma_j(p,q) = \sum_{|\alpha|+|\beta|=j} \frac{(-1)^{|\beta|}}{\alpha!\beta!} p^{(\alpha)}_{(\beta)}(x,\xi) q^{(\beta)}_{(\alpha)}(x,\xi) \in S^{m-(\rho-\delta)j}_{\rho,\delta},$$

$r^w_N(p,q) \in S^{m-(\rho-\delta)N}_{\rho,\delta}$ and there exist ℓ_0 and C such that the following estimate holds for any ℓ:

$$|r^w_N|^{(m-(\rho-\delta)N)}_\ell \leq C \sum_{|\alpha|+|\beta|=N} |p^{(\alpha)}_{(\beta)}|^{(m_1-\rho|\alpha|+\delta|\beta|)}_{\ell+\ell_0} |q^{(\beta)}_{(\alpha)}|^{(m_2+\delta|\alpha|-\rho|\beta|)}_{\ell+\ell_0}.$$

Remark 15.3.3. Pseudo-differential operators of Weyl symbols are pseudo-differential operators. In fact, we have

$$p^w(x,D) = q(x,D)$$

if

$$q(x,\xi) = Os - (2\pi)^{-n} \int_{\mathbb{R}^n \times \mathbb{R}^n} e^{-iy\cdot\eta} p\left(x + \frac{y}{2}, \xi + \eta\right) dy\, d\eta$$

and

$$p(x,\xi) = Os - (2\pi)^{-n} \int_{\mathbb{R}^n \times \mathbb{R}^n} e^{iy\cdot\eta} q\left(x + \frac{y}{2}, \xi + \eta\right) dy\, d\eta.$$

This shows that the condition $p(x,\xi) \in S^m_{\rho,\delta}$ is equivalent to $q(x,\xi) \in S^m_{\rho,\delta}$.

For the terms $\sigma_j(p,q)$ given in Theorem 15.3.2, we have the following remark.

Remark 15.3.4. It is clear that

$$\sigma_j(p,q) = (-1)^j \sigma_j(q,p) \text{ for any integer } j.$$

So we have

$$\sigma_j(p,p) = 0 \text{ if } j \text{ is an odd integer.}$$

Remark 15.3.5. In particular, $\sigma_j(p,q)$, $j = 1, 2$, have the following nice representation that is used in the construction of the fundamental solution:

$$\sigma_1(p,q) = < J\nabla p, \nabla q >,$$
$$\sigma_2(p,q) = -\frac{1}{2}\text{tr}(JH_pJH_q),$$

where

$$\nabla p =^t \Big(\frac{\partial}{\partial x_1}p, \dots, \frac{\partial}{\partial x_n}p, \frac{\partial}{\partial \xi_1}p, \dots, \frac{\partial}{\partial \xi_n}p\Big),$$

J is the $2n \times 2n$ matrix defined by

$$J = \begin{pmatrix} 0 & I \\ -I & 0 \end{pmatrix},$$

and H_p denotes the Hessian matrix. That is,

$$H_p = \begin{pmatrix} \partial_x \partial_x p & \partial_x \partial_\xi p \\ \partial_\xi \partial_x p & \partial_\xi \partial_\xi p \end{pmatrix}.$$

Proof of Theorem 15.3.2: In order to show the formula for $p \circ_w q$, we shall start by writing

$$\begin{aligned} & p^w(x,D)q^w(x,D)u(x) \\ &= Os - (2\pi)^{-2n} \int_{\mathbb{R}^n \times \mathbb{R}^n} \int_{\mathbb{R}^n \times \mathbb{R}^n} e^{i(x-x^1)\cdot\xi^1 + i(x^1-x^2)\cdot\xi^2} \\ &\quad \times p\Big(\frac{x+x^1}{2}, \xi^1\Big) q\Big(\frac{x^1+x^2}{2}, \xi^2\Big) u(x^2)\, dx^1 d\xi^1 dx^2 d\xi^2. \end{aligned}$$

The integrand can be written as an oscillatory integral:

$$p\Big(\frac{x+x^1}{2}, \xi^1\Big) = Os - (2\pi)^{-n} \int_{\mathbb{R}^n \times \mathbb{R}^n} e^{-i(x^1-w)\cdot\eta^2} p\Big(\frac{x+w}{2}, \xi^1\Big) dw\, d\eta^2.$$

By the change of variables

$$w = x^2 - y^2, \quad x^1 = x + y^1,$$

we have

$$\begin{aligned}&(x-x^1)\cdot\xi^1+(x^1-x^2)\cdot\xi^2-(x^1-w)\cdot\eta^2\\&\quad=(x-x^2)(\xi^2-\eta^2)-y^1(\eta^2+\xi^1-\xi^2)-y^2\eta^2\\&\quad=(x-x^2)\zeta-y^1\eta^1-y^2\eta^2\end{aligned}$$

if we set

$$\zeta=\xi^2-\eta^2,\quad \eta^1=\eta^2+\xi^1-\xi^2.$$

Then we obtain the following expression:

$$\begin{aligned}&p^w(x,D)q^w(x,D)u(x)\\&\quad=Os-(2\pi)^{-3n}\int_{\mathbf{R}^n\times\mathbf{R}^n}\int_{\mathbf{R}^n\times\mathbf{R}^n}\int_{\mathbf{R}^n\times\mathbf{R}^n}e^{i(x-x^2)\cdot\zeta-i(y^1\cdot\eta^1+y^2\cdot\eta^2)}\\&\quad\times p\Big(\frac{x+x^2-y^2}{2},\zeta+\eta^1\Big)q\Big(\frac{x+x^2+y^1}{2},\zeta+\eta^2\Big)u(x^2)\,dy^1d\eta^1dy^2d\eta^2dx^2d\zeta.\end{aligned}$$

So we have

$$\begin{aligned}(p\circ_w q)(x,\xi)&=Os-(2\pi)^{-2n}\int_{\mathbf{R}^n\times\mathbf{R}^n}\int_{\mathbf{R}^n\times\mathbf{R}^n}e^{-i(y^1\cdot\eta^1+y^2\cdot\eta^2)}\\&\quad\times p\left(x-\frac{y^2}{2},\xi+\eta^1\right)q\left(x+\frac{y^1}{2},\xi+\eta^2\right)dy^1d\eta^1dy^2d\eta^2.\end{aligned}$$

Using the Taylor expansion, we can write

$$\begin{aligned}(p\circ_w q)(x,\xi)&=Os-(2\pi)^{-2n}\int_{\mathbf{R}^n\times\mathbf{R}^n}\int_{\mathbf{R}^n\times\mathbf{R}^n}e^{-i(y^1\cdot\eta^1+y^2\cdot\eta^2)}\\&\quad\times p\left(x-\frac{y^2}{2},\xi+\eta^1\right)q\left(x+\frac{y^1}{2},\xi+\eta^2\right)dy^1d\eta^1dy^2d\eta^2\\&=Os-(2\pi)^{-2n}\int_{\mathbf{R}^n\times\mathbf{R}^n}\int_{\mathbf{R}^n\times\mathbf{R}^n}e^{-i(y^1\cdot\eta^1+y^2\cdot\eta^2)}\\&\quad\times\sum_{|\alpha|<N}\frac{1}{\alpha!}p^{(\alpha)}\left(x-\frac{y^2}{2},\xi\right)(\eta^1)^\alpha q\left(x+\frac{y^1}{2},\xi+\eta^2\right)dy^1d\eta^1dy^2d\eta^2\\&\quad+Os-(2\pi)^{-2n}\int_{\mathbf{R}^n\times\mathbf{R}^n}\int_{\mathbf{R}^n\times\mathbf{R}^n}e^{-i(y^1\cdot\eta^1+y^2\cdot\eta^2)}\sum_{|\gamma|=N}\frac{1}{\gamma!}(\eta^1)^\gamma\\&\quad\times\int_0^1 N(1-\theta)^{N-1}p^{(\gamma)}\left(x-\frac{y^2}{2},\xi+\theta\eta^1\right)d\theta\\&\quad\times q\left(x+\frac{y^1}{2},\xi+\eta^2\right)dy^1d\eta^1dy^2d\eta^2.\end{aligned}$$

The first term of the above equation can be written as

$$Os-(2\pi)^{-2n}\int_{\mathbb{R}^n\times\mathbb{R}^n}\int_{\mathbb{R}^n\times\mathbb{R}^n}e^{-i(y^1\cdot\eta^1+y^2\cdot\eta^2)}$$
$$\times\sum_{|\alpha|<N}\frac{1}{\alpha!}p^{(\alpha)}\left(x-\frac{y^2}{2},\xi\right)(\eta^1)^\alpha q\left(x+\frac{y^1}{2},\xi+\eta^2\right)dy^1d\eta^1dy^2d\eta^2$$
$$=Os-(2\pi)^{-n}\int_{\mathbb{R}^n\times\mathbb{R}^n}e^{-iy^2\cdot\eta^2}$$
$$\times\sum_{|\alpha|<N}(\frac{1}{2i})^{|\alpha|}\frac{1}{\alpha!}p^{(\alpha)}\left(x-\frac{y^2}{2},\xi\right)q_{(\alpha)}\left(x,\xi+\eta^2\right)dy^2d\eta^2$$
$$=Os-(2\pi)^{-n}\int_{\mathbb{R}^n\times\mathbb{R}^n}e^{-iy^2\cdot\eta^2}$$
$$\times\sum_{|\alpha|<N}\left(\frac{1}{2i}\right)^{|\alpha|}\sum_{|\beta|<N-|\alpha|}\frac{1}{\alpha!\beta!}p^{(\alpha)}\left(x-\frac{y^2}{2},\xi\right)q^{(\beta)}_{(\alpha)}(x,\xi)(\eta^2)^\beta dy^2d\eta^2$$
$$+Os-(2\pi)^{-n}\int_{\mathbb{R}^n\times\mathbb{R}^n}e^{-iy^2\cdot\eta^2}\sum_{|\alpha|<N}\left(\frac{1}{2i}\right)^{|\alpha|}\sum_{|\gamma|=N-|\alpha|}\frac{1}{\alpha!\gamma!}p^{(\alpha)}\left(x-\frac{y^2}{2},\xi\right)$$
$$\times\int_0^1(N-|\alpha|)(1-\theta)^{N-|\alpha|-1}q^{(\gamma)}_{(\alpha)}(x,\xi+\theta\eta^2)d\theta(\eta^2)^\gamma dy^2d\eta^2$$
$$=\sum_{|\alpha|+|\beta|<N}\left(\frac{1}{2i}\right)^{|\alpha|+|\beta|}(-1)^\beta\frac{1}{\alpha!\beta!}p^{(\alpha)}(x,\xi)q^{(\beta)}_{(\alpha)}(x,\xi)+r^w_N(1)(x,\xi),$$

where

$$r^w_N(1)(x,\xi)$$
$$=Os-(2\pi)^{-n}\int_{\mathbb{R}^n\times\mathbb{R}^n}e^{-iy^2\cdot\eta^2}\sum_{|\alpha|<N}\left(\frac{1}{2i}\right)^{|\alpha|}\sum_{|\gamma|=N-|\alpha|}\frac{1}{\alpha!\gamma!}p^{(\alpha)}\left(x-\frac{y^2}{2},\xi\right)$$
$$\times\int_0^1(N-|\alpha|)(1-\theta)^{N-|\alpha|-1}q^{(\gamma)}_{(\alpha)}(x,\xi+\theta\eta^2)d\theta(\eta^2)^\gamma dy^2d\eta^2.$$

So we have

$$(p\circ_w q)(x,\xi)=\sum_{j=1}^{N-1}\left(\frac{1}{2i}\right)^j\sigma_j(p,q)+r^w_N(1)(x,\xi)+r^w_N(2)(x,\xi),$$

where

$$r^w_N(2)(x,\xi)=\ Os-(2\pi)^{-2n}\int_{\mathbb{R}^n\times\mathbb{R}^n}\int_{\mathbb{R}^n\times\mathbb{R}^n}e^{-i(y^1\cdot\eta^1+y^2\cdot\eta^2)}\sum_{|\gamma|=N}\frac{1}{\gamma!}(\eta^1)^\gamma$$

$$\times \int_0^1 N(1-\theta)^{N-1} p^{(\gamma)}\left(x - \frac{y^2}{2}, \xi + \theta\eta^1\right) d\theta$$
$$\times q\left(x + \frac{y^1}{2}, \xi + \eta^2\right) dy^1 d\eta^1 dy^2 d\eta^2.$$

Applying Lemma 15.1.3 for $r_N^w(x,\xi) = r_N^w(1)(x,\xi) + r_N^w(2)(x,\xi)$, we get the desired assertion. ■

We have the following theorem for the multi-product of pseudo-differential operators of Weyl symbols.

Theorem 15.3.6. *Let $p_j \in S^{m(j)}_{\rho,\delta}$ for $j = 1, \ldots, \nu$, and consider ν pseudo-differential operators of Weyl symbols $p_1^w(x,D), \ldots, p_\nu^w(x,D)$. Then the product $p_1^w(x.D) \cdots p_\nu^w(x,D)$ is also a pseudo-differential operator of Weyl symbols $p(x,\xi) = \sigma^w(p_1^w(x,D) \cdots p_\nu^w(x,D)) \in S^m_{\rho,\delta}$, with $m = \Sigma_{j=1}^\nu m(j)$, such that the following estimate is satisfied:*

$$|p|_\ell^{(m)} \leq C^\nu \prod_{j=1}^{\nu} |p_j|_{\ell+\ell_0}^{m(j)}, \qquad \forall \ell,$$

with the constants C and ℓ_0 independent of ν.

Proof. Since the pseudo-differential operators of Weyl symbols are represented as pseudo-differential operators of usual symbols, applying Theorem 15.1.4, we obtain the desired assertion by transforming it into a pseudo-differential operator of Weyl symbols. ■

Another variant of the proof is based on direct computation. The reader can find the details in the appendix of reference [72]. In this case $p(x,\xi)$ is given by

$$p(x,\xi) = Os - (2\pi)^{-n\nu} \int_{\mathbb{R}^n} \int_{\mathbb{R}^n} \cdots \int_{\mathbb{R}^n} \exp\left(-i \sum_{j=1}^{\nu} y^j \cdot \eta^j\right)$$
$$\times \prod_{j=1}^{\nu} p_j \left(x + \frac{1}{2}\sum_{k=1}^{j-1} y^k - \frac{1}{2} \sum_{k=j+1}^{\nu} y^k, \xi + \eta^j\right) dV,$$

where

$$dV = dy^1 d\eta^1 dy^2 d\eta^2 \cdots dy^\nu d\eta^\nu.$$

15.4 Fundamental Solution by Symbolic Calculus: The Degenerate Case

This section deals with the construction of the fundamental solution for degenerate heat equations via symbolic calculus of Weyl symbols. We shall provide the construction of the following fundamental solution $E(t)$ as a pseudo-differential operator of Weyl symbol $e(t;x,\xi)$:

$$\frac{d}{dt}e^w(t;x,D) + p^w(x,D)e^w(t;x,D) = 0 \quad \text{on} \quad (0,T)\times\mathbb{R}^n,$$
$$e^w(0;x,D) = I,$$

where T is a finite positive number, and $p^w(x,D)$ is a pseudo-differential operator of Weyl symbol $p(x,\xi)$. Let $p(x,\xi) = \sum_{j=1}^m p_j(x,\xi)$, where $p_j(x,\xi)$ are homogeneous functions of order j with respect to ξ. In this section we consider a degenerate operator whose sub-elliptic estimation was provided in Melin [90].

15.4.1 Construction of the Symbol

We shall provide in the following proof sketch for the construction of the fundamental solution of the heat equation according to [72].

Theorem 15.4.1. *Suppose that $p(x,\xi) \in S^m_{1,0}$ satisfies the following conditions:*

1. $p_m(x,\xi) \geq 0,$
2. $p_{m-1} + \frac{1}{2}\mathrm{tr}^+(A) \geq c|\xi|^{m-1}$ *for $|\xi|$ large and some positive constant c on Σ,*

where Σ is the characteristic set of $p_m(x,\xi)$, $A = iJH_{p_m}$, $\mathrm{tr}^+(A)$ is the sum of positive eigenvalues of A, the Hessian matrix is given by

$$H_{p_m} = \begin{pmatrix} \partial_x\partial_x p_m & \partial_x\partial_\xi p_m \\ \partial_\xi\partial_x p_m & \partial_\xi\partial_\xi p_m \end{pmatrix},$$

and

$$J = \begin{pmatrix} 0 & I \\ -I & 0 \end{pmatrix}.$$

Then we can construct the symbol

$$e(t;x,\xi) \in \begin{cases} S^0_{1/2,1/2}, & \text{for } t \geq 0, \\ S^{-\infty}, & \text{for } t > 0, \end{cases}$$

which satisfies the following expansion for any integer N:

$$e(t;x,\xi) - \sum_{j=0}^{N-1} e_j(t;x,\xi) \in S^{-N/2}_{1/2,1/2},$$

$$e_0(t;x,\xi) = \exp\Big\{-\varphi(t;x,\xi)\Big\},$$

$$e_j(t;x,\xi) = f_j(t;x,\xi)\exp\Big\{-\varphi(t;x,\xi)\Big\} \in S^{-j/2}_{1/2,1/2},$$

with

$$\varphi(t;x,\xi) = p_m t + p_{m-1} t + \frac{1}{2}\mathrm{tr}\Big[\log\cosh\Big(\frac{At}{2}\Big)\Big] - \frac{it^2}{4} < G(At/2)J\nabla p_m, \nabla p_m >,$$

where

$$G(x) = (1 - x^{-1}\tanh x)/x.$$

Remark 15.4.2. We note that

$$e_0(t;x,\xi) = \frac{e^{-p_m t + it^2 < G(At/2)J\nabla p_m, \nabla p_m>/4}}{\sqrt{\det\{\cosh(At/2)\}}},$$

$$\nabla p_m = 0 \text{ on the characteristic set } \Sigma$$

and

$$\left|\frac{1}{\sqrt{\det\{\cosh(At/2)\}}}\right| \le C e^{-\mathrm{tr}^+ At/2} \text{ on } \Sigma.$$

Proof. The proof is divided into four steps as in the following:

In step 1 the construction of the main part of the fundamental solution is discussed.

In step 2 we provide the estimation of the main part of all $(x,\xi) \in \mathbb{R}^n \times \mathbb{R}^n$.

In step 3 we show that the symbol has the expansion $\sum_{j=0} e_j(t;x,\xi)$ with the terms of the form $e_j(t;x,\xi) = f_j(t;x,\xi)e^{-\varphi}$, where the f_j are approximate solutions of (15.4.22).

In the final step we discuss the construction of the fundamental solution by solving an integral equation.

Step 1: *Construction of the main part*

Assume the fundamental solution is a pseudo-differential operator of Weyl symbols, with $e(t;x,\xi) = e^{-\varphi(t;x,\xi)}$. Applying Theorem 15.3.2, we have

$$\frac{\partial}{\partial t}e(t) + \sum_{j=0}^{\infty}\Big(\frac{1}{2i}\Big)^j \sigma_j(p, e(t)) = 0, \quad e(0) = 1.$$

The following equations can be easily shown

$$\begin{aligned}
\sigma_1(p, e(t)) &= < J\nabla p, \nabla e(t) > = - < J\nabla p, \nabla\varphi > e(t),\\
\sigma_2(p, e(t)) &= \sigma_2(p, -\varphi)e(t) + \frac{1}{2} < J\nabla\varphi, H_p J\nabla\varphi > e(t)\\
&= \frac{1}{2}\{\mathrm{tr}(JH_p JH_\varphi) + < J\nabla\varphi, H_p J\nabla\varphi >\}e(t).
\end{aligned}$$

So we have

$$\begin{aligned}
&\sum_{j=0}^{2}\left(\frac{1}{2i}\right)^j \sigma_j(p, e(t))\\
&= \left\{p - \frac{1}{2i} < J\nabla p, \nabla\varphi > - \frac{1}{8}\mathrm{tr}(JH_p JH_\varphi) - \frac{1}{8} < J\nabla\varphi, H_p J\nabla\varphi >\right\} e(t).
\end{aligned}$$

Neglecting the terms $\sigma_j(p, e(t))$ for $j \geq 3$, we get the following equation for φ with $\varphi(0) = 0$:

$$\frac{\partial}{\partial t}\varphi - p + \frac{1}{2i} < J\nabla p, \nabla\varphi > + \frac{1}{8}\mathrm{tr}(JH_p JH_\varphi) + \frac{1}{8} < J\nabla\varphi, H_p J\nabla\varphi >= 0.$$

It might not be easy to find a solution of the above equation. But neglecting the derivatives of p and φ of order of greater than 3, we can find a suitable solution as in the following.

Using

$$\begin{aligned}
\nabla(< J\nabla p, \nabla\varphi >) &= -H_p J\nabla\varphi + H_\varphi J\nabla p,\\
\nabla(< J\nabla\varphi, H_p J\nabla\varphi >) &= -\nabla(< \nabla\varphi, JH_p J\nabla\varphi >)\\
&= -2H_\varphi JH_p J\nabla\varphi,
\end{aligned}$$

we have

$$\frac{\partial}{\partial t}\nabla\varphi - \nabla p + \frac{1}{2i}\{-H_p J\nabla\varphi + H_\varphi J\nabla p\} - \frac{1}{4}H_\varphi JH_p J\nabla\varphi = 0$$

and

$$\frac{\partial}{\partial t}H_\varphi - H_p + \frac{1}{2i}\{-H_p JH_\varphi + H_\varphi JH_p\} - \frac{1}{4}H_\varphi JH_p JH_\varphi = 0.$$

The above equation means that $X = iJH_\varphi$ satisfies

$$\frac{\partial}{\partial t}X - A + \frac{1}{2}(AX - XA) + \frac{1}{4}XAX = 0, \quad X|_{t=0} = 0,$$

with $A = iJH_p$. The unique solution of the equation is

$$X = 2\tanh(At/2).$$

So we can write

$$\frac{\partial}{\partial t} J\nabla\varphi - J\nabla p + \frac{1}{2}\{AJ\nabla\varphi - XJ\nabla p\} + \frac{1}{4} XAJ\nabla\varphi = 0.$$

Put

$$y = J\nabla\varphi, \quad b = J\nabla p.$$

Then we have

$$\frac{\partial}{\partial t} y - b + \frac{1}{2}\{Ay - Xb\} + \frac{1}{4} XAy = 0, \quad y|_{t=0} = 0.$$

The unique solution of the above equation is

$$y = A^{-1}Xb.$$

Now the equation for φ is written as

$$\frac{\partial}{\partial t}\varphi - p + \frac{i}{2} < \nabla p, y > -\frac{1}{8}\mathrm{tr}(AX) + \frac{1}{8} < y, H_p y >= 0, \quad \varphi|_{t=0} = 0.$$

So we have

$$\frac{\partial}{\partial t}\varphi - p - \frac{i}{2} < Jb, A^{-1}Xb > -\frac{1}{8}\mathrm{tr}(AX) + \frac{i}{8} < A^{-1}Xb, JXb > = 0.$$

Using the antisymmetry

$$< JAu, v > = - < Ju, Av >,$$

we have

$$\begin{aligned} < Jh(A)u, v > &= - < Ju, h(A)v > \text{ if } h(x) \text{ is an odd function,} \\ < Jh(A)u, v > &= < Ju, h(A)v > \text{ if } h(x) \text{ is an even function.} \end{aligned}$$

If $h(x)$ is an even function, we have

$$\begin{aligned} < Jh(A)u, u > &= < Ju, h(A)u > \\ &= < J^2u, Jh(A)u > = - < Jh(A)u, u >. \end{aligned}$$

So we have

$$< Jb, A^{-1}Xb > = 0,$$
$$< A^{-1}Xb, JXb > = - < A^{-1}X^2b, Jb > = < JA^{-1}X^2b, b>.$$

φ is obtained by the following formula:

$$\varphi = pt + \frac{1}{8}\int_0^t \mathrm{tr}(AX)ds - \frac{i}{8}\int_0^t < JA^{-1}X^2b, b > ds.$$

The integrals can be computed as

$$\begin{aligned}\frac{1}{8}\int_0^t X^2 ds &= \frac{1}{2}\int_0^t \tanh^2(As/2)ds \\ &= \frac{1}{2}\int_0^t \left(1 - \frac{1}{\cosh^2(As/2)}\right) ds \\ &= \frac{t}{2}\Big(1 - (At/2)^{-1}\tanh(At/2)\Big)\end{aligned}$$

and

$$\begin{aligned}\frac{1}{8}\int_0^t AX ds &= \frac{1}{2}\int_0^t A/2\tanh(As/2)ds \\ &= \frac{1}{2}\Big[\log\{\cosh(As/2)\}\Big]_0^t \\ &= \frac{1}{2}\log\{\cosh(At/2)\}.\end{aligned}$$

Then we have

$$\begin{aligned}\varphi &= pt + \frac{1}{2}\mathrm{tr}[\log\{\cosh(At/2)\}] - \frac{it}{2} < JA^{-1}\Big(1 - (At/2)^{-1}\tanh(At/2)\Big)b, b > \\ &= pt + \frac{1}{2}\mathrm{tr}[\log\{\cosh(At/2)\}] - \frac{it^2}{4} < G(At/2)J\nabla p, \nabla p > .\end{aligned}$$

The main part of φ is

$$p_m t + p_{m-1}t + \frac{\mathrm{tr}}{2}\left\{\log\left(\cosh\left(\frac{At}{2}\right)\right)\right\} - \frac{it^2}{4} < G(At/2)J\nabla p_m, \nabla p_m >$$

with $A = iJH_{p_m}$.

Step 2: *The estimation of the main part*
We note first that all the eigenvalues of A are real. If A has a nonzero eigenvalue λ, then $-\lambda$ is also an eigenvalue of A. These facts are shown in the following. Set

$$((u, v)) = < iJu, \bar{v}>.$$

Then we have

$$((Au, v)) = -((u, \bar{A}v)) = ((u, Av)),$$

and

$$((Au, u)) = ((H_p u, u)) \geq 0.$$

Assume $Au = \lambda u$, with $\lambda \neq 0$. Then we have

$$0 < ((H_p u, u)) = ((Au, u)) = \lambda((u, u)).$$

This means that $((u, u)) \neq 0$. On the other hands the following holds:

$$((Au, u)) = ((u, Au)) = ((u, \lambda u)) = \bar{\lambda}((u, u)),$$

and hence we have $\lambda = \bar{\lambda}$. We also have

$$A\bar{u} = -\lambda \bar{u},$$

and then $-\lambda$ is an eigenvalue of A.

By Remark 15.4.2, we have the estimation

$$\left| \frac{1}{\sqrt{\det\{\cosh(At/2)\}}} \right| \leq Ce^{-\mathrm{tr}^{+} At/2},$$

and by the assumptions on the characteristic set, we have

$$\varphi(x, \xi) \geq C < \xi >^m t \quad \text{if } p_m(x, \xi) \geq c|\xi|^m,$$
$$\varphi(x, \xi) \geq c < \xi >^{m-1} t \quad \text{in } \Sigma.$$

We need precise arguments to obtain an estimate for φ near the set Σ. According to [72], we have

$$e^{-\varphi} \in S^0_{1/2,1/2}.$$

Step 3: *The construction of the asymptotic term*
We shall give an idea of the method of construction of $e_j(t; x, \xi)$. The main difficulty of the construction appears on the characteristic set Σ. I this step we study how to construct $e_j(t; x, \xi)$ near Σ. For the precise proof, see Iwasaki [72]. Let

$$\varphi(t; x, \xi) = pt + \frac{\mathrm{tr}}{2}\left\{\log\left(\cosh\left(\frac{At}{2}\right)\right)\right\} - \frac{it^2}{4} < G(At/2)J\nabla p, \nabla p >,$$

where

$$G(x) = (1 - x^{-1} \tanh x)/x.$$

Proposition 15.4.3. *We have*

$$\frac{\partial}{\partial t}e^{-\varphi} + \sum_{j=0}^{2}\left(\frac{1}{2i}\right)^j \sigma_j(p, e^{-\varphi})$$
$$= \frac{i}{2} < (1 - \tanh(At/2))J\nabla p, \psi > + \frac{i}{8}tr(AY) - \frac{i}{8} < \psi, AJ\psi >,$$

where

$$\psi = \frac{1}{2}\nabla\left(\mathrm{tr}\left\{\log\left(\cosh\left(\frac{At}{2}\right)\right)\right\}\right) - \frac{it^2}{4} <\nabla(G(At/2))J\nabla p_m, \nabla p_m>, \quad (15.4.12)$$

$$Y = 2\nabla(A^{-1}\tanh(At/2))J\nabla p + J\nabla\psi. \quad (15.4.13)$$

Proof. By the definition of φ, (15.4.12) and (15.4.13), we have

$$J\nabla\varphi = 2A^{-1}\tanh\left(\frac{At}{2}\right)J\nabla p + J\psi, \quad (15.4.14)$$

$$JH_\varphi = -2i\tanh\left(\frac{At}{2}\right) + Y. \quad (15.4.15)$$

Equations (15.4.14) and (15.4.15) imply the following identities:

$$-\frac{1}{2i} < J\nabla p, \nabla\varphi > = -i < A^{-1}\tanh(At/2)J\nabla p, \nabla p > -\frac{i}{2} < \nabla p, J\psi >$$
$$= \frac{i}{2} < J\nabla p, \psi >, \quad (15.4.16)$$

where we used that $A^{-1}\tanh(At/2)$ is an even function of A. Then

$$-\frac{1}{8}\mathrm{tr}(JH_pJH_\varphi) = \frac{1}{2}\mathrm{tr}(A\tanh(At/2)/2) + \frac{i}{8}\mathrm{tr}(AY) \quad (15.4.17)$$

and

$$-\frac{1}{8} < J\nabla\varphi, H_pJ\nabla\varphi >$$
$$= -\frac{1}{2} < A^{-1}\tanh(At/2)J\nabla p, H_pA^{-1}\tanh(At/2)J\nabla p >$$
$$-\frac{1}{2} < A^{-1}\tanh(At/2)J\nabla p, H_pJ\psi > -\frac{1}{8} < J\psi, H_pJ\psi >$$

$$= -\frac{i}{2} < A^{-1}\tanh^2(At/2)J\nabla p, \nabla p >$$
$$-\frac{i}{2} < \tanh(At/2)J\nabla p, \psi > -\frac{i}{8} < \psi, AJ\psi >. \tag{15.4.18}$$

On the other hand, we have

$$\frac{\partial}{\partial t}\varphi = p + \frac{1}{2}\mathrm{tr}(A\tanh(At/2)/2)$$
$$-\frac{it}{2} < \left\{G(At/2) + \frac{1}{2}(At/2)G'(At/2)\right\} J\nabla p, \nabla p >$$
$$= p + \frac{1}{2}\mathrm{tr}(A\tanh(At/2)/2)$$
$$-\frac{i}{2} < A^{-1}\tanh^2(At/2)J\nabla p, \nabla p >, \tag{15.4.19}$$

where we used

$$G(x) + \frac{x}{2}G'(x) = \frac{x}{2}\frac{\tanh^2 x}{x}.$$

By (15.4.16)–(15.4.19), we obtain the assertion. ■

Now fix an integer $N \geq 3$. Since

$$p \in S^m_{1,0}, \quad e^{-\varphi} \in S^0_{1/2,1/2},$$

we have

$$\left(\frac{1}{2i}\right)^j \sigma_j(p, e^{-\varphi}) \in S^{m-\frac{j}{2}}_{1/2,1/2}$$

and

$$\sum_{j=3}^{N}\left(\frac{1}{2i}\right)^j \sigma_j(p, e^{-\varphi}) \in S^{m-\frac{3}{2}}_{1/2,1/2}.$$

Then $e_0(t; x, \xi) = e^{-\varphi}$ satisfies the equation

$$\frac{\partial}{\partial t}e^{-\varphi} + \sum_{j=0}^{N}\left(\frac{1}{2i}\right)^j \sigma_j(p, e^{-\varphi}) = g_0 e^{-\varphi},$$

with

$$g_0 e^{-\varphi} \in S^{m-\frac{3}{2}}_{1/2,1/2},$$

because

$$g_0 = \frac{i}{2} < (1 - \tanh(At/2))J\nabla p, \psi > + \frac{i}{8}\mathrm{tr}(AY) - \frac{i}{8} < \psi, AJ\psi >$$
$$+ \sum_{j=3}^{N} \left(\frac{1}{2i}\right)^j \sigma_j(p, e^{-\varphi})e^{\varphi}. \tag{15.4.20}$$

Note that g_0 is a polynomial of $J\nabla p$ whose coefficients are represented by functions of A and its derivatives are given by (15.4.12) and (15.4.13).

For any smooth function $h(x, \xi)$, we have

$$\frac{\partial}{\partial t}(he^{-\varphi}) + \sum_{j=0}^{N} \left(\frac{1}{2i}\right)^j \sigma_j(p, he^{-\varphi})$$
$$= \left(\frac{\partial}{\partial t}h\right) e^{-\varphi} + \sum_{j=1}^{N} \left(\frac{1}{2i}\right)^j (\sigma_j(p, he^{-\varphi}) - h\sigma_j(p, e^{-\varphi}))$$
$$+ \left\{\frac{\partial}{\partial t}e^{-\varphi} + \sum_{j=0}^{N} \left(\frac{1}{2i}\right)^j \sigma_j(p, e^{-\varphi})\right\} h$$
$$= \left(\frac{\partial}{\partial t}h + hg_0\right) e^{-\varphi} + \sum_{j=1}^{N} \left(\frac{1}{2i}\right)^j (\sigma_j(p, he^{-\varphi}) - h\sigma_j(p, e^{-\varphi})).$$

The main part of $\sum_{j=1}^{N} \left(\frac{1}{2i}\right)^j (\sigma_j(p, he^{-\varphi}) - h\sigma_j(p, e^{-\varphi}))$ is also given by

$$\sum_{j=1}^{2} \left(\frac{1}{2i}\right)^j (\sigma_j(p, he^{-\varphi}) - h\sigma_j(p, e^{-\varphi})).$$

The following equations can easily be shown:

$$\begin{aligned}\sigma_1(p, he^{-\varphi}) - h\sigma_1(p, e^{-\varphi}) &= \sigma_1(p, h)e^{-\varphi} \\ &= < J\nabla p, \nabla h > e^{-\varphi}, \\ \sigma_2(p, he^{-\varphi}) - h\sigma_2(p, e^{-\varphi}) &= \sigma_2(p, h)e^{-\varphi} - < J\nabla h, H_p J\nabla\varphi > e^{-\varphi} \\ &= \left\{-\frac{1}{2}\mathrm{tr}(JH_pJH_h) - < J\nabla h, H_pJ\nabla\varphi >\right\} e^{-\varphi}.\end{aligned}$$

Then we have

$$\sum_{j=1}^{2} \left(\frac{1}{2i}\right)^j (\sigma_j(p, he^{-\varphi}) - h\sigma_j(p, e^{-\varphi}))$$
$$= \frac{1}{2i} < J\nabla p, \nabla h > e^{-\varphi}$$

$$
\begin{aligned}
&+\left(\frac{1}{2i}\right)^2\left(-\frac{1}{2}\mathrm{tr}(JH_pJH_h)e^{-\varphi}- < J\nabla h, H_pJ\nabla\varphi > e^{-\varphi}\right)\\
&= (Qh)e^{-\varphi},
\end{aligned}
$$

where Q is a second-order partial differential operator defined by

$$
Qh = \frac{1}{2i} < J\nabla p, \nabla h > +\left(\frac{1}{2i}\right)^2\left(-\frac{1}{2}\mathrm{tr}(JH_pJH_h) - < J\nabla h, H_pJ\nabla\varphi >\right).
$$

So we have

$$
\begin{aligned}
&\frac{\partial}{\partial t}(he^{-\varphi}) + \sum_{j=0}^{N}\left(\frac{1}{2i}\right)^j\sigma_j(p, he^{-\varphi})\\
&= \left(\frac{\partial}{\partial t}h + hg_0 + Qh\right)e^{-\varphi} + \sum_{3\le j\le N}\left(\frac{1}{2i}\right)^j(\sigma_j(p, he^{-\varphi}) - h\sigma_j(p, e^{-\varphi})).
\end{aligned}
$$

Next we shall approximate Q. Using

$$
J\nabla\varphi = 2A^{-1}\tanh(At/2)J\nabla p + J\psi,
$$

we have

$$
\begin{aligned}
Qh &= -\frac{1}{2} < iJ\nabla p, \nabla h > +\frac{1}{2} < iJH_pA^{-1}\tanh(At/2)iJ\nabla p, \nabla h >\\
&\quad - \frac{i}{8}\mathrm{tr}(AJH_h) + \frac{1}{4} < AJ\psi, i\nabla h >\\
&= \frac{1}{2} < \{\tanh(At/2) - 1\}iJ\nabla p, \nabla h > + \frac{i}{8}\mathrm{tr}(AJH_h) + \frac{1}{4} < AJ\psi, i\nabla h >\\
&= Q_0h + \frac{1}{4} < AJ\psi, i\nabla h >,
\end{aligned}
$$

where

$$
Q_0h = \frac{1}{2} < \{\tanh(At/2) - 1\}iJ\nabla p, \nabla h > - \frac{i}{8}\mathrm{tr}(AJH_h).
$$

If $f_0 = 1$, we formally have

$$
\begin{aligned}
&\frac{\partial}{\partial t}\left(\sum_{k=0}^{\infty} f_ke^{-\varphi}\right) + \sum_{j=0}^{N}\left(\frac{1}{2i}\right)^j\sigma_j\left(p, \sum_{k=0}^{\infty} f_ke^{-\varphi}\right)\\
&= \left(\sum_{k=1}^{\infty}\frac{\partial}{\partial t}f_k + \sum_{k=0}^{\infty} f_kg_0 + \sum_{k=1}^{\infty} Q_0f_k\right)e^{-\varphi}, + \sum_{k=1}^{\infty}\frac{1}{4} < AJ\psi, i\nabla f_k > e^{-\varphi}\\
&\quad + \sum_{k=1}^{\infty}\sum_{3\le j\le N}\left(\frac{1}{2i}\right)^j(\sigma_j(p, f_ke^{-\varphi}) - f_k\sigma_j(p, e^{-\varphi})).
\end{aligned}
$$

Put

$$n_\ell(f_1, f_2, \ldots, f_\ell) = \frac{1}{4} < AJ\psi, i\nabla f_\ell > + \sum_{\substack{j+k=\ell+3 \\ 1\le k\le \ell,\, 3\le j\le N}} \times\left(\frac{1}{2i}\right)^j \Big(\sigma_j(p, f_k e^{-\varphi}) - f_k\sigma_j(p, e^{-\varphi})\Big)e^{\varphi}.$$

Then we have

$$\frac{\partial}{\partial t}\Big(\sum_{k=0}^{\infty} f_k e^{-\varphi}\Big) + \sum_{j=0}^{N}\left(\frac{1}{2i}\right)^j \sigma_j\Big(p, \sum_{k=0}^{\infty} f_k e^{-\varphi}\Big)$$
$$= \Big(\sum_{k=1}^{\infty}\frac{\partial}{\partial t} f_k + \sum_{k=0}^{\infty} f_k g_0 + \sum_{k=1}^{\infty} Q_0 f_k\Big)e^{-\varphi} + \sum_{\ell=1}^{\infty} n_\ell(f_1, f_2, \ldots, f_\ell)e^{-\varphi}.$$

It is natural to try to find solutions f_k for the following equations:

f_1 is the solution of

$$\frac{\partial}{\partial t} f_1 + Q_0 f_1 + g_0 = 0, \quad f_1|_{t=0} = 0. \tag{15.4.21}$$

f_k, with $k \geq 2$, is the solution of

$$\frac{\partial}{\partial t} f_k + f_{k-1}g_0 + Q_0 f_k + n_{k-1}(f_1, \ldots, f_{k-1}) = 0, \quad f_k|_{t=0} = 0. \tag{15.4.22}$$

However, we cannot solve the aforementioned equations exactly. We can only find the approximate solution of the aforementioned equations.

We prepare the ground for solving (15.4.21) by the following proposition.

Let $K(t)$ be a smooth function valued in an $\ell \times \ell$ matrix. Set $\theta(t, s)$ be the unique solution of

$$\Big(\frac{d}{dt} + {}^tK(t)\Big)\,{}^t\theta(t, s) = 0, \qquad \theta(s, s) = 0. \tag{15.4.23}$$

Proposition 15.4.4. *Let $g(t, \zeta)$ be a homogeneous polynomial of degree k in the variable $\zeta = {}^t(\zeta_1, \zeta_2, \ldots, \zeta_\ell)$, with coefficients smooth functions of t. Then the following polynomial defined by*

$$h(t, \zeta) = \int_0^t g(s, \theta(t, s)\zeta)ds \tag{15.4.24}$$

is also a homogeneous polynomial of degree k with respect to ζ, and satisfies

$$\frac{\partial}{\partial t}h(t,\zeta)+ < K(t)\zeta, \frac{\partial}{\partial \zeta}h(t,\zeta) > = g(t,\zeta),$$
$$h(0,\zeta) = 0,$$

where $\frac{\partial}{\partial \zeta}h(t,\zeta) = {}^t\Big(\frac{\partial}{\partial \zeta_1}h(t,\zeta), \frac{\partial}{\partial \zeta_2}h(t,\zeta), \dots, \frac{\partial}{\partial \zeta_\ell}h(t,\zeta)\Big)$.

Proof. Set $G(t,\zeta) = \frac{\partial}{\partial \zeta}g(t,\zeta)$. Then by (15.4.23), we have

$$\begin{aligned}
\frac{\partial}{\partial t}h(t,\zeta) &= g(t,\zeta) + \int_0^t < \frac{d}{dt}\theta(t,s)\zeta, G(s,\theta(t,s)\zeta) > ds \\
&= g(t,\zeta) - \int_0^t < \theta(t,s)K(t)\zeta, G(s,\theta(t,s)\zeta) > ds \\
&= g(t,\zeta) - < K(t)\zeta, \int_0^t {}^t\theta(t,s)G(s,\theta(t,s)\zeta)ds > \\
&= g(t,\zeta) - < K(t)\zeta, \frac{\partial}{\partial \zeta}h(t,\zeta) >,
\end{aligned}$$

because

$$\frac{\partial}{\partial \zeta}h(t,\zeta) = \int_0^t {}^t\theta(t,s)G(s,\theta(t,s)\zeta)ds$$

holds by (15.4.24). ∎

Corollary 15.4.5. *For an $\ell \times \ell$ constant matrix B, set*

$$\theta(t,s) = (1+e^{-Bs})(1+e^{-Bt})^{-1}$$

and

$$h(t,\zeta) = \int_0^t g(s,\theta(t,s)\zeta)ds.$$

Then we have

$$\frac{\partial}{\partial t}h(t,\zeta) + \frac{1}{2} < B(\tanh(Bt/2)-1)\zeta, \frac{\partial}{\partial \zeta}h(t,\zeta) > = g(t,\zeta),$$
$$h(0,\zeta) = 0.$$

Proof. Since in our case we have

$$K(t) = \frac{1}{2}B(\tanh(Bt/2)-1),$$

it suffices to show

$$\theta(t,s) = \exp\Big(-\int_s^t K(r)dr\Big).$$

Since it is easy to see that

$$-\int_s^t K(r)dr = \frac{1}{2}B(t-s) - \log[\cosh(Bt/2)] + \log[\cosh(Bs/2)],$$

$$\exp\Big(-\int_s^t K(r)dr\Big) = \exp\Big(\frac{t-s}{2}B\Big)\cosh(Bs/2)(\cosh(Bt/2))^{-1}$$
$$= (1+e^{-Bs})(1+e^{-Bt})^{-1},$$

we get the desired assertion. ■

Next we shall go back to solve (15.4.21).

Lemma 15.4.6. *We can obtain an approximate solution of (15.4.21), in the sense*

$$\frac{\partial}{\partial t} f_1 + Q_0 f_1 + g_0 = G_1, \quad f_1|_{t=0} = 0,$$

where G_1 is a polynomial of $b(=iJ\nabla p)$ with order less than g_0, in the sense that

$$G_1 e^{-\varphi} \in S^{m-2}_{1/2,1/2}.$$

Moreover, $e_1(t;x,\xi) = f_1(t,x,\xi,b)e^{-\varphi} \in S^{-1/2}_{1/2,1/2}$.

Proof. Set $b = iJ\nabla p$. We will prove the assertion by induction over the degree of g_0 with respect to b. Assume that $g_0(t,x,\xi,b)$ is a homogenous polynomial of degree k with respect to b. Then

$$f_1(t,x,\xi,b) = -\int_0^t g_0(s,x,\xi,\theta(t,s)b)\,ds$$

is a homogeneous polynomial of degree k and satisfies

$$\nabla\Big(f_1(t,x,\xi,b)\Big) = (\nabla f_1)(t,x,\xi,b) +^t A\frac{\partial}{\partial b} f_1(t,x,\xi,b),$$

where

$$\theta(t,s) = (1+e^{-As})(1+e^{-At})^{-1}.$$

Using Corollary 15.4.5 yields

$$\frac{\partial}{\partial t} f_1(t,x,\xi,b) + \frac{1}{2} < (\tanh(At/2)-1)b, \nabla\Big(f_1(t,x,\xi,b)\Big) >$$
$$= -g_0 + \frac{1}{2} < (\tanh(At/2)-1)b, (\nabla f_1)(t,x,\xi,b) >,$$
$$f_1(0,x,\xi,b) = 0,$$

and we can show that $f_1(t,x,\xi,b)e^{-\varphi} \in S^{-1/2}_{1/2.1/2}$. We have the following fact:

$$\mathrm{tr}(AJH_{f_1}) = \mathrm{tr}\left(AJ^tA\frac{\partial^2}{\partial b^2}f_1(t,x,\xi,b)A\right) + \tilde{G}_1$$

with $\tilde{G}_1 e^{-\varphi} \in S^{m-2}_{1/2.1/2}$. Then we have

$$\frac{\partial}{\partial t}f_1 + Q_0 f_1 + g_0 = -\frac{i}{8}\tilde{G}_1 - \frac{i}{8}\mathrm{tr}\left(AJ^tA\frac{\partial^2}{\partial b^2}f_1(t,x,\xi,b)A\right) + \frac{1}{2} < (\tanh(At/2)-1)b, (\nabla f_1)(t,x,\xi,b)>.$$

We can prove that $\langle(\tanh(At/2)-1)b, (\nabla f_1)(t,x,\xi,b)\rangle e^{-\varphi}$ belongs to $S^{m-2}_{1/2,1/2}$, and that $\mathrm{tr}(AJ^tA\frac{\partial^2}{\partial b^2}f_1(t,x,\xi,b)A)$ is a homogenous polynomial of degree $k-2$ with respect to b. This leads to the desired conclusion. ■

Instead of (15.4.22) for f_k $(k \geq 2)$, we need to solve the following equations step by step:

$$\frac{\partial}{\partial t}f_k + Q_0 f_k + f_{k-1}g_0 + n_{k-1}(f_1,\ldots,f_{k-1}) + G_{k-1} = G_k, \quad f_k|_{t=0} = 0, \tag{15.4.25}$$

under the condition

$$G_k e^{-\varphi} \in S^{m-3/2-k/2}_{1/2,1/2}.$$

We can obtain the solution f_k of (15.4.25) and also show that $e_j(t;x,\xi) = f_j(t,x,\xi,b)e^{-\varphi} \in S^{-j/2}_{1/2,1/2}$ by a similar method to the construction of f_1.

Step 4: *The construction of the fundamental solution*

We have constructed $e_j(t;x,\xi)$, for $j = 1,2,\ldots,N-1$, such that

$$\left(\frac{\partial}{\partial t} + p^w(x,D)\right)\left(\sum_{j=0}^{N-1} e_j^w(t;x,D)\right) = r_N^w(t;x,D),$$

with $r_N(t;x,\xi) \in S^{m-N/2}_{1/2,1/2}$. So we can construct the symbol of the fundamental solution of the form

$$e(t) = \sum_{j=0}^{N-1} e_j(t) + \int_0^t \sum_{j=0}^{N-1} e_j(t-s) \circ_w \psi(s)ds,$$

with $\psi(t) \in S^{m-N/2}_{1/2,1/2}$ if $m - N/2 \leq 0$. In fact, we can construct $\psi(t)$ as a unique solution of the following equation by the similar method as in Theorem 15.2.1:

$$r_N(t) + \psi(t) + \int_0^t r_N(t-s) \circ_w \psi(s)ds = 0.$$

In this case we apply the estimate of symbols of the multi-product of the pseudo-differential operators of Weyl symbols. Then we have the solution $\psi(t) \in S^{m-N/2}_{1/2,1/2}$ of the previous integral equation. So $e(t)$ has the expansion $e(t) - \sum_{j=0}^{N-1} e_j(t) \in S^{m-N/2}_{1/2,1/2}$ for any N. Since each $e_j(t)$ belongs to $S^{-j/2}_{1/2,1/2}$, we can show that $e(t) - \sum_{j=0}^{N-1} e_j(t) \in S^{-N/2}_{1/2,1/2}$. This completes the proof. ■

15.4.2 The Symbol $p_m = \frac{1}{2}\sum_{j=1}^{\ell} q_j^2$

If the principal symbol p_m has the exactly the double characteristic Σ, that is, $p_m(x,\xi) = \frac{1}{2}\sum_{j=1}^{\ell} q_j^2(x,\xi)$, then the following theorem provides a closed-form expression for φ.

Theorem 15.4.7. *If $p_m(x,\xi) = \frac{1}{2}\sum_{j=1}^{\ell} q_j^2(x,\xi)$, then we can choose*

$$\varphi = p_{m-1}t + \frac{1}{2}\mathrm{tr}\Big[\log\{\cosh(Mt/2)\}\Big] + \frac{t}{2} < F(Mt/2)\mathbf{q}, \mathbf{q} >,$$

where

$$F(x) = \frac{\tanh x}{x}, \quad \mathbf{q} = {}^t(q_1, \dots, q_\ell),$$

and M is an $\ell \times \ell$ Hermitian matrix defined by

$$M = (M_{jk}), \quad M_{jk} = i < \nabla q_j, J\nabla q_k >.$$

We note that if the q_j are symbols of vector fields, then the M_{jk} are symbols for the corresponding commutators.

Proof. The previous formula can be obtained in two ways. One is to repeat the proof under the assumption

$$\nabla p = \sum_{j=1}^{\ell} q_j \nabla q_j, \quad H_p = \sum_{j=1}^{\ell} \nabla q_j^t (\nabla q_j).$$

The other way is to use the result obtained in Theorem 15.4.1:

$$\begin{aligned}\frac{it^2}{4} < G(At/2)J\nabla p, \nabla p > &= \frac{t}{2}\sum_{j,k=1}^{\ell} < G(At/2)(iJ\nabla q_j t/2)q_j, q_k \nabla q_k > \\ &= \frac{t}{2}\sum_{j,k=1}^{\ell} q_j q_k (\tilde{F}(Mt/2))_{jk},\end{aligned}$$

where $\tilde{F}(x) = xG(x) = 1 - F(x)$. Then the following identity holds:

$$pt - \frac{it^2}{4} < G(At/2)J\nabla p, \nabla p > = \frac{1}{2}\sum_{j=1}^{\ell} q_j^2 t - \frac{t}{2} < \tilde{F}(Mt/2)\mathbf{q}, \mathbf{q} >$$
$$= \frac{t}{2} < F(Mt/2)\mathbf{q}, \mathbf{q}>.$$

On the other hand, we have

$$AS = SM,$$

where S is a $(2n \times \ell)$ matrix defined by

$$S = (iJ\nabla q_1, iJ\nabla q_2, \ldots, iJ\nabla q_\ell).$$

So we get

$$\det\{\cosh(At/2)\} = \det\{\cosh(Mt/2)\}.$$

This completes the proof. ∎

15.4.3 The Special Case of Quadratic Symbols

Let $p(x,\xi)$ be a polynomial in (x,ξ) of degree at most 2. In this case the fundamental solution $E(t)$ is obtained as a pseudo-differential operator of Weyl symbol $e(t;x,\xi) = \exp\big(-\varphi(t,x,\xi)\big)$, where

$$\varphi(t;x,\xi) = pt + \frac{\mathrm{tr}}{2}\left\{\log\left(\cosh\left(\frac{At}{2}\right)\right)\right\} - \frac{it^2}{4} < G(At/2)J\nabla p, \nabla p >,$$

because $\varphi(t;x,\xi)$ is the exact solution of the equation

$$-\frac{d\varphi}{dt} + p - \frac{1}{2i}\sigma_1(p,\varphi) + \frac{1}{4}\sigma_2(p,\varphi) - \frac{1}{8} < J\nabla\varphi, H_p J\nabla\varphi >= 0.$$

More precisely, we have the following result. This is the key theorem for the construction of the fundamental solution for the heat equation with polynomial coefficients. It is also useful for the construction of the fundamental solution of degenerate parabolic equations.

Theorem 15.4.8. *If $p(x,\xi)$ is a quadratic polynomial with respect to $X = {}^t(x,\xi) \in \mathbb{R}^d \times \mathbb{R}^d$,*

$$p = \frac{1}{2}\langle X, HX\rangle + i\langle X, p_0\rangle + b,$$

then $E(t) = e^w(t; x, D)$ *is given by*

$$e(t; x, \xi) = \frac{e^{-bt}}{\sqrt{\det\cosh(At/2)}} \exp\Big[-i\Big\{\langle J\tanh(At/2)X, X\rangle + t\langle J\tanh(At/2)(At/2)^{-1}X, Jp_0\rangle + \frac{t^2}{4}\langle JG(At/2)Jp_0, Jp_0\rangle\Big\}\Big],$$

where $A = iJH$ *is a* $2d \times 2d$ *matrix,*

$$J = \begin{pmatrix} 0 & I \\ -I & 0 \end{pmatrix},$$

and

$$G(x) = (1 - x^{-1}\tanh x)/x.$$

Proof. Since

$$\nabla p = HX + ip_0, \quad H_p = H,$$

we have

$$\begin{aligned}\langle G(At/2)Jp, p\rangle &= \langle G(At/2)JHX, HX\rangle + \langle G(At/2)JHX, p_0\rangle \\ &\quad + i\langle G(At/2)Jp_0, HX\rangle - \langle G(At/2)Jp_0, p_0\rangle \\ &= \langle JG(At/2)JHX, JHX\rangle + i\langle JG(At/2)JHX, Jp_0\rangle \\ &\quad + i\langle JG(At/2)Jp_0, JHX\rangle - \langle JG(At/2)Jp_0, Jp_0\rangle,\end{aligned}$$

and hence

$$\begin{aligned}\frac{t^2}{4}\langle G(At/2)Jp, p\rangle &= -\langle JG(At/2)(At/2)X, (At/2)X\rangle \\ &\quad + \frac{t}{2}\langle JG(At/2)(At/2)X, Jp_0\rangle \\ &\quad + \frac{t}{2}\langle JG(At/2)Jp_0, (At/2)X\rangle - \frac{t^2}{4}\langle JG(At/2)Jp_0, Jp_0\rangle.\end{aligned}$$

Using that

$${}^tAJ = -JA$$

and

$${}^t(JG(At/2)) = {}^tG(At/2){}^tJ = -{}^tG(At/2)J = JG(At/2),$$

for the odd function $G(x)$, we have

$$\begin{aligned}\frac{t^2}{4}\langle G(At/2)Jp, p\rangle &= \langle JG(At/2)(At/2)^2X, X\rangle + t\langle JG(At/2)(At/2)X, Jp_0\rangle \\ &\quad - \frac{t^2}{4}\langle JG(At/2)Jp_0, Jp_0\rangle.\end{aligned}$$

Since

$$x^2 G(x) = x - \tanh x, \quad xG(x) = 1 - x^{-1} \tanh x,$$

it holds that

$$\begin{aligned}\langle JG(At/2)(At/2)^2 X, X\rangle &= \langle J(At/2 - \tanh(At/2))X, X\rangle \\ &= -\frac{it}{2}\langle HX, X\rangle - \langle J \tanh(At/2)X, X\rangle\end{aligned}$$

and

$$\begin{aligned}\langle JG(At/2)(At/2)X, Jp_0\rangle &= \langle JX, Jp_0\rangle - \langle J(At/2)^{-1} \tanh(At/2)X, Jp_0\rangle \\ &= \langle X, p_0\rangle - \langle J(At/2)^{-1} \tanh(At/2)X, Jp_0\rangle.\end{aligned}$$

Therefore, we have

$$\begin{aligned}i\frac{t^2}{4}\langle G(At/2)Jp, p\rangle &= \frac{t}{2}\langle HX, X\rangle - i\langle J \tanh(At/2)X, X\rangle \\ &\quad + it\langle X, p_0\rangle - it\langle J(At/2)^{-1} \tanh(At/2)X, Jp_0\rangle \\ &\quad -i\,\frac{t^2}{4}\langle JG(At/2)Jp_0, Jp_0\rangle \\ &= pt - bt - i\langle J \tanh(At/2)X, X\rangle \\ &\quad - it\langle J(At/2)^{-1} \tanh(At/2)X, Jp_0\rangle \\ &\quad - i\frac{t^2}{4}\langle JG(At/2)Jp_0, Jp_0\rangle.\end{aligned}$$

Finally, we have

$$\begin{aligned}Q = bt &+ \frac{\mathrm{tr}}{2}\left\{\log\left(\cosh(\frac{At}{2})\right)\right\} + i\langle J \tanh(At/2)X, X\rangle \\ &+ it\langle J(At/2)^{-1} \tanh(At/2)X, Jp_0\rangle \\ &+ i\frac{t^2}{4}\langle JG(At/2)Jp_0, Jp_0\rangle.\end{aligned}$$

Noting that

$$\exp\left[-\frac{\mathrm{tr}}{2}\left\{\log\left(\cosh(\frac{At}{2})\right)\right\}\right] = \frac{1}{\sqrt{\det \cosh(At/2)}},$$

we get the assertion. ∎

15.4.4 A Key Theorem for Eigenfunction Expansion

In a forthcoming section we shall show a method of obtaining eigenfunctions using the fundamental solution represented by pseudo-differential operators of Weyl symbols. The following result is the key of a future proof of the expansion for the kernel of the fundamental solution obtained as a pseudo-differential operator.

Proposition 15.4.9. *If the symbol of a pseudo-differential operator is of the form* $p(x,\xi) = h(x,\xi)g(x,\xi)$, *then the kernel of operator* $p^w(x,D)$ *in* $\mathbb{R}^d$ *is given by*

$$(2\pi)^{-d}\int_{\mathbb{R}^d} e^{i(x-x')\cdot\xi}\, p\Big(\frac{x+x'}{2},\xi\Big)d\xi = h\Big(\frac{r}{2},-i\frac{\partial}{\partial_q}\Big)\tilde{g}\Big(\frac{r}{2},q\Big)\Big|_{r=x+x',q=x-x'},$$

where

$$\tilde{g}\Big(\frac{r}{2},q\Big) = (2\pi)^{-d}\int_{\mathbb{R}^d} e^{iq\cdot\xi} g\Big(\frac{r}{2},\xi\Big)d\xi.$$

Proof. If the symbol has the expansion $h(x,\xi) = \sum_\alpha a_\alpha(x)\xi^\alpha$, then with $r = x+x'$ and $q = x-x'$, we formally have

$$\begin{aligned}(2\pi)^{-d}\int_{\mathbb{R}^d} e^{i(x-x')\cdot\xi}\, p\Big(\frac{x+x'}{2},\xi\Big)d\xi &= (2\pi)^{-d}\int_{\mathbb{R}^d}\sum_\alpha a_\alpha\Big(\frac{r}{2}\Big)\xi^\alpha e^{iq\cdot\xi} g\Big(\frac{r}{2},\xi\Big)\,d\xi\\ &= \sum_\alpha a_\alpha\Big(\frac{r}{2}\Big)\Big(-i\frac{\partial}{\partial_q}\Big)^\alpha (2\pi)^{-d}\int_{\mathbb{R}^d} e^{iq\cdot\xi} g\Big(\frac{r}{2},\xi\Big)d\xi\\ &= h\Big(\frac{r}{2},-i\frac{\partial}{\partial_q}\Big)\tilde{g}\Big(\frac{r}{2},q\Big)\Big|_{r=x+x',q=x-x'}.\end{aligned}$$

■

15.5 The Hermite Operator

In the rest of this section we will provide symbols of the fundamental solution as pseudo-differential operators according to Theorem 15.4.8 for several examples that have been studied in previous chapters by different methods.

First we will study the heat equation corresponding to the Hermite operator. We will also provide its eigenfunction expansion according to Proposition 15.4.9.

The one-dimensional Hermite operator considered in this section is given by

$$P = \frac{1}{2}\Big(-\frac{\partial^2}{\partial x^2} + c^2x^2\Big), \tag{15.5.26}$$

where c is a positive constant.

15.5.1 Exact Form of the Symbol of the Fundamental Solution

The next result deals with the fundamental solution $E(t)$ of the operator (15.5.26) as a pseudo-differential operator. We note that its kernel $K(t;x,x')$ is given by

$$K(t;x,x') = \frac{1}{2\pi}\int_{-\infty}^{\infty} e^{i(x-x')\xi} e\left(t;\frac{x+x'}{2},\xi\right) d\xi.$$

Theorem 15.5.1. *The fundamental solution is $E(t) = e^{w}(t;x,D)$, with the Weyl symbol*

$$e(t;x,\xi) = \frac{1}{\cosh(ct/2)}\exp\left[-\frac{\tanh(ct/2)}{c}(c^2x^2+\xi^2)\right].$$

The kernel $K(t;x,x')$ is given by

$$K(t;x,x') = \sqrt{\frac{c}{2\pi\sinh(ct)}}\exp\left[-\frac{c}{2\sinh(ct)}\left\{(x^2+x'^2)\cosh(ct) - 2xx'\right\}\right].$$

Proof. The symbol of the operator (15.5.26) is given by

$$\sigma(P) = \frac{1}{2}(\xi^2 + c^2x^2).$$

Then choosing in Theorem 15.4.8

$$H = \begin{pmatrix} c^2 & 0 \\ 0 & 1 \end{pmatrix}$$

and

$$X = {}^t(x,\xi),$$

we have

$$e(t;x,\xi) = \frac{1}{\sqrt{\det\cosh(At/2)}}\exp\{-i\langle J\tanh(At/2)X, X\rangle\}.$$

If $c = 0$, then

$$e(t;x,\xi) = e^{-|\xi|^2t/2}.$$

Assume that $c \neq 0$. The 2×2 matrix $A = iJH$ has eigenvalues

$$c, -c$$

and eigenvectors

$$U^+ = \begin{pmatrix} 1 \\ -ic \end{pmatrix}, \quad U^- = \begin{pmatrix} 1 \\ ic \end{pmatrix}.$$

It is easy to see that

$$X = c_1 U^+ + c_2 U^-,$$

with

$$c_1 = \frac{x}{2} + i\frac{\xi}{2c}, \quad c_2 = \frac{x}{2} - i\frac{\xi}{2c}.$$

Then we have

$$\begin{aligned} iJ\tanh(At/2)X &= i\tanh(ct/2)(c_1 JU^+ - c_2 JU^-) \\ &= \tanh(ct/2)W, \end{aligned}$$

where

$$W = {}^t\left(cx, \frac{\xi}{c}\right).$$

We have

$$\begin{aligned} \langle W, X\rangle &= \frac{1}{c}(\xi^2 + c^2x^2), \\ \sqrt{\det\cosh(At/2)} &= \cosh(ct/2), \\ i\langle J\tanh(At/2)X, X\rangle &= \frac{\tanh(ct/2)}{c}(\xi^2 + c^2x^2). \end{aligned}$$

Thus the kernel of the fundamental solution is given by

$$\begin{aligned} K(t; z, s, z', s') &= \frac{1}{2\pi}\int_{\mathbb{R}} e^{i(x-x')\xi} e\left(t; \frac{x+x'}{2}, \xi\right) d\xi \\ &= \frac{1}{2\pi}\int_{\mathbb{R}} e^{i(x-x')\xi} \frac{1}{\cosh(ct/2)} \\ &\quad \times \exp\left[-\frac{\tanh(ct/2)}{c}\left(\frac{1}{4}c^2(x+x')^2 + \xi^2\right)\right] d\xi \\ &= \frac{1}{2\sqrt{\pi}\cosh(ct/2)}\sqrt{\frac{c}{\tanh(ct/2)}} \\ &\quad \times \exp\left[-\frac{c}{4}\left(\tanh(ct/2)(x+x')^2 + \frac{(x-x')^2}{\tanh(ct/2)}\right)\right]. \end{aligned}$$

Using the equalities

$$\begin{aligned} \tanh(ct/2) + \frac{1}{\tanh(ct/2)} &= \frac{2}{\tanh(ct)}, \\ \tanh(ct/2) - \frac{1}{\tanh(ct/2)} &= -\frac{2}{\sinh(ct)}, \\ \sinh(ct/2)\cosh(ct/2) &= \frac{1}{2}\sinh(ct) \end{aligned}$$

yields the following expression for the heat kernel:

$$K(t;x,x') = \sqrt{\frac{c}{2\pi\sinh(ct)}}\exp\left[-\frac{c}{2\sinh(ct)}\left\{(x^2+x'^2)\cosh(ct) - 2xx'\right\}\right].$$

■

15.5.2 Eigenfunction Expansion

We shall apply Proposition 15.4.9 to $e(t;x,\xi)$ given by the previous theorem. The Weyl symbol in the case of the Hermite operator can be written as

$$\begin{aligned} e(t;x,\xi) &= \frac{1}{\cosh(ct/2)}\exp\left[-\frac{\tanh(ct/2)}{c}(c^2x^2+\xi^2)\right] \\ &= h(t,x,\xi)e_\infty(t;x,\xi), \end{aligned}$$

where

$$e_\infty(t;x,\xi) = 2e^{-ct/2}\exp\left[-\frac{1}{c}(c^2x^2+\xi^2)\right],$$

and

$$h(t;x,\xi) = \frac{1}{1+e^{-ct}}\exp\left[\frac{1}{c}\left\{1-\tanh(ct/2)\right\}(c^2x^2+\xi^2)\right].$$

With the substitution

$$w = e^{-ct},$$

we have

$$\begin{aligned} h\left(t;r/2,-i\frac{\partial}{\partial q}\right) &= \frac{1}{1+w}\exp\left[\frac{2w}{c(1+w)}\left(c^2r^2/4-\frac{\partial^2}{\partial q^2}\right)\right] \\ &= \frac{1}{1+w}\exp\left[\frac{2w}{c(1+w)}\left(cr/2+\frac{\partial}{\partial q}\right)\left(cr/2-\frac{\partial}{\partial q}\right)\right]. \end{aligned}$$

If we let $r = x + x'$ and $q = x - x'$, we have

$$cr/2+\frac{\partial}{\partial q} = \frac{1}{2}\left\{\left(\frac{\partial}{\partial x}+cx\right)+\left(-\frac{\partial}{\partial x'}+cx'\right)\right\}$$

and

$$cr/2-\frac{\partial}{\partial q} = \frac{1}{2}\left\{\left(-\frac{\partial}{\partial x}+cx\right)+\left(\frac{\partial}{\partial x'}+cx'\right)\right\}.$$

Then the following equation holds:

$$h\Big(t;r/2,-i\frac{\partial}{\partial q}\Big)=\frac{1}{1+w}\exp\Big[\frac{w}{(1+w)}\Big(B_x+(B_{x'})^*\Big)\Big((B_x)^*+B_{x'}\Big)\Big],$$

where

$$B_x=\frac{1}{\sqrt{2c}}\Big(\frac{\partial}{\partial x}+cx\Big),\quad B_{x'}=\frac{1}{\sqrt{2c}}\Big(\frac{\partial}{\partial x'}+cx'\Big).$$

Note that

$$[B_x,(B_x)^*]=1,\quad [B_x+(B_{x'})^*,(B_x)^*+B_{x'}]=0.$$

Set

$$\psi(x)=\Big(\frac{c}{\pi}\Big)^{\frac{1}{4}}\exp\Big(-\frac{cx^2}{2}\Big). \tag{15.5.27}$$

We conclude with the following result.

Proposition 15.5.2. *Let* $\tilde{e}_\infty(t;x,x')$ *be the kernel of* $e^{\mathrm{w}}_\infty(t;x,D)$. *Then we have*

$$\tilde{e}_{t,\infty}(t;x,x')=\sqrt{w}\,\psi(x)\psi(x').$$

Proof. We have the following computation:

$$\begin{aligned}\tilde{e}_\infty(t;x,x')&=\frac{1}{2\pi}\int_{\mathbb{R}}e^{i(x-x')\xi}e_\infty\left(t;\frac{x+x'}{2},\xi\right)d\xi\\&=\frac{2\sqrt{w}\sqrt{c\pi}}{2\pi}\exp\left\{-\frac{c(x-x')^2}{4}-\frac{c(x+x')^2}{4}\right\}\\&=\sqrt{w}\psi(x)\psi(x').\end{aligned}$$

■

We shall deal next with the following expansion regarding the operator $h\big(t;r/2,-i\frac{\partial}{\partial q}\big)$.

Proposition 15.5.3. *(1) Let* $C_j(S)$ *be the coefficients defined by the following expansion:*

$$\frac{1}{1+w}\exp\Big[\frac{w}{(1+w)}S\Big]=\sum_{j=0}^{\infty}\frac{C_j(S)}{j!}w^j.$$

Then the following recursive formula holds:

$$\begin{aligned}C_{\ell+1}(S)&=SC_\ell(S)-(2\ell+1)C_\ell(S)-\ell^2C_{\ell-1}(S),\quad \ell\ge 0,\\C_0(S)&=1.\end{aligned}$$

(2) If $S=\Big(B_x+(B_{x'})^*\Big)\Big((B_x)^*+B_{x'}\Big)$, *then*

$$C_j(S)\psi(x)\psi(x')=\psi_j(x)\psi_j(x'),$$

where

$$\psi_j(x) = (B^*)^j \psi(x).$$

Proof. *(1)* Consider the analytic function $f(z)$ defined by

$$f(z) = \frac{1}{(1+z)} \exp\Big(\frac{z}{1+z} S\Big).$$

Since we have

$$(1+z) f(z) = \exp\Big(\frac{z}{1+z} S\Big),$$

it follows that

$$f'(z)(1+z)^2 + (1+z) f(z) = f(z) S, \quad f(0) = 1.$$

Differentiating ℓ times, we obtain

$$\begin{aligned}&(1+z)^2 f^{(\ell+1)}(z) + 2\ell(1+z) f^{(\ell)}(z) + \ell(\ell-1) f^{(\ell-1)}(z)\\ &\quad + (1+z) f^{(\ell)}(z) + \ell f^{(\ell-1)}(z) = f^{(\ell)}(z) S.\end{aligned}$$

Making $z = 0$, we obtain the desired equation for $C_\ell(S) = f^{(\ell)}(0)$.

(2) We note that the following equations hold:

(i)

$$B_x \psi(x) = 0, \quad B_x \psi_\ell(x) = \ell \psi_{\ell-1}(x),$$

(ii)

$$(B_x)^k (B_x^*)^k \psi(x) = k! \psi(x),$$

(iii)

$$\begin{aligned}S\big(\psi_j(x)\psi_j(x')\big) &= \psi_{j+1}(x)\psi_{j+1}(x') + (2j+1)\psi_j(x)\psi_j(x')\\ &\quad + j^2 \psi_{j-1}(x)\psi_{j-1}(x').\end{aligned}$$

We shall proceed by induction with respect to j. The assertion obviously holds for $j = 0$, because $C_0(S) = 1$ for all S. Note that for any operator S, the operators $C_j(S)$ satisfy part *(1)*. Assume that formula *(2)* holds for all $\ell \leq j$. Then we have

$$\begin{aligned}C_{j+1}(S)\Big(\psi(x)\psi(x')\Big) &= S C_j(S)\Big(\psi(x)\psi(x')\Big) - (2j+1) C_j(S)\Big(\psi(x)\psi(x')\Big)\\ &\quad - j^2 C_{j-1}(S)\Big(\psi(x)\psi(x')\Big)\\ &= S\big(\psi_j(x)\psi_j(x')\big) - (2j+1)\psi_j(x)\psi_j(x')\\ &\quad - j^2 \psi_{j-1}(x)\psi_{j-1}(x')\\ &= \psi_{j+1}(x)\psi_{j+1}(x'),\end{aligned}$$

where the last equality is obtained by (iii). Thus the induction method provides the desired conclusion.

We still have to prove formulas (i), (ii), and (iii). We shall next give the sketch of the proof for (i).

$$\begin{aligned} B_x\psi_\ell(x) &= [B_x, (B_x^*)^\ell]\psi(x) + (B_x^*)^\ell B_x\psi(x) \\ &= \ell(B_x^*)^{\ell-1}\psi(x) \\ &= \ell\psi_{\ell-1}(x). \end{aligned}$$

Formula (ii) is proved by using the following equation inductively:

$$\begin{aligned} B_x^k(B_x^*)^k\psi(x) &= B_x^k\psi_k(x) \\ &= B_x^{k-1}k\psi_{k-1}(x) \\ &= kB_x^{k-1}(B_x^*)^{k-1}\psi(x). \end{aligned}$$

We have

$$\begin{aligned} \Big((B_x)^* + B_{x'}\Big)(\psi_j(x)\psi_j(x')) &= \Big((B_x)^*\psi_j(x)\Big)\psi_j(x') + \psi_j(x)\Big(B_{x'}\psi_j(x')\Big) \\ &= \psi_{j+1}(x)\psi_j(x') + j\psi_j(x)\psi_{j-1}(x'), \end{aligned}$$

noting

$$(B_x)^*\psi_j(x) = \psi_{j+1}(x).$$

Then we have

$$\begin{aligned} &S(\psi_j(x)\psi_j(x')) \\ &\quad = \Big(B_x + (B_{x'})^*\Big)(\psi_{j+1}(x)\psi_j(x') + j\psi_j(x)\psi_{j-1}(x')) \\ &\quad = (B_x\psi_{j+1}(x))\psi_j(x') + j(B_x\psi_j(x))\psi_{j-1}(x') \\ &\qquad + \psi_{j+1}(x)((B_{x'})^*\psi_j(x')) + j\psi_j(x)((B_{x'})^*\psi_{j-1}(x')) \\ &\quad = (j+1)\psi_j(x)\psi_j(x') + j^2\psi_{j-1}(x)\psi_{j-1}(x') \\ &\qquad + \psi_{j+1}(x)\psi_{j+1}(x') + j\psi_j(x)\psi_j(x') \\ &\quad = \psi_{j+1}(x)\psi_{j+1}(x') + (2j+1)\psi_j(x)\psi_j(x') + j^2\psi_{j-1}(x)\psi_{j-1}(x'). \end{aligned}$$

The proof of (iii) is complete.

Theorem 15.5.4. *The kernel $K(t;x,x')$ of $e^w(t;x,D)$ has the following expansion:*

$$K(t;x,x') = \sum_{j=0}^{\infty}\frac{1}{j!}e^{-ct(j+1/2)}\psi_j(x)\psi_j(x'),$$

with

$$\psi_j(x) = \Big\{\frac{1}{\sqrt{2c}}\Big(-\frac{\partial}{\partial x} + cx\Big)\Big\}^j\psi(x),$$

with $\psi(x)$ defined by (15.5.27).

Proof. By Propositions 15.4.9, 15.5.2, and 15.5.3, we have

$$K(t;x,x') = h\left(t;r/2,-i\frac{\partial}{\partial q}\right)\tilde{e}_\infty(t;x,x') = \sum_{j=0}^{\infty}\frac{1}{j!}w^j C_j(S)\tilde{e}_\infty(t;x,x')$$
$$= \sum_{j=0}^{\infty}\frac{1}{j!}w^{j+1/2}C_j(S)\psi(x)\psi(x') = \sum_{j=0}^{\infty}\frac{1}{j!}w^{j+1/2}\psi_j(x)\psi_j(x').$$

■

The above theorem tells us that the Hermite operator has the eigenvalues $\{c(j+1/2)\}_{j=0}^{\infty}$ and the corresponding eigenfunctions $\{\psi_j(x)/\sqrt{j!}\}_{j=0}^{\infty}$.

These results are useful in the study of the following Grushin operator. Taking the Fourier transform, we obtain symbol similar to that of the Hermite operator.

15.6 The Grushin Operator

In this section we shall provide the exact form for the symbol of the fundamental solution of the Grushin operator. This is a degenerate operator on $\mathbb{R}^2$ defined as

$$P = -\frac{1}{2}\left(\frac{\partial^2}{\partial x^2} + x^2\frac{\partial^2}{\partial y^2}\right).$$

Theorem 15.6.1. *We have $E(t) = e^w(t;x,y,D_x,D_y)$ with*

$$e(t;x,y,\xi,\eta) = \frac{1}{\cosh(\eta t/2)}\exp\left[-\frac{\tanh(\eta t/2)}{\eta}(\xi^2 + x^2\eta^2)\right],$$

and with its kernel given by

$$K(t;x,y,x',y') = \frac{1}{2\pi t^{\frac{3}{2}}}\int_{\mathbb{R}} e^{i(y-y')\frac{v}{t}}\sqrt{\frac{v}{2\pi\sinh(v)}}$$
$$\times\exp\left[-\frac{v}{2t\sinh(v)}\left\{(x^2+x'^2)\cosh(v) - 2xx'\right\}\right]dv.$$

Proof. Since we have

$$\sigma^w(P) = \frac{1}{2}(\xi^2 + x^2\eta^2),$$

using Theorem 15.5.1 we easily get the symbol of the fundamental solution. The kernel of the fundamental solution is obtained as in the following:

$$K(t;x,y,x',y') = \left(\frac{1}{2\pi}\right)^2\int_{\mathbb{R}^2} e^{i(x-x')\xi+i(y-y')\eta}e\left(t;\frac{x+x'}{2},\frac{y+y'}{2},\xi,\eta\right)d\xi d\eta$$

$$= \frac{1}{2\pi} \int_{\mathbb{R}} e^{i(y-y')\eta} \sqrt{\frac{\eta}{2\pi \sinh(\eta t)}}$$

$$\times \exp\left[-\frac{\eta}{2\sinh(\eta t)}\left\{(x^2 + x'^2)\cosh(\eta t) - 2xx'\right\}\right] d\eta$$

$$= \frac{1}{2\pi t^{\frac{3}{2}}} \int_{\mathbb{R}} e^{i(y-y')\frac{v}{t}} \sqrt{\frac{v}{2\pi \sinh(v)}}$$

$$\times \exp\left[-\frac{v}{2t\sinh(v)}\left\{(x^2 + x'^2)\cosh(v) - 2xx'\right\}\right] dv.$$

■

15.7 Exact Form of the Symbol of the Fundamental Solution for the Sub-Laplacian

Let $x, y, s \in \mathbb{R}$, $z = x + iy \in \mathbb{C}$ and consider the following vector fields:

$$Z = \frac{\partial}{\partial z} + ia\bar{z}\frac{\partial}{\partial s}, \quad \bar{Z} = \frac{\partial}{\partial \bar{z}} - iaz\frac{\partial}{\partial s}, \quad T = \frac{\partial}{\partial s}.$$

This section is concerned with the sub-Laplacian on $\mathbb{R}^3$ defined by

$$P = -\frac{1}{2}\left(Z\bar{Z} + \bar{Z}Z\right) + i\alpha T.$$

Theorem 15.7.1. *Let $a \in \mathbb{R}$ and $\alpha \in \mathbb{C}$ be constants such that $|\alpha| < a$ and $a > 0$. Then $E(t) = e^w(t; x, y, s, D_x, D_y, D_s)$ is obtained with*

$$e(t; x, y, s, \xi, \eta, \sigma) = \frac{e^{\alpha\sigma t}}{\cosh(a\sigma t)} \exp\left[-\frac{\tanh(a\sigma t)}{4a\sigma}\{(\xi + 2a\sigma y)^2 + (\eta - 2a\sigma x)^2\}\right]$$

and its kernel is given by

$$K(t; z, s, z', s') = \int_{-\infty}^{\infty} \frac{a\sigma}{2\pi^2 \sinh(a\sigma t)} \exp\left[i\{s - s' + 2a(\Im(z\bar{z}'))\}\sigma\right.$$

$$\left. - \frac{a\sigma}{\tanh(a\sigma t)}|z - z'|^2 + \alpha\sigma t\right] d\sigma.$$

Proof. We have

$$\sigma^w(L) = \frac{1}{4}\{(\xi + 2a\sigma y)^2 + (\eta - 2a\sigma x)^2\} - \alpha\sigma,$$

by Theorem 15.4.8, with

$$H = \begin{pmatrix} 2a^2\sigma^2 & 0 & 0 & -a\sigma \\ 0 & 2a^2\sigma^2 & a\sigma & 0 \\ 0 & a\sigma & \frac{1}{2} & 0 \\ -a\sigma & 0 & 0 & \frac{1}{2} \end{pmatrix}$$

and

$$X = {}^t(x, y, \xi, \eta);$$

this yields the symbol

$$e(t; x, \xi) = \frac{e^{\alpha\sigma t}}{\sqrt{\det \cosh(At/2)}} \exp\{-i \langle J \tanh(At/2)X, X\rangle\}.$$

Assume that $\sigma \neq 0$. The 4×4 matrix $A = iJH$ has the eigenvalues

$$2a\sigma, \quad -2a\sigma, \quad 0, \quad 0$$

and the corresponding eigenvectors

$$U^+ = \begin{pmatrix} 1 \\ -i \\ -2ia\sigma \\ -2a\sigma \end{pmatrix}, \quad U^- = \begin{pmatrix} 1 \\ i \\ 2ia\sigma \\ -2a\sigma \end{pmatrix}, \quad U_0 = \begin{pmatrix} 1 \\ 0 \\ 0 \\ 2a\sigma \end{pmatrix}, \quad \tilde{U_0} = \begin{pmatrix} 0 \\ 1 \\ -2a\sigma \\ 0 \end{pmatrix}.$$

It is easy to see that we have the decomposition

$$X = c_0 U_0 + \tilde{c_0}\tilde{U_0} + c_1 U^+ + c_2 U^-,$$

with

$$c_0 = \frac{x}{2} + \frac{\eta}{4a\sigma}, \quad \tilde{c_0} = \frac{y}{2} - \frac{\xi}{4a\sigma},$$

$$c_1 = \frac{x + iy}{4} + i\frac{(\xi + i\eta)}{8a\sigma}, \quad c_2 = \frac{x - iy}{4} - i\frac{(\xi - i\eta)}{8a\sigma}.$$

Then we have

$$\begin{aligned} iJ \tanh(At/2)X &= i \tanh(a\sigma t)(c_1 JU^+ - c_2 JU^-) \\ &= \tanh(a\sigma t)W, \end{aligned}$$

where

$$W = {}^t\Big(a\sigma x - \frac{\eta}{2}, \quad a\sigma y + \frac{\xi}{2}, \quad \frac{y}{2} + \frac{\xi}{4a\sigma}, \quad -\frac{x}{2} + \frac{\eta}{4a\sigma}\Big).$$

By computation we obtain

$$\begin{aligned}
\langle W, X\rangle &= \frac{1}{4a\sigma}\{(\xi + 2a\sigma y)^2 + (\eta - 2a\sigma x)^2\},\\
\sqrt{\det\cosh(At/2)} &= \cosh(a\sigma t),\\
i\langle J\tanh(At/2)X, X\rangle &= \frac{\tanh(a\sigma t)}{4a\sigma}\{(\xi + 2a\sigma y)^2 + (\eta - 2a\sigma x)^2\}.
\end{aligned}$$

Then the kernel of the fundamental solution is given by

$$\begin{aligned}
K(t; z, s, z', s') &= \left(\frac{1}{2\pi}\right)^3 \int_{\mathbb{R}^3} e^{i(s-s')\sigma + i(x-x')\xi + i(y-y')\eta}\\
&\quad \times e\left(t; \frac{x+x'}{2}, \frac{y+y'}{2}, \xi, \eta, \sigma\right) d\xi\, d\eta\, d\sigma\\
&= \left(\frac{1}{2\pi}\right)^3 \int_{\mathbb{R}^3} e^{i\{s-s'+2a(x'y-xy')\}\sigma + i(x-x')\xi + i(y-y')\eta}\\
&\quad \times \frac{e^{\alpha\sigma t}}{\cosh(a\sigma t)} \exp\left[-\frac{\tanh(a\sigma t)}{4a\sigma}(|\xi|^2 + |\eta|^2)\right] d\xi d\eta d\sigma\\
&= \int_{-\infty}^{\infty} \frac{a\sigma}{2\pi^2\sinh(a\sigma t)} \exp\left[i\{s - s' + 2a(x'y - xy')\}\sigma\right.\\
&\quad \left. - \frac{a\sigma}{\tanh(a\sigma t)}|z - z'|^2 + \alpha\sigma t\right] d\sigma.
\end{aligned}$$

■

15.8 The Sub-Laplacian on Step-2 Nilpotent Lie Groups

We apply the last construction of the fundamental solution to a case of the degenerate operators, that is, to the case of the sub-Laplacian on two-step free nilpotent Lie groups.

So let $F_{N+N(N-1)/2} \cong \mathbb{R}^N \oplus \mathbb{R}^{N(N-1)/2}$ be a connected and simply connected free two-step nilpotent Lie group with the Lie algebra $\mathfrak{f}_{N+N(N-1)/2}$ (it is also identified with $\mathbb{R}^N \oplus \mathbb{R}^{N(N-1)/2}$). We fix a basis $\{X_j, Z_{j,k} \mid 1 \le j, k \le N, j < k\}$ of the Lie algebra $\mathfrak{f}_{N+N(N-1)/2}$. Their bracket relation is assumed to be

$$[X_j, X_k] = 2Z_{jk}$$

for $1 \leq j < k \leq N$, and the group multiplication $* : F_{N+N(N-1)/2} \times F_{N+N(N-1)/2} \to F_{N+N(N-1)/2}$ is given by

$$\begin{aligned}
&\mathbb{R}^N \oplus \mathbb{R}^{N(N-1)/2} \times \mathbb{R}^N \oplus \mathbb{R}^{N(N-1)/2} \\
&\ni \big\langle \big(\sum x_j X_j \oplus \sum z_{jk} Z_{jk}\big), \big(\sum \tilde{x}_j X_j \oplus \sum \tilde{z}_{jk} Z_{jk}\big)\big\rangle \\
&\longmapsto \big(\sum x_j X_j \oplus \sum z_{jk} Z_{jk}\big) * \big(\sum \tilde{x}_j X_j \oplus \sum \tilde{z}_{jk} Z_{jk}\big) \\
&= \sum (x_j + \tilde{x}_j) X_j \oplus \sum \big(z_{jk} + \widetilde{z}_{jk} + x_j \tilde{x}_k - x_k \tilde{x}_j\big) Z_{jk}.
\end{aligned}$$

Let $\tilde{X}_j$ be the left invariant vector fields on $F_{N+(N-1)/2}$:

$$\begin{aligned}
\tilde{X}_j(f)g &= \frac{d}{dt} f(g * e^{tX_j})|_{t=0} \\
&= \frac{\partial f}{\partial x_j} + \sum_{k<j} x_k \frac{\partial f}{\partial z_{kj}} - \sum_{k>j} x_k \frac{\partial f}{\partial z_{jk}},
\end{aligned}$$

where $g = (x, z) \in \mathbb{R}^N \oplus \mathbb{R}^{N(N-1)/2} \cong F_{N+N(N-1)/2}$.

Let $(x, z; \xi, \zeta) = (x_j, z_{jk}\ ;\ \xi_j, \zeta_{jk}) \in T^*(F_{N+N(N-1)/2}) \cong \mathbb{R}^N \oplus \mathbb{R}^{N(N-1)/2} \times \mathbb{R}^N \oplus \mathbb{R}^{N(N-1)/2}$ be the dual coordinates on the cotangent bundle; then we understand the symbol of vector fields $\tilde{X}_j$ and their Weyl symbol as

$$\begin{aligned}
\sigma^w(\tilde{X}_j) = \sigma(\tilde{X}_j) &= \sqrt{-1}\left(\xi_j + \sum_{k<j} x_k \zeta_{kj} - \sum_{k>j} x_k \zeta_{jk}\right) \\
&= \sqrt{-1}(\xi - \Omega(\zeta)x)_j,
\end{aligned}$$

where $\Omega = \Omega(\zeta)$ is an $N \times N$ skew-symmetric matrix defined by

$$\Big(\Omega(\zeta)\Big)_{jk} = \zeta_{jk} \quad (1 \leq j < k \leq N).$$

Let P be the sub-Laplacian

$$P = -\frac{1}{2} \sum_{j=1}^{N} \tilde{X}_j^2;$$

then its Weyl symbol is given by

$$\begin{aligned}
\sigma^w(P) &= -\frac{1}{2} \sum_{j=1}^{N} \sigma^w(\tilde{X}_j)^2 \\
&= \frac{1}{2} < X, HX >, \ X = {}^t(x, \xi),
\end{aligned}$$

with a $2N \times 2N$ matrix H defined by

$$H = \begin{pmatrix} -(\Omega(\zeta))^2 & \Omega(\zeta) \\ -\Omega(\zeta) & I \end{pmatrix}.$$

We consider the pseudo-differential operator $p^w(x, z, D_x, D_z)$ with the Weyl symbol $\sigma^w(P)$ and construct the following fundamental solution $E(t)$ as a pseudo-differential operator of the Weyl symbol $e(t; x, z, \xi, \zeta)$:

$$\frac{d}{dt} e^w(t; x, z, D_x, D_z) + p^w(x, z, D_x, D_z) e^w(t; x, z, D_x, D_z) = 0 \quad \text{in } (0, T) \times \mathbb{R}^{N+N(N-1)/2},$$

$$e^w(0; x, z, D_x, D_z) = I.$$

Theorem 15.8.1. *The Weyl symbol $e(t; x, z, \xi, \zeta)$ is obtained as follows:*

$$e(t; x, z, \xi, \zeta) = \frac{1}{\sqrt{\det \cosh(it\Omega(\zeta))}} \exp\left[-\frac{t}{2} \left\langle \frac{\tanh(i\,t\Omega_0)}{i\,t\Omega_0} H\,X,\ X \right\rangle \right],$$

where

$$\Omega_0 = \begin{pmatrix} \Omega(\zeta) & 0 \\ 0 & \Omega(\zeta) \end{pmatrix}.$$

Then

$$\begin{aligned} E(t)u(x, z) &= e^w(t; x, z, D_x, D_z) u(x, z) \\ &= (2\pi)^{-N-N(N-1)/2} \int_{\mathbb{R}^N \times \mathbb{R}^N} \int_{\mathbb{R}^{N(N-1)/2} \times \mathbb{R}^{N(N-1)/2}} e^{i<x-\tilde{x},\xi>+i<z-\tilde{z},\zeta>} \times \\ &\quad \times e(t; (x+\tilde{x})/2,\ (z+\tilde{z})/2,\ \xi,\ \zeta)\, u(\tilde{x},\ \tilde{z})\, d\zeta\, d\tilde{z} d\xi\, d\tilde{x}. \end{aligned}$$

Corollary 15.8.2. *The kernel function of the above fundamental solution $K(t; x, z, \tilde{x}, \tilde{z})$ is given by*

$$\begin{aligned} K(t; x, z, \tilde{x}, \tilde{z}) &= (2\pi t)^{-N/2-N(N-1)/2} \int_{\mathbb{R}^{N(N-1)/2}} e^{i<x,\Omega(\zeta)\tilde{x}>/t+i<z-\tilde{z},\zeta>/t} \\ &\quad \times \exp\left\{ -\frac{1}{2t} < x-\tilde{x},\ \frac{i\Omega(\zeta)}{\tanh(i\Omega(\zeta))}(x-\tilde{x}) > \right\} \\ &\quad \times \sqrt{\det\left[\frac{i\Omega(\zeta)}{\sinh(i\Omega(\zeta))}\right]} d\zeta, \end{aligned}$$

where $< z, \zeta > = \sum_{1 \le j < k \le N} z_{jk}\zeta_{jk}$.

From the above expression of the kernel function $K(t)$, its value on the diagonal is given by

Corollary 15.8.3.

$$K(t;x,z,x,z) = (2\pi t)^{-N/2-N(N-1)/2} \int_{\mathbb{R}^{N(N-1)/2}} \sqrt{\det\Big[\frac{i\Omega(\zeta)}{\sinh(i\Omega(\zeta))}\Big]}d\zeta.$$

We provide now the proof of Theorem 15.8.1. Let $A = iJH$ be a $2N\times 2N$ matrix. Then the power of the matrix A is given by

$$A^n = A(-2i\Omega_0)^{n-1} = (-2i\Omega_0)^{n-1}A. \tag{15.8.28}$$

Since

$$A = i\begin{pmatrix} -\Omega(\zeta) & I \\ \Omega(\zeta)^2 & -\Omega(\zeta) \end{pmatrix},$$

formula (15.8.28) is proved by the induction with respect to n if we note

$$A^2 = -2\begin{pmatrix} \Omega(\zeta)^2 & -\Omega(\zeta) \\ -\Omega(\zeta)^3 & \Omega(\zeta)^2 \end{pmatrix} = -2iA\Omega_0.$$

The next result will be used shortly.

Lemma 15.8.4. *If $h_0(x)$ is an entire function, then for the entire function $h(x) = xh_0(x)$, we have*

$$h(tA/2) = t\,A\,h_0(-i\Omega_0 t)/2. \tag{15.8.29}$$

In the case $f(x) = \cosh x$, we have

$$\det\{\cosh(t\,A/2)\} = \det\{\cosh(i\Omega t)\}. \tag{15.8.30}$$

Proof. Relation (15.8.29) is clear by the previous formula (15.8.28). Also, the formula

$$\cosh(t\,A/2) = At\,h(-i\Omega_0 t)/2 + I_{2N}$$

is derived from (15.8.28), where

$$I_{2N} = \text{identity matrix of size } (2N)\times(2N)$$

and $h(x) = (\cosh x - 1)/x$. We also have

$$\begin{aligned}\cosh(At/2) &= Ath(-i\Omega_0 t)/2 + I_{2N} \\ &= i\begin{pmatrix} \frac{t}{2}\Omega(\zeta)h(it\Omega(\zeta)) & -\frac{t}{2}h(it\Omega(\zeta)) \\ -\frac{t}{2}(\Omega(\zeta))^2 h(it\Omega(\zeta)) & \frac{t}{2}\Omega(\zeta)h(it\Omega(\zeta)) \end{pmatrix} + I_{2N} \\ &= \begin{pmatrix} \frac{it}{2}\Omega(\zeta)h(it\Omega(\zeta)) + I & -\frac{it}{2}h(it\Omega(\zeta)) \\ -\frac{it}{2}(\Omega(\zeta))^2 h(it\Omega(\zeta)) & \frac{it}{2}\Omega(\zeta)h(it\Omega(\zeta)) + I \end{pmatrix}\end{aligned}$$

and

$$\begin{aligned}\det(\cosh(tA/2)) &= \det\begin{pmatrix} \frac{it}{2}\Omega(\zeta)h(it\Omega(\zeta)) + I & -\frac{it}{2}h(it\Omega(\zeta)) \\ -\frac{it}{2}(\Omega(\zeta))^2 h(it\Omega(\zeta)) & \frac{it}{2}\Omega(\zeta)h(it\Omega(\zeta)) + I \end{pmatrix} \\ &= \det\begin{pmatrix} I & -\frac{it}{2}h(it\Omega(\zeta)) \\ \Omega(\zeta) & \frac{it}{2}h(it\Omega(\zeta)) + I \end{pmatrix} \\ &= \det\begin{pmatrix} I & -\frac{it}{2}h(it\Omega(\zeta)) \\ 0 & it\Omega(\zeta)h(it\Omega(\zeta)) + I \end{pmatrix} \\ &= \det(it\Omega(\zeta)h(it\Omega(\zeta)) + I) \\ &= \det\cosh(it\Omega(\zeta)).\end{aligned}$$

■

Theorem 15.8.1 is obtained by Lemma 15.8.4 and Theorem 15.4.8:
The kernel $K(t; x, z, \tilde{x}, \tilde{z})$ is given by the following formula:

$$\begin{aligned}K(t; x, z, \tilde{x}, \tilde{z}) = (2\pi)^{-N-N(N-1)/2} \int_{\mathbb{R}^N}\int_{\mathbb{R}^{N(N-1)/2}} e^{i<x-\tilde{x},\xi>+i<z-\tilde{z},\zeta>} \\ \times e(t; (x+\tilde{x})/2,\ (z+\tilde{z})/2,\ \xi,\ \zeta)\, d\zeta\, d\xi.\end{aligned}$$

The following equation is clear for a symmetric nonsingular matrix M:

$$\begin{aligned}&\int_{\mathbb{R}^N} \exp(-<M\xi, \xi> + i <a, \xi>)\, d\xi \\ &= (\det M)^{-1/2}(\pi)^{N/2} \exp(-<a, M^{-1}a>/4).\end{aligned}$$

Applying the above equation for $M = \frac{t}{2}\frac{\tanh(it\Omega(\zeta))}{it\Omega(\zeta)}$ and

$$a = (x - \tilde{x}) - \tanh(it\Omega(\zeta))(x + \tilde{x})/2,$$

we get the assertion of Corollary 15.8.2.

15.9 The Kolmogorov Operator

Let P be an operator on $\mathbb{R}^2$ defined by

$$P = -\frac{1}{2}\frac{\partial^2}{\partial x^2} + x\frac{\partial}{\partial y}. \tag{15.9.31}$$

The next result deals with the exact form of a symbol of the fundamental solution of the Kolmogorov operator.

Theorem 15.9.1. *The fundamental solution for the operator (15.9.31) is* $E(t) = e^w(t; x, y, D_x, D_y)$, *with the symbol*

$$e(t; x, y, \xi, \eta) = \exp\left[-\frac{t}{2}\xi^2 - ix\eta t - \frac{t^3}{24}\eta^2\right].$$

The heat kernel of (15.9.31) is given by

$$K(t; x, y, x', y') = \frac{\sqrt{3}}{\pi t^2}\exp\left[-\frac{1}{2t}(x-x')^2 - \frac{3}{2t}\left\{2(y-y')\frac{1}{t} - (x+x')\right\}^2\right].$$

Proof. We have

$$\sigma^w(P) = \frac{1}{2}\xi^2 + ix\eta.$$

In this case

$$H = \begin{pmatrix} 0 & 0 \\ 0 & 1 \end{pmatrix}, \quad p_0 = \begin{pmatrix} \eta \\ 0 \end{pmatrix}, \quad b = 0$$

and

$$X = {}^t(x, \xi).$$

Note that

$$A^2 = 0.$$

So we have

$$\tanh(At/2) = At/2,$$
$$G(At/2) = \frac{1}{6}At$$

because

$$G(x) = \frac{1}{3}x + \mathcal{O}(x^5).$$

By Theorem 15.4.8, we have

$$\begin{aligned}
&e(t;x,y,\xi,\eta)\\
&= \exp\left[-i\left\{\langle J(At/2)X,X\rangle + t\langle JX,Jp_0\rangle + \frac{t^2}{4}\langle J(At/6)Jp_0,Jp_0\rangle\right\}\right]\\
&= \exp\left[-\frac{t}{2}\langle HX,X\rangle - it\langle X,p_0\rangle - \frac{t^3}{24}\langle HJp_0,Jp_0\rangle\right]\\
&= \exp\left[-\frac{t}{2}\xi^2 - ix\eta t - \frac{t^3}{24}\eta^2\right].
\end{aligned}$$

The kernel of the fundamental solution is given by

$$\begin{aligned}
&K(t;x,y,x',y')\\
&= \left(\frac{1}{2\pi}\right)^2 \int_{\mathbb{R}^2} e^{i(x-x')\xi+i(y-y')\eta} e\left(t;\frac{x+x'}{2},\xi,\eta\right) d\xi\, d\eta\\
&= \left(\frac{1}{2\pi}\right)^2 \int_{\mathbb{R}^2} \exp\left[-\left\{\sqrt{\frac{t}{2}}\xi - \frac{i(x-x')}{\sqrt{2t}}\right\}^2 - \frac{t^3}{24}\eta^2 + i\beta\eta - \frac{(x-x')^2}{2t}\right] d\xi\, d\eta\\
&= \left(\frac{1}{2\pi}\right)^2 \frac{\sqrt{2\pi}}{\sqrt{t}} e^{-\frac{(x-x')^2}{2t}} \int_{\mathbb{R}} \exp\left\{-\frac{t^3}{24}\eta^2 + i\beta\eta\right\} d\eta,
\end{aligned}$$

with

$$\beta = y - y' - \frac{x+x'}{2}t.$$

So we have

$$\begin{aligned}
K(t;x,y,x',y') &= \frac{\sqrt{2}}{4\pi t^2}\sqrt{24}\exp\left\{-\frac{(x-x')^2}{2t} - \frac{6\beta^2}{t^3}\right\}\\
&= \frac{\sqrt{3}}{\pi t^2}\exp\left[-\frac{1}{2t}(x-x')^2 - \frac{3}{2t}\left\{2(y-y')\frac{1}{t} - (x+x')\right\}^2\right].
\end{aligned}$$

■

To conclude, the method of pseudo-differential operators is feasible in several important cases, including degenerate operators (the Kolmogorov case), operators with potential (the Hermite case), and sub-elliptic Laplacians on two-step nilpotent Lie groups.

Appendix

Relationship Between the Heat and Wave Kernels

This appendix presents the relationship between the wave kernel and the heat kernel of an operator. Since finding wave kernels is more difficult than finding heat kernels, this is not always an efficient method for computing heat kernels. On the other hand, the number of examples which can be worked out explicitly is limited. Here we provide just the example of the operator ∂_x^2. A similar method is used in [62] to provide the explicit formula of the heat kernel on the hyperbolic space.

Let L be a second-order operator defined on $C_0^\infty(\mathbb{R}^n)$ with a self-adjoint non-positive definite extension on $L^2(\mathbb{R}^n)$. One defines the *wave kernel* of L to be the distribution u on $\mathbb{R}^n \times (-\infty, \infty)$ which satisfies the following Cauchy problem:

$$\begin{aligned} \partial_t^2 u &= Lu, \\ u_{|t=0} &= 0, \\ \partial_t u_{|t=0} &= \delta_0, \end{aligned}$$

where δ_0 is the Dirac distribution centered at $x = 0$. The solution $u(x,t)$ is a wave which is flat at $t = 0$ and has the initial velocity given by an impulse function centered at the origin. A formal computation shows that

$$u(x,t) = \frac{1}{\sqrt{-L}} \sin\left(t\sqrt{-L}\right), \tag{A.1}$$

where $-L$ is assumed to be positive definite. On the other side, the heat kernel of L is the distribution v on $\mathbb{R}^n \times (0, \infty)$ such that

$$\begin{aligned} \partial_t v &= Lv, \\ v_{|t=0} &= \delta_0. \end{aligned}$$

One may formally write

$$v(x,t) = e^{tL}, \qquad t > 0. \tag{A.2}$$

In order to establish relationship between formulas (A.1) and (A.2), we consider the following Fourier transform identity, where λ can be either a number or an operator:

$$e^{-t\lambda^2} = \frac{1}{\sqrt{4\pi t}}\int e^{-\frac{s^2}{4t}} e^{is\lambda}\,ds = \frac{1}{\sqrt{4\pi t}}\int e^{-\frac{s^2}{4t}}\cos(s\lambda)\,ds$$
$$= \frac{1}{\sqrt{4\pi t}}\int e^{-\frac{s^2}{4t}}\,\partial_s\left(\frac{1}{\lambda}\sin(s\lambda)\right)ds.$$

Substituting $\sqrt{-L}$ for λ yields

$$e^{tL} = \frac{1}{\sqrt{4\pi t}}\int_{\mathbb{R}} e^{-\frac{s^2}{4t}}\,\partial_s\left(\frac{1}{\sqrt{-L}}\sin(s\sqrt{-L})\right)ds.$$

To conclude, we have

Theorem 15.9.2. *Let $v(x,t)$ be the heat kernel and $u(x,t)$ the wave kernel for L. Then*

$$v(x,t) = \frac{1}{\sqrt{4\pi t}}\int_{\mathbb{R}} e^{-\frac{s^2}{4t}}\,\partial_s\big(u(x,s)\big)\,ds, \qquad t>0. \tag{A.3}$$

As an immediate application of this formula, one may consider the operator $L=-\partial_x^2$ with the wave kernel $u(x,t) = H(t-|x|)$, where $H(u) = \begin{cases} 1, & \text{if } u>0 \\ 0, & \text{if } u\le 0\end{cases}$ is the Heaviside function. In this case the derivative in the distribution sense $\partial_s\big(u(x,s)\big) = \partial_s\big(H(s-|x|)\big) = \delta(s-|x|)$ yields the Dirac distribution. Then formula (A.3) provides the well-known expression for the heat kernel

$$v(x,t) = \frac{1}{\sqrt{4\pi t}}\int_{\mathbb{R}} e^{-\frac{s^2}{4t}}\,\delta(s-|x|)\,ds = \frac{1}{\sqrt{4\pi t}}e^{-\frac{|x|^2}{4t}}, \qquad t>0.$$

Other computations of heat kernels starting from the associated wave kernel constitute a launching ground for future research. A good reference for wave kernels is the paper [59].

Conclusions

At the end of this journey through several techniques of finding explicit formulas for heat kernels, we shall conclude with a few final remarks. Some of the methods presented in this monograph cover only certain operators in most of the cases and are not well suited for the others. We shall discuss in the following the experience gained from using these methods. More precisely, we shall be concerned with what method is suited to which operator.

Elliptic and sub-elliptic operators via the geometric method. The heat kernel for a given elliptic operator has a determined physical meaning; i.e., it is the amount of heat transferred between two given points on a Riemannian manifold in a given time. It is physically reasonable to assume that the diffusion of heat occurs along the geodesics of the Riemannian geometry associated with the elliptic operator considered initially. There are a few classes of complete manifolds where the connectivity by geodesics holds; i.e., there is only one geodesic joining any two given points on the manifold. For instance, Hadamard manifolds (negatively curved Riemannian manifolds) and compact manifolds are just a couple of examples. However, the local connectivity by geodesics property always holds on any manifold (Whitehead's theorem). For the elliptic operators associated with these types of Riemannian geometries, the expression of the heat kernel can be obtained by a closed-form formula that involves a product between the diffusion coefficient, called the volume function, and an exponential term of the action function along the geodesic. This action function, in the case of Riemannian geometry, is obtained by dividing the square of the Riemannian distance by $2t$, where t denotes the time parameter. When the geodesics have conjugate points, the expression of the heat kernel is more complicated, as in the case of the circle S^1. In other cases there might not be any explicit formulas, as the case of the sphere S^2.

We have extended this method to elliptic operators with potential. For these operators the geodesics are replaced by the solutions of the Euler–Lagrange equations associated with the Lagrangian obtained as a difference between the kinetic energy induced by the metric and the potential function. Explicit formulas exist only in the cases when the potential is linear or quadratic. Higher powers lead to unsolved problems, such as the exact solution for the quartic oscillator problem.

If the operator is sub-elliptic, the appropriate geometry is sub-Riemannian. Unfortunately, in this case the problem of connectivity by geodesics is a very complicated one, since the "local" and "global" behaviors in this case coincide. Since the sub-elliptic operators have always at least one missing direction, it was proved successful in several cases to integrate along the characteristic variety (the space where the principal symbol vanishes) a product between a volume function and the exponential of a modified complex action. The latter function has the important property that it contains information about the lengths of all the geodesics joining any two points. In this case the expression of the heat kernel has an integral representation that in general cannot be reduced to a function. However, in the case of the Kolmogorov operator this integral can be computed explicitly, and the result obtained is a function-type heat kernel.

Despite its robustness, the geometric method has its own difficulties. It is not always easy to find the action function, since this is a solution of a nonlinear equation, called the Hamilton–Jacobi equation. In the Riemannian case one can use other equivalent formalisms to find the action function, such as the Lagrangian formalism and the Hamiltonian formalism. In either case, one ends up solving a system of ODEs with boundary conditions. In the case of sub-Riemannian geometry the aforementioned formalisms are not always equivalent, since they produce possibly distinct geodesics (regular and normal geodesics, respectively). It is believed that the normal geodesics are the ones useful in computing the heat kernel.

Even if the action function has a reasonable expression, the transport equation, whose solution is the volume function, might not be easily solved explicitly, unless it is of a very particular type. The geometric method can be successfully applied in a number of cases when the Riemannian metric has a familiar expression and the transport equation has separable variables.

The role of the Fourier transform. This method is used whenever one wants to eliminate a variable. One of the situations where the Fourier method was proved efficient was in computing the heat kernels for sub-elliptic operators with one missing direction. In this case the partial Fourier transform transforms the sub-elliptic operator into an elliptic one, for which the heat kernel can be obtained by any other method. Then an application of the inverse partial Fourier transform on the heat kernel of the elliptic operator provides the heat kernel for the sub-elliptic operator.

The method of eigenfunction expansion. This method can be applied in the case when the operator has a discrete spectrum. For instance, the Laplace–Beltrami operator on a compact Riemannian manifold has this property. The formula for the heat kernel provided by this method involves infinite series, unless some generating formulas exist. Unfortunately, these are rare, and their occurrence deserves to be cherished. The limited number of cases when generating formulas can reduce the heat kernel series to a familiar function includes the generating formulas of Mehler, Hille–Hardy, and Poisson. In general, the method of eigenfunction expansion does not provide a closed-form expression for the kernel. In the case of sub-elliptic operators this method was not as successful as the geometric method.

The method of path integrals. In spite of its obscurity, this is the most remarkable of the methods, since it provides an integral formula over all the paths joining the endpoints, reminding us of the way radiation propagates between two given states, making the relationship with quantum mechanics. This path integral can be computed explicitly in some particular cases, by either van Vleck's or Feynman–Kac's formula. The first one provides the heat kernel as a function-type formula, reminding us of the geometric method, where the volume function is the square root of the van Vleck determinant. However, this formula is not successfully applicable to sub-elliptic operators. Explicitly working out heat kernels for sub-elliptic operators by path integrals, even for simple cases like the Grushin or Heisenberg operators, has yet to be accomplished at the moment.

The stochastic method. Each of the differential operators considered is a generator for an associated Ito diffusion processes. The success of this method is based on the ability to compute the probability density function of the associated Ito diffusion. This method can be applied for both elliptic and sub-elliptic operators. The main difficulty of the method is solving the associated stochastic differential equation. This can be done easily for simple operators such as the Laplace, Grushin or Kolmogorov. In the last two of these cases the probability density is obtained by evaluating an expectation integral using Feynman–Kac's formula. The stochastic methods have proved useful not only in the case of parabolic operators but also in solving Dirichlet problems.

The method of Laguerre calculus. This is the symbolic tensor calculus induced by the Laguerre functions on the Heisenberg group. These functions have been used in the study of the twisted convolution, or equivalently, the Heisenberg convolution. The Laguerre calculus plays a role similar to the Fourier series for a reasonable function defined on the unit circle. In order to invert a sub-elliptic operator (not necessarily second-order) on the Heisenberg group, we need to invert the Laguerre tensor of the operator. The method was successful in computing the heat kernels for the Kohn Laplacian $\Box_b = \bar{\partial}_b \bar{\partial}_b^* + \bar{\partial}_b^* \bar{\partial}_b$ and Paneitz operator. The advantage of this method over the others is its applicability to finding heat kernels to powers of sub-Laplacians, wave operators and higher-order operators as long as the Laguerre tensor is diagonalizable. However, the method is somewhat limited since it is heavily based on the group structure and orthogonality of the Laguerre functions.

The method of pseudo-differential operators. Using the symbolic calculus of pseudo-differential operators, we retrieved the heat kernels of several operators, such as the Hermite, Grushin, Kolmogorov and step-2 sub-Laplacian $-\frac{1}{2}(Z\bar{Z} + \bar{Z}Z) + i\alpha T$ operators. The disadvantage of this method is its limitation to strongly elliptic operators and some sub-elliptic operators defined on two-step Heisenberg manifolds. The technique cannot be applied to sub-Laplacians of step higher than 2.

The sum-of-squares operators. One of the special classes of operators treated in this book is the sum of squares of n linear independent vector fields. If n is the dimension of the space, then the operator is elliptic. Otherwise, it is sub-elliptic. In the latter case, if the vector fields and their iterated brackets generate the tangent space at each point, then we say the bracket generating property holds. If this holds, two things happen: the sum-of-squares operator is hypoelliptic (Hörmander's theorem), and the distribution generated by the vector fields has the global connectivity property (Chow's theorem). If exactly one bracket is needed to generate the tangent space, then the operator is said to be of step 2. The prototype operator in this case is the Heisenberg operator. If exactly $k-1$ brackets are needed to generate the tangent space, then the operator is said to be of step k. Step $k=1$ operators correspond to the elliptic case. The methods and techniques discussed in the present book apply for steps $k=1,2$. Even if most of the methods work theoretically for any step, we did not encounter any explicit example where the computation produces exact solutions. There are a few examples of superior step operators treated in the literature by some adhoc methods, which do not belong to any of the methods treated in this book. One of the further developments is to create techniques to approach these types of operators.

List of Frequently Used Notations and Symbols

$\mathbb{R}^n$	The n-dimensional Euclidean space
$\vert\alpha\vert$	The length of the muti-index $\alpha_1 + \cdots + \alpha_n$, where $\alpha = (\alpha_1, \ldots, \alpha_n)$
∂_{x_k}	Partial derivative with respect to x_k
∂_x^α	$\partial_{x_1}^{\alpha_1} \cdots \partial_{x_n}^{\alpha_n}$
$\vert x\vert$	$\left(\sum_{k=1}^n x_k^2\right)^{1/2}$ if $x = (x_1, \ldots, x_n) \in \mathbb{R}^n$
$\mathfrak{L}^2(I)$	Measurable and square integrable functions on I
S^n	n-dimensional unit sphere in $\mathbb{R}^{n+1}$
∇_x	The gradient in the x-variable
$\operatorname{div} X$	The divergence of the vector field X
Γ_{ij}^k	The Christoffel symbols
$\Gamma(x)$	The Gamma function
δ_{x_0}	The Dirac distribution centered at x_0
$\theta_1, \theta_2, \theta_3, \theta_4$	The theta-functions
X_t	One-dimensional stochastic process
$X(t)$	n-dimensional stochastic process $(X_t^1, \ldots, X_t^n)$
$E(X_t)$	The expected value operator of the random variable X_t
$\operatorname{Cov}(X_t, X_u)$	The covariance function
dX_t	The increment of the stochastic process X_t within time dt
$F(x_1, x_2; t_1, t_2)$	The joint distribution function of random variables X_1 and X_2
$P(A\vert B)$	The conditional probability of A given B
W_t	The one-dimensional Brownian motion
W(t)	The n-dimensional Brownian motion $(W_{t_1}, \ldots, W_{t_n})$
$p_t(x_0, x)$	The probability density function
$G_t(x)$	Gaussian distribution of parameter t
$I_k(x)$	The Bessel function of first kind of order k
$L\big(x(t), \dot{x}(t)\big)$	Lagrangian function $L : TM \to \mathbb{R}$
S_{cl}	The action associated with the classical Lagrangian
$S(x(t))$	The action evaluated along the path $x(t)$
$d(x_0, x)$	The Riemannian distance between x_0 and x
$H(x, p)$	Hamiltonian function $H : T^*M \to \mathbb{R}$
$\hat{K}$	The integral kernel of the integral operator K

$\mathfrak{P}_{x,y;t}$	The space of continuous paths from x_0 to x within time t
$d\mathfrak{m}(\phi)$	The Weyl measure on the path space $\mathfrak{P}_{x,y;t}$
Δ	The Laplace operator $\frac{1}{2}(\partial_{x_1}^2 + \cdots + \partial_{x_n}^2)$
H_{2n+1}	The $(2n+1)$-dimensional Heisenberg group
$\mathfrak{h}_{2n+1}$	The Lie algebra of the Heisenberg group H_{2n+1}
Δ_H	The Heisenberg Laplacian $\sum_{k=1}^{2n} X_k^2$ on the Heisenberg group H_{2n+1}
$\Delta_{\mathcal{G}}$	The Grushin operator
Δ_X	The sub-elliptic Laplacian $\sum_{k=1}^{m} X_k^2$ on a manifold of dimension $n > m$
$K(x, y;t)$	Heat kernel—the amount of heat transferred from x_0 to x within time $t > 0$
$\mathcal{F}_y$	The partial Fourier transform with respect to y
i	The imaginary number $\sqrt{-1}$
$[X, Y]$	The bracket $XY - YX$, with X, Y vector fields or operators
$\langle x, y \rangle$	The inner product $\sum_{k=1}^{n} x_k y_k$ if $x = (x_1, \dots, x_n), y = (y_1, \dots, y_n) \in \mathbb{R}^n$
$H_n(x)$	The Hermite polynomial of degree n
$L_n^{(a)}$	The Laguerre polynomial of degree n and parameter a
TM	The tangent bundle of the manifold M
$\mathcal{H}$	A nonintegrable distribution—a sub-bundle of TM
$C^\infty(M)$	The set of smooth functions on the manifold M
$\mathbb{H}$	The set quaternion numbers
Id	The identity map
$\mathfrak{E}_4$	The Engel group
$\mathfrak{e}_4$	The Lie algebra of the Engel group
$\partial\Omega$	The boundary of the set Ω
$G *_\tau F$	The noncommutative twisted product of G and F
$\mathcal{W}_k^{(p)}$	The Cauchy–Szegö kernel
$\mathcal{P}_\alpha$	The Paneitz operator
$\Box_b$	The Kohn Laplacian
$e(t; x, \xi)$	The Weyl symbol
$\mathcal{M}_+(F)$	The positive Laguerre matrix of F
$\mathcal{M}(F)$	The Laguerre tensor of the operator F
δ_i^j	The Kronecker delta function

References

1. J. Aarão. *Fundamental solutions for some partial differential operators from fluid dynamics and statistical physics*. SIAM Rev., **49**, no. 2, pp. 303–314, 2007.
2. J. Aarão. *A transport equation of mixed type*. J. Differ. Equ. **150**, pp. 188–202, 1998.
3. A.A. Agrachev, U. Boscain, and M. Sigalotti. *A Gauss-Bonnet-like formula on two-dimensional almost – Riemannian manifolds*. preprint, ArXiv:math/0609566v2.
4. M.Y. Antimirov, A.A. Kolyshkin, and R. Vaillancourt. *Complex Variables*. Academic Press, San Diego, CA, 1997.
5. M.F. Atiyah, R. Bott, and A. Shapiro. *Clifford modules*. Topology, **3**, Suppl. 1, pp. 3–38, 1964.
6. V.I. Arnold. *Mathematical Methods of Classical Mechanics*. GTM 60, Springer-Verlag, Berlin, 1989.
7. R.O. Bauer. *Analysis of the horizontal Laplacian for the Hopf fibration*. Forum Math. **17**, no. 6, pp. 903–920, 2005.
8. W. Bauer and K. Furutani. *Quantization operators on quadrics*. Kyushu J. Math. **62**, no. 1, pp. 221–258, 2008.
9. W. Bauer and K. Furutani. *Spectral Analysis and Geometry of a Sub-Riemannian Structure on S^3 and S^7*. J. Geom. Phys. **58**, no. 12, pp. 1693–1738, 2008.
10. W. Bauer and K. Furutani. *Zeta regularized determinant of the Laplacian for a class of spherical space forms*. J. Geom. Phys. **58**, pp. 64–88, 2008.
11. R. Beals. *Analysis and geometry on the Heisenberg Group*. Conference on Inverse Spectral Geometry, University of Kentucky, June 2002.
12. R. Beals. *A note on fundamental solutions*. Comm. PDE, **24**, pp. 369–376, 1999.
13. R. Beals, B. Gaveau, and P.C. Greiner. *On a geometric formula for the fundamental solution of subelliptic Laplacians*. Math. Nachr. **181**, pp. 81–163, 1996.
14. R. Beals, B. Gaveau, and P.C. Greiner. *Complex Hamiltonian mechanics and parametrices for subelliptic Laplacians I, II, III*. Bull. Sci. Math. **121**, pp. 1–36, 97–149, 195–259, 1997.
15. R. Beals, B. Gaveau, and P.C. Greiner. *The Green function of model step two hypoelliptic operators and the analysis of certain tangential cauchy riemannian complexes*. Adv. Math. **121**, 1996.
16. R. Beals, B. Gaveau, and P.C. Greiner. *Hamilton–Jacobi theory and the heat kernel on Heisenberg groups*. J. Math. Pures Appl. **79**, pp. 633–689, 2000.
17. R. Beals, B. Gaveau, P. Greiner, and J. Vauthier. *The Laguerre calculus on the Heisenberg group, II*. Bull. Sci. Math. **110**, pp. 255–288, 1986.
18. R. Beals and P.C. Greiner. *Calculus on Heisenberg manifolds*. Ann. Math. Studies, no. 119. Princeton University Press, Princeton, NJ, 1988.
19. C. Berenstein, D.C. Chang, and J. Tie. *Laguerre Calculus and Its Application on the Heisenberg Group*. AMS/IP Series in Advanced Mathematics, **22**, International Press, Cambridge, MA, 2001.
20. A. Boggess and A. Raich. *A simplified calculation for the fundamental solution to the heat equation on the Heisenberg group*. Proc. Amer. Math. Soc. **137**, pp. 937–944, 2009.

21. O. Calin and D.C. Chang. *Geometric mechanics on a step 4 subRiemannian manifold.* Taiwanese J. Math. **9**, no. 2, pp. 261–280, 2005.
22. O. Calin and D.C. Chang. *The geometry on a step 3 Grushin operator*. Appl. Anal. **84**, no. 2, pp. 111–129, 2005.
23. O. Calin and D.C. Chang. *On a Class of Nilpotent Distributions*, to appear in Taiwanese J. Math. 2010.
24. O. Calin and D.C. Chang. *Geometric Mechanics on Riemannian Manifolds*. Birkhäuser, Boston, 2004.
25. O. Calin and D.C. Chang. *Heat Kernels for Differential Operators with Radical Function Coefficients*, to appear in Taiwanese J. Math. 2010.
26. O. Calin and D.C. Chang. *Heat kernels for operators with gravitational potential operators in one and two variables*. Pure Appl. Math. Quart. **6**, no. 3, pp. 677–692, 2008.
27. O. Calin and D.C. Chang. *Sub-Riemannian Geometry, General Theory and Examples*. Cambridge University Press, Encyclopedia of Mathematics and Its Applications, **126**, Cambridge, 2009.
28. O. Calin, D.C. Chang, and P.C. Greiner. *Geometric Analysis on the Heisenberg Group and Its Generalizations*. AMS/IP Series in Advanced Mathematics, **40**, International Press, Cambridge, MA, 2007.
29. G. Campolieti and R. Makarov. *Path integral pricing of Asian options on state-dependent volatility models*. Quant. Fin. **8**, pp. 147–161, 2008.
30. R. Camporesi. *Harmonic analysis and propagators on homogeneous spaces*. Phys. Rep. **196**, no. 1, pp. 1–134, 1990.
31. C.H. Chang, D.C. Chang, B. Gaveau, P. Greiner, and H.P. Lee. *Geometric analysis on a step 2 Grushin operator*. Bull. Inst. Math. Academia Sinica (New Series), **4**, no. 2, pp. 119–188, 2009.
32. D.C. Chang, S.C. Chang, and J. Tie. *Laguerre calculus and Paneitz operator on the Heisenberg group*. Sci. China Ser. A 2009.
33. D.C. Chang and I. Markina. *Anisotropic quaternion Carnot groups: geometric analysis and Green's function*. Adv. Appl. Math. **39**, pp. 345–394, 2007.
34. S.C. Chang, J. Tie, and C.T. Wu. *Subgradient estimate and Liouville-type theorems for the CR heat equation on Heisenberg groups*, to appear in Asian J. Math. 2011.
35. W.L. Chow. *Über Systeme von linearen partiellen Differentialgleichungen erster Ordnung*. Math. Ann. **117**, no. 1, pp. 98–105, 1939.
36. M. Christ and D. Geller. *Singular integral characterizations of Hardy spaces on homogeneous groups*. Duke Math. J. **51**, pp. 547–598, 1984.
37. L.C. Cox, J.E. Ingersoll, and S.A. Ross. *A theory of the term structure of interest rates*. Econometrica, **53**, pp. 385–407.
38. M. de Gosson. *The Principles of Newtonian and Quantum Mechanics*. Imperial College Press, London, 2001.
39. A. Debiard, B. Gaveau, and E. Mazet. *Théorémes de comparison in géométrie riemanniene*. Publ. Kyoto University **12**, pp. 391–425, 1976.
40. A.H. Dolley, C. Benson, and G. Ratcliff. *Fundamental solutions for the power of the Heisenberg sub-Laplacian*. Illinois J. Math. **37**, pp. 455–476, 1993.
41. A. Erdélyi. *Higher Transcendental Functions*, vols. I, II, III, McGraw-Hill, Inc., Springer, 1953.
42. R. Feynman. *Space-time approach to nonrelativistic quantum mechanics*. Rev. Mod. Phys. **20**, pp. 367–387, 1948.
43. G.B. Folland. *Harmonic analysis in phase space*. Ann. Math. Studies, **122**, Princeton University Press, Princeton, NJ, 1989.
44. G.B. Folland. *Introduction to Partial Differential Equations*, 2nd ed. Princeton University Press, Princeton, NJ, 1995.
45. G.B. Folland and E.M. Stein. *Estimates for the $\bar{\partial}_b$ complex and analysis on the Heisenberg group*. Comm. Pure Appl. Math. **27**, pp. 429–522, 1974.
46. K. Furutani. *The heat kernels and the spectrum of a class of nilmanifolds*. Comm. Partial Differ. Equ. **21**, nos. 3&4, pp. 423–438, 1996.

47. K. Furutani and C. Iwasaki. *Grushin operator and heat kernel on milpotent Lie groups.* RIMS Kôkyûroku 1502, Developments of Cartan Geometry and Related Mathematical Problems, July 2006, Research Inst. Math. Sci., Kyoto University, Kyoto, Japan.
48. B. Gaveau. *Principe de moindre action, propagation de la chaleur et estimées sous elliptiques sur certains groupes nilpotents.* Acta Math. **139**, pp. 95–153, 1977.
49. B. Gaveau. *Systémes dynamiques associés à certains opérateurs hypoelliptiques.* Bull. Sci. Math. **102**, pp. 203–229, 1978.
50. I.M. Gelfand and G.E. Shilov. *Generalized Functions I.* Academic, New York, 1964.
51. D. Geller. *Fourier analysis on the Heisenberg group.* Proc. Natl. Acad. Sci. USA, **74**, pp. 1328–1331, 1977.
52. M. Giaquinta and S. Hildebrand. *Calculus of Variations, I, II*, vol. 310. Springer, New York, 1977.
53. P.B. Gilkey. *Invariance Theory, The Heat Equation, and the Atiyah-Singer Index Theorem.* Publish or Perish, Inc., Wilmington, Delaware (USA) 1984, Studies in Adv. Math. CRC Press, Boca Raton, Ann Arbor, London, Tokyo 1994 (Second Edition).
54. H. Goldstein. *Classical Mechanics.* Addison-Wesley, Reading, MA, 1959.
55. C. Gordon and J. Dodziuk. *Integral structures on H-type Lie algebras.* J. Lie Theory, **12**, no. 1, pp. 69–79, 2002.
56. C.R. Graham and J.M. Lee. *Smooth solutions of degenerate Laplacians on strictly pseudo-convex Domains.* Duke Math. J. **57**, pp. 697–720, 1988.
57. P.C. Greiner. *On the Laguerre calculus of left-invariant convolution operators on the Heisenberg group.* Seminaire Goulaouic-Meyer-Schwartz, exp. **XI**, pp. 1–39, 1980–81.
58. P. Greiner and O. Calin. *On subRiemannian geodesics.* Analysis and Applications, **1**, no. 3, pp. 289–350, 2003.
59. P.C. Greiner, D. Holcman, and Y. Kannai. *Wave kernels related to second-order operators.* Duke Math. J. **114**, no. 2, pp. 329–387, 2002.
60. P.C. Greiner, J.J. Kohn, and E.M. Stein. *Necessary and sufficient conditions for solvability of the Lewy equation.* Proc. Natl. Acad. Sci. USA, **72**, pp. 3287–3289, 1975.
61. P.C. Greiner and E.M. Stein. *On the solvability of some differential operators of type* $\Box_b$. Proc. International Conf., Cortona 1976–77, Ann. Scuola Norm. Sup. Pisa Cl. Sci. **4**, pp. 106–165, 1978.
62. A. Grigor'yan and Masakazu Noguchi. *The heat kernel on hyperbolic space.* Bull. Lond. Math. Soc. **30**, pp. 643–650, Cambridge University Press, 1998.
63. S.J. Gustavson and I.M. Sigal. *Mathematical Concepts of Quantum Mechannics.* Springer, Universitext, 2003.
64. K. Hirachi. *Scalar pseudo-Hermitian Invariants and the Szegö kernel on 3-dimensional CR manifolds. Lecture Notes in Pure and Appl. Math.*, Marcel-Dekker, **143**, pp. 67–76, 1992.
65. L. Hörmander. *Hypoelliptic second order differential equations.* Acta Math. **119**, pp. 147–171, 1967.
66. L. Hörmander. *Pseudo-differential operators and hypoelliptic equations.* Proc. Symposium on Singular Integrals, Am. Math. Soc. **10**, pp. 138–183, 1967.
67. L. Hörmander. *The Weyl calculus of pseudo-differential operators.* Comm. Pure Appl. Math. **32**, pp. 359–443, 1979.
68. A. Hulanicki. *The distribution of energy in the Brownian motion in the Gausssian field and analytic hypoellipticity of certain subelliptic operators on the Heisenberg group.* Stud. Math. **56**, pp. 165–173, 1976.
69. C. Iwasaki. *The fundamental solution for pseudo-differential operators of parabolic type.* Osaka J. Math. **14**, pp. 569–592, 1977.
70. C. Iwasaki. *A proof of the Gauss–Bonnet–Chern theorem by the symbol calculus of pseudo-differential operators.* Jpn. J. Math. **21**, pp. 235–285, 1995.
71. C. Iwasaki. *Symbolic calculus for construction of the fundamental solution for a degenerate equation and a local version of Riemann-Roch theorem* in Geometry, Analysis and Applications *(R.S. Pathak, ed.)*. World Scientific, pp. 83–92, 2000.

72. C. Iwasaki and N. Iwasaki. *Parametrix for a degenerate parabolic equation and its application to the asymptotic behavior of spectral functions for stationary problems*. Publ. Res. Inst. Math. Sci. **17**, pp. 557–655, 1981.
73. G. Johnson and M. Lapidus. *The Feynman Integral and Feynman's Operational Calculus*. Oxford Science Publications, Oxford, 2000.
74. M. Kac. *Integration in Function Spaces and Some of Its Applications*. Accademia Nazionale dei Lincei, Scuola Normale Superioare, Pisa, 1980.
75. M. Kac. *On distributions of certain Wiener functionals*. Trans. Am. Math. Soc. **65**, pp. 1–13, 1949.
76. R.P. Kanwal. *Generalized Functions: Theory and Applications*, 3rd ed. Birkhäuser, Boston, 2004.
77. A. Kaplan. *On the geometry of groups of Heisenberg type*. Bull. Lond. Math. Soc. **15**, no. 1, pp. 35–42, 1983.
78. A. Klinger. *New derivation of the Heisenberg kernel*. Comm. Partial Differ. Equ. **22**, pp. 2051–2060, 1997.
79. S. Kobayashi and K. Nomizu. *Foundations of Differential Geometry*, vols. I–II. Wiley Interscience, New York, 1996.
80. A.N. Kolmogorov. *Uber die analytischen Methoden in der Wahrscheinlichkeitsrechnung*. Math. Ann. **104**, pp. 415–458, 1931.
81. A.N. Kolmogorov. *Zufällige Bewegungen (Zur Theorie der Brownschen Bewegung)*. Ann. Math. **35**, no. 1, pp. 116–117, 1934.
82. H. Kumano-go. *Pseudo-Differential Operators*. MIT, Cambridge, MA, 1981.
83. H.H. Kuo. *Introduction to Stochastic Integration*. Springer, Universitext, New York, 2006.
84. J. Lafferty and G. Lebanon. *Diffusion kernels on statistical manifolds*. Jo. Mach. Learning Res. 2006 (http://www-2.cs.cmu.edu).
85. J. Lamperti. *Probability*. W.A. Benjamin, Inc. Reading, MA, 1966.
86. D.F. Lawden. *Elliptic Functions and Applications*, vol. 80. Springer, New York, 1989.
87. J.M. Lee. *Pseudo-Einstein Structure on CR-Manifolds*. Am. J. Math. **110**, pp. 157–178, 1988.
88. J.F. Lingevitch and A.J. Bernoff. *Advection of a passive scalar by a vortex couple in the small-diffusion limit*. J. Fluid Mech. **270**, pp. 219–249, 1994.
89. H.P. McKean. *An upper bound to the spectrum of Δ on a manifold of negative curvature*. J. Diff. Geom. **4**, pp. 359–366, 1970.
90. A. Melin. *Lower bounds for pseudo-differential operators*. Ark. Mat. **9**, pp. 117–140, 1971.
91. S.G. Mikhlin. *Compounding of double singular integrals*. Doklady Akad. Nauk. USSR, **2**, pp. 3–6, 1936.
92. R. Montgomery. *A tour of subriemannian geometries, their geodesics and applications*. AMS, Math. Surv. Monogr. **91**, Providence, RI, 2002.
93. D. Müller and E.M. Stein. *L^p-estimates for the wave equation on the Heisenberg group*. J. Math. Pures Appl. **73**, pp. 413–440, 1994.
94. S. Nishikawa. *Variational Problems in Geometry*. A.M.S. Translations of Mathematical Monographs, Iwanami Series in Modern Mathematics, **205**, 2002.
95. B. Oksendal. *Stochastic Differential Equations, An Introduction with Applications*, 6th ed. Springer, Berlin, 2003.
96. J. Peetre. *The Weyl transform and Laguerre polynomials*. Le Matematiche, **27**, pp. 301–323, 1972.
97. J.H. Rawnsley. *A non-unitary pairing of polarization for the Kepler problem*. Trans. Amer. Math. Soc. **250**, pp. 167–180, 1979.
98. D.B. Ray and I.M. Singer. *R-Torsion and the Laplacian on Riemannian manifolds*. Adv. Math. **7**, pp. 145–210, 1971.
99. S.M. Ross. *Stochastic Processes*. Wiley, New York, 1996.
100. L.P. Rothschild and E.M. Stein. *Hypoelliptic differential operators and nilpotent Lie groups*. Acta Math. **137**, nos. 3/4, pp. 247–320, 1976.
101. L.S. Schulman. Ph.D. thesis. Phys. Rev. **176**, p. 1558, 1968.
102. L.S. Schulman. *Techniques and Applications of Path Integration*. Dover Publications, Inc., Mineola, NY, 2005.

103. E.M. Stein. *Harmonic Analysis – Real Variable Methods, Orthogonality, and Oscillatory Integrals*. 43, Princeton Mathematical Series, Princeton University Press, Princeton, NJ, 1993.
104. R. Strichartz. *L^p harmonic analysis and Radon transforms on the Heisenberg group*. J. Funct. Anal. **96**, pp. 350–406, 1991.
105. R. S. Strichartz. *Sub-Riemannian geometry*. J. Diff. Geom. **24**, no. 2, pp. 221–263, 1986.
106. M. Taylor. *Noncommutative Harmonic Analysis*. Am. Math. Soc., Providence, RI, 1986.
107. S. Thangavelu. *Lectures on Hermite and Laguerre Expansions*. Mathematical Notes, no. **42**, Princeton University Press, Princeton, NJ, 1993.
108. C. Tsutsumi. *The fundamental solution for a degenerate parabolic pseudo-differential operator*. Proc. Jpn. Acad. **50**, pp. 11–15, 1974.
109. J.H. van Vleck. *The correspondence principle in the statistical interpretation of quantum mechanics*. Proc. Natl. Acad. Sci. USA, **14**, no. 176, pp. 178–188, 1928.
110. N. Wallach. *Symplectic Geometry and Fourier Analysis in Lie Groups: History, Frontiers and Applications*. vol. V, Math. Sci. Press, 1977.
111. D.V. Widder. *The Heat Equation*. Academic, London, 1975.
112. S. Wolfgang. *Wiener path integrals and the fundamental solution for the Heisenberg Laplacian*. J. d'Analyse Math. **91**, pp. 389–400, 2003.

Index

Applied and Numerical Harmonic Analysis

J.M. Cooper: *Introduction to Partial Differential Equations with MATLAB* (ISBN 978-0-8176-3967-9)

C.E. D'Attellis and E.M. Fernández-Berdaguer: *Wavelet Theory and Harmonic Analysis in Applied Sciences* (ISBN 978-0-8176-3953-2)

H.G. Feichtinger and T. Strohmer: *Gabor Analysis and Algorithms* (ISBN 978-0-8176-3959-4)

T.M. Peters, J.H.T. Bates, G.B. Pike, P. Munger, and J.C. Williams: *The Fourier Transform in Biomedical Engineering* (ISBN 978-0-8176-3941-9)

A.I. Saichev and W.A. Woyczyński: *Distributions in the Physical and Engineering Sciences* (ISBN 978-0-8176-3924-2)

R. Tolimieri and M. An: *Time-Frequency Representations* (ISBN 978-0-8176-3918-1)

G.T. Herman: *Geometry of Digital Spaces* (ISBN 978-0-8176-3897-9)

A. Procházka, J. Uhlíř, P.J.W. Rayner, and N.G. Kingsbury: *Signal Analysis and Prediction* (ISBN 978-0-8176-4042-2)

J. Ramanathan: *Methods of Applied Fourier Analysis* (ISBN 978-0-8176-3963-1)

A. Teolis: *Computational Signal Processing with Wavelets* (ISBN 978-0-8176-3909-9)

W.O. Bray and C.V. Stanojević: *Analysis of Divergence* (ISBN 978-0-8176-4058-3)

G.T Herman and A. Kuba: *Discrete Tomography* (ISBN 978-0-8176-4101-6)

J.J. Benedetto and P.J.S.G. Ferreira: *Modern Sampling Theory* (ISBN 978-0-8176-4023-1)

A. Abbate, C.M. DeCusatis, and P.K. Das: *Wavelets and Subbands* (ISBN 978-0-8176-4136-8)

L. Debnath: *Wavelet Transforms and Time-Frequency Signal Analysis* (ISBN 978-0-8176-4104-7)

K. Gröchenig: *Foundations of Time-Frequency Analysis* (ISBN 978-0-8176-4022-4)

D.F. Walnut: *An Introduction to Wavelet Analysis* (ISBN 978-0-8176-3962-4)

O. Bratteli and P. Jorgensen: *Wavelets through a Looking Glass* (ISBN 978-0-8176-4280-8)

H.G. Feichtinger and T. Strohmer: *Advances in Gabor Analysis* (ISBN 978-0-8176-4239-6)

O. Christensen: *An Introduction to Frames and Riesz Bases* (ISBN 978-0-8176-4295-2)

L. Debnath: *Wavelets and Signal Processing* (ISBN 978-0-8176-4235-8)

J. Davis: *Methods of Applied Mathematics with a MATLAB Overview* (ISBN 978-0-8176-4331-7)

G. Bi and Y. Zeng: *Transforms and Fast Algorithms for Signal Analysis and Representations* (ISBN 978-0-8176-4279-2)

J.J. Benedetto and A. Zayed: *Sampling, Wavelets, and Tomography* (ISBN 978-0-8176-4304-1)

E. Prestini: *The Evolution of Applied Harmonic Analysis* (ISBN 978-0-8176-4125-2)

O. Christensen and K.L. Christensen: *Approximation Theory* (ISBN 978-0-8176-3600-5)

L. Brandolini, L. Colzani, A. Iosevich, and G. Travaglini: *Fourier Analysis and Convexity* (ISBN 978-0-8176-3263-2)

W. Freeden and V. Michel: *Multiscale Potential Theory* (ISBN 978-0-8176-4105-4)

O. Calin and D.-C. Chang: *Geometric Mechanics on Riemannian Manifolds* (ISBN 978-0-8176-4354-6)

Applied and Numerical Harmonic Analysis (Cont'd)

J.A. Hogan and J.D. Lakey: *Time-Frequency and Time-Scale Methods* (ISBN 978-0-8176-4276-1)

C. Heil: *Harmonic Analysis and Applications* (ISBN 978-0-8176-3778-1)

K. Borre, D.M. Akos, N. Bertelsen, P. Rinder, and S.H. Jensen: *A Software-Defined GPS and Galileo Receiver* (ISBN 978-0-8176-4390-4)

T. Qian, V. Mang I, and Y. Xu: *Wavelet Analysis and Applications* (ISBN 978-3-7643-7777-9)

G.T. Herman and A. Kuba: *Advances in Discrete Tomography and Its Applications* (ISBN 978-0-8176-3614-2)

M.C. Fu, R.A. Jarrow, J.-Y. J. Yen, and R.J. Elliott: *Advances in Mathematical Finance* (ISBN 978-0-8176-4544-1)

O. Christensen: *Frames and Bases* (ISBN 978-0-8176-4677-6)

P.E.T. Jorgensen, K.D. Merrill, and J.A. Packer: *Representations, Wavelets, and Frames* (ISBN 978-0-8176-4682-0)

M. An, A.K. Brodzik, and R. Tolimieri: *Ideal Sequence Design in Time-Frequency Space* (ISBN 978-0-8176-4737-7)

S.G. Krantz: *Explorations in Harmonic Analysis* (ISBN 978-0-8176-4668-4)

G.S. Chirikjian: *Stochastic Models, Information Theory, and Lie Groups, Volume 1* (ISBN 978-0-8176-4802-2)

C. Cabrelli and J.L. Torrea: *Recent Developments in Real and Harmonic Analysis* (ISBN 978-0-8176-4531-1)

B. Luong: *Fourier Analysis on Finite Abelian Groups* (ISBN 978-0-8176-4915-9)

M.V. Wickerhauser: *Mathematics for Multimedia* (ISBN 978-0-8176-4879-4)

P. Massopust and B. Forster: *Four Short Courses on Harmonic Analysis* (ISBN 978-0-8176-4890-9)

O. Christensen: *Functions, Spaces, and Expansions* (ISBN 978-0-8176-4979-1)

J. Barral and S. Seuret: *Recent Developments in Fractals and Related Fields* (ISBN 978-0-8176-4887-9)

O. Calin, D.-C. Chang, K. Furutani, and C. Iwasaki: *Heat Kernels for Elliptic and Sub-elliptic Operators* (ISBN 978-0-8176-4994-4)

Printed in the United States
By Bookmasters